INTELLIGENT AUTONOMOUS SYSTEMS 10

Intelligent Autonomous Systems 10

IAS-10

Edited by

Wolfram Burgard

University of Freiburg

Rüdiger Dillmann

University of Karlsruhe

Christian Plagemann

University of Freiburg

and

Nikolaus Vahrenkamp

University of Karlsruhe

IOS Press

Amsterdam • Berlin • Oxford • Tokyo • Washington, DC

ISBN 978-1-58603-887-8
Library of Congress Control Number: 2008929931

Publisher
IOS Press
Nieuwe Hemweg 6B
1013 BG Amsterdam
Netherlands
fax: +31 20 687 0019
e-mail: order@iospress.nl

Distributor in the UK and Ireland
Gazelle Books Services Ltd.
White Cross Mills
Hightown
Lancaster LA1 4XS
United Kingdom
fax: +44 1524 63232
e-mail: sales@gazellebooks.co.uk

Distributor in the USA and Canada
IOS Press, Inc.
4502 Rachael Manor Drive
Fairfax, VA 22032
USA
fax: +1 703 323 3668
e-mail: iosbooks@iospress.com

PRINTED IN THE NETHERLANDS

Intelligent Autonomous Systems 10
W. Burgard et al. (Eds.)
IOS Press, 2008

Preface

Welcome to the 10th International Conference on Intelligent Autonomous Systems (IAS-10). The International Conference on Intelligent Autonomous Systems belongs to the most traditional robotics events and we are proud to host it in Baden Baden, Germany, this year. The goal of the IAS conference is to bring together leading researchers interested in all aspects of autonomy and adaptivity of artificial systems. One of the driving forces of this conference is the observation that intelligence and autonomy is best studied and demonstrated using mobile robots acting autonomously in real-world environments and under challenging conditions.

This year, 80 papers have been submitted to IAS-10. Each paper was evaluated by between two and six reviewers and 49 were accepted for presentation at the conference. IAS-10 features technical presentations of papers with high scientific quality, invited talks, demonstrations, and workshops. The papers contained in the final program cover a wide spectrum of research in autonomous intelligent systems including agent technology, walking robots, motion planning, robot control, multi-robot systems, navigation, perception, applications, learning and adaptation, and humanoid robots, just to mention some of them.

We are especially proud that the keynote presentation will be given by Professor Roland Siegwart from ETH Zurich. He belongs to the most outstanding European researchers in the area of autonomous intelligent robots.

The proceedings include all accepted papers and reflect a variety of topics concerning intelligent autonomous systems. The organizers would like to express their gratitude to all contributors in the preparation phase as well as during the meeting. Without additional assistance, IAS-10 would have never been a success. We would especially like to thank the program committee members for their valuable support and for the preparation of the reviews, which allowed to make a proper selection of high-quality papers. Many thanks also go to the additional reviewers.

The staff at the Autonomous Intelligent Systems lab of the University of Freiburg and at the Institute for Computer Science and Engineering of the University of Karlsruhe took a great part in planning and organizing the conference.

We wish all participants that they enjoy IAS-10 and the beautiful area of Baden Baden. We hope that IAS-10 will provide you with new ideas, allow you to exchange knowledge, and be a prosperous event for you.

Enjoy IAS-10.

Wolfram Burgard
Rüdiger Dillmann
Christian Plagemann
Nikolaus Vahrenkamp

IAS-10 Conference Organization

General Chair

Rüdiger Dillmann
University of Karlsruhe (TH), Germany

Program Chair

Wolfram Burgard
Albert-Ludwigs-Univ. Freiburg, Germany

Workshop Chair

Paul Levi
University of Stuttgart, Germany

Publication Co-Chairs

Christian Plagemann
Albert-Ludwigs-Univ. Freiburg, Germany

Nikolaus Vahrenkamp
University of Karlsruhe (TH), Germany

Local Arrangements Co-Chairs

Nela Redzovic
University of Karlsruhe (TH), Germany

Nikolaus Vahrenkamp
University of Karlsruhe (TH), Germany

Steering Committee

Tamio Arai
Maria Gini
Anthony Stentz
Nancy Amato
Frans Groen
Enrico Pagello

Program Committee

Paul Levi (Program Co-Chair Europe)
Nicholas Roy (Program Co-Chair America)
Hajime Asama (Program Co-Chair Asia)

Maren Bennewitz
Panos Trahanias
Tom Duckett
Emanuele Menegatti
Patrick Jensfelt
Danica Kragic
Michael Montemerlo
Dave Ferguson
Joseph Modayil
Kristian Kersting
Thomas Henderson
Javier Minguez
Achim Lilienthal
Gisbert Lawitzky
Gregorz Cielniak
Christoph Stiller
Karsten Berns
Gamini Dissanayake
Norihiro Hagita
Tamio Arai
Hiroaki Yamaguchi
Ryuichi Ueda

Advisory Committee

Minoru Asada
James L. Crowley
Toshio Fukuda
Gerd Hirzinger
Katsushi Ikeuchi
Masatoshi Ishikawa
Sukhan Lee
Arthur C. Sanderson
Yoshiaki Shirai

Contents

Intelligent Autonomous Systems 10
W. Burgard et al. (Eds.)
IOS Press, 2008

doi:10.3233/978-1-58603-887-8-1

Agent Standard Protocols for Building Multi-Robot Applications

Enric CERVERA [a] and Veysel GAZI [b]
[a] *Robotic Intelligence Lab, Jaume-I University, Spain*
[b] *TOBB University of Economics and Technology, Ankara, Turkey*

Abstract. A practical solution based on multi-agent protocols for the development of real-world multi-robot applications is presented. FIPA standard protocols implemented by the JADE library provide the standard functionality for a number of tasks. Robot behaviors are built upon the Player middleware. Such components provide off-the-shelf tools which allow a straightforward implementation of indoor localization and navigation tasks for a team of mobile robots. Such integration combines proven mobile robot algorithms with a distributed infrastructure, and extends the capabilities from a robot alone to a whole team of robots, thus allowing the development of cooperative applications. As a proof of concept, an auction-like goal assignment task is presented: the robot team is given a goal, and each robot proposes an estimated cost for achieving it, then the best proposal is selected. Most of the control flow is automated by the standard interaction protocols. Experimental evaluation demonstrates the advantages of combining both frameworks, for a practical yet sound development of multi-robot applications.

Keywords. Networked Robots, Multi-Robot Systems

1. Introduction

The aim of our research is the integration of a robot middleware and a multi-agent system, which paves the way for the straightforward development of multi-robot cooperative applications. Both components are complementary: the robot middleware provides generic interfaces with the robot hardware (base and sensors) and a handful of sound algorithms for single robot localization, planning and navigation; the multi-agent system adds the necessary infrastructure to extend the robot capabilities to a team of robots, based on standard protocols.

The efficacy of the integrated system undoubtedly derives from the choice of its components, Player [7] and JADE [2], whose capabilities and weaknesses will be highlighted in the context of the proposed system architecture. Besides their philosophy or design, they have been chosen mainly due to *practical* reasons: their widespread use, rich community of users, constant development and support, and overall usefulness in our previous experience in teaching and research of Robotics and AI. Last but not least, both platforms are open-source and free.

As a result, we develop two simple yet effective multi-robot applications which showcase the benefits of both platforms: extending the functionality of the sophisticated planning and navigation robot algorithms to the team of robots, thus allowing the realization of cooperative tasks. The synergy between the components is emphasized by their seamless and elegant integration, as well as for the complementary strengths.

2. System architecture

The proposed architecture consists of two platforms: a middleware for robot control (Player[1]) and a multi-agent development environment (JADE[2]).

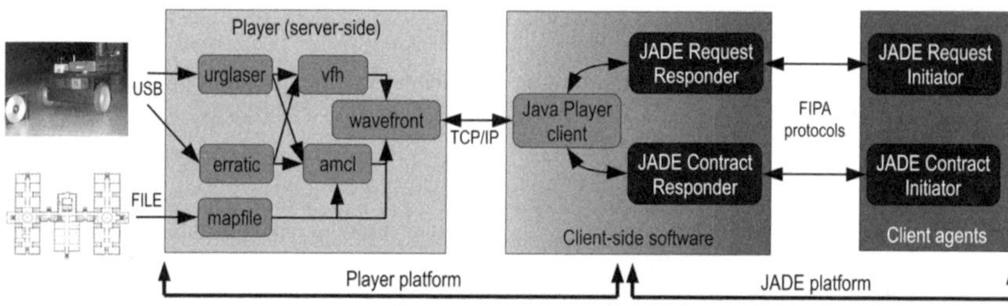

Figure 1. System architecture.

Robotic middlewares are playing an increasingly important role in robotics research, particularly for the development of architectures for mobile robots. They allow portability of code and enhance reusability, thus making the application independent of the robot hardware. Many popular middlewares are available, see e.g. [4] for a recent survey. However, such middlewares focus on solutions for *single* robot applications, and the extension to multi-robot applications, while possible, is not straightforward, since it may require significant communication and programming resources.

On the other hand, multi-agent platforms have been widely studied in the last decade [5], and successfully applied to many domains, including cooperative robotics [8]. But they are not designed with robots in mind, and again, a significant effort to interface with the robot hardware may be needed to perform a successful application.

We claim that the fusion of such complementary platforms may boost the development of cooperative multi-robot tasks. In the following, we introduce our choices of components, Player and JADE, and we describe their synergy and ease of use for practical multi-robot application development.

Figure 1 depicts the system architecture, namely Player and JADE platforms, with the different modules used in each of them, which are thoroughly described in the following.

2.1. Player robot platform

The Player framework interfaces with the underlying hardware (mobile robot base and sensors). In addition, it provides ready-to-use software drivers which implement localization and (local and global) navigation functionalities [6,3,1].

[1]http://playerstage.sf.net
[2]http://jade.tilab.com

2.2. Multi-agent protocols

JADE is a software framework which simplifies the implementation of multi-agent systems through a middleware that complies with the FIPA[3] specifications and through a set of graphical tools that supports the debugging and deployment phases.

Two FIPA-standard protocols [2] are used. In the first case (FIPA-Request) the initiator will ask for one or some robots to attain a given goal. In the second case (FIPA-Contract-Net) the initiator will ask for proposals of the cost estimated by each robot for achieving the goal, then selecting one or some of them to proceed to the goal.

It should be taken into account that the JADE framework provides the domain-independent source code for implementing the above protocols, thus freeing the developer from all the flow control and message processing burden. She only needs to concentrate on the particular details of the application, i.e. in our case multi-robot planning and navigation.

2.3. Interfacing Player and JADE

The client/server design of Player decouples the implementation language of each side, as long as they can communicate by TCP/IP. In our architecture, the localization and navigation drivers run natively on the server-side of Player. The client-side is programmed in the Java language[4] (see Fig. 1), thus allowing a seamless integration with JADE, which uses the same language.

As a result, both components are highly decoupled, since the agent will communicate with the wavefront planner driver. Interaction with the other drivers is indeed possible, yet not needed in our applications.

3. Applications

In the experiments, we have used three mobile robots, as shown in Fig. 2. Each one is endowed with a small laser rangefinder for localization and obstacle avoidance, and an onboard PC with wireless networking. Besides the operating system, each computer is running the Player drivers for controlling the hardware elements, those drivers which implement the localization, planning and navigation algorithms, and the Player/JADE responder agents which listen either for requests or contracts.

3.1. Multi-robot navigation based on request

The first example is a straightforward extension from a single-robot behavior (go to "goal") to a team of robots. The user asks some robots to go to a particular location in the map. Those robots which are available, will agree; then each robot will compute its path, and move towards the goal. If any robot fails to achieve the goal, it will send a failure message to the user; otherwise, an inform message will be sent.

Figure 3(a) depicts a screenshot of a special JADE agent, named *sniffer*, which allows the traffic of messages in the application to be monitored. In this example, all the robots have been requested to go towards the goal, they all have agreed, and successfully informed of the accomplishment.

[3] http://www.fipa.org

[4] http://java-player.sf.net/

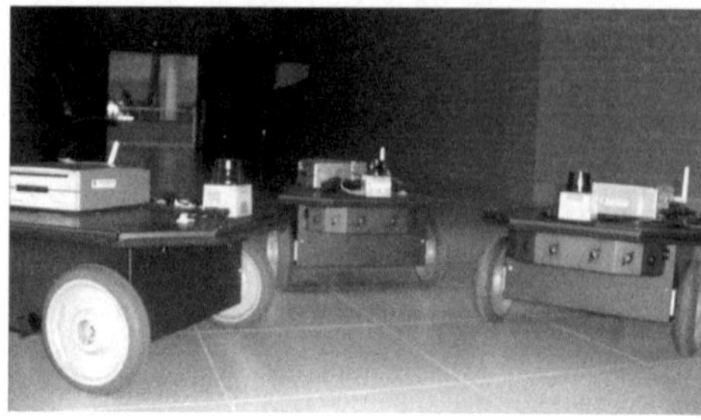

Figure 2. Team of robots used in the experiments. On top of each one, a compact PC is running all the control software, and a Hokuyo URG laser detects obstacles and provides range data for localization.

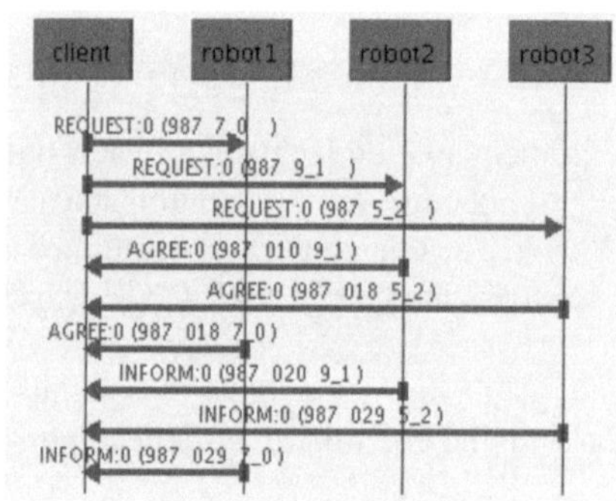

(a) Request Protocol.

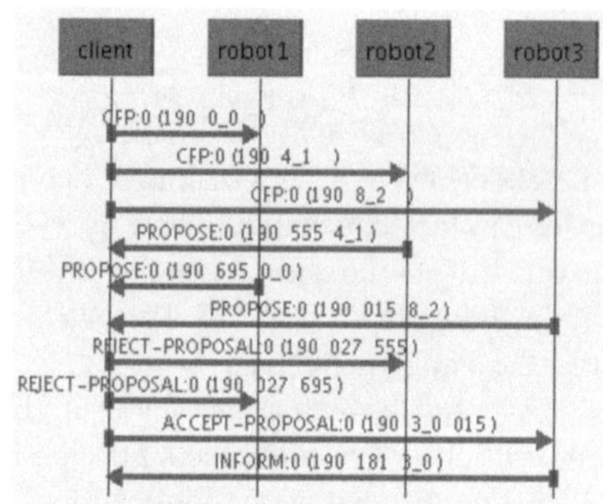

(b) ContractNet Protocol.

Figure 3. Traffic monitoring of messages in the 3-robot application using the each protocol (vertical axis is not scaled in time).

3.2. *Multi-robot navigation based on contract net*

In this second example, the robots will not move right away to the goal, but compute a cost (the length of the path) and answer to the initiator with a proposal. The initiator will select some of the robots, usually the shortest path, and will send an acceptance message to that robot, and reject messages to the rest, as presented in the protocol before. Figure 3(b) depicts another screenshot of the JADE *sniffer* agent when asking for proposals to attain a goal. In this example, the proposal of robot 3 is accepted and it successfully informed the initiator of the accomplishment; the others are rejected.

4. Conclusion

To this end, what makes the combination of Player and JADE so appealing for cooperative robotics? JADE is a mature and widely-used agent platform, which lacks the interface to *real things* like e.g. mobile robots. Player is an extremely popular software for mobile robots, a de-facto standard; despite its distributed client-server model, it is by no means designed for multi-robot applications.

Combining Player and JADE brings together the best of both, resulting in an extremely useful framework for the development of real, cooperative robotics tasks. The examples presented, though simple, demonstrate on one side how straightforward is the translation from a generic protocol to a real robot application, and on the other side how simply behaviors can be shared and combined by the robot team.

Comparison with equivalent frameworks is our next target in the near future. From the developer's point of view, one possible measurement of the strength of each platform could be the length of the source code for the same task. Our responder agent based on the contract-net protocol has only 134 lines of Java code, distributed almost at 50% between generic agent code and specific robot application code. This is indeed a very short program, yet it performs remarkably well and it scales to any number of robots (as long as communication infrastructure allows).

More complex protocols (iterated contract net, english auction, dutch auction, brokering, recruiting, subscribe) do exist which could suit well to other robotic tasks. Simple yet effective problems could form the basis for a benchmark in cooperative robotics.

Acknowledgements

This work has been partially funded by the EU-VI Framework Programme under grant IST-045269 - "GUARDIANS" of the EC Cognitive Systems initiative, and by the Spanish Ministry of Science and Education (MEC) under grant DPI2005-08203-C02-01.

References

[1] J. Barraquand, B. Langlois, and J. C. Latombe. Numerical potential field techniques for robot path planning. *IEEE Transactions on Systems, Man and Cybernetics*, 22(2):224–241, 1992.

[2] Fabio Bellifemine, Giovanni Caire, Agostino Poggi, and Giovanni Rimassa. JADE: A software framework for developing multi-agent applications. lessons learned. *Information and Software Technology*, 50(1–2):10–21, 2007.

[3] Dieter Fox, Wolfram Burgard, Frank Dellaert, and Sebastian Thrun. Monte Carlo Localization: Efficient position estimation for mobile robots. In *Proceedings of the AAAI/IAAI*, pages 343–349, 1999.

[4] James Kramer and Matthias Scheutz. Robotic development environments for autonomous mobile robots: A survey. *Autonomous Robots*, 22(2):101–132, 2007.

[5] A. Sloman. What's an AI toolkit. In *Software Tools for Developing Agents: Papers from the 1998 Workshop. AAAI Press, Technical Report WS–98–10*, pages 1–10, 1998.

[6] I. Ulrich and J. J. Borenstein. VFH+: reliable obstacle avoidance for fast mobile robots. In *Proceedings of the IEEE International Conference on Robotics and Automation (ICRA)*, volume 2, pages 1572–1577, 1998.

[7] Richard T. Vaughan and Brian P. Gerkey. Really reusable robot code and the Player / Stage project. In D. Brugali, editor, *Software Engineering for Experimental Robotics*, Springer Tracts on Advanced Robotics. Springer, 2007.

[8] K. Wieczerzak, G.; Kozlowski. Agents that live in robots: how are successful applications built? In *Proceedings of the Fourth International Workshop on Robot Motion and Control, 2004. RoMoCo'04.*, pages 97–102, 2004.

Intelligent Autonomous Systems 10
W. Burgard et al. (Eds.)
IOS Press, 2008

doi:10.3233/978-1-58603-887-8-6

Control of A Trident Steering Walker

Hiroaki YAMAGUCHI [a,1]

[a] *Department of Integrated Information Technology,*
College of Science and Engineering,
Aoyama Gakuin University

Abstract. This paper introduces and describes a new type of wheeled locomotor, which we refer to as a "trident steering walker." The wheeled locomotor is a nonholonomic mechanical system, which consists of an equilateral triangular base, three joints, three links and four steering systems. The equilateral triangular base has a steering system at its center of mass. At each apex of the base is a joint which connects the base and a link. The link has a steering system at its midpoint. The wheeled locomotor transforms driving of the three joints into its movement by operating the four steering systems. This means that the wheeled locomotor achieves undulatory locomotion in which changes in its own shape are transformed into its net displacement. We assume that there is a virtual joint at the end of the first link. The virtual joint connects the first link and a virtual link which has a virtual axle at its midpoint and a virtual steering system at its end. We prove that, by assuming the presence of such virtual mechanical elements, it is possible to convert the kinematical equation of the trident steering walker into five-chain, single-generator chained form in a mathematical framework, differential geometry. Based on chained form, we derive a path following feedback control method which causes the trident steering walker to follow a straight path. The validity of the mechanical design of the trident steering walker, the conversion of its kinematical equation into chained form, and the straight path following feedback control method has been verified by computer simulation.

Keywords. nonholonomic system, wheeled locomotor, undulatory locomotion, differential geometry, chained form

Introduction

Establishing design and control methodologies of mobile mechanisms are important issues in the field of mechanical engineering. In this paper, we describe a novel mobile mechanism performing undulatory locomotion, which will be hereinafter referred to as a "trident steering walker," and its straight path following feedback control method.

Heretofore, there have been developed mobile mechanisms which move by undulatory locomotion, such as a snake-like robot [1], a snake board [2], a roller walker [3] and a rollerblading robot [4]. The snake-like robot [1] is constructed by connecting a plurality of links on which passively rotating wheels are attached. Movement of the snake-like robot is achieved by driving the joints between the links. The undulatory locomotion is motion achieved by transforming changes in the shape of a mobile mechanism into its displacement. The snake board [2] has a rigid body with moment of inertia which rotates

[1]Corresponding Author: Associate Professor, Dr., Hiroaki Yamaguchi, Department of Integrated Information Technology, College of Science and Engineering, Aoyama Gakuin University, 5-10-1 Fuchinobe, Sagamihara-shi, Kanagawa 229-8558, Japan; E-mail: yamaguchi@it.aoyama.ac.jp.

around the center of mass of the vehicle body. It has, in addition, front and rear passively rotating wheels attached on the vehicle body. Both the front and rear wheels are provided with a steering function. Movement of the snake board is achieved by rotating the rigid body with moment of inertia while operating the two steering systems. The roller walker [3] is a mechanism in which passively rotating wheels are attached at the ends of the legs of a four-legged robot. Movement of the roller walker is achieved by driving the joints of the legs. The rollerblading robot [4] has a base mounted on four omnidirectional casters. The base is provided with two legs. At each of the ends of the two legs is attached a passively rotating wheel. The rollerblading robot moves and rotates by driving the joints of its two legs.

In order to achieve stable operation of the snake-like robot, the snake board, the roller walker and the rollerblading robot, a closed loop control method, such as a feedback control method for causing their positions and postures to converge to desired ones asymptotically or a feedback control method for causing them to follow desired paths (or desired trajectories) asymptotically, is required.

As shown in Figure 1, a trident steering walker described in this paper has an equilateral triangular base. At the center of mass of the base is attached a steering system. At each apex of the equilateral trinagular base is provided a link which is connected to it by a joint. The link has a steering system attached at its midpoint. Movement of the trident steering walker is achieved by driving the three joints while operating the four steering systems. In this paper, we describe a feedback control method for causing the trident steering walker to follow a straight path based on its kinematical equation. It should be particularly noted that the method can cause the trident steering walker to follow a straight path smoothly. The trident steering walker is a kind of multiple-connected vehicle system. Other types of multiple-connected vehicle systems include a mobile robot towing multiple trailers [5], a mobile robot towing multiple trailers with multiple steering systems [6] [7] and two car-like mobile robots cooperatively transporting a carrier [8]. While the multiple-connected vehicle systems mentioned above are moved by driving wheels, the movement of the trident steering walker we describe here is achieved by driving joints. In other words, driving of the joints is transformed into the movement of the trident steering walker. In each of the multiple-connected vehicle systems mentioned above, a kinematical equation can be converted into chained form. Chained form is a kind of canonical form, based on which effective control methods have been developed. We will convert the kinematical equation of the trident steering walker into chained form and develop a control method, in particular, a feedback control method for the trident steering walker based on this canonical form in a similar manner. However, it is not simple to convert the kinematical equation of the trident steering walker into chained form, since it has an off-hook structure in which the steering systems are spaced apart from the joints connecting the respective links to the corresponding apexes of the equilateral triangular base, as shown in Figure 1.

In view of this, in this paper we assume that there are a virtual joint, a virtual link, a virtual axle and a virtual steering system as shown in Figure 2, whereby we will demonstrate that the kinematical equation of the trident steering walker can be converted into five-chain, single-generator chained form in spite of its off-hook type structure, by showing calculation results. In addition, we propose a feedback control method based on chained form which enables the trident steering walker to follow a straight path. The validity of the design of the trident steering walker, the conversion of its kinematical equation into chained form and its feedback control method enabling it to follow a straight path has been verified by computer simulation.

1. Structure of the Trident Steering Walker

In this section, the structure of the trident steering walker shown in Figure 1 will be described. The mobile mechanism which performs undulatory locomotion described in this paper is intended to carry objects. For this purpose, the trident steering walker is provided with an equilateral triangular base which can bear objects to be carried. At the center of mass of the equilateral triangular base are attached wheels. The wheels are passive. The rotating direction of the wheels relative to the orientation of the base can be changed positively. This means that the equilateral triangular base is provided with a steering system at the center of mass of the base. Three links are connected to the equilateral triangular base by three joints at the respective apexes of the base. Steering systems are attached on the respective links at their midpoints. As per the above, the equilateral triangular base is supported at its center of mass by the steering system and the base is also supported at its three apexes by the three links having the steering systems attached thereon respectively. Thus, the trident steering walker can bear objects to be carried placed on the base with sufficient stability. In the trident steering walker, it is possible 1) to move and 2) to rotate the equilateral triangular base by periodically driving the three joints and periodically operating the four steering systems. In this paper, we propose a feedback control method for causing the equilateral triangular base to follow a straight path.

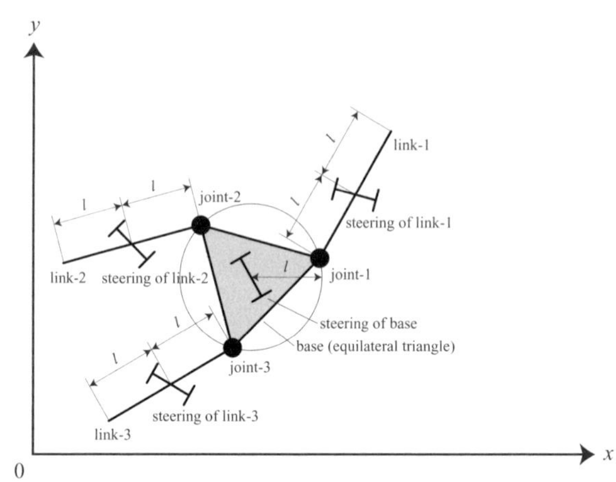

Figure 1. A Trident Steering Walker

The trident steering walker has singular postures in which it is not possible to control its motion by driving the three joints. It should be particularly noted that the trident steering walker can avoid such singular postures by practicing the steering functions of the wheels attached on the three links and on the base.

2. Kinematical Equation of the Trident Steering Walker with Virtual Mechanical Elements

In this paper, we assume that there is a virtual link which is connected to the end of the first link by a virtual joint as shown in Figure 2. We also assume that on the virtual link are attached a virtual axle at its midpoint and a virtual steering system at its end. These virtual mechanical elements are imaginarily supposed in order to convert the kinematical equation of the trident steering walker into five-chain, single-generator chained form, and they do not exist actually. Therefore, these virtual mechanical elements do not impose any physical constraint on the motion of the trident steering walker. In the control method described here, the angular velocities of driving of the first, second and third joints and the angular velocities of operating of the steering systems on the first, second and third links and on the equilateral triangular base are determined according to the angle of the virtual joint and the steering angle of the virtual steering system on the virtual link, which means that the trident steering walker moves while satisfying the constraints imaginarily imposed by the virtual mechanical elements.

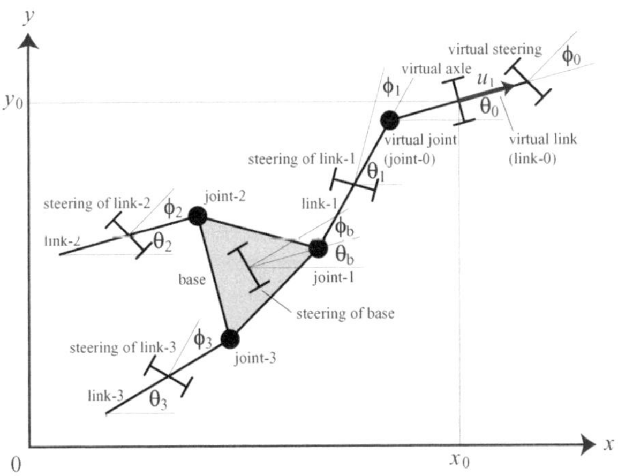

Figure 2. A Trident Steering Walker with Virtual Mechanical Elements

Let u_1 be the displacement velocity of the virtual link, then the kinematical equation of the trident steering walker shown in Figure 2 is derived as:

$$\dot{x} = g_1(x)u_1 + g_2(x)u_2 + g_3(x)u_3 + g_4(x)u_4 + g_5(x)u_5 + g_6(x)u_6, \tag{1}$$

$$x = (x_0, y_0, \phi_0, \theta_0, \phi_1, \theta_1, \phi_b, \theta_b, \phi_2, \theta_2, \phi_3, \theta_3)^T,$$

$$g_1 = (g_{11}, g_{12}, g_{13}, g_{14}, g_{15}, g_{16}, g_{17}, g_{18}, g_{19}, g_{110}, g_{111}, g_{112})^T,$$

$$g_2 = (0,0,1,0,0,0,0,0,0,0,0,0)^T,$$

$$g_3 = (0,0,0,0,1,0,0,0,0,0,0,0)^T,$$

$$g_4 = (0,0,0,0,0,0,1,0,0,0,0,0)^T,$$

$$g_5 = (0,0,0,0,0,0,0,0,1,0,0,0)^T,$$

$$g_6 = (0,0,0,0,0,0,0,0,0,0,1,0)^T,$$

$$\begin{cases} g_{11} = \cos\theta_0,\ g_{12} = \sin\theta_0,\ \ g_{13} = 0,\ g_{14} = \frac{\alpha_{0n}}{l}, \quad g_{15} = 0,\ \ g_{16} = \frac{\alpha_{1n}}{l\cos\phi_1}, \\ g_{17} = 0, \quad g_{18} = \frac{\alpha_{b1n}}{l\cos\phi_b},\ g_{19} = 0,\ g_{110} = \frac{\alpha_{2n}}{l\cos\phi_2},\ g_{111} = 0,\ g_{112} = \frac{\alpha_{3n}}{l\cos\phi_3}, \end{cases}$$

$$\begin{cases} \alpha_{0t} = 1 \\ \alpha_{0n} = \tan\phi_0, \end{cases} \tag{2}$$

$$\begin{cases} \alpha_{1t} = \alpha_{0t}\cos(\theta_0 - \theta_1 - \phi_1) + \alpha_{0n}\sin(\theta_0 - \theta_1 - \phi_1) - 2l\sin\phi_1 \frac{\alpha_{1n}}{l\cos\phi_1} \\ \alpha_{1n} = \alpha_{0t}\sin(\theta_0 - \theta_1 - \phi_1) - \alpha_{0n}\cos(\theta_0 - \theta_1 - \phi_1), \end{cases} \tag{3}$$

$$\begin{cases} \alpha_{b1t} = \alpha_{1t}\cos(\theta_1+\phi_1-\theta_b-\phi_b)+\alpha_{1n}\sin(\theta_1+\phi_1-\theta_b-\phi_b) \\ \alpha_{b1n} = \alpha_{1t}\sin(\theta_1+\phi_1-\theta_b-\phi_b)-\alpha_{1n}\cos(\theta_1+\phi_1-\theta_b-\phi_b), \end{cases} \tag{4}$$

$$\begin{cases} \alpha_{b2t} = \alpha_{b1t} - l\sin\phi_b \frac{\alpha_{b1n}}{l\cos\phi_b} - l\cos(\phi_b-\pi/6)\frac{\alpha_{b1n}}{l\cos\phi_b} \\ \alpha_{b2n} = -l\sin(\phi_b-\pi/6)\frac{\alpha_{b1n}}{l\cos\phi_b}, \end{cases} \tag{5}$$

$$\begin{cases} \alpha_{b3t} = \alpha_{b1t} - l\sin\phi_b \frac{\alpha_{b1n}}{l\cos\phi_b} - l\cos(\phi_b-5\pi/6)\frac{\alpha_{b1n}}{l\cos\phi_b} \\ \alpha_{b3n} = -l\sin(\phi_b-5\pi/6)\frac{\alpha_{b1n}}{l\cos\phi_b}, \end{cases} \tag{6}$$

$$\begin{cases} \alpha_{2t} = \alpha_{b2t}\cos(\theta_b+\phi_b-\theta_2-\phi_2)+\alpha_{b2n}\sin(\theta_b+\phi_b-\theta_2-\phi_2) \\ \qquad\qquad -2l\sin\phi_2\frac{\alpha_{2n}}{l\cos\phi_2} \\ \alpha_{2n} = \alpha_{b2t}\sin(\theta_b+\phi_b-\theta_2-\phi_2)-\alpha_{b2n}\cos(\theta_b+\phi_b-\theta_2-\phi_2), \end{cases} \tag{7}$$

$$\begin{cases} \alpha_{3t} = \alpha_{b3t}\cos(\theta_b+\phi_b-\theta_3-\phi_3)+\alpha_{b3n}\sin(\theta_b+\phi_b-\theta_3-\phi_3) \\ \qquad\qquad -2l\sin\phi_3\frac{\alpha_{3n}}{l\cos\phi_3} \\ \alpha_{3n} = \alpha_{b3t}\sin(\theta_b+\phi_b-\theta_3-\phi_3)-\alpha_{b3n}\cos(\theta_b+\phi_b-\theta_3-\phi_3). \end{cases} \tag{8}$$

A vector, $(x_0, y_0)^T$, represents the position of the midpoint of the virtual link. Angles, ϕ_0 and θ_0, respectively represent the steering angle of the virtual steering system and the orientation of the virtual link. Five control inputs, u_2, u_3, u_4, u_5 and u_6, respectively represent the angular velocities of the virtual steering system on the virtual link and the steering systems on the first link, on the equilateral triangular base and on the second and third links. The displacement velocity, u_1, of the virtual link is not a control input, as a matter of course. Given angular velocity inputs, v_1, v_2 and v_3, for the angular velocities, $\dot{\psi}_1$, $\dot{\psi}_2$ and $\dot{\psi}_3$, of driving of the first, second and third joints, the displacement velocity, u_1, is determined as the value which satisfies the following equation as:

$$\begin{cases} \dot{\psi}_1 = \dot{\theta}_1 - \dot{\theta}_b = (g_{16}-g_{18})\,u_1 = v_1 \\ \dot{\psi}_2 = \dot{\theta}_2 - \dot{\theta}_b = (g_{110}-g_{18})\,u_1 = v_2 \\ \dot{\psi}_3 = \dot{\theta}_3 - \dot{\theta}_b = (g_{112}-g_{18})\,u_1 = v_3. \end{cases} \tag{9}$$

The angular velocity, $\dot{\psi}_0$, of driving of the virtual joint, namely, the angular velocity input, v_0, is calculated as:

$$\dot{\psi}_0 = \dot{\theta}_0 - \dot{\theta}_1 = (g_{14}-g_{16})\,u_1 = v_0. \tag{10}$$

3. Conversion into Chained Form

In this paper, it is assumed that there is a virtual link which is connected to the end of the first link by a virtual joint. On this virtual link are attached a virtual axle at its midpoint and a virtual steering system at its end. In this section, we will demonstrate, by assuming the presence of these virtual mechanical elements, that the kinematical equation of the trident steering walker can be converted into five-chain, single-generator chained form by showing calculation results. This conversion is based on differential geometry [9]. First, six vector fields, g_1, g_2, g_3, g_4, g_5 and g_6, in Eq.(1) are converted as:

$$\begin{cases} f_1 = g_1/\cos\theta_0 \\ f_i = g_i, \quad i = 2, 3, \cdots, 6. \end{cases} \tag{11}$$

Secondly, u_1, u_2, u_3, u_4, u_5 and u_6 are converted as:

$$\begin{cases} \tilde{u}_1 = u_1\cos\theta_0 \\ \tilde{u}_i = u_i, \quad i = 2, 3, \cdots, 6. \end{cases} \tag{12}$$

Thus, the kinematical equation of the trident steering walker, Eq.(1), is converted as:

$$\dot{x} = f_1(x)\tilde{u}_1 + f_2(x)\tilde{u}_2 + f_3(x)\tilde{u}_3 + f_4(x)\tilde{u}_4 + f_5(x)\tilde{u}_5 + f_6(x)\tilde{u}_6. \tag{13}$$

Using these six vector fields, f_1, f_2, f_3, f_4, f_5 and f_6, the variables in the kinematical equation of the trident steering walker, Eq.(1), are converted as:

$$\left\{\begin{array}{llll} z_{11} = & h_1 & = x_0 \\ z_{21} = & L_{f_1}^2 h_2 & = \frac{1}{\cos^3\theta_0}\frac{\tan\phi_0}{l} \\ z_{22} = & L_{f_1} h_2 & = \tan\theta_0 \\ z_{23} = & h_2 & = y_0 \\ z_{31} = & L_{f_1} h_3 & = \frac{1}{\cos\theta_0} g_{14} - \frac{1}{\cos\theta_0} g_{16} \\ z_{32} = & h_3 & = \theta_0 - \theta_1 \\ z_{41} = & L_{f_1} h_4 & = \frac{1}{\cos\theta_0} g_{16} - \frac{1}{\cos\theta_0} g_{18} \\ z_{42} = & h_4 & = \theta_1 - \theta_b \\ z_{51} = & L_{f_1} h_5 & = \frac{1}{\cos\theta_0} g_{110} - \frac{1}{\cos\theta_0} g_{18} \\ z_{52} = & h_5 & = \theta_2 - \theta_b \\ z_{61} = & L_{f_1} h_6 & = \frac{1}{\cos\theta_0} g_{112} - \frac{1}{\cos\theta_0} g_{18} \\ z_{62} = & h_6 & = \theta_3 - \theta_b. \end{array}\right. \tag{14}$$

A Lie derivative, $L_p q$, represents an inner product of a vector whose components are partial derivatives with respect to state variables, x_i, $i = 1, 2, \cdots, n$, of a scalar quantity, q, namely, a vector, $(\partial q/\partial x_1, \partial q/\partial x_2, \cdots, \partial q/\partial x_n)^T$, and a vector, $p=(p_1, p_2, \cdots, p_n)^T$. That is, $L_p q = \frac{\partial q}{\partial x_1}p_1 + \frac{\partial q}{\partial x_2}p_2 + \cdots + \frac{\partial q}{\partial x_n}p_n$. The derivative with respect to timc of Eq.(14) is five-chain, single-generator chained form as:

$$\begin{array}{llll} \dot{z}_{11} = w_1 & \dot{z}_{21} = w_2 & \dot{z}_{31} = w_3 & \dot{z}_{41} = w_4 \\ & \dot{z}_{22} = z_{21}w_1 & \dot{z}_{32} = z_{31}w_1 & \dot{z}_{42} = z_{41}w_1 \\ & \dot{z}_{23} = z_{22}w_1 & & \\ & & & \\ & \dot{z}_{51} = w_5 & \dot{z}_{61} = w_6 & \\ & \dot{z}_{52} = z_{51}w_1 & z_{62} = z_{61}w_1, & \end{array} \tag{15}$$

$$\begin{array}{l} w_1 = \tilde{u}_1 \\ w_2 = L_{f_1}^3 h_2\tilde{u}_1 + L_{f_2}L_{f_1}^2 h_2\tilde{u}_2 + L_{f_3}L_{f_1}^2 h_2\tilde{u}_3 + L_{f_4}L_{f_1}^2 h_2\tilde{u}_4 + L_{f_5}L_{f_1}^2 h_2\tilde{u}_5 + L_{f_6}L_{f_1}^2 h_2\tilde{u}_6 \\ w_3 = L_{f_1}^2 h_3\tilde{u}_1 + L_{f_2}L_{f_1} h_3\tilde{u}_2 + L_{f_3}L_{f_1} h_3\tilde{u}_3 + L_{f_4}L_{f_1} h_3\tilde{u}_4 + L_{f_5}L_{f_1} h_3\tilde{u}_5 + L_{f_6}L_{f_1} h_3\tilde{u}_6 \\ w_4 = L_{f_1}^2 h_4\tilde{u}_1 + L_{f_2}L_{f_1} h_4\tilde{u}_2 + L_{f_3}L_{f_1} h_4\tilde{u}_3 + L_{f_4}L_{f_1} h_4\tilde{u}_4 + L_{f_5}L_{f_1} h_4\tilde{u}_5 + L_{f_6}L_{f_1} h_4\tilde{u}_6 \\ w_5 = L_{f_1}^2 h_5\tilde{u}_1 + L_{f_2}L_{f_1} h_5\tilde{u}_2 + L_{f_3}L_{f_1} h_5\tilde{u}_3 + L_{f_4}L_{f_1} h_5\tilde{u}_4 + L_{f_5}L_{f_1} h_5\tilde{u}_5 + L_{f_6}L_{f_1} h_5\tilde{u}_6 \\ w_6 = L_{f_1}^2 h_6\tilde{u}_1 + L_{f_2}L_{f_1} h_6\tilde{u}_2 + L_{f_3}L_{f_1} h_6\tilde{u}_3 + L_{f_4}L_{f_1} h_6\tilde{u}_4 + L_{f_5}L_{f_1} h_6\tilde{u}_5 + L_{f_6}L_{f_1} h_6\tilde{u}_6. \end{array}$$

4. Control Inputs and Control Stability

In this section, we describe a feedback control method in which the virtual joint and the first, second and third joints are driven, and the virtual steering system on the virtual link and the steering systems on the first, second and third links and on the equilateral triangular base are operated in such a manner that the midpoint of the virtual link moves along the x-axis of the static coordinate system, in other words, in such a manner as to cause the equilateral triangular base to follow the x-axis. In order to achieve the aforementioned movement following the x-axis, it is necessary to cause the variables, z_{23}, z_{22} and z_{21}, to converge to zero. To this end, under the following condition as:

$$w_1 = a_0, \tag{16}$$

the control input, w_2, is given as:

$$w_2 = p_{21}z_{21} + p_{22}\frac{z_{22}}{a_0} + p_{23}\frac{z_{23}}{a_0^2}. \tag{17}$$

In the above equation, Eq.(16), a_0 is a constant value not equal to zero, and it physically means the displacement velocity, $\tilde{u}_1$, of the midpoint of the virtual link with respect to the x-axis. That is,

$$\dot{x}_0 = \tilde{u}_1 = w_1 = a_0 \neq 0.$$

Therefore, the displacement velocity, $u_1(=\tilde{u}_1/\cos\theta_0)$, of the virtual link is also not equal to zero, and the angular velocities, $\dot{\psi}_1$, $\dot{\psi}_2$ and $\dot{\psi}_3$, of driving of the first, second and third joints, namely, the angular velocity inputs, v_1, v_2 and v_3, are uniquely determined by Eq.(9). The angular velocity, $\dot{\psi}_0$, of driving of the virtual joint, namely, the angular velocity input, v_0, is uniquely determined by Eq.(10). When a three dimensional vector, ζ_2, is defined as Eq.(18), its derivative with respect to time, $\dot{\zeta}_2$, is given as Eq.(19).

$$\zeta_2 = (\zeta_{21}, \zeta_{22}, \zeta_{23})^T = \left(z_{21}, \frac{z_{22}}{a_0}, \frac{z_{23}}{a_0^2} \right)^T. \tag{18}$$

$$\dot{\zeta}_2 = \begin{pmatrix} p_{21} & p_{22} & p_{23} \\ 1 & 0 & 0 \\ 0 & 1 & 0 \end{pmatrix} \zeta_2 = \Lambda_2 \zeta_2. \tag{19}$$

Giving coefficients, p_{21}, p_{22} and p_{23}, which make all the real parts of the eigenvalues of a matrix, Λ_2, in Eq.(19) negative causes the vector, ζ_2, to converge exponentially to zero, whereby the movement following the x-axis is achieved.

To follow the x-axis, it is necessary for the trident steering walker to continue to move, as a matter of course. In other words, it is necessary that at least one of the first, second and third joints is driven and the condition, $w_1=a_0\neq 0$, is always satisfied. Therefore, it is necessary to change the angles, $\psi_1(=z_{42})$, $\psi_2(=z_{52})$ and $\psi_3(=z_{62})$, of the first, second and third joints periodically while preventing the trident steering walker from having singular postures in which its motion cannot be controlled. The angle, $\psi_0(=z_{32})$, of the virtual joint should also be changed periodically if need be. To this end, the following control inputs, w_3, w_4, w_5 and w_6, are given as:

$$\begin{cases} w_3 = p_{31} z_{31} + p_{32} \frac{z_{32}}{a_0} + e_3 \\ w_4 = p_{41} z_{41} + p_{42} \frac{z_{42}}{a_0} + e_4 \\ w_5 = p_{51} z_{51} + p_{52} \frac{z_{52}}{a_0} + e_5 \\ w_6 = p_{61} z_{61} + p_{62} \frac{z_{62}}{a_0} + e_6. \end{cases} \tag{20}$$

In the control inputs in Eq.(20), variables, e_3, e_4, e_5 and e_6, are oscillatory functions as:

$$\begin{cases} e_3 = A_3 \sin(\omega_3 t + \varphi_3) \\ e_4 = A_4 \sin(\omega_4 t + \varphi_4) \\ e_5 = A_5 \sin(\omega_5 t + \varphi_5) \\ e_6 = A_6 \sin(\omega_6 t + \varphi_6). \end{cases} \tag{21}$$

When a two dimensional vector, ζ_3, is defined as Eq.(22), its derivative with respect to time, $\dot{\zeta}_3$, is given as Eq.(23).

$$\zeta_3 = (\zeta_{31}, \zeta_{32})^T = \left(z_{31}, \frac{z_{32}}{a_0} \right)^T. \tag{22}$$

$$\dot{\zeta}_3 = \begin{pmatrix} p_{31} & p_{32} \\ 1 & 0 \end{pmatrix} \zeta_3 + \begin{pmatrix} A_3 \sin(\omega_3 t + \varphi_3) \\ 0 \end{pmatrix} = \Lambda_3 \zeta_3 + \begin{pmatrix} 1 \\ 0 \end{pmatrix} A_3 \sin(\omega_3 t + \varphi_3). \tag{23}$$

Similarly, when two dimensional vectors, ζ_4, ζ_5 and ζ_6, are defined as Eqs.(24), (26) and (28), their derivatives with respect to time, $\dot{\zeta}_4$, $\dot{\zeta}_5$ and $\dot{\zeta}_6$, are given as Eqs.(25), (27) and (29), respectively.

$$\zeta_4 = (\zeta_{41}, \zeta_{42})^T = \left(z_{41}, \frac{z_{42}}{a_0}\right)^T. \tag{24}$$

$$\dot{\zeta}_4 = \begin{pmatrix} p_{41} & p_{42} \\ 1 & 0 \end{pmatrix} \zeta_4 + \begin{pmatrix} A_4 \sin(\omega_4 t + \varphi_4) \\ 0 \end{pmatrix} = \Lambda_4 \zeta_4 + \begin{pmatrix} 1 \\ 0 \end{pmatrix} A_4 \sin(\omega_4 t + \varphi_4). \tag{25}$$

$$\zeta_5 = (\zeta_{51}, \zeta_{52})^T = \left(z_{51}, \frac{z_{52}}{a_0}\right)^T. \tag{26}$$

$$\dot{\zeta}_5 = \begin{pmatrix} p_{51} & p_{52} \\ 1 & 0 \end{pmatrix} \zeta_5 + \begin{pmatrix} A_5 \sin(\omega_5 t + \varphi_5) \\ 0 \end{pmatrix} = \Lambda_5 \zeta_5 + \begin{pmatrix} 1 \\ 0 \end{pmatrix} A_5 \sin(\omega_5 t + \varphi_5). \tag{27}$$

$$\zeta_6 = (\zeta_{61}, \zeta_{62})^T = \left(z_{61}, \frac{z_{62}}{a_0}\right)^T. \tag{28}$$

$$\dot{\zeta}_6 = \begin{pmatrix} p_{61} & p_{62} \\ 1 & 0 \end{pmatrix} \zeta_6 + \begin{pmatrix} A_6 \sin(\omega_6 t + \varphi_6) \\ 0 \end{pmatrix} = \Lambda_6 \zeta_6 + \begin{pmatrix} 1 \\ 0 \end{pmatrix} A_6 \sin(\omega_6 t + \varphi_6). \tag{29}$$

Giving coefficients, p_{31}, p_{32}, p_{41}, p_{42}, p_{51}, p_{52}, p_{61} and p_{62}, which make all the real parts of the eigenvalues of the matrices, Λ_3, Λ_4, Λ_5 and Λ_6, in Eqs.(23), (25), (27) and (29) negative causes the angles, $\psi_0(=z_{32})$, $\psi_1(=z_{42})$, $\psi_2(=z_{52})$ and $\psi_3(=z_{62})$, of the virtual joint and the first, second and third joints to converge respectively to corresponding periodic functions, r_3, r_4, r_5 and r_6, as:

$$\begin{cases} \psi_0 = z_{32} = r_3 = \hat{A}_3 \sin(\hat{\omega}_3 t + \hat{\varphi}_3) \\ \psi_1 = z_{42} = r_4 = \hat{A}_4 \sin(\hat{\omega}_4 t + \hat{\varphi}_4) \\ \psi_2 = z_{52} = r_5 = \hat{A}_5 \sin(\hat{\omega}_5 t + \hat{\varphi}_5) \\ \psi_3 = z_{62} = r_6 = \hat{A}_6 \sin(\hat{\omega}_6 t + \hat{\varphi}_6). \end{cases} \tag{30}$$

The angular frequencies, $\hat{\omega}_3$, $\hat{\omega}_4$, $\hat{\omega}_5$ and $\hat{\omega}_6$, of these periodic functions, r_3, r_4, r_5 and r_6, are ω_3, ω_4, ω_5 and ω_6 respectively, namely, $\hat{\omega}_3{=}\omega_3$, $\hat{\omega}_4{=}\omega_4$, $\hat{\omega}_5{=}\omega_5$ and $\hat{\omega}_6{=}\omega_6$. The amplitudes, $\hat{A}_3$, $\hat{A}_4$, $\hat{A}_5$ and $\hat{A}_6$, and the phases, $\hat{\phi}_3$, $\hat{\phi}_4$, $\hat{\phi}_5$ and $\hat{\phi}_6$, of these periodic functions, r_3, r_4, r_5 and r_6, are functions of the amplitudes, A_3, A_4, A_5 and A_6, and the phases, ϕ_3, ϕ_4, ϕ_5 and ϕ_6, of the oscillatory functions, e_3, e_4, e_5 and e_6, of the control inputs in Eq.(20). Therefore, the amplitudes, $\hat{A}_3$, $\hat{A}_4$, $\hat{A}_5$ and $\hat{A}_6$, the angular frequencies, $\hat{\omega}_3$, $\hat{\omega}_4$, $\hat{\omega}_5$ and $\hat{\omega}_6$, and the phases, $\hat{\phi}_3$, $\hat{\phi}_4$, $\hat{\phi}_5$ and $\hat{\phi}_6$, of the periodic functions, r_3, r_4, r_5 and r_6, are designed by means of the parameters, A_3, A_4, A_5, A_6, ω_3, ω_4, ω_5, ω_6, ϕ_3, ϕ_4, ϕ_5 and ϕ_6, in such a manner that the trident steering walker continues to move while avoiding singular postures in which its motion cannot be controlled.

5. Simulation

In this section, the validity of the described control method which causes the equilateral triangular base of the trident steering walker to follow the x-axis will be verified by computer simulation. In the simulation, all of the virtual link and the first, second and third links are set to have a length of $0.15[\mathrm{m}]$(l=$0.075[\mathrm{m}]$). The displacement velocity, $\tilde{u}_1$, of the midpoint of the virtual link of the trident steering walker with respect to the x-axis is set as:

$$\dot{x}_0 = \tilde{u}_1 = w_1 = a_0 = 0.4[\mathrm{m/s}].$$

The coefficients of the control inputs in Eqs.(17) and (20) are set as:

$$\begin{aligned} &p_{21} = -3.0, \quad p_{22} = -\tfrac{p_{21}^2}{3.0} = -3.0, \quad p_{23} = \tfrac{p_{21}^3}{27.0} = -1.0, \\ &p_{31} = -2.0, \quad p_{32} = -p_{31}^2/4.0 = -1.0, \\ &p_{41} = -2.0, \quad p_{42} = -p_{41}^2/4.0 = -1.0, \\ &p_{51} = -2.0, \quad p_{52} = -p_{51}^2/4.0 = -1.0, \\ &p_{61} = -2.0, \quad p_{62} = -p_{61}^2/4.0 = -1.0. \end{aligned}$$

In this case, all the eigenvalues of the matrices, Λ_2, Λ_3, Λ_4, Λ_5 and Λ_6, in Eqs.(19), (23), (25), (27) and (29) become equal to -1.0. The amplitudes, $\hat{A}_3$, $\hat{A}_4$, $\hat{A}_5$ and $\hat{A}_6$, the angular frequencies, $\hat{\omega}_3$, $\hat{\omega}_4$, $\hat{\omega}_5$ and $\hat{\omega}_6$, and the phases, $\hat{\phi}_3$, $\hat{\phi}_4$, $\hat{\phi}_5$ and $\hat{\phi}_6$, of the periodic functions, r_3, r_4, r_5 and r_6, in Eq.(30), to which the angles, $\psi_0(=z_{32})$, $\psi_1(=z_{42})$, $\psi_2(=z_{52})$ and $\psi_3(=z_{62})$, of the virtual joint and the first, second and third joints converge respectively are set as:

$$\begin{aligned}
&\hat{A}_3 = \pi/6[\text{rad}], \hat{\omega}_3 = 2\pi/5[\text{rad/sec}], \hat{\varphi}_3 = 0[\text{rad}],\\
&\hat{A}_4 = \pi/6[\text{rad}], \hat{\omega}_4 = 2\pi/5[\text{rad/sec}], \hat{\varphi}_4 = \pi[\text{rad}],\\
&\hat{A}_5 = \pi/6[\text{rad}], \hat{\omega}_5 = 2\pi/5[\text{rad/sec}], \hat{\varphi}_5 = \pi/3[\text{rad}],\\
&\hat{A}_6 = \pi/6[\text{rad}], \hat{\omega}_6 = 2\pi/5[\text{rad/sec}], \hat{\varphi}_6 = 2\pi/3[\text{rad}].
\end{aligned}$$

Then, the amplitudes, A_3, A_4, A_5 and A_6, the angular frequencies, ω_3, ω_4, ω_5 and ω_6, and the phases, ϕ_3, ϕ_4, ϕ_5 and ϕ_6, of the oscillatory functions, e_3, e_4, e_5 and e_6, of the control inputs in Eq.(20) are given as:

$$\begin{aligned}
A_3 &= \hat{A}_3\left(4\hat{\omega}_3^2 + p_{31}^2\right)/\left(4a_0\right), \omega_3 = \hat{\omega}_3, \varphi_3 = \hat{\varphi}_3 - \text{atan2}\left\{4\hat{\omega}_3 p_{31}/\left(-4\hat{\omega}_3^2 + p_{31}^2\right)\right\},\\
A_4 &= \hat{A}_4\left(4\hat{\omega}_4^2 + p_{41}^2\right)/\left(4a_0\right), \omega_4 = \hat{\omega}_4, \varphi_4 = \hat{\varphi}_4 - \text{atan2}\left\{4\hat{\omega}_4 p_{41}/\left(-4\hat{\omega}_4^2 + p_{41}^2\right)\right\},\\
A_5 &= \hat{A}_5\left(4\hat{\omega}_5^2 + p_{51}^2\right)/\left(4a_0\right), \omega_5 = \hat{\omega}_5, \varphi_5 = \hat{\varphi}_5 - \text{atan2}\left\{4\hat{\omega}_5 p_{51}/\left(-4\hat{\omega}_5^2 + p_{51}^2\right)\right\},\\
A_6 &= \hat{A}_6\left(4\hat{\omega}_6^2 + p_{61}^2\right)/\left(4a_0\right), \omega_6 = \hat{\omega}_6, \varphi_6 = \hat{\varphi}_6 - \text{atan2}\left\{4\hat{\omega}_6 p_{61}/\left(-4\hat{\omega}_6^2 + p_{61}^2\right)\right\}.
\end{aligned}$$

The initial conditions are given as:

$$\begin{aligned}
&(x_0, y_0)^T|_{t=0} = (0.6[\text{m}], 0.0[\text{m}])^T,\\
&\theta_0|_{t=0} = 0.0[\text{rad}], \quad \phi_0|_{t=0} = 0.0[\text{rad}],\\
&\theta_1|_{t=0} = 0.0[\text{rad}], \quad \phi_1|_{t=0} = 0.1227498[\text{rad}],\\
&\theta_b|_{t=0} = 0.0[\text{rad}], \quad \phi_b|_{t=0} = 0.2419082[\text{rad}],\\
&\theta_2|_{t=0} = 0.4534498[\text{rad}], \phi_2|_{t=0} = -0.2693041[\text{rad}],\\
&\theta_3|_{t=0} = 0.4534498[\text{rad}], \phi_3|_{t=0} = -0.1523114[\text{rad}].
\end{aligned}$$

In the initial conditions above, the midpoint of the virtual link is located on the x-axis, and the orientation of the virtual link is parallel to the x-axis, and the steering angle of the virtual steering system on the virtual link is equal to zero. In addition, the angles, $\psi_0(=z_{32})$, $\psi_1(=z_{42})$, $\psi_2(=z_{52})$ and $\psi_3(=z_{62})$, of the virtual joint and the first, second and third joints have already converged to the periodic functions, r_3, r_4, r_5 and r_6, respectively.

A simulation result is shown in Figure 3. It is understood from Figure 3 that the end of the first link always moves along the x-axis. This means that the virtual link connected to the end of the first link by the virtual joint moves along the x-axis. The displacement velocity of the virtual link with respect to the x-axis is $\tilde{u}_1=w_1=a_0=0.4[\text{m/s}]$. In other words, the movement of the equilateral triangular base following the x-axis is achieved. It is also understood from Figure 3 that the equilateral triangular base always keeps a posture parallel to the x-axis. This is because the amplitude, $\hat{A}_3$, and the angular frequency, $\hat{\omega}_3$, of the periodic function, r_3, to which the angle, $\psi_0(=z_{32})$, of the virtual joint converges are designed to be equal respectively to the amplitude, $\hat{A}_4$, and the angular velocity, $\hat{\omega}_4$, of the periodic function, r_4, to which the angle, $\psi_1(=z_{42})$, of the first joint converges, and the phase difference, $\hat{\phi}_3-\hat{\phi}_4$, between the periodic functions, r_3 and r_4, is designed to be $-\pi[\text{rad}]$. As per the above, from the simulation result shown in Figure 3, the validity of the control method described here which makes it possible to cause the equilateral triangular base of the trident steering walker to follow the x-axis has been verified.

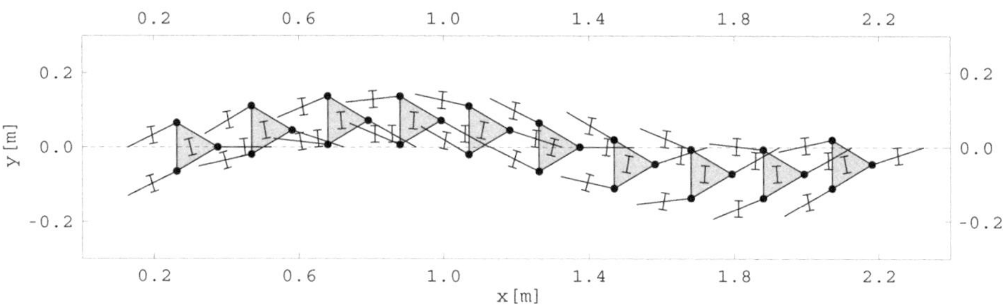

$t = 0.0 - 4.5[s]$

Figure 3. A Simulation Result

Conclusion

In this paper, a novel mobile mechanism which is referred to as a trident steering walker and its feedback control method have been described. It has been demonstrated, by showing calculation results, that the kinematical equation of the trident steering walker can be converted into five-chain, single-generator chained form by assuming the presence of a virtual joint, a virtual link, a virtual axle and a virtual steering system. The feedback control method which enables the trident steering walker to follow a straight path based on chained form has been proposed. Especially, it is achieved that the trident steering walker follows a straight path smoothly by driving its three joints and operating its four steering systems. The validity of the mechanical design of the trident steering walker, the conversion of its kinematical equation into chained form, and the straight path following feedback control method has been verified by computer simulation.

Acknowledgements

This research has been supported by Category C of Grant-in-Aid for Scientific Research (19560238), Japan Society for the Promotion of Science.

References

[1] S. Hirose: Biologically Inspired Robots (Snake-like Locomotor and Manipulator), *Oxford University Press*, (1993).

[2] J.P. Ostrowski and J.W. Burdick: The Geometric Mechanics of Undulatory Robotic Locomotion, *International Journal of Robotics Research*, **17(7)** (1998), 683-701.

[3] S. Hirose and H. Takeuchi: Study on Roller-Walk (Basic Characteristics and Its Control), Proceedings of the 1996 IEEE International Conference on Robotics and Automation, (1996), 3265-3270.

[4] S. Chitta and V. Kumar: Dynamics and Generation of Gaits for a Planar Rollerblader, *Proceedings of the 2003 IEEE/RSJ International Conference on Intelligent Robots and Systems*, (2003), 860-865.

[5] C. Samson: Control of Chained Systems: Application to Path Following and Time-Varying Point-Stabilization of Mobile Robots, *IEEE Transactions on Automatic Control*, **40(1)** (1995), 64-77.

[6] D.M. Tilbury, O.J. Sordalen, L.G. Bushnell, and S.S. Sastry: A Multisteering Trailer System: Conversion into Chained Form using Dynamic Feedback, *IEEE Transactions on Robotics and Automation*, **11(6)** (1995), 807-818.

[7] D.M. Tilbury and S.S. Sastry: The Multi-Steering N-Trailer System: A Case Study of Goursat Normal Forms and Prolongations, *International Journal of Robust and Nonlinear Control*, **5(4)** (1995), 343-364.

[8] H. Yamaguchi and T. Arai: A Path Following Feedback Control Method for A Cooperative Transportation System with Two Car-Like Mobile Robots, *Transactions on Society of Instruments and Control Engineers (SICE in Japanese)*, **39(6)**, (2003), 575-584.

[9] A. Isidori: Nonlinear Control Systems, Second Edition, *Springer-Verlag*, New York (1989).

Intelligent Autonomous Systems 10
W. Burgard et al. (Eds.)
IOS Press, 2008

doi:10.3233/978-1-58603-887-8-16

Novelty Detection and Online Learning for Vibration-based Terrain Classification

Christian WEISS [a,1] and Andreas ZELL [a]
[a] *Computer Science Department, University of Tübingen, Germany*

Abstract. In outdoor environments, a great variety of ground surfaces exists. To ensure safe navigation, a mobile robot should be able to identify the current terrain so that it can adapt its driving style. If the robot navigates in known environments, a terrain classification method can be trained on the expected terrain classes in advance. However, if the robot is to explore previously unseen areas, it may face terrain types that it has not been trained to recognize. In this paper, we present a vibration-based terrain classification system that uses novelty detection based on Gaussian mixture models to detect if the robot traverses an unknown terrain class. If the robot has collected a sufficient number of examples of the unknown class, the new terrain class is added to the classification model online. Our experiments show that the classification performance of the automatically learned model is only slightly worse than the performance of a classifier that knows all classes beforehand.

Keywords. vibration-based terrain classification

Introduction

In outdoor environments, a mobile robot faces many different types of ground surfaces. Each surface poses different hazards to the robot. One type of hazards are positive and negative obstacles, e.g. rocks or ditches. Other dangers originate from the properties of the ground surface itself, for example that it is very rough or slippery. Such dangers are called *non-geometric hazards* [1]. To avoid accidents or damage, a robot should be able to identify the current terrain in order to adapt its driving style. A common way to determine the terrain type is to group the terrain into classes that have known properties, e.g. grass or gravel. These classes are learned from training examples.

The most common methods for terrain classification use laser scanners (e.g. [2,3]) or cameras (e.g. [4,3,5]). A way to detect non-geometric hazards is vibration-based terrain classification, which was first suggested by Iagnemma and Dubowsky [6]. The main idea is based on the observation that traversing different terrain types induces different kinds of vibrations in the body or the wheels of the robot. Commonly, the vibrations are measured by accelerometers and the characteristic vibrations of different terrain classes are learned from training examples. A method based on probabilistic neural networks has been presented by DuPont et al. [7]. An approach based on linear discriminant analysis

[1]Corresponding Author: Christian Weiss, Computer Science Department, University of Tübingen, Sand 1, 72076 Tübingen, Germany; E-mail: c.weiss@uni-tuebingen.de.

was suggested by Brooks and Iagnemma [8]. We proposed a method that uses a support vector machine [9,10]. Stavens et al. focus on assessing the roughness of the terrain instead of grouping the ground surface into classes [11]. There also exist systems that combine vision and vibration sensing [12,13,14].

These approaches can only identify terrain classes that are contained in the training set. If the robot explores unknown areas, it is likely to encounter terrain it has never seen before. In this case, the new terrain will be classified wrongly as a class from the training set. Unlike the other approaches, the one presented by Brooks [8] can classify terrain as "unknown". This occurs, however, if a test example is equally similar to two terrain classes, and does not indicate that the current terrain is new. A possibility to detect new terrain is novelty detection which detects if a test sample is dissimilar to the training set (and therefore novel). Novelty detection approaches have been presented, for example, based on nearest-neighbor techniques [15], string matching [16], multi-layer perceptrons [17], one-class support vector machines (SVM) [18] or habituation [19]. An overview over different novelty detection methods can be found in [20,21].

In the paper at hand, we present a vibration-based terrain classification system that is able to detect if the robot traverses a ground surface it has not been trained to identify. We achieve this by novelty detection based on Gaussian mixture models (GMM) which have been sucessfully used for novelty detection in other domains, e.g. the detection of masses in mammograms [22] or sensor fault detection [23]. We chose GMMs because they are relatively simple and fast. In initial experiments, they also performed better than one-class SVMs. If in our system the robot has collected enough samples of a new class, this class is added to the model online. Our experimental results show that the classification performance of automatically learned models is only slightly worse than the performance of models that were trained manually on the same classes.

The rest of this paper is organized as follows. Section 1 gives an overview of our system and Section 2 describes the components in more detail. Section 3 presents our experimental results. Finally, Section 4 concludes the paper and suggests future work.

1. System Overview

The first step in our system is an offline training phase. As training data, we use vibration data of a number of terrain classes that are known to the robot right from the start. We represent vibration data by accelerations measured in three directions: perpendicularly to the ground floor (z-acceleration), left-to-right (y-acceleration) and front-to-back (x-acceleration) [10]. We split these x-, y- and z- vibration signals into smaller segments corresponding to a period of 1 s. At a measurement frequency of 100 Hz, this leads to three 1×100 vectors per terrain segment. We then transform each vector individually to the frequency domain using a log-scaled power spectral density (PSD), as suggested by Brooks [8]. After this, we concatenate the PSDs of the x-, y- and z- vibrations of a terrain segment to create the feature vector of the segment. Next, we normalize the feature vectors in the training set such that each feature has mean 0 and standard deviation 1. Additionally to such a feature vector, our system holds two other representations for a terrain segment. The first one is the feature vector reduced by principal component analysis (PCA) to dimension 1×20. The second one is formed by the raw acceleration data of the segment. In each stage of the system, the appropriate representation is used.

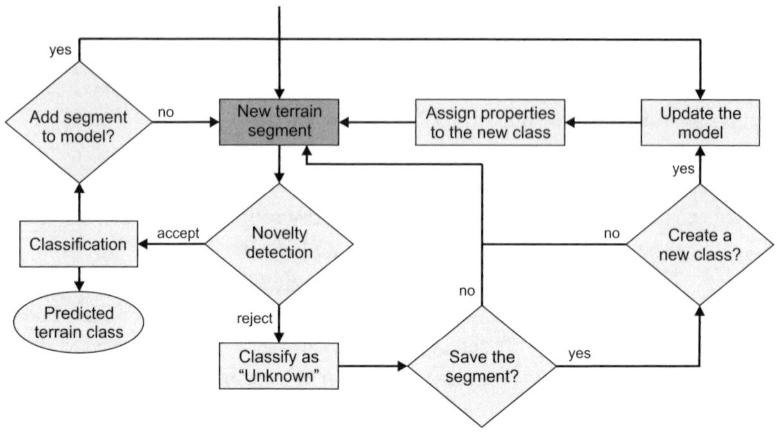

Figure 1. Overview of the classification phase of our system.

To complete the training phase, we train a k-nearest-neighbor (kNN) classifier on the initial training set. Additionally, we create an initial set of Gaussian mixture models for novelty detection.

The classification phase is depicted in Fig. 1. This phase forms a repeating cycle starting with the collection of a new one-second terrain segment and the computation of the corresponding feature vector $\mathbf{f}$. In the following step, the novelty detection method checks if $\mathbf{f}$ belongs to one of the known classes ($\mathbf{f}$ is "accepted") or if the class of $\mathbf{f}$ seems to be unknown so far ($\mathbf{f}$ is "rejected"). In case of acceptance, the kNN classifier predicts the segment's terrain type. Next, the system decides whether to add $\mathbf{f}$ to the training set. Being able to update existing classes is especially useful for increasing the number of training examples of newly created classes until the training set is balanced.

If the new terrain segment $\mathbf{f}$ is "rejected", the system decides if $\mathbf{f}$ should be stored in a buffer for further use, or if $\mathbf{f}$ is discarded. This step is an attempt to separate rejected outliers of known classes from truly unknown terrain segments. The next stage checks if the number of stored feature vectors in the buffer exceeds a given threshold. This indicates that enough examples of the unknown class are available to create a new class. In this case, the classifier must be retrained and new GMMs for novelty detection must be computed. Finally, the system tries to assign some properties like hardness or bumpiness to the new class.

2. System Components

2.1. Novelty Detection Using GMMs

The purpose of the novelty detection stage is to predict whether or not a newly aquired terrain segment $\mathbf{f}$ belongs to a class that is contained in the training set. For this purpose, we follow a novelty detection approach described in [24]. The main idea is to model the distribution of the available training data by a Gaussian mixture model. If a test example cannot be explained by the model, i.e. has a low probability given the model, it is likely to belong to a class that is not present in the training set. Our terrain classification system works on the feature vectors whose size have been reduced by PCA to dimension $1 \times d$.

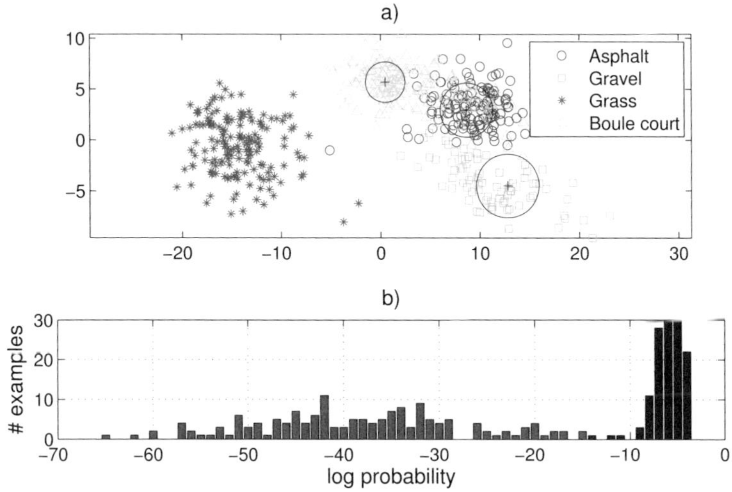

Figure 2. a) GMM with three spherical Gaussians fitted to vibration data (100 examples per class) of asphalt, gravel and boule court. Grass is used for testing. The centers of the Gaussians are indicated by a blue plus sign. Blue circles show one standard deviation error bars. The data dimension was reduced to 2 by applying PCA. b) Histogram of log probabilities on the training set (dark blue) and on the test set (light purple). Values larger than 30 were cut for clearer visibility.

In experiments, we found $d = 20$ to be a good trade-off between novelty detection performance and computation time. Smaller d generally led to decreasing novelty detection performance, while for larger d, generating GMMs becomes to slow.

In a GMM, a probability density function is expressed as a linear combination of Gaussian basis functions. If there are M Gaussians, the model is given by

$$p(\mathbf{x}) = \sum_{j=1}^{M} P(j)p(\mathbf{x}|j), \tag{1}$$

where the *mixing coefficient* $P(j)$ can be viewed as the prior probability of the j-th component, and the $p(\mathbf{x}|j)$ are the Gaussian component density functions. By constraining the mixing coefficients by $\sum_{j=1}^{M} P(j) = 1$ and $0 \leq P(j) \leq 1 \forall j = 1, \ldots, M$ and having $\int p(\mathbf{x}|j)d\mathbf{x} = 1$, the model represents a probability density function.

Each of the Gaussian densities is represented by a mean vector μ_j of dimension d and a covariance matrix $\mathbf{\Sigma}_j$. We use spherical Gaussians with $\mathbf{\Sigma}_j = \sigma_j^2\mathbf{I}$, where $\mathbf{I}$ is the identity matrix. Fig. 2 a) shows an example of a GMM with three Gaussians fitted to vibration data. In practice, we use 16 Gaussians to form a GMM. For fitting a GMM to given data in an unsupervised fashion, one must determine the parameters of the GMM, i.e. the means and standard deviations of the Gaussians. This is done by maximizing the data likelihood or, equivalently, by minimizing the negative log likelihood $-\sum_{n=1}^{N} \log p(\mathbf{x}_n)$ of the data, where N is the number of data points. To solve this problem we follow the common approach to use the *expectation-maximization* (EM) algorithm [25]. The EM algorithm is initialized by the result of five iterations of K-means clustering. For both the initialization and the fitting of GMMs to data, we use the Netlab Matlab toolbox [24].

For novelty detection, we first use the GMM to compute the log probabilities $\log p(\mathbf{x}_i)$ of all training examples. To identify a test vector $\mathbf{f}$ as novel, its log probability $\log p(\mathbf{f})$ must be lower than the log probability of most training examples. Fig. 2 b) illustrates this method by comparing two histograms: the histogram of log probability of the training set and the histogram of log probability of test examples whose class is not in the training set. In practice, our terrain classification system computes a log probability threshold t_p below which test examples will be identified as novel. The higher t_p, the more samples belonging to unknown classes are be recognized as unknown, but also more (outliers) of known classes. The smaller t_p, the more test samples of unknown classes are misclassified as being known, but also more known samples are correctly accepted. Experimentally we found that good results can be obtained by setting t_p such that the log probabilities of 99% of the training data are above the threshold.

The K-means clustering, which initializes the EM algorithm, is initialized randomly. Depending on the random starting values, the minima found by the EM algorithm may be suboptimal, which can lead to suboptimal GMMs. Therefore, we use ten GMMs for novelty detection instead of a single one. Each of these GMMs gives an individual novelty prediction for a test example $\mathbf{f}$. A voting scheme then marks $\mathbf{f}$ as novel if more than half of the GMMs rejected $\mathbf{f}$.

2.2. *Classification using kNN*

As our system must be able to frequently update its classifier online, we use the k-nearest-neighbor algorithm, because its training is extremely fast. Additionally, despite its simplicity, we found that kNN performs very well for vibration-based terrain classification [26]. Training the kNN simply means to store all training vectors $\mathbf{x}_i$. To classify a test vector $\mathbf{f}$, the distances d_i of $\mathbf{f}$ to all $\mathbf{x}_i$ are computed and the k training vectors with the smallest distances are selected. The predicted class is the one that is most frequent among the selected vectors. If there is a draw, a random label is chosen from the winners. We use $k = 10$ in our experiments, because in [26], we found $k = 10$ to work best among $k \in \{1, 2, 5, 10, 15\}$. As source data, we use the PSD feature vectors of full dimension.

2.3. *Update of a Known Class*

Having classified a terrain segment $\mathbf{f}$ as a known class k_i, the system decides whether to update the training set with $\mathbf{f}$. Currently, our system has a fixed upper limit of 100 training examples per class. When a new class is created, however, this class typically has fewer members. If the robot traverses further examples of the new class, they are added to the training set until the new class reaches the maximum number of members. In order not to update the model by wrongly classified test examples, we only add test vectors whose predecessor was classified as the same class. If one changes the condition so that more than only the preceding terrain segment must be of the same class, less samples are added. These samples, however, are also more likely not to be outliers. Each time we update a known class, we must recompute the GMMs on the updated training set. Thus to save computation time, we do not update the model after each selected test sample, but whenever 10 such samples have been stored in a buffer.

2.4. Creation of a New Class

Before the system adds a new terrain class, a sufficiently large number q of examples of the unknown class must be available. Thus, a rejected terrain segment $\mathbf{f}$ is first stored in a buffer. To prevent outliers of known classes from being stored, $\mathbf{f}$ is only added if either the preceding or the successive example is also added. For the same reason, the system additionally possesses a forget mechanism. If no terrain segment is added to the buffer during a fixed time t (currently, we use $t = 10$ s), the oldest entry of the buffer is deleted.

As soon as the number of buffer entries exceeds q, a new terrain class is created from the buffer entries. In our experiments, we use $q = 20$, which enables fast learning of new classes. If there is a frequent change of terrain classes, this relatively small q also reduces the chance of creating new classes containing different terrain types. To update the model with the new class, three steps are necessary. First, the kNN must be retrained by simply adding the samples of the new class to the training set. Second, the PCA must be recomputed on the updated training set. Finally, new GMMs for novelty detection are created on the version of the updated training set that has been processed by PCA.

2.5. Estimation of Class Properties

When a new class is created, its physical properties are unknown, because there is no label like "asphalt" or "grass" indicating the properties. Therefore, our system tries to assign the properties "hardness" and "bumpiness" to the new class based on its raw vibration data. This section only shortly describes this procedure, because this is not the focus of this paper and our research in this direction is still in the beginning.

In experiments on raw vibration signals of different terrain surfaces, we identified some values computed from the signals that can serve as a hint for hardness and bumpiness. For example, the absolute value of the maximum and minimum of the y- and z-vibration typically is smaller the softer the surface, because soft surfaces dampen the vibration. An example of an indicating feature for the bumpiness is the number of sign changes in the x- or y-vibration signal. Lower numbers indicate bumpier surfaces. During big bumps, the acceleration stays positive or negative for a longer period, whereas for flat surfaces, the acceleration changes more rapidly between being positive and negative. These values are, however, often not characteristic enough when looking at a single terrain segment but only when averaged over a sufficiently large number of examples of a class.

Our current system computes a number of such values that measure hardness and bumpiness for each class. By comparison to a threshold, each measure casts a vote for hard or soft (or bumpy or flat). The largest number of votes per property finally leads to a label for the class, for example "hard and bumpy".

3. Experimental Results

We acquired experimental data with our RWI ATRV-JR outdoor robot (Fig. 3 a). The robot is equipped with an Xsens MTi three-axis accelerometer working at 100 Hz. At speeds of around 0.5 m/s, the robot traversed four different ground surfaces: asphalt, gravel, grass and the surface of a boule court (Fig. 3 b). The number of terrain segments

Figure 3. a) Our RWI ATRV-JR robot, equipped with an Xsens MTi sensor. b) Terrain types we used in the experiments: asphalt (1), gravel (2), grass (3), boule court (4).

	Evaluation data	Initial model	Test run
Known class k_1	100	100	Rest
Known class k_2	100	100	Rest
Known class k_3	100	100	Rest
Unknown class u_1	100	Rest	

Figure 4. Number of samples per class in the evaluation dataset, the initial model and the test run for the experiment with one unknown class.

is 513 for asphalt, 323 for gravel, 572 for grass and 579 for the boule court, which corresponds to about 33 minutes of robot drive. As depicted in Fig. 2 a), grass and gravel are relatively well seperated from the other classes. However, a considerable overlap exists between asphalt and boule court. This overlap does decrease, but it does not vanish, when more than two dimensions are used.

3.1. Experiments with One Unknown Class

In the first experiments we present in this paper, we trained the system on three classes and left one class unknown. Before starting an experiment, we split the data into parts (Fig. 4). Firstly, we put aside 100 evaluation examples per class. We classified the evaluation data at the end of the experiment using the final, automatically updated model, and compared the results to the predictions of a manually trained classifier which knows all classes from the start. Secondly, we use 100 examples of each known class to train the initial model. With the remaining data, we simulated a test run of the robot. In such a run, the system is first presented half the examples of each known class. Then, the system encounters half of the unknown examples. After that, the robot traverses the remaining half of all classes. In the first block of the test run, i.e. from the beginning until a new class is created from the unknown examples, most known examples should be accepted, and most unknown examples should be rejected. In the subsequent second block, all classes are known, so only a small number of test examples should be rejected. We executed experiments for all possible combinations of known and unknown classes. Additionally, we repeated each experiment five times with a permuted data order such that the training, test and evaluation sets are different each time.

Table 1. Average rejection rates (%) of the experiment with one unknown class ± standard deviation.

	Block 1	Block 2
Known examples	2.74 ± 0.34	3.51 ± 0.89
Unknown examples	71.01 ± 23.22	(no unknown class exists)

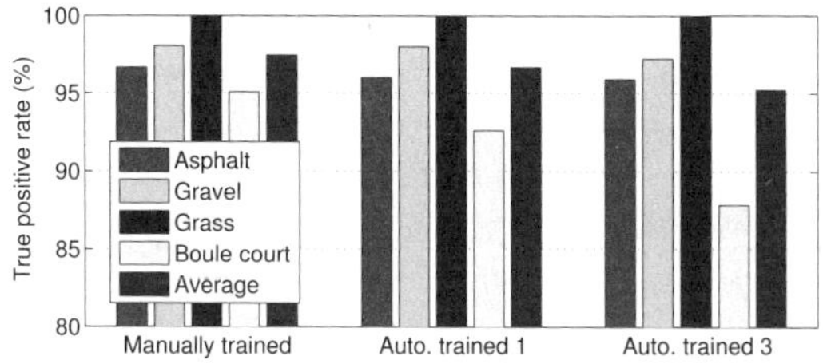

Figure 5. True positive rates on the evaluation dataset: manually trained (left) and automatically trained classifiers started with one (middle) or three (right) unknown classes.

Tab. 1 summarizes the percentage of test examples per block that were rejected by the novelty detector. From the known examples, only a small fraction was rejected. The rejection rate in the second block is a little larger than in the first block. This is because newly created classes consist of only 20 examples in the beginning. These 20 examples are not enough to explain almost all of the new classes's examples. For the examples belonging to unknown classes, the average rejection rate is over 70%. Without novelty detection, all of these examples would be accepted and classified wrongly. However, the large standard deviation of this value shows that the rejection rate differs greatly depending on which class is unknown. If there is little overlap with the known classes, the rejection rate is high, e.g. over 99% for grass. If there is a considerable overlap with one or more known classes, the rejection rate is lower, e.g. about 45% for asphalt.

Fig. 5 compares the mean true positive rate (TPR) on the evaluation dataset, achieved by the automatically updated system (96.65%), to the TPR of a manually trained kNN classifier (97.43%). These TPRs differ only slightly, as well as the rates for the individual classes. Only for the boule court, which has a considerable overlap with asphalt, there is a gap of about 2.5%. Note that in the automatically generated training set, newly added classes will only have little overlap with the other classes. The reason why the TPR on asphalt is better than on boule court is that some boule court test examples are spread in the main asphalt area, whereas much fewer asphalt test examples are in the main boule court region (Fig. 2 a). Note that all classes in our evaluation dataset have the same number of test samples. Therefore, the mean true positive rates correspond to the accuracy measure.

3.2. Experiments with Three Unknown Classes

In the second experiment, the system started with a single known class k_1 and three unknown classes u_1, u_2, u_3. We used the same data split and setup as in the first experiment, but with the following order of appearance of classes in the test run: $(k_1, u_1, k_1, \hat{u}_1, u_2, k_1, \hat{u}_1, \hat{u}_2, u_3, k_1, \hat{u}_1, \hat{u}_2, \hat{u}_3)$. $\hat{u}_i$ denotes that the i-th unknown class already has been added to the model, which should be the case if the system works well. We repeated the experiment for all 24 possible combinations of known and unknown classes, and repeated this five times for each combination with a permuted data order.

Table 2. Rejection rates (%) of the experiment with three unknown classes ± standard deviation.

	Block 1	Block 2	Block 3	Block 4
Known examples	6.83 ± 2.62	9.06 ± 2.11	7.73 ± 1.85	6.61 ± 1.21
Unknown examples	92.07 ± 10.56	85.65 ± 13.14	77.80 ± 15.83	-

Compared to the first experiment, the rejection rates for known classes are larger now (Tab. 2). The main reason is that during the experiment, the average number of training examples is smaller, because the three new classes start with only 20 examples. Thus, the explanatory power of the training set is lower on average. On the other hand, the rejection rates for unknown examples are also larger than in the first experiment. Again, the rejection rates vary depending on overlaps of the unknown class with known classes.

The mean TPR on the evaluation dataset using the final, automatically trained model is 95.22%, which is about 2% less than the TPR using a manually trained kNN (Fig. 5). Again, the boule court is classified significantly worse than the other classes.

On a 3 GHz Pentium 4 PC with 1 GHz RAM, we measured the following computation times for our Matlab code. When using the maximum number of training examples (400), computing the PCA took about 0.085 s. On average, novelty detection for one vector only required about 0.34 ms. kNN classification took about 4.9 ms per test vector. The time for training the kNN is negligible. Creating the novelty detector is the computationally most expensive part, because 10 Gaussian mixture models must be fitted to the data using the EM algorithm. When using maximally 20 iterations of the EM algorithm, fitting one GMM took about 0.160 s if the training set contains 400 examples. In this case, creating a new class (computing the PCA + creating the novelty detector) would take about 1.685 s. Using a C++ implementation, this should be possible in less time.

At the end of all experiments, the system updated all unknown classes to 100 examples. It assigned the following properties to the classes: asphalt is hard and flat, gravel is hard and bumpy, grass is soft and bumpy, and the boule court is hard and flat. Depending on the weather, the boule court could also be soft. As we recorded our data on a dry summer day, however, the boule court actually was hard and the assignment is correct.

4. Conclusion

We presented a vibration-based terrain classification system that is capable of identifying terrain classes that are not included in the training set. When the robot has collected a sufficiently large number of examples of the unknown terrain class, it adds a new terrain class online. Our experiments showed that a classifier trained by our system in an online fashion has a classification performance which is only slightly worse compared to a manually trained classifier to which all classes were known beforehand.

A problem for the current system appears if different unknown terrain types follow each other closely. Then, a new class will be possibly created containing more than one physical terrain type. We are currently searching for methods to detect such situations. Additionally, we will further improve our methods to assign properties to new classes. We will also investigate alternative novelty detection methods to GMMs. Furthermore, we want to implement the system in C++ instead of Matlab to achieve accelerated computation times for the model update.

References

[1] B. H. Wilcox, Non-Geometric Hazard Detection for a Mars Microrover, in *Proc. AIAA Conf. Intell. Robot. Field, Factory, Service, Space*, Houston, TX, USA, 1994, 675–684

[2] N. Vandapel, D. Huber, A. Kapuria and M. Hebert, Natural Terrain Classification using 3-D Ladar Data, in *Proc. IEEE Int. Conf. on Robotics and Automation (ICRA)*, New Orleans, LA, USA, 2004

[3] R. Manduchi, A. Castano, A. Talukder and L. Matthies, Obstacle Detection and Terrain Classification for Autonomous Off-Road Navigation, *Autonomous Robots* **18**, 2005, 81–102

[4] P. Bellutta, R. Manduchi, L. Matthies, K. Owens and A. Rankin, Terrain Perception for Demo III, in *Proc. IEEE Intelligent Vehicles Symposium*, Dearborn, MI, USA, 2000, 326–332

[5] A. Angelova, L. Matthies, D.M. Helmick and P. Perona, Fast Terrain Classification Using Variable-Length Representation for Autonomous Navigation, in *Proc. IEEE Conf. on Computer Vision and Pattern Recognition (CVPR)*, Minneapolis, USA, 2007, 1–8

[6] K. Iagnemma and A. Dubowsky, Terrain Estimation for High-Speed Rough-Terrain Autonomous Vehicle Navigation, in *Proc. SPIE Conf. on Unmanned Ground Vehicle Technology IV*, 2002

[7] E.M. DuPont, R.G. Roberts and C. Moore, The Identification of Terrains for Mobile Robots Using Eigenspace and Neural Network Methods, in *Proc. Florida Conf. on Recent Advances in Robotics*, Miami, Florida, USA, 2006

[8] C.A. Brooks and K. Iagnemma, Vibration-Based Terrain Classification for Planetary Exploration Rovers, *IEEE Transactions on Robotics* **21**(6), 2005, 1185–1191

[9] C. Weiss, H. Fröhlich and A. Zell, Vibration-based Terrain Classification Using Support Vector Machines, in *Proc. Int. Conf. on Intelligent Robots and Systems (IROS)*, Beijing, China, 2006, 4429–4434

[10] C. Weiss, M. Stark, A. Zell, SVMs for Vibration-based Terrain Classification, in *Proc. Autonome Mobile Systeme (AMS)*, Kaiserslautern, Germany, 2007, 1–7

[11] D. Stavens, G. Hoffmann and S. Thrun, Online Speed Adaptation Using Supervised Learning for High-Speed, Off-Road Autonomous Driving, in *Proc. Int. Joint Conf. on Artificial Intelligence (IJCAI)*, Hyderabad, India, 2007

[12] R.E. Karlsen and G. Witus, Terrain Understanding for Robot Navigation, in *Proc. Int. Conf. on Intelligent Robots and Systems (IROS)*, San Diego, CA, USA, 2007, 895–900

[13] C.A. Brooks and K. Iagnemma, Self-supervised Terrain Classification for Planetary Rovers, in *Proc. NASA Science Technology Conference*, 2007

[14] I. Halacti, C.A. Brooks and K. Iagnemma, Terrain Classification and Classifier Fusion for Planetary Exploration Rovers, in *Proc. IEEE Aerospace Conference*, Big Sky, MT, USA, 2007

[15] D.M.J. Tax and R.P.W. Duin, Data Description in Subspaces, in *Proc. Int. Conf. on Pattern Recognition (ICPR)*, Barcelona, Spain, 2000, 672–675

[16] D. Dasgupta and F. Nino, A Comparison of Negative and Positive Selection Algorithms in Novel Pattern Detection, in *Proc. Int. Conf. on Systems, Man, and Cybernetics (SMC)*, Nashville, USA, 2000, 125–130

[17] G.C. Vasconcelos, M.C. Fairhurst and D.L. Bisset, Investigating Feedforward Neural Networks with Respect to the Rejection of Spurious Patterns, *Pattern Recognition Letters* **16**(2), 1995, 207–212

[18] B. Schölkopf, R. Williamson, A. Smola, J. Shawe-Taylor and J. Platt, Support Vector Method for Novelty Detection, in *Proc. Advances in Neural Information Processing Systems*, 2000, 582–588

[19] S. Marsland, U. Nehmzow and J. Shapiro, Environment-Specific Novelty Detection, in *Proc. Int. Conf. on Simulation of Adaptive Behaviour (SAB)*, Edinburgh, UK, 2002, 36–45

[20] M. Markou and S. Singh, Novelty Detection: a Review - Part 1: Statistical Approaches, *Signal Processing* **83**(12), 2003, 2481–2497

[21] M. Markou and S. Singh, Novelty Detection: a Review - Part 2: Neural Network Based Approaches, *Signal Processing* **83**(12), 2003, 2499–2521

[22] L. Tarassenko, P. Hayton, N. Cerneaz and M. Brady, Novelty Detection for the Identification of Masses in Mammograms, in *Proc. Int. Conf. on Artificial Neural Networks (ICANN)*, Paris, France, 1995, 442–447

[23] S.J. Hickinbotham and J. Austin, Neural networks for novelty detection in airframe strain data, in *Proc. IEEE Int. Joint Conf. on Neural Networks (IJCNN)*, Como, Italy, 2000, 375–380

[24] I.T. Nabney, *NETLAB: Algorithms for Pattern Recognition*, Springer, 2002

[25] A.P. Dempster, N.M. Laird and D.B. Rubin, Maximum Likelihood From Incomplete Data via the EM Algorithm, *Journal of the Royal Statistical Society* **39**(1), 1977, 1–38

[26] C. Weiss, N. Fechner, M. Stark and A. Zell, Comparison of Different Approaches to Vibration-based Terrain Classification, in *Proc. European Conf. on Mobile Robots (ECMR)*, Freiburg, Germany, 2007

Intelligent Autonomous Systems 10
W. Burgard et al. (Eds.)
IOS Press, 2008

doi:10.3233/978-1-58603-887-8-26

Comparison of Data Amount for Representing Decision Making Policy

Ryuichi UEDA [1],
Dept. of Precision Engineering, School of Engineering, Univ. of Tokyo

Abstract. This paper deals with the problem of how to implement software controller for a robot with a small amount of random access memory (RAM) on its computer. This problem is essentially different from how to solve it with a small amount of RAM. This paper purely compares the trade-off between memory use and performance of a controller. We found that policies that are compressed by vector quantization have an efficient representation.

Keywords. optimal control, value iteration, compression of policies

1. Introduction

How to generate decision making policies of robots with a limited amount of resources has been one of the most essential problems in robotics. The amount of memory, numbers of CPUs and their power, the I/O bus bandwidth, and time will be included in the resources. When a special hardware is not considered, advantage and disadvantage of a method is evaluated by the balance of use of the above resources, and performance of the controller.

We have studied some methods only for reducing the amount of memory for representing random access policies [1,2,3]. This specialization is very useful when there are large differences of resources between a computer for planning and some intelligent products, which will be robots or other machines. At calculations of a policy, calculations are generally stable if there are abundant computing resources. Once we can obtain a reliable policy, it can be compressed in consideration of the trade-off between the performance and the amount of memory on a product.

In this paper, we compare some popular memory reduction methods for decision making policies. They are applied to policies for a puddle world task [4], which are obtained by precise calculations.

2. Related Work

In this paper, methods are evaluated in the context of the optimal control theory, which is the most general framework for solving or evaluating policies. We consider

[1] 7-3-1 Hongo, Bunkyo-ku, Tokyo, Japan; E-mail: ueda@robot.t.u-tokyo.ac.jp; url: http://www.robot.t.u-tokyo.ac.jp/~ueda/.

the following state equation and functional:

$$\dot{\boldsymbol{x}}(t) = \boldsymbol{f}[\boldsymbol{x}(t), \boldsymbol{u}(t)], \quad t \in [0, t_{\mathrm{f}}] \tag{1}$$

$$\text{and } J[\boldsymbol{u}] = \int_0^{t_{\mathrm{f}}} g[\boldsymbol{x}(t), \boldsymbol{u}(t)]dt. \tag{2}$$

$\boldsymbol{x}(t)$ denotes a state, which can be represented by some continuous variables. $\boldsymbol{u}(t)$ is called a control input vector. In the field of artificial intelligence, it can be called an action. Time t_{f} is defined as the time at the end of control. It does not have to be fixed. Control in a state is evaluated by a scalar: $g[\boldsymbol{x}(t), \boldsymbol{u}(t)] \in \Re$.

The decision making from the initial state $t = 0$ to $t = t_{\mathrm{f}}$ is totally evaluated by the functional in Eq.(2). When a decision making from $t = 0$ to $t = t_{\mathrm{f}}$ maximizes the value of the functional, the decision is called the optimal control. The optimal control can be represented by a function from a state to a control input $\boldsymbol{\pi}^* : \mathcal{X} \to \mathcal{U}$ when the state equation is time-invariant. $\mathcal{X}$ and $\mathcal{U}$ are the sets (or space) of states and controls respectively. $\boldsymbol{\pi}$ is called a (decision making) policy in this paper.

When a policy $\boldsymbol{\pi}$ is given, the function from $\forall \boldsymbol{x}$ to the value of the functional can be calculated. This function is called a state-value function $V^{\boldsymbol{\pi}} : \mathcal{X} \to \Re$. The state-value function related to the optimal policy $\boldsymbol{\pi}^*$ is called the optimal state-value function V^*, which fulfills the Hamilton-Jacobi-Bellman equation:

$$\frac{\partial V(\boldsymbol{x})}{\partial t} = \max_{\boldsymbol{u} \in \mathcal{U}} \left[g[\boldsymbol{x}, \boldsymbol{u}] + \frac{\partial V(\boldsymbol{x})}{\partial \boldsymbol{x}} \boldsymbol{f}[\boldsymbol{x}, \boldsymbol{u}] \right]. \tag{3}$$

V and $\boldsymbol{\pi}$ are represented as functions with some parameters in a computer as

$$V^{\boldsymbol{\pi}}(\boldsymbol{x}; \theta_1, \theta_2, \dots, \theta_{N_\theta}), \text{ or } \boldsymbol{\pi}(\boldsymbol{x}; \theta_1, \theta_2, \dots, \theta_{N'_\theta}). \tag{4}$$

This representation contains many suggestions. How to represent the parameters is one of the most important problems for representing the above equation.

If an optimal control problem is complicated, discontinuous, or defined with many boundary conditions, numerical calculations must be done on computers. Dynamic programming (DP) [5] is a representative method for computation. When the dynamics is unknown, reinforcement learning [6] is required.

By the number of parameters, we can classify representation methods into two types: global one and local one. In the case of a global function approximator (e.g. typical artificial neural networks [7]), a change of a parameter makes an impact all over the state space. Since the global method needs heuristics for setting parameters, local methods are argued in this paper.

In the case of local methods, each parameter has its own area in space. The simplest one is discretization with a multi-dimensional lattice. In other methods, areas of parameters partially overlap with each other. It makes a function smooth. Techniques with tile-coding [8,4] and radial basis functions [9,10,11] utilize this property.

Interpolation techniques also realize the smoothness [12]. Munos and Moore have presented a complex interpolation method with a tree structure [13]. They have indicated that this structure is useful not only for reinforcement learning but also for DP.

In the first part of [13], they have presented some criteria of where nodes should be added to for representing V^{π}. However, as they have mentioned, representation of V^{π} never relates to representation of π. Parameters for a value function and a policy are different in general cases. In the center part of [13], a criterion for creating the tree structure for representing π accurately is indicated. In this criterion, a cell surrounded edges is divided when its representative states cannot give suitable control inputs at any state in the cell. According to them, this lack of resolution can be found by the comparison of the value obtained from the Hamilton-Jacobi-Bellman equation shown in Eq. (3).

3. Example Task for Evaluation: Puddle World Task

The puddle world task is used for studies about learning algorithms [4]. Figure 1 illustrates the puddle world. An agent, whose state is defined as a point on the xy-coordinate system, aims to go to the goal area. The puddle in the figure prevents the agent from reaching the goal directly. If the agent enters the puddle, penalty is given. The agent should minimize the sum of the penalty and the number of steps from any point to the goal. Though this task is simple, there are some characteristic parts in this world such as a narrow path, a broad area, the area where the agent should detour, edge of the world, and the puddle area. Therefore, a method examined in this world must work well in the various areas assuredly.

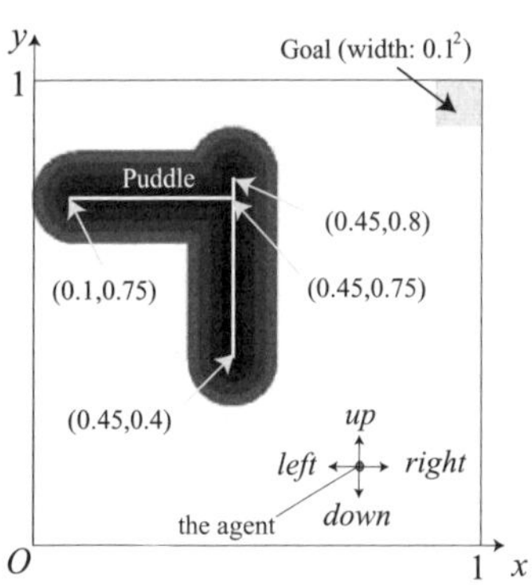

Figure 1. Puddle World

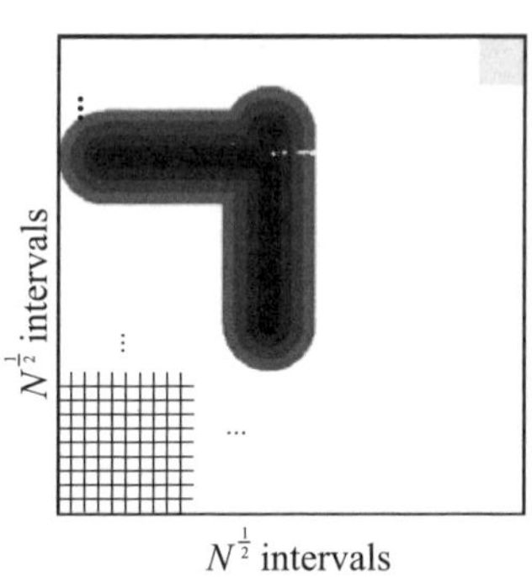

Figure 2. Discretization

3.1. Definition of Environment

As shown in Fig. 1, the shape of the world is a square: $\mathcal{X} = \{(x, y)|x \in (0, 1), y \in (0, 1)\}$, which can be regarded as the state space. In this case, x and y are the state variables of the agent. The agent does not have its orientation.

The center of the puddle is composed of the two line segments as shown in the figure. The puddle spreads 0.1 in width from the center lines. The goal is the square area: $\mathcal{X}_{\mathrm{f}} = \{(x, y)|x \in (0.9, 1), y \in (0.9, 1)\}$ at the upper right of the figure.

The control vector can be defined as $\boldsymbol{u} = (\delta_x\ \delta_y)^T$. δ_x and δ_y denote target displacements of the agent along x-axis and y-axis respectively. Random Gaussian noise, whose standard deviation is 0.01, is added to the displacement in the direction of each axis. In this case, a probability that the agent enters an area X' from a state $\boldsymbol{x}$ by $\boldsymbol{u}$ becomes

$$P(X'|\boldsymbol{x}, \boldsymbol{u}) = 5000/\pi \int_{\boldsymbol{x}' \in X'} \exp\left(-5000|\boldsymbol{x}' - \boldsymbol{x} - \boldsymbol{u}|^2\right) d\boldsymbol{x}'. \quad (5)$$

The world is surrounded by a wall. The agent cannot go beyond the wall. When the posterior state $\boldsymbol{x}'$ is not in the state space $\mathcal{X}$, $\boldsymbol{x}'$ is replaced at the intersecting point of the wall and the line segment between $\boldsymbol{x}$ and $\boldsymbol{x}'$.

We give -1[step] to each step as an evaluation. If it enters there, $-400\{0.1 - \ell(\boldsymbol{x})\}$ is also given. $\ell(\boldsymbol{x})$ is the distance from $\boldsymbol{x}$ to the nearest edge of the puddle. The evaluation is defined as

$$r_{\boldsymbol{x}} = -1 - 400 \max\{0.1 - \ell(\boldsymbol{x}), 0\}. \quad (6)$$

This evaluation is given just after the agent steps away from $\boldsymbol{x}$.

3.2. Calculation of Policies on Simple Lattice

We use a lattice that divides the puddle world into N square discrete states for calculating a policy. Each axis is divided into $\sqrt{N}$ intervals, and each square area is regarded as a discrete state s_i $(i = 1, 2, \ldots, N)$. We then consider only four kinds of the motion of the agent: $\boldsymbol{u} = (\pm 0.05\ 0)^T, (0\ \pm 0.05)^T$. They are given symbols: $a_{\text{right}}, a_{\text{left}}, a_{\text{up}}$, and a_{down} as shown in Fig.2. Each discrete state s has a pair of value $V(s)$ and action $\pi(s)$.

Since a discrete state is not a point in $\mathcal{X}$, but a subset of $\mathcal{X}$, Eq.(5) and (6) must be rewritten. When we cannot know which state the agent is in s, the transition probability from s to s' is calculated by

$$\mathcal{P}^a_{ss'} = P(s'|s, a) = \int_{\boldsymbol{x} \in s} P(s'|\boldsymbol{x}, a) d\boldsymbol{x} \Big/ \int_{\boldsymbol{x} \in s} d\boldsymbol{x} = N \int_{\boldsymbol{x} \in s} P(s'|\boldsymbol{x}, a) d\boldsymbol{x}. \quad (7)$$

For precise calculations, however, we have expanded this equation by Mathematica and have implemented some calculations for obtaining values. The evaluation for the discrete states is also calculated as

$$\mathcal{R}_s = N \int_{\boldsymbol{x} \in s} r_{\boldsymbol{x}} d\boldsymbol{x}. \quad (8)$$

It is calculated by using a Monte Carlo integral in our implementation.

3.3. Computing Result

State-value functions $V(s)$ and policies $\pi(s)$ are calculated by a value iteration algorithm with $N = 10^2, 20^2, 40^2, 100^2, 200^2, 400^2$. Value iteration is an implementation of DP. The policy, its state-value function, and some examples of behavior of the agent with $N = 400^2$ are shown in Fig.3.

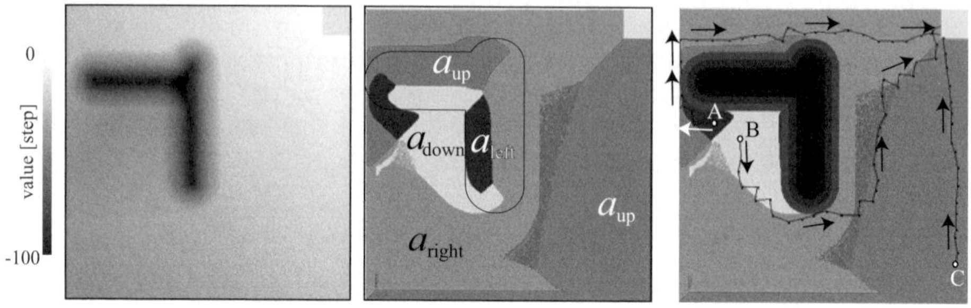

Figure 3. State-Value Function, Policy, and Examples of Behavior

Table 1. Relation between Size and Performance of The State-Action Maps

N	10^2	20^2	40^2	100^2	200^2	400^2
size[bit]	200	800	3200	20000	80000	320000
avg.[step]	21.29	20.87	20.51	20.35	20.30	20.29

We choose 1000^2 initial states uniformly from the state space, and measure the average performance (the sum of evaluations) of the agent with each policy. The result is shown in Table 1. There is no case where the agent cannot go to the goal. In the case of the puddle world task, the trade-off between memory use and the performance is clear. The amount of memory is calculated as $L = 2N$[bit], where 2 denotes the number of bits to distinguish the four kinds of actions.

4. Memory Reduction

Once policies with fine discretization can be obtained, it seems to be easy that the policy is abstracted and represented with a small amount of memory. Though this problem is basically not easy, it is easier than the problem of how to create a small policy from scratch.

To change a policy for memory reduction, the following equation:

$$d^{\pi}(s,a) = V^{\pi}(s) - \sum_{s' \in \mathcal{S}} \mathcal{P}^{a}_{ss'} \left[\mathcal{R}^{a}_{ss'} + V^{\pi}(s')\right] \tag{9}$$

is proposed [3]. This is called a state-value distortion at a change of the action in a discrete state. The larger this value is, the more performance is sacrificed. In fact, this equation is an approximation, it can be successfully applied to various problems [2,14,3].

4.1. Vector Quantization

Since we have published many papers of the vector quantization (VQ) method of policies [2,14,3], we state only the final product of VQ, which is called *a VQ policy*. A sequence of actions in a policy: $\pi(s_i)$ $(i = 1, 2, \ldots, N)$ can be transformed to a two-dimensional format: $\pi(i_c, i_\varepsilon)$. The indexes i_c, i_ε are obtained from the following transforms:

$$\big(\nu(i)\ \epsilon(i)\big) \mapsto \big(i_v\ i_\varepsilon\big) \tag{10}$$

$$\omega(i_v) \mapsto i_c. \tag{11}$$

Equation (10) means that the action $\pi(s_i)$ is located to i_εth element of i_vth vector. i_ε and i_v are called an element index and a vector index respectively. Their names indicate that a sequence of actions are arranged to some blocks that are called vectors.

The vector index i_v is changed to i_c by the transform ω, which is a key of the compression. When different vector indexes can be safely associated with an identical index i_c, the number of vectors can be reduced. An index i_c is called a representative vector index. It means that the i_cth representative vector represents some vectors that are associated by the transform ω. The state-value distortion is utilized for choosing ω and every element of every representative vector.

A transform ω is an array that contains each representative vector index to every vector. $\pi(i_c, i_\varepsilon)$ contains all representative vectors. In the case of the puddle world task, which requires 2[bit] for identifying actions, the number of bits of the compressed policy is

$$L_{\mathrm{VQ}} = N/N_\varepsilon \lceil \log_2 N_c \rceil + 2N_\varepsilon N_c, \tag{12}$$

where the first term of the right side is the size of ω, and the second term of the right side is that of $\pi(i_c, i_\varepsilon)$. N_ε and N_c are the numbers of elements in a vector and the number of representative vectors. N/N_ε is the number of vectors before compression.

In Fig.6, we show a VQ policy. In this case, we divided the $N = 400^2$ policy into 400 rows, which are regarded as vectors. They are classified into 32 groups. 32 representative vectors are allocated to the groups.

4.2. Representation by Binary-Trees

We have implemented a code for rewriting a state-value function on a lattice to a tree structure. As shown in Fig. 4, a binary-tree has nodes, branches, and leaves. This code starts from a node, which has two branches: e^-,e^+. They cover $0 \le x < 1/2$ area and $1/2 \le x < 1$ area respectively. Each branch has a leaf in which an action is written. Every state-value distortion $d^\pi(s, a)$ has been calculated from a state-value function on a fine look-up table. From that state, the tree grows with a procedure that changes a leaf into a node. The leaf that is changed is chosen from all leaves based on the state-value distortion. To decide the leaf that is changed, the method calculates every sum of $d^\pi(s, a)$ in the area of every branch that has a leaf. A branch that has the largest sum is attached a node instead of the leaf. When the area of a branch is not larger than the size of a discrete state on the state-value function, however, the branch cannot have a node. This restriction can prevent excess splitting of the state-space.

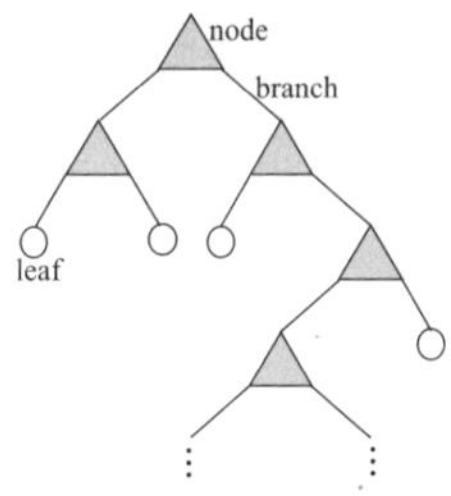

Figure 4. Binary-tree structure

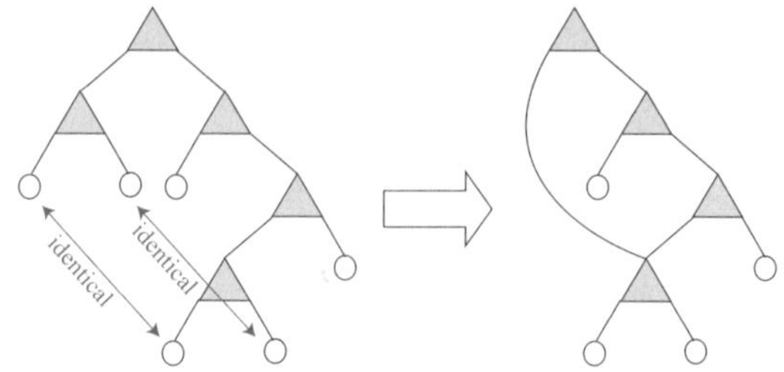

Figure 5. Reconnection of a Branch

When a node is added to the tree, the area that is covered by the node must be split and allocated to both leaves. If the area can be split arbitrarily, the node must use some bits to memory the way of splitting. We choose to fix a splitting rule so as to save the amount of memory. The rule is as follows: when the area is a square, it is split evenly along y-axis; otherwise it is split evenly perpendicular to long edges. By this rule, the area that is covered by every leaf can be calculated only with the structure of the tree.

After that, unnecessary nodes are cut. A node that has an identical action on its both branches can be erased. They can be replaced by leaves. This procedure is iterated until unnecessary nodes exist. The policy that is obtained by this process is called *a tree policy* in this paper. Moreover, a node can be erased when another node has identical contents. In this case, it can be erased when the branch that points the node is reconnected to the other node as shown in Fig. 5. This procedure is also iterated until unnecessary nodes exist.

We show two tree policies as examples in Fig. 7. The $N = 400^2$ state-value function was used for computing $d^{\pi}(s, a)$. The tree policy in Fig. (a) is composed of 277 leaves ($N_{\text{node}} = 276$). Incidentally, there were 300 nodes before the cut of unnecessary nodes. Moreover, the number of the nodes can be reduced to $N_{\text{node}} = 172$ by the method in Fig. 5. Figure (b) shows a tree policy with 1913 nodes, which are reduced from $2,000$.

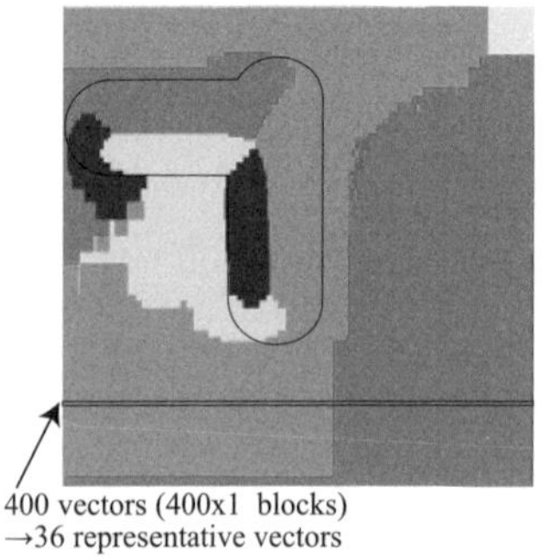

Figure 6. VQ Policy

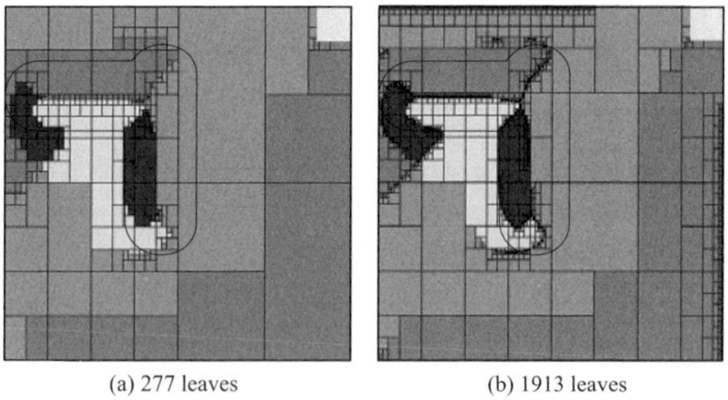

Figure 7. Policies on Binary-Trees

In a computer, the branches are implemented as pointers (address variables). A node has two pointers and each of them points the address of another node or a leaf. Since an address variable consumes 32 bits on a 32-bit computer, address variables of a system should not be used for addressing. If we prepare original minimal pointers for searching a binary-tree, the size of a tree becomes

$$L_{\mathrm{tree}} = 2N_{\mathrm{node}}\lceil \log_2(N_{\mathrm{node}} + M)\rceil, \tag{13}$$

where N_{node} and M denote the numbers of nodes and actions respectively.

4.3. Implementation of Interpolation

We utilize the interpolation technique shown in [12] to enhance pseudo resolution of the state-value functions shown in Fig. 3. In Fig. 8, a part of a state-value function is drawn. The value at $\boldsymbol{x}$ is calculated as the weighed mean of values of states around $\boldsymbol{x}$. The weight is in proportion to the square measure of the overlapping area between each discrete state and a square whose center is $\boldsymbol{x}$. The size of the square is identical with a discrete state. The interpolated state-value functions consume no more memory than the size of the original $V(s)$.

5. Comparison Result

5.1. Size of State-Value Functions and Their Performance

At first, some types of state-value functions are evaluated on the trade-off between memory consumption and performance. A part of this result is also shown in [14]. In [14], we have mentioned a problem of tile-coding in the context of value iteration. Here we add the result of the interpolation technique.

To obtain an appropriate action from a state-value function, we have used a Monte Carlo method. Before the agent moves, 100 simulations are tried for each action and the most valuable action is chosen.

Figure 9 illustrates the results of evaluation. If there are trials in which the agent cannot go to the goal, the result is removed from this figure. In this figure, data of the policies in Table 1 are also written as *normal policies.* In Fig.(a) and (b), the number of bits for representing a value is different for each other (8[bit] and 16[bit]). Since only 2[bit] are required for representing an action, a normal policy is essentially advantageous in this comparison. Nonetheless, normal functions and interpolated functions are not far behind from normal policies in the cases of 8[bit].

However, the more important thing is that interpolated functions are not better than the simple lattice function. Since the Monte Carlo method can interpolate the function with a nonlinear way, such a linear interpolation shown in Fig.8 is not effective, especially when the discretization is coarse.

This result would change if we could not know the state equation in Eq.(1). When we use a set of trials for calculating a state-value function, the smoothing effect with function approximation is sometimes effective. It is related to easy implementations. However, it is not directly related to saving the amount of memory. Though this puddle world task is an example, it is enough to insist that the simple lattice discretization should not be discarded. When state transitions are calculated as streams of probability as shown in Eq.(7), there is a possibil-

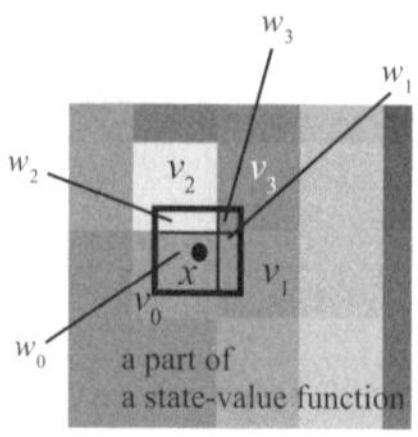

Figure 8. Interpolation

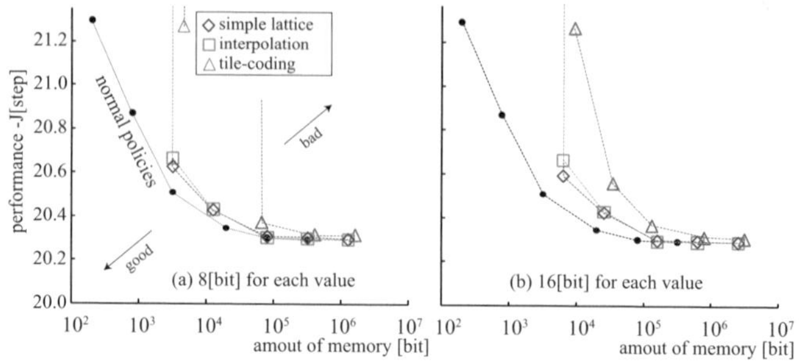

Figure 9. Performance of State-Value Functions and Policies

ity that a very coarse discretization is still better than a complicated function approximation.

5.2. Size of Policies and Their Performance

We have compressed all of the policies in Table 1 to VQ policies and tree policies. As the numbers of representative vectors, $N_c = 2, \ldots, 9, 10, 16, 32, 64, 128, \ldots$ are chosen. N_{node} is chosen from $1, 2, \ldots, 10, 20, \ldots, 100, 200, \ldots$ The numbers of nodes are reduced from the chosen numbers by the technique shown in Fig.5. VQ policies are compressed again with lossless VQ.

The results with the evaluation are shown in Fig.10(a). There are many compressed policies that are smaller and higher performance than the normal policies. In the area between 10^3 and 10^4, we can obtain both VQ and tree policies. However, we cannot obtain a tree policy that can bring the agent in all trials when the available memory is less than 10^3[bit]. Though there is a possibility that our implementation of tree creation is imperfect, Fig.10(b) shows another reason. The horizontal axis of this graph indicates how much elements (elements in representative vectors or leaves in a tree) are contained in a policy. In the viewpoint from the number of elements, tree policies are much more efficient than VQ policies. Instead, this efficiency is supported by memory for the pointers. If we want to obtain a tree policy smaller than 10^3[bit], at most 100 leaves can be used for discretization.

6. Conclusion

We have evaluated performance and memory consumption of decision making policies obtained by some memory reduction techniques. Though only one task is examined in this paper, the following things should be described as conclusions.

- We should not think that simple discretization with a lattice is inefficient, especially for DP.
- The above statement is not for solving feasible state-value functions but for representing accurate state-value functions. However, discretization with a coarse lattice should be tried when we can obtain accurate probabilities of state transitions.

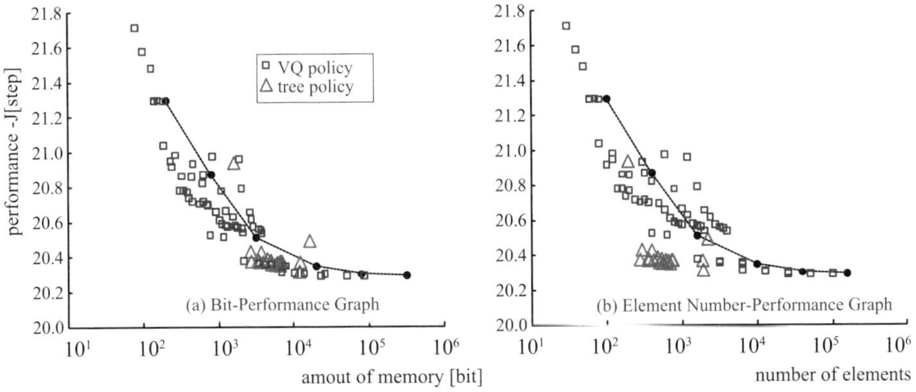

Figure 10. Efficiency of VQ Policies and Tree Structured Policies

• Tree policies can split state space into much smaller numbers of elements than VQ policies. However, tree policies consume memory for representing their structure. On the other hand, a VQ policy has a simple form that is a kind of hash table and its representation is compact.

References

[1] T. Fukase *et al.*: "Real-time Decision Making under Uncertainty of Self-Localization Results," RoboCup 2002: Robot Soccer World Cup VI, pp. 375–383, 2003.
[2] R. Ueda *et al.*: "Vector Quantization for State-Action Map Compression," In *Proc. of ICRA*, pp. 2356–2361, 2003.
[3] R. Ueda and T. Arai: "Creation and Compression of Global Control Policy for Swinging up Control of the Acrobot," In *Proc. of IROS*, pp. 2232–2237, 2006.
[4] R. Sutton: "Generalization in Reinforcement Learning: Successful Examples Using Space Coarse Coding," In *Neural Information Processing Systems*, pp. 1038–1044, 1996.
[5] R. Bellman: *Dynamic Programming*, Princeton University Press, 1957.
[6] R. Sutton and A. Barto: *Reinforcement Learning: An Introduction*, The MIT Press, 1998.
[7] G. Tesauro: "Temporal Difference Learning and TD-Gammon," *Communications of the ACM*, 38(3), pp. 58–68, 1995.
[8] J. Albus: "A New Approach to Manipulator Control: The Cerebellar Model Articulation Controller (CMAC)," *Journal of Dynamic Systems, Measurement and Control*, 97(3), pp. 220–227, 1975.
[9] D. Broomhead and D. Lowe: "Multivariable Functional Interpolation and Adaptive Networks," *Complex Systems*, 2(3), pp. 321–355, 1988.
[10] J. Moody and C. Darken: "Fast Learning in Networks of Locally-Tuned Processing Units," *Neural Computation*, 1(2), pp. 281–294, 1989.
[11] K. Samejima and T. Omori: "Adaptive internal state space construction method for reinforcement learning," *Neural Networks*, 12(7-8), pp. 1143–1155, 1999.
[12] Y. Takahashi *et al.*: "Improvement Continuous Valued Q-learning and Its Application to Vision Guided Behavior Acquisition," *RoboCup 2000: Robot Soccer World Cup IV*, pp. 385–390, 2001.
[13] R. Munos and A. Moore: "Variable Resolution Discretization in Optimal Control," *Machine Learning*, 49(2-3), pp. 291–323, 2002.
[14] R. Ueda *et al.*: "Vector Quantization for State-Action Map Compression –Comparison with Coarse Discretization Techniques and Efficiency Enhancement," In *Proc. of IROS*, pp. 166–171, 2005.

Intelligent Autonomous Systems 10
W. Burgard et al. (Eds.)
IOS Press, 2008

doi:10.3233/978-1-58603-887-8-36

A Multiple Walking Person Tracker for Laser-Equipped Mobile Robots

Nicolas A. TSOKAS and Kostas J. KYRIAKOPOULOS
National Technical University of Athens, Greece

Abstract. The presented algorithm keeps track of people walking within the field of view of a laser rangefinder while constructing a short-term model of the static environment. Background detection does not require a learning phase and is based on the extraction of simple geometric shapes and on the notion of their *vitality* introduced in this study. Object tracking is based on Multiple Hypothesis Tracking and Kalman Filtering, whereas separate Kalman Filters follow people's trajectories. The proposed method is experimentally evaluated.

Keywords. People tracking, Multiple Hypothesis Tracking, Background subtraction

Introduction

Autonomous robotic systems are likely to increasingly cooperate and closely interact with people in everyday life in the near future. Absolute prerequisite for collision-free navigation in populated places is the robots awareness of the people walking by.

In [1] the authors propose a method that makes use of Rao-Blackwellized Particle Filters (PFs) for implementing Multiple Hypothesis Tracking (MHT). The method combines the accuracy of laser rangefinders with the identification certainty of id-sensors, however, trackable objects need to carry special equipment. In [2] MHT is used and a walking model switching between the left and right foot is presented. In [3] Probabilistic Data Association Filters are used and a combination of a laser rangefinder and a camera is proposed. In [4] the authors use PFs and have the tracking algorithm guessing people's intentions. In [5] a system for tracking pedestrians that models the swinging movement of human legs is presented. Tracking involves Kalman Filtering (KF) and data processing by a central server. [6] extends this method for the case of moving robotic platforms by maintaining a background grid image. In both cases the model of the background has to be acquired *a priori*. In [7] the motion of human legs is compared to a pendulum. In [8] the authors use Parzen window and the distribution of accumulating scan points in consecutive frames for point clustering and leg detection.

This paper tackles the problem of tracking walking people with a moving robot that has not any knowledge of the surrounding environment. The approach proposes a tracking algorithm based on Reid's MHT [9] for data association and a novel on-the-fly background extraction procedure, namely Dynamic Background Detection (DBD). The latter, unlike most background subtraction methods, does not require an explicit learning phase and constructs a short-term model of the environment particularly suitable to a moving robot. Humans are implied by their walking legs positions and are assigned sep-

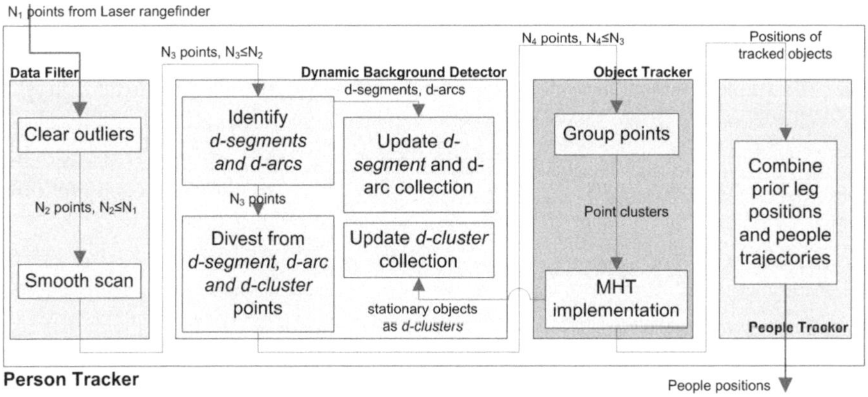

Figure 1. Overview of the walking person tracking algorithm ($N_1 = 361$)

arate Kalman Filters, which can alleviate long occlusions of legs. The proposed method is evaluated through an experiment that verifies the accuracy of the produced trajectories.

1. Proposed Method

As outlined in Fig. 1, the method presented is roughly divided in the following four parts.

1.1. Data Filtering

In the case of reflective surfaces, laser scanners are often prone to false measurements. The Data Filter considers scan points with range greater than 10 meters as outliers and eliminates them. The scan data is smoothed with linear interpolation.

1.2. Dynamic Background Detection

Background extraction is highly important to tracking as it alleviates the tracking algorithm from significant computational burden. DBD is based on the extraction of geometric shapes, namely *d-shapes*, from the laser scan data and on the notion of their *vitality*. d-shapes are currently divided into: *d-segments* (line segments), *d-arcs* (circle arcs) and *d-clusters* (clusters of points). Background subtraction involves the absorbing of scan points by the d-shapes, so that the remaining laser scan data only depict moving objects.

d-shapes have an area of ownership and they can absorb data points falling within this area. Equality and compatibility between d-shapes are explicitly defined. Equality may be straightforward, whereas compatibility refers to whether two d-shapes originate from two adjacent or overlapping parts of the same real-world shape. Compatible d-shapes can be fused resulting in a new d-shape.

Weight is a property of d-shapes whose initial value is equal to the number of scan points that made each one of them up. The value of weight can change through time as shapes are combined with other shapes and new shapes emerge. *Life* represents the number of the algorithm's processing cycles that have passed since a d-shape was constructed. *Power* holds the total number of points absorbed by a d-shape and is thus increased every time a d-shape absorbs data points lying within its area of ownership.

Vitality is a property assigned to all d-shapes as a metric assessing the confidence in their existence. Vitality is represented by a positive real number whose initial value is 1.0 or slightly higher (depending on the geometric definition of the area-of-ownership, the inital value of power might be greater than the initial weight). A value of vitality close to 1.0 implies that the belief in the existence of the corresponding real-world shape is high. For each processing cycle and for each d-shape vitality is calculated by the equation:

$$Vitality = \frac{Power}{Weight \times Life} \tag{1}$$

When the vitality of a d-shape gets too low that shape is eliminated from the background collection. Vitality threshold is a trade-off between background completeness and robust moving/static objects distinction. Low vitality threshold induces a complete background but also causes a slight confusion between static and moving objects. High vitality threshold makes moving shapes to be forgotten quickly, and thus not to be considered as part of the background, but the actual background may end up with some missing parts.

The extraction of d-segments from scan points is carried out with the use of linear least squares regression, whereas R. Israel's *Best Fit Circles* [10] method has been applied in order for the d-arcs to be figured. d-clusters are formed when groups of points not absorbed by any d-shape show up in consecutive scans within a certain area.

In the context of DBD, the static background is defined as a structure representing the collection of all d-shapes. Fig. 1 illustrates the data flow within DBD. After the d-segments and d-arcs have been extracted from the scan, the corresponding collections are updated. The scan is then divested from the points that fall within the ownership areas of the background d-shapes. The remaining scan points feed the Object Tracker, which in turn updates the background d-clusters with stationary point clusters observed.

1.2.1. Object Tracking

Points surviving DBD, i.e. points that have possibly emerged by legs of humans, are grouped into clusters. The centers of gravity of these clusters form the set of measurements, which are supplied to the rest of the tracking procedure.

MHT has been implemented based on [9]. In the method presented here, as measurements represent clusters of 3 or more points and given the sensor resolution the probability of false alarms has been considered negligible. After all, in the unlikely case that a false alarm initiates an object, the latter will quickly be eliminated after the next sets of measurements have been processed.

Human walking being relatively stable (usually 1.0 to $1.5m/s$), and given the periodic motion of human legs, in the context of Kalman Filtering their acceleration has been modelled as white noise. When a moving object is lost from the sensor sight, a counter is initiated and the object is assigned the *occluded* status. The position of objects in *occluded* status, is updated by using the posterior position of the corresponding KF. When the occlusion counter reaches a number of steps, the object is discarded.

The algorithm feeds the objects considered as stationary back to the DBD and forwards the positions of the moving objects implied by the associations of the hypotheses involved to the People Tracker.

1.2.2. People Tracking

The People Tracker exploits the object tracking inference in order to calculate the trajectories of people walking by. The assumption involved here is that, during normal walking, the relative positions of the human CoG and legs satisfy the following two arguments: a) the triangle defined by the human CoG and the feet tends to stay isosceles with its base parallel to the ground and b) the plane defined by that triangle is quasi-normal to the ground. This way a human is considered to be making less effort to keep his/her balance. These two arguments are valid for the scanning plane of the laser rangefinder as well, as the latter is usually parallel to the ground.

The people tracking algorithm is maintaining a Kalman Filter for every people detected. By combining this KF with the two KFs maintained by the Object Tracker for the two legs, the above assumption and some basic geometry, People Tracker can accommodate for periods of occlusion of one or even of both legs. It can also identify a human even in the case when, due to occlusion, one or both legs have been mistakenly considered by the Object Tracker as new objects and their trajectories have been reset.

2. Experimental Evaluation

The proposed method was evaluated through an experiment that measured the accuracy of the produced trajectories. For the needs of the conducted experiment three people walked within an outdoor area of about $50m^2$. A robot equipped with a laser-rangefinder scanning at the height of the calves, and without any knowledge of the surrounding environment, was observing their motion while moving with a low constant speed. Random walking included an instance of multiple occlusion as the trajectories of two people (4 legs) crossed inside the field of view of the rangefinder. A camera mounted at a height of $11m$ recorded the experiment through a sequence of frames.

The trajectories calculated by the People Tracker refer to the global coordinate system that has its center on the initial position of the robot. The X-axis coincides with the initial direction of the robot motion, whereas the positive part of the Y-axis lies on its left. The positions of the walking people implied are first transformed from the pixel-based image coordinate system to the millimeter-based "World" coordinate system of the tracking algorithm. It should be noted that the latter infers the position of the human CoG by observing the legs walking, whereas the trajectories extracted by the processed images refer to the human heads.

The first three graphs of Fig. 2 illustrate people paths. Dotted lines refer to those paths implied by the recorded image sequence, whereas solid lines refer to the ones calculated by the proposed algorithm. Graph (d) depicts the position of human (B) implied by both methods in 10 distinct consecutive seconds. Each pair of circle (o) and cross (+) refers to a distinct moment in time. Trajectory synchronization for human (B) reveals an average distance of $77,6mm$ between points corresponding to the same moment.

3. Conclusion and Future Work

The presented approach allows a moving robot to successfully track a group of walking people by observing their legs with a laser rangefinder. Unlike other approaches, the

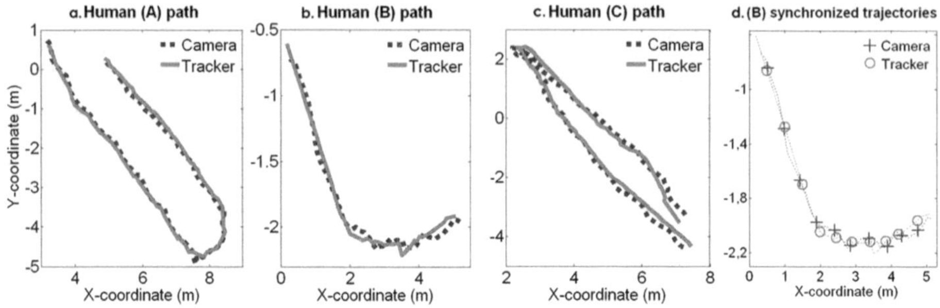

Figure 2. a,b,c: Paths of humans (A), (B) and (C) as seen by the camera (dotted lines) and as calculated by the proposed tracking method (solid line). d: Synchronized trajectories of human (B)

robot does not need to have any prior knowledge of the environment as it constructs and updates an on-the-fly model of its surrounding space.

Object tracking is currently focused on people tracking, so all objects not qualifying as human legs are ignored. A general moving-object Tracker is under developement, that will deal with those measurements currently rejected.

The presented framework for background detection and object tracking will be adapted, tested and evaluated in the context of a multi-mobile-robot setup. The related work to be undertaken will focus on developing a language or high-level protocol for inter-robot communication. Apart from exchanging tracked object trajectories, the latter will allow for effective exchange of information related to d-shapes and thus will contribute to acquiring a global model for the static environment used by all team robots.

References

[1] D. Schulz, D. Fox and J. Hightower, "People Tracking with Anonymous and ID-Sensors Using Rao-Blackwellised Particle Filters", *Proceedings of the International Joint Conference on Artificial Intelligence (IJCAI)*, 2003.

[2] G. Taylor and L. Kleeman, "A Multiple Hypothesis Walking Person Tracker with Switched Dynamic Model", *Proceedings of the Australasian Conference on Robotics and Automation*, Australia, 2004.

[3] M. Kobilarov, J. Hyams, P. Batavia and G. S. Sukhatme, "People tracking and following with mobile robot using an omnidirectional camera and a laser", *In IEEE International Conference on Robotics and Automation*, pp. 557-562, 2006.

[4] A. Bruce and G. Gordon, "Better motion prediction for people tracking", *In Proc. of the International Conference on Robotics and Automation (ICRA)*, New Orleans, USA, 2004.

[5] H. Zhao and R. Shibasaki, "A Novel System for Tracking Pedestrians using Multiple Single-row Laser Range Scanners", *IEEE Trans. SMC Part A: Systems and Humans*, A 35(2): 283-291, 2005.

[6] H. Zhao, Y. Chen, X. Shao, K. Katabira and R. Shibasaki, "Monitoring a populated environment using single-row laser range scanners from a mobile platform", *IEEE International Conference on Robotics and Automation*, pp. 4739-4745, Rome, Italy, 2007.

[7] J. H. Lee, T. Tsubouchi, K. Yamamoto and S. Egawa, "People Tracking Using a Robot in Motion with Laser Range Finder", *Proceedings of the 2006 IEEE/RSJ Int. Conf. on Intelligent Robots and Systems*, pp. 2936 - 2942, Beijing, China, 2006.

[8] J. Cui, H. Zha, H. Zhao and R. Shibasaki, "Laser-based Interacting People Tracking Using Multi-level Observations", *Proceedings of the 2006 IEEE/RSJ Int. Conf. on Intelligent Robots and Systems*, Beijing, China, 2006.

[9] D. B. Reid, "An Algorithm for Tracking Multiple Targets", *IEEE Transactions on Automatic Control*, AC-24(6):843-854, 1979.

[10] http://hdebruijn.soo.dto.tudelft.nl/jaar2006/kromming.pdf

Intelligent Autonomous Systems 10
W. Burgard et al. (Eds.)
IOS Press, 2008

doi:10.3233/978-1-58603-887-8-41

Nonlinear Model Predictive Control of Omnidirectional Mobile Robot Formations

Kiattisin KANJANAWANISHKUL [a,1], Xiang LI [a] and Andreas ZELL [a]
[a] *Department of Computer Architecture, University of Tübingen, Tübingen, Germany*

Abstract. In this paper we focus on solving a path following problem and keeping a geometrical formation. The problem of formation control is divided into a leader agent subproblem and a follower agent subproblem such that a leader agent follows a given path, while each follower agent tracks a trajectory, estimated by using the leader's information. In this paper, we exploit nonlinear model predictive control (NMPC) as a local control law due to its advantages of taking the robot constraints and future information into account. For the leader agent, we propose to explicitly integrate the rate of progression of a virtual vehicle into the local cost function of NMPC. This strategy can overcome stringent initial constraints in a path following problem. Our approach was validated by experiments using three omnidirectional mobile robots.

Keywords. Omnidirectional mobile robots, multi-robot systems, formation control, nonlinear model predictive control, path-following problems

Introduction

The main objective of formation control is that every mobile robot in the group follows a given path (or in case of a given trajectory, tracks a time parameterized reference trajectory) and all mobile robots keep a desired spatial formation at any time [1,2]. Solutions of formation control can be applied in a wide range of applications, such as security patrols, search and rescue in hazardous environments, area coverage and reconnaissance in military missions.

Various strategies have been investigated for solving the formation control problem. These approaches can be roughly categorized as the leader-following approach [3], the virtual structure approach [4], and the behavior-based approach [5]. In this paper, we want to steer a group of omnidirectional mobile robots along a reference path while keeping a desired flexible formation pattern. We use the leader-following strategy because the main advantage of the leader-following approach is its simplicity in that formation maneuvers can be completely specified in terms of the leader's path or trajectory, and the leader-following problem can be reduced to a tracking problem. Consequently, the task of keeping formation pattern can be divided into two subtasks: the leader agent follows

[1]Corresponding Author: Wilhelm-Schickard-Institute, Department of Computer Architecture, University of Tübingen, Sand 1, 72076 Tübingen, Germany; E-mail: kiattisin.kanjanawanishkul@uni-tuebingen.de.

a given reference path and each follower agent tracks its *estimated* trajectory. To control the leader agent, we employ nonlinear model predictive control (NMPC) to solve the path following problems by integrating the rate of progression of a virtual vehicle, $\dot{s}$, into the local cost function. This local controller not only controls the leader agent's motions, but also produces an optimal predicted trajectory. This information is broadcasted to all follower agents.

This paper is organized in the following way: Section 1 describes how to combine the formation configuration with the path following problem. In Section 2, the omnidirectional mobile robots, which have been used in our experiments, are modeled. Then NMPC is presented in Section 3. Section 4 shows the experimental results. Finally, our conclusions and future work are given in Section 5.

1. Formation Configuration

Let the path be parameterized by the path-length s. As proposed in [6], when the formation is turning, the coordinates in a curvilinear coordinate system instead of a rectilinear coordinate system are applied, since this allows us to slightly modify the formation's shape (see Figure 1). In our method, only the path that the leader follows is generated, while each individual follower agent, F_i, in the group has a pre-specified offset $(p_i(s), q_i(s))$ in curvilinear coordinates relative to the reference point *C*, which the leader agent follows, as shown in Figure 1.

In some situations, the collision-free path does not always guarantee the safety for the whole formation. For example, the width of the path could be too narrow to allow for more than one robot. Thus, the formation must be changed to a column (see Figure 2(b)). However, as stated in [6], the width of the formation (q-offset) can only be changed if the second derivative d^2q/ds^2 exists, i.e., offset q_i must be adequately smooth with respect to the corresponding progenitor path during the transient from one configuration to another. To solve this problem, we propose to use a fifth-order (quintic) polynomial to join two path segments with different offset as follows[2]:

$$q(s) = q_{start} + (q_{end} - q_{start})(6s_d^5 - 15s_d^4 + 10s_d^3), \tag{1}$$

where (s, q) is the position on the offset curve at the path length s, $s_d = \frac{s - s_{start}}{s_{end} - s_{start}}$, (s_{start}, q_{start}) and (s_{end}, q_{end}) are the starting and end points of the quintic curve, respectively.

Let u_c be the translational velocity of point C, which the leader agent follows. In other words, u_c is the rate of progression of a virtual vehicle. Once the $(p_i(s), q_i(s))$ coordinates of a follower agent *i* have been determined, the path length of a follower agent, s_i, can be obtained by $s_i = s_c + p_i$, where s_c is the path length at point C. Then its velocity profiles can be obtained by[2]

$$u_i = H u_p, \quad \omega_i = k_i u_i, \tag{2}$$

where $k_i = sign(b)\frac{\sqrt{a^2+b^2}}{H^2}$, $H = \sqrt{(1 - k_p q)^2 + (\frac{dq}{ds})^2}$,
$a = -2k_p\frac{dq}{ds} - q\frac{dk_p}{ds} - (1 - k_p q)\frac{G}{H^2}$, $b = k_p - k_p^2 q + \frac{d^2q}{ds^2} - \frac{dq}{ds}\frac{G}{H^2}$,

[2]The derivations can be found at: http://www.ra.cs.uni-tuebingen.de/mitarb/kanjana/IASderiv.pdf.

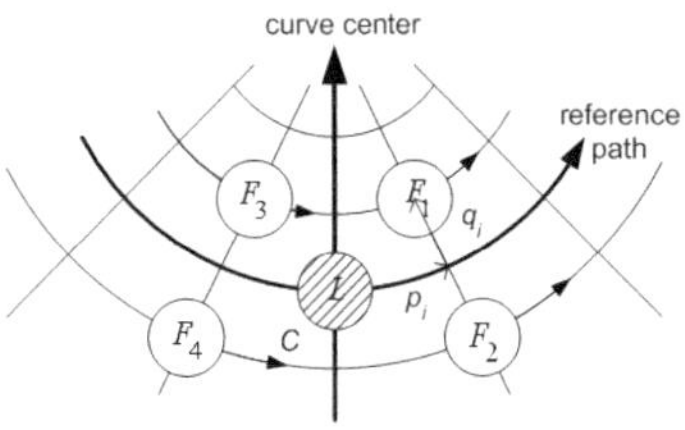

Figure 1. Graphical depiction of a mobile robot path and accompanying offset quantities [6] when the formation is turning. L denotes a leader agent and F_1 - F_4 denote follower agents.

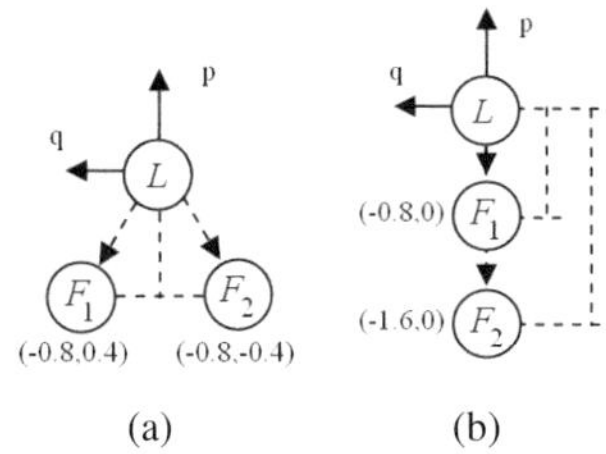

Figure 2. Graphical description of formation configurations: (a) a triangle and (b) a column. L, F_1, and F_2 denote the leader agent, the follower agent 1 and the follower agent 2, respectively. Units are given in meter.

$G = (1 - k_p q)(-k_p \frac{dq}{ds} - q\frac{dk_p}{ds}) + \frac{dq}{ds}\frac{d^2q}{ds^2}$,
u_i, ω_i and k_i are the translational velocity, the rotational velocity and the curvature of the follower agent i, respectively, u_p is the translational velocity at s_i, which is usually equal to u_c. k_p is the curvature at s_i on the reference path, and $q_i(s)$ is the offset at s_i.

2. Omnidirectional Mobile Robot Model

Omnidirectional mobile robots have simultaneously and independently controlled rotational and translational motion capabilities. The annual RoboCup competition is an example of a highly dynamic environment where omnidirectional mobile robots have been exploited highly successfully (see RoboCup Official Site: *http://www.robocup.org*).

There are two coordinate frames used in the modeling: the body frame (X_m, Y_m) and the world frame (X_w, Y_w). The body frame is fixed on the moving robot with the origin at the center of the robot, whereas the world frame is fixed on the ground, as shown in Figure 3. The kinematic model of an omnidirectional mobile robot is given by

$$\dot{\mathbf{x}}(t) = \mathbf{f}(\mathbf{x}(t), \mathbf{u}(t)), \mathbf{x}(0) = \mathbf{x}_0,$$
$$\begin{bmatrix} \dot{x} \\ \dot{y} \\ \dot{\theta} \end{bmatrix} = \begin{bmatrix} \cos\theta & -\sin\theta & 0 \\ \sin\theta & \cos\theta & 0 \\ 0 & 0 & 1 \end{bmatrix} \begin{bmatrix} u \\ v \\ \omega \end{bmatrix}, \tag{3}$$

where $\mathbf{x}(t) = [x, y, \theta]^T$ is the state vector in the world frame and $\mathbf{u}(t) = [u, v, \omega]^T$ is the vector of robot velocities observed in the body frame. θ denotes the robot orientation,

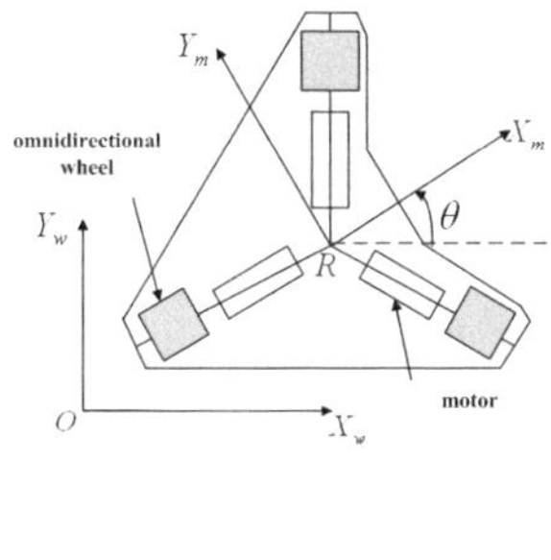

Figure 3. Coordinate frames of an omnidirectional mobile robot.

which is the angle of the axis X_m in the world coordinate system. u and v are the robot translational velocities and ω is the robot rotational velocity. The velocity control inputs are subject to input constraints: $\mathbf{u}_{min} \leq \mathbf{u} \leq \mathbf{u}_{max}$, with constant vectors $\mathbf{u}_{min}, \mathbf{u}_{max}$.

Motion control of omnidirectional robot has been well studied. However, the linearized model in [7,8] cannot fully handle the nonlinear properties of the original system. Liu et al. [7] implemented trajectory linearization control (TLC), which is based on linearization along the desired trajectory and inversion of the dynamics. Watanabe [8] introduced the PID control, self-tuning PID control, and fuzzy control of omni-directional mobile robot. In our proposed approach, NMPC combined with path following has been carried out for the leader agent, and is explained in details in the next section. NMPC is an attractive approach, because it can explicitly account for system constraints and easily handle nonlinear and time-varying systems, since the controller is a function of the model that can be modified in real-time.

3. Nonlinear Model Predictive Control

Nonlinear model predictive control (NMPC) is based on a finite-horizon continuous time minimization of nonlinear predicted tracking errors with constraints on the control inputs and the state variables. It predicts system outputs based on current states and the system model, finds an open loop control profile by numerical optimization, and applies the first control signal in the optimized control profile to the system [9]. However, due to the use of a finite predictive control horizon, control stability becomes one of the main problems. To guarantee control stability, many approaches have been investigated, e.g., using so called terminal region constraints and/or a terminal penalty term [10]. More explanations regarding NMPC can be found in [9,10].

In the centralized system of formation control, the complete system is modeled and all the control inputs are computed in one optimization problem. Using this strategy, the size of the state variables depends typically on the number of mobile robots. When the control horizon becomes larger, the number of variables, of which the agent has to find the value, increases rapidly. Thus, our research has been directed at extending the single-agent MPC framework to multiple agents by means of decomposing the centralized system into smaller subsystems. This can be achieved by using distributed/decentralized control or hierarchical design (see [11,12,13]). The main difference of these control approaches is the kind of interaction between two subsystems via state variables, constraints or objectives. In this paper, we divide a group of robot agents into a leader agent as a downstream stage and a set of follower agents as an upstream stage. Each agent computes a solution to its local problem. The leader agent communicates the most recent variables to its followers, each of which re-solves its optimization problem with the updated values.

Each omnidirectional mobile robot can be described by a nonlinear differential equation as given in Eq. (3). In NMPC, the input applied to the system is usually given by the solution of the following finite horizon open-loop optimal control problem, which is solved at every sampling instant:

$$\min_{\mathbf{u}(\cdot)} \int_{t}^{t+T_p} F(\mathbf{x}(\tau), \mathbf{u}(\tau))\, d\tau, \tag{4}$$

$$\text{subject to:} \quad \dot{\mathbf{x}}(\tau) = \mathbf{f}(\mathbf{x}(\tau), \mathbf{u}(\tau)), \\ \mathbf{x}(\tau) \in \mathcal{X}, \;\; \mathbf{u}(\tau) \in \mathcal{U}, \;\; (\tau \in [t, t + T_p]), \tag{5}$$

where T_p is the prediction horizon, $F(\mathbf{x}(\tau), \mathbf{u}(\tau))$ is the cost function, and $\mathcal{U}$ and $\mathcal{X}$ denote the set of feasible input values and the set of feasible states, respectively.

Furthermore, a simple delay compensation approach is applied in order to overcome the computational delay problem. The optimal control problem at time t_k is solved for the predicted state at $\mathbf{x}(t_k + \delta^r)$, where δ^r is the computational delay. At initial time the internal clock of each agent has to be synchronized using clock synchronization, and in our implementation, the sampling time is calculated based on the average past results [14]. Thus sampling time is varying. This leads us to work with asynchronous agents with different sampling time. This strategy allows agents to proceed in parallel, each one at its own speed.

The main tasks for the leader agent are to steer itself to the reference path, to produce an optimal predicted trajectory of itself at each time instant, and to send out its information to all follower agents via broadcast communication. In this paper, we propose to add $\dot{s}$ into the cost function in order to overcome stringent initial condition constraints that are present in a number of path following control strategies described in the literature as suggested in [15]. The local cost function becomes:

$$F_{leader}(\mathbf{x}, \mathbf{u}) = (\mathbf{x} - \mathbf{x}_s)^T \mathbf{Q}_l (\mathbf{x} - \mathbf{x}_s) + (\mathbf{u} - \mathbf{u}_s)^T \mathbf{R}_l (\mathbf{u} - \mathbf{u}_s) + (\dot{s} - u_o)^2 p_l \quad , \tag{6}$$

where $\mathbf{x}_s = [x_r, y_r, \theta_r]^T$ is the vector of the desired pose in the world frame, $\mathbf{u}_s = [u_r, v_r, \omega_r]^T$ is the vector of the desired velocity in the body frame. θ_r denotes the desired robot orientation, which is the angle of the axis X_m in the world coordinate system. u_o is the desired translational velocity along the reference path and $\dot{s}$ is the rate of progression. The deviation from the desired values is weighted by the positive definite matrices $\mathbf{Q}_l$, $\mathbf{R}_l$, and the positive constant p_l.

After the optimization problem at time t_k is solved, the current reference state ($s_{l,k}$), the optimal predicted reference trajectory ($s_{l,k+1|k}$, ..., $s_{l,k+T_p|k}$), and the sampling time (δ_l) are transmitted to all follower agents. Each data packet is time-stamped, so that the *age* of the information can be extracted at a follower controller.

Then each follower agent has to track it own reference trajectory, estimated by using the leader agent's optimal predicted trajectory and the predefined formation configuration. In practice, some problems may arise, for example, the information time delay is not zero, the sampling time of the follower agent can be different (asynchronous timing conditions) from that of the leader agent or the data packet can be lost. To overcome these problems, we calculate the *estimated* reference trajectory based on the leader agent's most recent information, where the time stamp is taken into account and the missing information is filled if packet loss happens. For example, at time $t_{i,k}$, follower agent F_i receives an optimal predicted trajectory of its leader, which is generated at time $t_{l,k} < t_{i,k}$. Over the interval $[t_{i,k}, t_{l,k} + T_p]$, follower agent F_i has the optimal predicted trajectory of its leader agent. For the rest of the time interval, i.e., $[t_{l,k} + T_p, t_{i,k} + T_p]$, follower agent i has to estimate the optimal predicted trajectory of its leader agent.

The local cost function of the follower agent can be given as

$$F_{follower}(\mathbf{x}, \mathbf{u}) = (\mathbf{x} - \mathbf{x}_s)^T \mathbf{Q}_f (\mathbf{x} - \mathbf{x}_s) + (\mathbf{u} - \mathbf{u}_s)^T \mathbf{R}_f (\mathbf{u} - \mathbf{u}_s), \tag{7}$$

where $\mathbf{x}_s = [x_r, y_r, \theta_r]^T$ is the vector of the desired pose in the world frame, $\mathbf{u}_s = [u_r, v_r, \omega_r]^T$ is the vector of the desired velocity in the body frame and the deviation from the desired values is weighted by the positive definite matrices $\mathbf{Q}_f$, $\mathbf{R}_f$.

4. Experimental Results

We implemented our algorithm on omnidirectional mobile robots shown in Figure 4. Each has an omnidirectional camera as sole sensor, which is used for self localization. Thanks to [16], the self localization applied for the RoboCup field has been employed in our experiments. This self-localization algorithm is based on probabilistic Monte-Carlo localization (MCL). We use three omnidirectional mobile robots to make a formation, one is defined as a leader, following a circular reference path and the others as followers. The reference path is given by

$$x_o(s) = r\cos(s/r), \quad y_o(s) = r\sin(s/r), \quad r = 1.1m, \tag{8}$$

where r is the radius of the circle. A formation transition from a triangle (see Figure 2(a)) to a column (see Figure 2(b)) happens between $s = 2\pi r$ and $s = 2.5\pi r$, and the formation is switched back to the triangle at $s = 4\pi r$. The parameters for the leader and for both followers are selected as follows:

Leader: $\mathbf{Q}_l = 0.05\mathbf{I}_3$, $\mathbf{R}_l = 0.005\mathbf{I}_3$, $p_l = 0.001$, prediction steps = 3,
Followers: $\mathbf{Q}_f = 0.05\mathbf{I}_3$, $\mathbf{R}_f = 0.01\mathbf{I}_3$, prediction steps = 3,
where $\mathbf{I}_3 = diag(1, 1, 1)$.

By our implementation, the average sampling time used by the leader agent and by the follower agents are approximately 0.1 s and 0.07 s, respectively. A free package DONLP2 [17] has been used to solve the online optimization problem and PID controllers have been implemented for motor velocity control in our experiments. Figure 5

Figure 4. Omnidirectional mobile robots used in the formation control experiments.

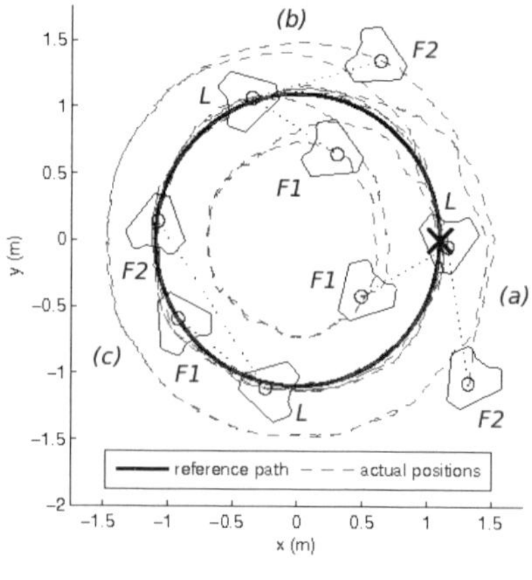

Figure 5. The snapshots are taken at the following time: (a) original configuration at t = 0 s, (b) triangle configuration at t = 8.0 s, and (c) column formation obtained at t = 31.8 s. × denotes the starting position.

shows the superimposed snapshots of three mobile robots keeping and switching the formation, while the leader follows the circular reference path with the translational velocity, u_o, of 0.4 m/s and the rotational velocity, ω_o, of $k_o u_o$, where k_o is the curvature at the reference point. L, $F1$, and $F2$ denote the leader, the follower 1, the follower 2, respectively. The pose errors of the leader, of the follower 1, and of the follower 2 are shown in Figure 6 and the velocities of the leader, of the follower 1, and of the follower 2, compared with their reference velocities, are shown in Figure 7.

5. Conclusions and Future Work

In this paper, the formation control problem of omnidirectional mobile robots has been solved by using NMPC as a local control law, in such a way that a leader agent follows a given reference path and each follower agent tracks an *estimated* reference trajectory. NMPC is a promising control method as can be seen in our experimental results. Not only can it handle control and input constraints, but it also utilizes future information to generate a trajectory of optimal control inputs at each time step. Two key points, which are employed to solve the path following problem and formation keeping problem, are proposed in this paper. First, the rate of progression of a virtual vehicle is integrated into

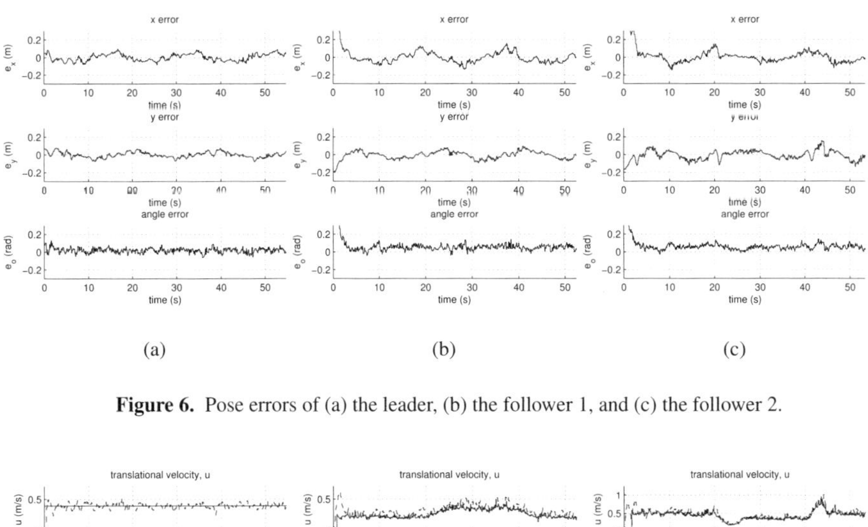

(a) (b) (c)

Figure 6. Pose errors of (a) the leader, (b) the follower 1, and (c) the follower 2.

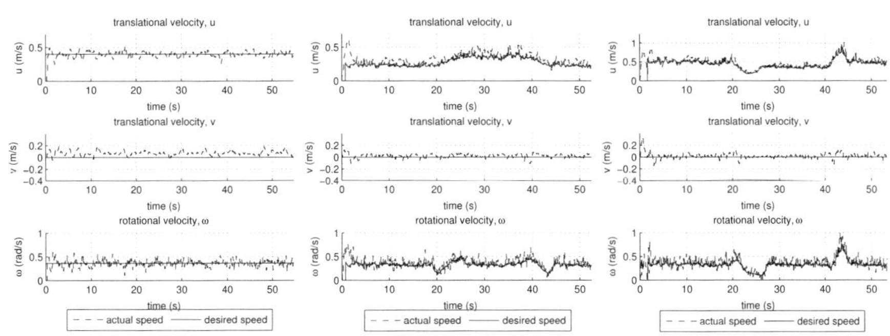

(a) (b) (c)

Figure 7. Velocities of (a) the leader, (b) the follower 1, and (c) the follower 2.

the local cost function of the leader agent. An optimal predicted trajectory, generated when the open-loop optimization problem is solved, is sent out to all follower agents. Second, with this information and a desired formation configuration, each follower agent can estimate its own reference trajectory with velocity profiles by taking the time stamps into account. However, in our problem, the constraint to enforce a degree of consistency between what the leader agent is actually doing and what the follower agents believe that the leader agent is doing is required, as suggested in [11]. This issue is currently under our investigation. Furthermore, in the future, we would like to integrate obstacle avoidance as coupling constraints and analyze the formation stability.

References

[1] Y.Q. Chen and Z.M. Wang, Formation control: a review and a new consideration, in *Proc. of IEEE/RSJ International Conference on Intelligent Robots and Systems (IROS '05)*, Edmonton, Canada, Aug. 2005, pp. 3181-3186.

[2] R.M. Murray, Recent research in cooperative-control of multivehicle systems, *Journal of Dynamics, Systems, Measurement and Control*, 129, pp. 571-583, 2007.

[3] A.K. Das, R. Fierro, V. Kumar, J.P. Ostrowski, J. Spletzer, and C.J. Taylor, A vision-based formation control framework, *IEEE Trans. on Robotics and Automation*, 18(5), pp. 813-825, 2002.

[4] M.A. Lewis and K.-H. Tan, High precision formation control of mobile robots using virtual structures, *Autonomous Robots*, 4(4), pp. 387-403, 1997.

[5] T. Balch and R.C. Arkin, Behavior-based formation control for multirobot teams, *IEEE Trans. on Robotics and Automation*, 14(6), pp. 1-15, 1998.

[6] T.D. Barfoot and C.M. Clark, Motion planning for formations of mobile robots, *Robotics and Autonomous Systems*, 46(2), pp. 65-78, 2004.

[7] Y. Liu, X. Wu , J. Jim Zhu, and J. Lew, Omni-directional mobile robot controller design by trajectory linearization, in *Proc. of American Control Conference*, Denver, Colorado, Jun. 2003, pp. 3423-3428.

[8] K. Watanabe, Control of omnidirectional mobile robot, in *Proc. of 2nd Int. Conf. on Knowledge-Based Intelligent Electronic Systems*, Adelaide, Australia, Apr. 1998, pp. 51-60.

[9] D.Q. Mayne, J.B. Rawlings, C.V. Rao, and P.O. M. Scokaert, Constrained model predictive control: stability and optimality, *Automatica*, 36, pp. 789-814, 2000.

[10] H. Chen and F. Allgöwer, A quasi-infinite horizon nonlinear model predictive control scheme with guaranteed stability, *Automatica*, 34, pp. 335-348, 1998.

[11] W.B. Dunbar and R.M. Murray, Distributed receding horizon control for multi-vehicle formation stabilization, *Automatica*, 42(4), pp. 549-558, 2006.

[12] T. Keviczky, F. Borrelli, and G. J. Balas, A study on decentralized receding horizon control for decoupled systems, in *Proc. of American Control Conference*, Boston, Massachusetts, June 30 - July 2, 2004, pp. 4921-4926.

[13] M.G. Earl and R. D'Andrea, A decomposition approach to multi-vehicle cooperative control, *Robotics and Autonomous Systems*, 55(4), pp. 276-291, 2007.

[14] R. Franz, M. Milam, and J. Hauser, Applied receding horizon control of the Calteth ducted fan, in *Proc. of American Control Conference*, Anchorage AK, May 2002, pp. 3735 - 3740.

[15] D. Soeanto, L. Lapierre, and A. Pascoal, Adaptive non-singular path-following, control of dynamic wheeled robots, in *Proc. of International Conference on Advanced Robotics*, Coimbra, Portugal, June 30 - July 3, 2003, pp. 1387-1392.

[16] P. Heinemann, J. Haase, and A. Zell, A Combined Monte-Carlo Localization and Tracking Algorithm for RoboCup, in *Proc. of IEEE/RSJ International Conference on Intelligent Robots and Systems (IROS '06)*, Beijing, China, Oct. 2006, pp. 1535-1540.

[17] P. Spellucci, An SQP method for general nonlinear programs using only equality constrained subproblems, *Mathematical Programming*, 82, pp. 413-448, 1998.

Intelligent Autonomous Systems 10
W. Burgard et al. (Eds.)
IOS Press, 2008

doi:10.3233/978-1-58603-887-8-49

Memory-based Software Integration for Development in Autonomous Robotics

Thorsten Peter SPEXARD, Frederic H. K. SIEPMANN and Gerhard SAGERER
Applied Computer Science, Bielefeld University,
Universitätstraße 25, 33615 Bielefeld, Germany;
E-mail: {tspexard, fsiepman, sagerer}@techfak.uni-bielefeld.de

Abstract. Focusing the development of non-industrial robotics in the last decade the growing impact of service and entertainment robots for daily life has emerged from pure science fiction to a serious scientific subject. But still many questions in how to solve everyday tasks like laying the table or even "simpler" detecting objects in unstructured areas with varying lighting conditions are unsolved. Hence the strong need to evaluate and exchange different approaches and abilities of multiple robotic demonstrators under real world conditions is also a crucial aspect in the development of system architectures. In this paper an architecture will be described providing strong support for simple exchange and integration of new robot abilities.

Keywords. autonomous robotics, human-robot interaction, cognitive architecture, system integration

1. Introduction

In combination with the increasing computational power both the amount of abilities for robots and the complexity of single abilities grew. Besides the impressive progress in the skills of robots and the concerning algorithms behind them, another issue becomes more and more important: Operating in human-oriented surroundings. This requires the combination of multiple different robot abilities. Based on our previous experience in building integrated robotic systems we present an advancement of our former System Infrastructure. This enhanced approach increases both the flexibility in integrating and changing components, as well as to efficiently combine information given from separate modules. After a short discussion on alternative proposals for integrated systems in Section 2, the robotic demonstrator acting in a human centered environment is presented in Section 3. Subsequently the new architecture with its benefits is discussed in Section 4. In Section 5 we present numbers for the decreased system complexity and practical experiences before closing with a brief summary.

2. Related Work

In research a broad variety of different architectures have been developed. Using classical AI strategies Nilsson [1] propagated the deliberative robot control. Based on a world model and a given goal plans are generated to achieve the goal. If the environment changes the planning has to be repeated. This is almost always too time consuming for embedded systems in dynamic human-centered environments. Therefore, reactive architectures [2] should overcome this burden, but the complexity of the tasks to be executed is strongly limited. Thus the combination of both deliberative and reactive components in a hybrid architecture [3] should enable a robot to generate plans for complex tasks and to react dynamically to a changing surrounding. Motivated by biological models behavior based architectures propose systems with separate abilities driven by reactive or deliberative components. These abilities can be seen as separate behaviors like following a human and avoiding obstacles. Hence behaviors may provide contradictory results like in this following task an arbitration mechanism has to select the final action.

We propose a system that is more deterministic than a weighted arbitration and therefore easier to design on the one hand, on the other hand it is sufficient dynamic to allow even online changes in the behavior modules *and* arbitration without redesigning source code or configuration files offline. Additionally the approach described in this paper is not depending on globally equal interface definitions as these often result in overhead and a less dynamic integration. This concerns especially behaviors, which are either exchanged with partners of different institutes or were not planned at the start of a project as their necessity emerged during experiments.

3. Applications in the Real World

Instead of artificial experiments under laboratory conditions an apartment is continously rent to test our demonstrator *BIRON* (**Bi**elefeld **Ro**bot Companio**n**) under real world conditions including small doorways, sticky grounds, and uncontrolled lighting. BIRON is equipped with several sensors that allow an assessment of the current situation as a basis for interaction. Its onboard computational equipment in combination with a wireless LAN notebook is sufficient to achieve a system running completely in real time which is necessary for HRI in the Home-Tour-Scenario. The scenario envisions a newly purchased robot being introduced to its new working area – usually an apartment – by an unexperienced user. Due to the huge variety of application areas especially in home environments only a small set of pre-programmed knowledge is useful. The main knowledge such as maps or objects in the new environment has to be obtained online during interaction with a person [4].

To succeed in this task a modular approach was chosen to easily add and exchange abilities. More than 30 asynchronous modules are running in parallel using an XML based Communication Framework (*XCF*, [5]). The communication was mainly realized in 1:n XML publish/subscribe data streams and remote method invocations (see Fig. 1(a)). The synchronization of events like switching the robot control from the person tracking and attention module (*PTA*) [6] to object tracking was realized by a finite state machine implementation (*ESV*) using the information sent via the data streams as transitions between its internal states. For each transition modules like the dialog system (*DLG*) [7] were reconfigured according to their new task. DLG, ESV, PTA, and Hardware Control (*HWC*) build a core system for basic HRI which is extended with modules for localization, navigation, gesture recognition and object attention.

With the growing number of modules the number of stream connections between them became a limitation: A module getting information from a data stream needed the data stream and therefore the stream generating module already running before being started itself. Thus modules depending on each other caused restarts of the whole system if a 1st-level module exited unexpectedly during development. Although this problem was caught by exception handling, the system could get into an inconsistent state when a module broke and was restarted, as during the reconnection process of the data streams information was lost in the meantime. The sender was not aware whether the information was successfully received or not. Furthermore the combination of information from different modules needed to either add a combining module with even more stream connections or the adaptation of an existing module, implementing worldknowledge concerning the software which could be adapted by source code manipulation only.

4. Integration

To overcome these limitations we propose that as less knowledge as possible concerning the robotic system is incorporated in the single modules depending not on the operation of another module but only on incoming data. This was realized by the active memory concept [8]. This system wide memory is able to hold all information of the modules and structure them semantically with regard to the expiration of validity:

Short term memory This memory contains information with a short duration of validity. A forgetting process removes all entries in the memory which are not updated within a certain period of time, e.g., person positions.

Scene memory As a representation of the current scene this memory keeps all information that are valid until changed or removed by a module. The memory contains, e.g., information about the current room the robot is in.

Long term memory This memory stores information which shall be kept permanently. This includes system parameters for BIRON's behavior as well as, e.g., object descriptions or maps.

Instead of stream connections a module registers database triggers which specify the kind of information the module needs by XPath and a database event. It is activated automatically when the information is available, independently from the information source and its current state. This reduces the communication complexity (see Fig. 1(b)) and encapsulates modules in terms of reducing system knowledge in all modules, increasing the extensibility and adaptivity of the system. Alternatively, remote method invocation (*RMI*) is used when an affirmation for information exchange is required. For continuous raw data (e.g laser data) the primary streaming approach was kept as partial drop outs of these data do not effect the system and memory flooding is avoided.

As a consequence of the decreased system knowledge within the modules a more powerful coordination module replaced the former ESV. The active control memory interface *ACMI* takes the current memory state with all its information into account. It observes the system by registering to the different memories to derive information like the current system state (e.g., following someone, recognizing new rooms) from combining stored information. ACMI is also capable of manipulating the current memory content by inserting, removing or replacing information. It takes responsibility for behavior arbitration and RMI for crucial information exchange, like the information which module gains control of the hardware in a certain situation. The necessary parameters are part of the ACMI configuration and represented in XML. The mayor and as simple as powerful advantage is that this configuration is also stored in the memory. Thus it can be modified during runtime either explicitly by a developer or by a software module. With ACMI

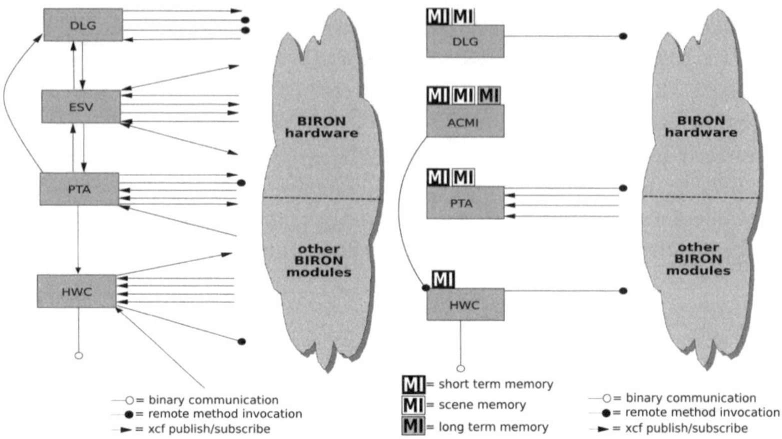

(a) Communication using mainly streams. (b) Advanced communication with memory.

Figure 1. Comparing former and memory based communication dependencies: The number of inter module connections was decreased by more than 70% from 30 to 8, now using streams for low level sensor data, only.

triggering its own configuration information the system is enabled to adapt e.g. its arbitration routine if a preprogrammed strategy does not succeed. An example of the use of triggers and adaptive memory use is provided by the HWC.

If the content of the memories indicates the necessity for a change in the hardware controlling module, ACMI uses an RMI on the HWC to send the name of the module, which is to control the hardware next. The RMI enables the system to estimate when the configuration is concluded and to propagate this back to the memory as well as the result of the HWC configuration. Both, the name of the controlling module, and the necessary memory content are defined in the ACMI configuration and thus stored in the memory, too. Using a standard prefix with the module name as suffix an XPath is generated. The XPath defines the location in the memories where the hardware commands of the controlling module are inserted. Subsequently a trigger is registered to newly inserted information under the given XPath and a potential former trigger is removed. If new information is inserted into the memory matching the specified XPath HWC actuates the hardware with the extracted information. This method proposed here is introduced as *memory pointing* comparable to programming languages not the information itself but where to find it is exchanged. The benefit of memory pointing is that it is flexible and extensible and no source code adaptation is needed to integrate new modules that need to control the hardware, which was necessary with the prior stream-based implementation. In addition memory pointing is robust towards missing or misbehaving modules as with missing information the system just idles.

5. Benefits

Regarding the core system, inter module connections were reduced by more than 70% from 30 to 8 streams as demonstrated in Figure 1. This reduces the number of possible communication failures as well as it decouples the modules. They now depend on the active memory module only, to interact with the whole system. Previously 8 layers of software modules had to be started causing modules in a higher layer to become unstable if a lower layer module broke, now the number of layers was reduced to 4. Taking into account that the first two layers are used to set up the communication framework the dependency depth of robot operating modules is finally reduced from 6 to 2. This results in modules restartable during development with only minimal influence to the system.

Besides the theoretical value of communication complexity the authors present a first test concerning both the robustness of the system but also the benefit of the architecture for integration of new modules. A module for localization from a project partner of a different institute[1] was integrated. Providing a sample source file for memory access methods and exchanging an XML-interface definition the whole practical integration was done within one day. During the integration the localization module had to be restarted several times. In contrast to previous approaches the remaining system did not need to be restarted but continued working stable, saving a huge amount of time. After the integration a first Home-Tour was performed. The robot worked robust and in real time for half a day limited only by battery life.

6. Summary

We presented a new architectural approach for autonomous mobile robots operating in natural, human centered environments. For this approach the usage of data stream was changed to a biologically inspired memory system consisting of short term, scene, and longterm memory. The memory approach allows the robotic system to store high level information as gestures or objects and combine the information with global access for each component. Moreover the new approach strongly supports the idea of loose coupling by reducing both the component dependencies on each other and the direct data stream connectivity by 70% The applicability of the presented work was proven by implementation of a core system and testing the demonstrator by a Home Tour in the apartment with a module for location learning from a different institute. The system run stable during both the integration and interaction for hours.

References

[1] N. J. Nilsson, *Principles of Artificial Intelligence.* Berlin, Germany: Springer Verlag, 1982.

[2] R. A. Brooks, "A robust control system for mobile robot," *IEEE Journal of Robotics an Automation*, vol. 2, no. 1, pp. 14–23, March 1986.

[3] R. C. Arkin, "Path planning for a vision-based autonomous robot," in *Proc. SPIE Conf. on Mobile Robots*, Cambridge, MA, October 1986, pp. 240–249.

[4] T. Spexard, S. Li, B. Wrede, J. Fritsch, G. Sagerer, O. Booij, Z. Zivkovic, B. Terwijn, and B. Kröse, "Biron, where are you? - enabling a robot to learn new places in a real home environment by integrating spoken dialog and visual localization." Beijing, P.R. China: IEEE, October 2006, inproceedings, pp. 934–940.

[5] S. Wrede, J. Fritsch, C. Bauckhage, and G. Sagerer, "An XML Based Framework for Cognitive Vision Architectures," in *Proc. Int. Conf. on Pattern Recognition*, no. 1, 2004, pp. 757–760.

[6] T. Spexard, M. Hanheide, and G. Sagerer, "Human-oriented interaction with an anthropomorphic robot," *IEEE Transactions on Robotics*, vol. 23, pp. 852–862, 2007.

[7] S. Li, B. Wrede, and G. Sagerer, "A computational model of multi-modal grounding." ACL Press, 2006, inproceedings, pp. 153–160.

[8] S. Wrede, M. Hanheide, C. Bauckhage, and G. Sagerer, "An Active Memory as a Model for Information Fusion," *Proc. 7th Int. Conference on Information Fusion*, 2004.

[1] Olaf Booij, Intelligent Systems Lab Amsterdam, University of Amsterdam

Intelligent Autonomous Systems 10
W. Burgard et al. (Eds.)
IOS Press, 2008

doi:10.3233/978-1-58603-887-8-54

An Experimental Environment for Optimal Spatial Sampling in a Multi-Robot System

ANSSI KEMPPAINEN [a], TONI MÄKELÄ [a], JANNE HAVERINEN [a] and JUHA RÖNING [a]

[a] *Department of Electrical and Information Engineering, University of Oulu, Finland*

Abstract.

In our research, we are concerned with sensing the environment using mobile robots. This enables selection of optimal sampling locations in order to produce maximum information about the environment. Selection of sampling locations plays a key role in hospital environments, for example, where humidity and temperature levels or carbon dioxide concentration may require regular monitoring. On the other hand, the accuracy of a regression model depends on the sampling locations, which is significant, for example, in planetary exploration.

In geostatistics optimal spatial sampling strategies search for sensor locations that produce minimal variance in estimates with a restricted number of sensors. Minimal variance is achieved, for example, by minimizing the conditional entropy of unobserved locations, where the environment is modelled using Gaussian processes.

In this paper, we propose an experimental environment for optimal spatial sampling using mobile sensors. Mobility reduces the reliability of sampling locations due to odometer failures, which again reduce the likelihood of the model. We have resolved this problem by building an experimental environment where a ceiling camera vision system provides multi-robot localization in an area measuring approximately 240 x 160 cm. Preliminary experiments compare optimal spatial sampling for both stationary and nonstationary models using scalar measurements of ambient light and magnetic flux density.

Keywords. Non-stationary Gaussian Processes, Distributed Sensing, Environment Modelling

Introduction

Gaussian processes have a long history in geostatistics in the context of interpolation, where the method is known as kriging [10]. Later, the machine learning community adopted Gaussian processes [16] for regression problems, where they enabled reasonable interpretations for predictions of models, yet provided a simple and sufficent choice of models. In the field of robotics, Gaussian processes have recently attracted attention in terrain modelling [9], for example, since they allow prior knowledge of the environment to be taken into account in the model.

Gaussian processes are also useful in distributed sensing scenarios, like in hospital environments, where humidity and temperature levels or carbon dioxide concentration may require regular monitoring with a restricted number of sensors. On the other hand, in spatial sampling, for example in planetary exploration, the accuracy of the regression model depends on the sampling locations. In our research, we are concerned with sensing the environment using mobile robots. This enables selection of optimal sampling locations in order to produce maximum information about the environment.

In geostatistics, optimal spatial sampling [3] searches for sensor locations that produce minimal variance in estimates with a restricted number of sensors. Correlations between the samples are modelled using a Gaussian process [14], which is a generalization from probability distributions to functions: a mean function describes the expected value and a covariance function, the variance in an estimate at a given location. Minimal variance is obtained by minimizing the conditional entropy of unobserved locations [15], which is analogous to maximizing the entropy of sensor locations. An alternative approach is to maximize mutual information [8] between samples, which leads to samples focused closer to the centre of the sampling area.

In this paper we propose an experimental environment for a study on optimal spatial sampling strategies using mobile robots. The experimental environment is a restricted area where mobile robots can measure ambient light and magnetic flux density and place the measurements onto maps. Mapping and optimal sampling use a centralized control system aided by roof camera-based localization. Control and sensor information between the robots and the control system is supplied through radio communication. Section 1 briefly reviews the theory of Gaussian process regression and Section 2 introduces information about theoretical criteria in optimal spatial sampling. Section 3 presents the experimental environment with some preliminary results of optimal spatial sampling with two different data sets: one from ambient light and the other from magnetic flux density measurements.

1. Gaussian processes

A Gaussian process [14] can be seen as a collection of candidate functions that model the environment. The range of functions is restricted by a covariance function and the most probable function is described by a mean function. Formally, a Gaussian process is defined by a mean function (1)

$$\mu(\boldsymbol{x}) = \mathbb{E}[f(\boldsymbol{x})] \tag{1}$$

and a covariance function (2)

$$c(\boldsymbol{x}, \boldsymbol{x}') = \mathbb{E}[(f(\boldsymbol{x}) - \mu(\boldsymbol{x}))(f(\boldsymbol{x}') - \mu(\boldsymbol{x}'))] \tag{2}$$

where $f(x)$ is a random function obeying the Gaussian process (3). [1]

$$f(\boldsymbol{x}) \sim \mathcal{GP}(\mu(\boldsymbol{x}), c(\boldsymbol{x}, \boldsymbol{x}')) \tag{3}$$

We can imagine a function $f(\boldsymbol{x})$ as a finite dimensional vector $\boldsymbol{y}$, where each random variable y^i in some orthonormal basis $\boldsymbol{e}_i$ corresponds to a function value $f(\boldsymbol{x}_i)$. Thus,

[1] Bold lowercase characters denote vectors and uppercase matrices. Not bolded italic characters stand for scalars and constants.

a Gaussian process defines a joint distribution of random variables $p(y^1, y^2, \cdots, y^n)$ with a mean value vector $\boldsymbol{\mu} = (\mu_1, \mu_2, \cdots, \mu_n)$ and a covariance matrix $\boldsymbol{C}_{nn}$ with elements $c_{ij} = cov(y^i, y^j)$. In order to define a continuous covariance function, we can use integral operators to define the correlation between bases. Thus, the covariance among function values is given by

$$\begin{aligned} cov(y^i, y^j) &= \mathbb{E}[(f(\boldsymbol{x}_i) - \mu(\boldsymbol{x}_i))(f(\boldsymbol{x}_j) - \mu(\boldsymbol{x}_j))] \\ &= \mathbb{E}[(f(\boldsymbol{x}_i) - \mu(\boldsymbol{x}_i))k(\boldsymbol{x}_i, \boldsymbol{x}_j)(f(\boldsymbol{x}_i) - \mu(\boldsymbol{x}_i))] \\ &= var(y^i)k(\boldsymbol{x}_i, \boldsymbol{x}_j) \end{aligned}$$

where the kernel function $k(\boldsymbol{x}_i, \boldsymbol{x}_j)$ must be symmetric, positive and definite.

1.1. Gaussian process regression

A Gaussian process defines a prior probability distribution over functions before any observations are made. This leads to a Bayesian inference, which in this context is known as a Gaussian process regression. In the Gaussian process regression we try to estimate the most likely posterior distribution over functions $f(\boldsymbol{x})$, conditioned on measurement function $z(\boldsymbol{x})$. For a multivariate Gaussian distribution we have that the conditional distribution is given by a conditional mean (4)

$$\hat{\boldsymbol{y}}_n = \boldsymbol{\mu}_n + \boldsymbol{C}_{nm}(\boldsymbol{C}_{mm} + \sigma_z^2 \boldsymbol{I}_{mm})^{-1}(\boldsymbol{z}_m - \boldsymbol{\mu}_m) \tag{4}$$

and a conditional variance (5)

$$var\,(\hat{\boldsymbol{y}}_n - \boldsymbol{y}_n) = \boldsymbol{C}_{nn} - \boldsymbol{C}_{nm}(\boldsymbol{C}_{mm} + \sigma_z^2 \boldsymbol{I}_{mm})^{-1}\boldsymbol{C}_{mn} \tag{5}$$

where σ_z^2 is the measurement variance (independent measurements) and m is the number of measurements.

1.2. Stationarity

A Gaussian process is said to be stationary if its mean and covariance functions are invariant to translations, i.e., $\mu(\boldsymbol{x}) = A$ and $c(\boldsymbol{x}, \boldsymbol{x}') = c(\boldsymbol{x} - \boldsymbol{x}')$. The stationarity assumption is somewhat artificial, since nature provides spatial information that is inherently irregular but yet continuous. In order to obtain a more accurate model of the environment, one could try to learn the parameters of the model based on experience. This leads to non-stationary Gaussian processes, where the mean function is defined by parameters and the covariance function, by hyperparameters. A covariance function restricts models to those that do not violate the definition of symmetric positive definiteness.

The literature [14] provides many non-stationary versions of covariance functions. In this study we adopted the *HSK* covariance function (6) [6], [13]

$$c(\boldsymbol{x}_i, \boldsymbol{x}_j) = \int_{\mathbb{R}^2} k(\boldsymbol{x}_i, \boldsymbol{u})k(\boldsymbol{x}_j, \boldsymbol{u})d\boldsymbol{u} \tag{6}$$

which uses local kernel functions $k(\boldsymbol{v}, \boldsymbol{u})$ centred on location $\boldsymbol{v}$. If each of these are given through Gaussian kernels, i.e,

$$k(\boldsymbol{v},\boldsymbol{u}) = (2\pi)^{-1/2}|\boldsymbol{\Sigma}_v|^{-1/2}\exp\left(-\frac{1}{2}(\boldsymbol{v}-\boldsymbol{u})^T\boldsymbol{\Sigma}_v^{-1}(\boldsymbol{v}-\boldsymbol{u})\right)$$

the covariance function can be given as (7)

$$c(\boldsymbol{x}_i,\boldsymbol{x}_j) = S(\boldsymbol{x}_i,\boldsymbol{x}_j)\exp\left(-\frac{1}{2}(\boldsymbol{x}_i-\boldsymbol{x}_j)^T(\boldsymbol{\Sigma}_i+\boldsymbol{\Sigma}_j)^{-1}(\boldsymbol{x}_i-\boldsymbol{x}_j)\right) \tag{7}$$

where $S(\boldsymbol{x}_i,\boldsymbol{x}_j) = (2\pi)^{-1/2}|\boldsymbol{\Sigma}_i+\boldsymbol{\Sigma}_j|^{-1/2}$.

2. Optimal spatial sampling

Optimal experimental design for spatial sampling [3] applies the theory of optimum experimental designs [1] to stochastic processes, typically to Gaussian processes. The problem is to select the sampling locations $\mathcal{X}^* = \{\boldsymbol{x}_1^*, \boldsymbol{x}_2^*, \ldots, \boldsymbol{x}_m^*\}$ that provide maximum information about the environment. In stationary Gaussian processes, optimal sampling is suggested to correspond to regular sampling with evenly distributed sampling locations [12]. In non-stationary processes one could intuitively argue that in optimal sampling the sampling locations should be distributed so that we get more information from areas where the data are less spatially correlated, and vice versa.

2.1. Maximum Entropy Sampling

In maximum entropy sampling [15] the objective is to minimize uncertainty in unobserved locations. Now we state that the variables $\boldsymbol{u}$ and $\boldsymbol{v}$ are members of observed $\mathcal{U}$ and unobserved $\mathcal{V}$ sets, respectively, and for the sets we have that $\mathcal{V} = \mathcal{X} \setminus \mathcal{U}$ where $\mathcal{X} = \{\boldsymbol{x}_1, \boldsymbol{x}_2, \ldots\}$. Minimal uncertainty is obtained by selecting m sampling locations $\mathcal{X}^*$ so that conditional entropy is minimized. This is equivalent to maximizing the entropy of a random function $f(\boldsymbol{x})$ in sampling locations (8)

$$\begin{aligned}\mathcal{X}^* &= \operatorname*{argmin}_{\mathcal{U}} H(f(\boldsymbol{v})|f(\boldsymbol{u}))\\ &= \operatorname*{argmin}_{\mathcal{U}} (H(f(\boldsymbol{x})) - H(f(\boldsymbol{u})))\\ &= \operatorname*{argmax}_{\mathcal{U}} H(f(\boldsymbol{u}))\end{aligned} \tag{8}$$

where $\mathcal{U} \subset \mathcal{X} : |\mathcal{U}| = m$.

For a Gaussian process $f(\boldsymbol{x}) \sim \mathcal{GP}(\mu(\boldsymbol{x}), c(\boldsymbol{x},\boldsymbol{x}'))$, the entropy at locations $\mathcal{U}$ is given by (9)

$$\begin{aligned}H(\boldsymbol{y}_m) &= -\int p(\boldsymbol{y}_m)\log p(\boldsymbol{y}_m)\mathrm{d}\boldsymbol{y}_m\\ &= \frac{1}{2}\log((2\pi e)^m|\boldsymbol{C}_{mm}|)\end{aligned} \tag{9}$$

where $\boldsymbol{y}_m = \{y^1, y^2, \ldots, y^m\}$ and $y^i = f(\boldsymbol{u}_i)$. Thus, maximal entropy in the sampling locations is obtained by maximizing the determinant of covariance matrix $\boldsymbol{C}_{mm}$, which is one form of *D*-optimal design [1], [2]. However, this is a *NP*-hard problem, for which reason the following greedy heuristic [8] is typically used:

Start with an empty optimal set $\mathcal{X}^* = \emptyset$ and a full retrieval set $\mathcal{U} = \mathcal{X}$
Add on each iteration i the location $\boldsymbol{x}_i^*$ that has the highest conditional entropy $\boldsymbol{x}_i^* = \operatorname{argmax}_{\boldsymbol{u}} H(f(\boldsymbol{u})|f(\mathcal{X}_{i-1}^*))$
Remove on each iteration the optimal location from the retrieval set $\mathcal{U} = \mathcal{U} \setminus \boldsymbol{x}_i^*$
Finish with an optimal set of locations $\mathcal{X}^* : |\mathcal{X}^*| = m$

which follows from the chain rule of entropy (conditional entropy), i.e,

$$H(f(\mathcal{X}_i^*)) = H(f(\boldsymbol{x}_i^*)|H(f(\mathcal{X}_{i-1}^*))) + H(f(\mathcal{X}_{i-1}^*)).$$

In the previous section we found that the conditional distribution of a Gaussian process is given by a multivariate normal distribution with a mean (4) and variance (5), which for a single random variable $y^i = f(\boldsymbol{u}_i)$ is given by a 1-dimensional normal distribution. As given by (9), we are now searching for the location $\boldsymbol{u}_i$ that maximizes the conditional variance $var(y^i|f(\mathcal{X}_{i-1}^*)) = var(y^i) - cov(y^i, f(\mathcal{X}_{i-1}^*))var(f(\mathcal{X}_{i-1}^*))cov(f(\mathcal{X}_{i-1}^*), y^i)$ on every iteration.

3. Experiments

In our experiments, a distributed multi-robot sensing system was used to collect scalar maps of ambient light and magnetic flux density. These data were further used in simulated optimal spatial sampling for both stationary and non-stationary models. The accuracy of the models was measured by the integrated mean squared error (IMSE) between the models and the maps. We assumed that optimal spatial sampling using non-stationary Gaussian processes would provide more accurate models than stationary processes. The following subsections present the experimental environment, data acquisition and simulation results, respectively.

3.1. Experimental Environment

In spatial modelling, mobile sensors provide an opportunity to select sampling locations that maximize the accuracy of the model. However, mobility reduces the reliability of the sampling locations due to odometer failures, which again reduce the likelihood of the model. Therefore, we built an experimental environment that provides an opportunity to study spatial modelling with known sensor locations. The experimental environment was realized using roof camera-based multi-robot localization, remote-operated robots and a central computer that performed data acquisition, robot control and environment modelling.

The multi-robot localization [11] was based on colour segmentation, which proved to be an efficient and reliable method. In order to localize and identify each robot from a roof camera image, localization discs (Figure 1) were mounted on top of the homogenous robots [4]. The segmentation algorithm exploits an HSV colour space with the V channel discarded for better tolerance against changes in lighting. The hue and saturation of each pixel was compared against preset minimum and maximum thresholds. The robot candidates with a large enough area were found using connected component labelling. The smaller rings under the robot area were sought in a similar manner in order to identify the

Figure 1. Three different localization discs attached to homogenous robots.

robots. The orientation of the robots was calculated as the angle between the centroids of the rings.

The segmentation algorithm provided multi-robot localization on image coordinates as presented in Figure 2. In order to find a transformation between image and metric coordinates, the camera was calibrated using Heikkilä's Camera calibration toolbox for Matlab [5]. Based on the calibration, one pixel in the image was judged to correspond to ca. 2.3 mm in the metric coordinate system. The remaining distortion, occurring especially near the image corners, was further reduced by nonlinear transformation resulting from curve fitting.

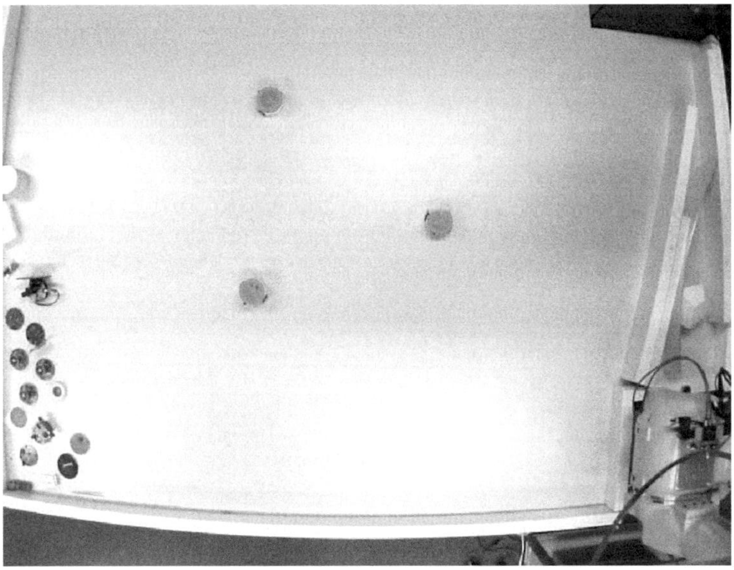

Figure 2. A Roof camera localization in a three-robot system

The accuracy of the localization system was evaluated for both position and orientation. Accuracy was measured on a plane in ten positions chosen on the basis of systematic unaligned sampling [7]. Table 1 presents the results of localization accuracy. The localization method reached a frame rate of 18 FPS with a resolution of 1024 x 768 pixels on a 3 GHz P4 desktop computer. The presented experimental environment can provide simultaneous localization for up to six robots with virtually unaffected performance.

Table 1. Observed localization errors

	Position, x axis	Position, y axis	Orientation
Maximum absolute error	3.8cm	4.4cm	13.7 deg.
Mean absolute error	1.6cm	1.7cm	4.7 deg.
Localization standard deviation	0.1cm	0.1cm	2.9 deg.

3.2. Data acquisition

The environmental data were acquired using a group of three robots. The robots moved arbitrarily in an area measuring approximately 180 x 160 cm while measuring ambient light and magnetic flux density. Furthermore, this data was stored and averaged on scalar maps containing 288 samples with a 10 cm grid. These maps are presented in Figure 3(a) and Figure 3(b). The units of ambient light and magnetic flux density were scaled and do not correspond to true luminance [cd/cm^2] or magnetic flux density [T].

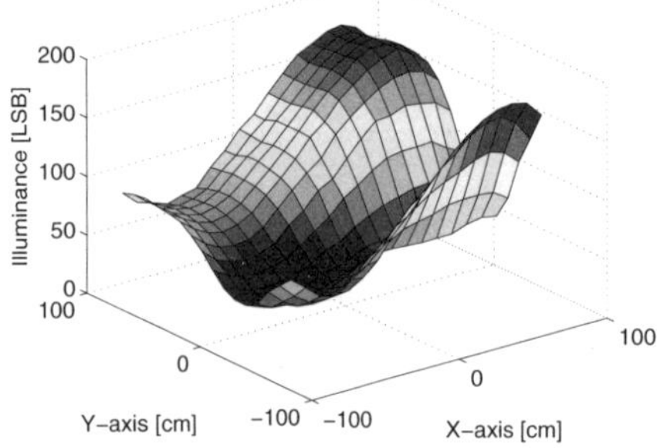

(a) Ambient light

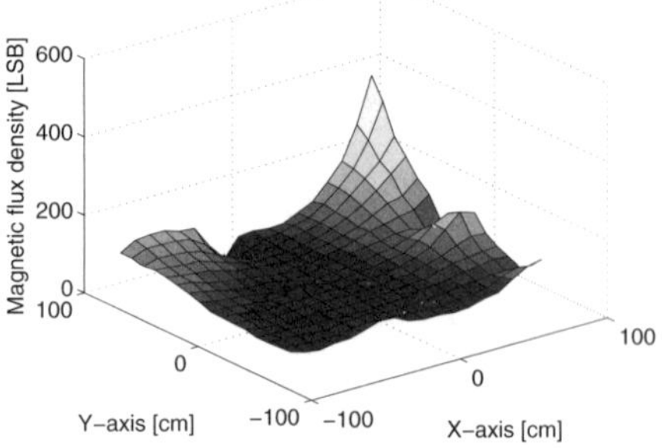

(b) Magnetic flux density

Figure 3. Scalar maps

3.3. Simulation results

Optimal spatial sampling simulations were performed for environmental data collected in advance using robots. The stationary models used a Gaussian kernel (10)

$$k(\boldsymbol{x}_i, \boldsymbol{x}_j) = (2\pi)^{-1/2}|\boldsymbol{\Sigma}|^{-1/2} \exp\left(-\frac{1}{2}(\boldsymbol{x}_i - \boldsymbol{x}_j)^T \boldsymbol{\Sigma}^{-1}(\boldsymbol{x}_i - \boldsymbol{x}_j)\right) \qquad (10)$$

where the variance hyperparameter was given a constant value

$$\boldsymbol{\Sigma} = \begin{bmatrix} 4000 & 0 \\ 0 & 4000 \end{bmatrix}.$$

This kernel was also an initial estimate for each local kernel in the non-stationary Gaussian processes. For both ambient light and magnetic flux density simulations, the mean value functions in the Gaussian processes were set to a constant $\mu(\boldsymbol{x}) = 100$ LSB.

For local kernel adaptation in the non-stationary Gaussian processes we used bounded linear adaptation with evaluation structure tensors [EST], as presented in [9]. However, in our simulation local gradients had to be calculated from interpolated values, in contrast to [9], who were able to use all the collected samples in the adaptation. Especially at the beginning of the sampling, this caused overestimated spatial smoothness in some interpolation areas. Thus, for maximum entropy sampling these areas seemed to provide a small amount of information and they were undersampled. To overcome these difficulties, the learning rate and the hyperparameter boundaries were given values $\eta = 0.2$, $\sigma_{min} = 3000$ and $\sigma_{max} = 5000$.

Figures 4(a) and 4(b) present 50 optimal samples in ambient light and magnetic flux density interpolation, respectively. These samples were selected by maximum entropy sampling for both stationary and non-stationary models. In the stationary models the optimal samples are distributed uniformly over the sampling area, whereas in the non-stationary models the sampling locations are focused more on the areas with the greatest gradients, i.e., on the areas that provide the most information about quantity.

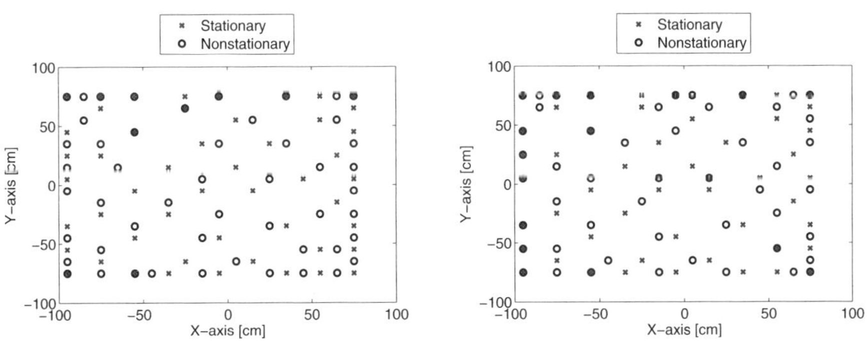

(a) Ambient light

(b) Magnetic flux density

Figure 4. Optimal sampling locations

Figures 5(a) and 5(b) present integrated mean squared errors as a function of the optimal samples up to 50 samples. The integrated mean squared errors were obtained by comparing the interpolation results with the maps. The IMSE of ambient light interpolation was nearly equal after 50 samples for both models which results from the smooth variation in spatial data. The IMSE of magnetic flux density interpolation was slightly smaller for the non-stationary model. However, the difference was not statistically significant. Figures 6(a) and 6(b) present stationary interpolation, and Figures 7(a) and 7(b) nonstationary interpolation in a 50-sample optimal spatial sampling. We believe larger data sets with more variation would bring out the advantages of the non-stationary models.

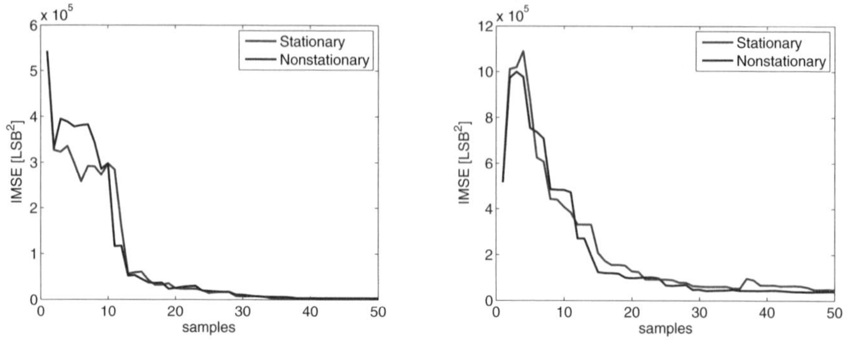

(a) Ambient light

(b) Magnetic flux density

Figure 5. Integrated mean squared errors

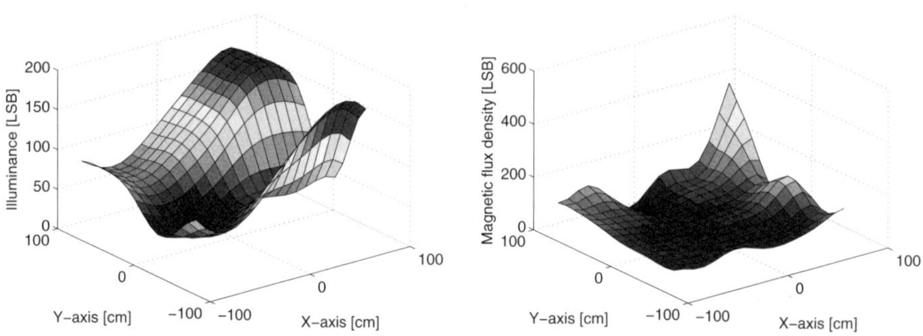

(a) Ambient light

(b) Magnetic flux density

Figure 6. Stationary interpolation

4. Conclusions

In this paper we presented an experimental environment for optimal spatial sampling in a multi-robot system. The environment enabled research on optimal spatial sampling using distributed and mobile sensors. The experimental environment was realized using roof camera-based multi-robot localization, remote-operated robots and a central computer for data acquisition, robot control and environment modelling.

In the simulations, the accuracy of optimal spatial sampling was measured for both the stationary and non-stationary models. Although the results did not show a clear advantage of the non-stationary models, authors believe that given larger and more variable data sets the difference would have been more significant. The simulation results encourage the authors to develop actual optimal spatial sampling methods for multi-robot systems. Such a research area contains not only optimization of information, but also optimization of measurement time, travelled distances, etc. Outdoors, aided with GPS localization, these methods could measure pH levels or the oxygen content of water, for example.

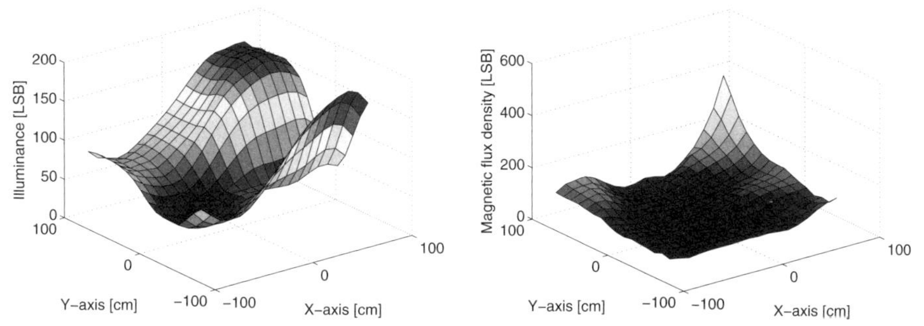

(a) Ambient light (b) Magnetic flux density

Figure 7. Nonstationary interpolation

Acknowledgements

This work was supported by the Academy of Finland and Infotech Oulu.

References

[1] A.C. Atkinson and A.N. Donev. *Optimum Experimental Designs*. Oxford University Press, New York, 1992.

[2] D.D.Cox, L.H. Cox, and K.B. Ensor. Spatial sampling and the environment: some issues and directions. *Environmental and Ecological Statistics*, 4:219–233, 1997.

[3] V.V. Fedorov. Optimal spatial design: Spatial sampling. *Calcutta Statistical Association Bulletin*, 44, 1994.

[4] J. Haverinen, M. Parpala, and J. Röning. A miniature mobile robot with a color stereo camera system for swarm robotics research. In *IEEE International Conference on Robotics and Automation (ICRA2005)*, pages 2494–2497, Barcelona, Spain, Apr 18 - 22 2005.

[5] J. Heikkilä. Geometric camera calibration using circular control points. *IEEE Transactions on Pattern Analysis and Machine Intelligence*, 22(10):1066–1077, 2000.

[6] D. Higdon, J. Swall, and J. Kern. Non-stationary spatial modeling. *Bayesian Statistics 6, eds. J.M. Bernardo et al., Oxford University Press*, pages 761–768, 1999.

[7] L. J. King. *Statistical Analysis in Geography*. Englewood Cliffs (N.J.) : Prentice-Hall, 1969.

[8] A. Krause, A. Singh, and C. Guestrin. Near-optimal sensor placements in gaussian processes: Theory, effecient algorithms and empirical studies. Technical report, Carnegie Mellon, Machine Learning Department, 2007.

[9] T. Lang, C. Plagemann, and W. Burgard. Adaptive non-stationary kernel regression for terrain modeling. In *Proceedings of Robotics: Science and Systems*, Atlanta, GA, USA, June 2007.

[10] G. Matheron. Kriging, or polynomial interpolation procedures? *CIM transactions*, LXX:240–244, 1967.

[11] T. Mäkelä, A. Tikanmäki, and J. Röning. A general-purpose vision system for diverse robots. In *Robotics and Applications and Telematics*, Wurzburg, Germany, Aug 29 - 8 2007.

[12] B. Mukherjee. A note on sampling designs for random processes with no quadratic mean derivative. *Australian & New Zealand Journal of Statistics*, 48(3):305–319, 2006.

[13] C. Paciorek. *Nonstationary Gaussian Processes for Regression and Spatial Modelling*. PhD thesis, Carnegie Mellon University, Pittsburgh, Pennsylvania, 2003.

[14] C.E. Rasmussen and C.K.I. Williams. *Gaussian Processes for Machine Learning*. The MIT Press, 2006.

[15] M.C. Shewry and H.P. Wynn. Maximum entropy sampling. *Applied Statistics*, 46:165–170, 1987.

[16] C.K.I. Williams and C.E. Rasmussen. Gaussian processes for regression. In David S. Touretzky, Michael C. Mozer, and Michael E. Hasselmo, editors, *Proc. Conf. Advances in Neural Information Processing Systems, NIPS*, volume 8. MIT Press, 1996.

Intelligent Autonomous Systems 10
W. Burgard et al. (Eds.)
IOS Press, 2008

doi:10.3233/978-1-58603-887-8-64

Sonar Scan Matching by Filtering Scans using Grids of Normal Distributions[1]

Antoni BURGUERA [a,2], Yolanda GONZÁLEZ [a] and Gabriel OLIVER [a]
[a] *Universitat de les Illes Balears, Spain*

Abstract. The NDT[3] was presented as an alternative to ICP-based scan matching algorithms. The problems of ICP-based algorithms derived of the establishment of correspondences do not appear in the NDT approach. However, the NDT requires accurate and dense sets of readings to work with. Consequently, the NDT is not well suited to perform scan matching with sensors that produce sparse sets of noisy readings, such as ultrasonic range finders. In this paper, an extension of the NDT to deal with sparse sets of noisy readings, the sNDT, is presented. The necessary processes to overcome the sparsity of sonar readings, as well as those involved in the scan filtering, are described in the paper. Experimental results compare the new approach to other well known scan matching methods.

Keywords. Sonar Scan Matching, Localization, Sonar

Introduction

Mobile robot localization is one of the most active areas in robotics research nowadays. It is broadly accepted that methods based on matching recent sensory information against prior knowledge of the environment constitute the best approach to long term localization. Although the use of geometric constraints in this methods is common, they render the process useful only in structured indoor environments. To give more generality to their localization methods, some authors determine the robot displacement by matching up successive sets of range measurements called *scans*. This technique is known as *scan matching*. Scan matching was introduced by Lu and Milios [9]. This work was inspired on the ICP (*Iterative Closest Points*) algorithm, which is broadly used in the computer vision community to perform image registration [1]. ICP scan matching consists in an iterative process to estimate the robot displacement and rotation that maximize the overlap between two consecutive sensor scans by establishing point to point correspondences. Due to its conceptual simplicity, the ICP algorithm was a great success, defining a *de facto* standard for scan matching localization. Different methods based on the ICP algorithm have been presented since then [11,12,10]. In the context of this document, the algorithms following the structure defined by ICP will be referred to as ICP-based algorithms.

[1]This work is partially supported by DPI 2005-09001-C03-02 and FEDER funding.
[2]Corresponding Author: Antoni Burguera, Dept. Matemàtiques i Informàtica, Universitat de les Illes Balears, 07122 Palma de Mallorca, Spain; E-mail: antoni.burguera@uib.es.

ICP-based algorithms have two main drawbacks. First, they implicitly assume that corresponding readings have been produced at the same position in the environment. This assumption is not correct, not only due to the spurious measurements but mainly because sensors sample the environment, providing a discrete view of the surroundings. The second drawback is that ICP-based algorithms do not hold information to guarantee coherence between iterations. After the execution of one iteration, a new set of correspondences is established. Consequently, a different function is optimized at each iteration. This leads the algorithm to a large number of iterations until convergence.

The NDT (*Normal Distributions Transform*) [3] approach to scan matching constitutes an interesting alternative to ICP-based algorithms. This approach does not rely on the establishment of correspondences. In consequence, none of the mentioned problems of the ICP based algorithms is present in the NDT.

Both the NDT and the most part of ICP-based algorithms are designed to match dense sets of readings provided by accurate sensors, such as laser range finders. Only a few works deal with sparse sets of noisy readings, such as the ones provided by ultrasonic range finders, using the ICP-based approach [4,7]. Sonar sensors are neither as accurate as laser sensors nor produce dense sets of readings. However, they are appealing in terms of cost, weight and power consumption. Moreover, their basic features are shared with underwater sonars, which are extensively used in underwater robotics.

In this paper we extend the NDT algorithm so that it can be used with sparse sets of noisy readings. This extended version is referred to as sNDT (*sonar Normal Distributions Transform*). In order to do this, the following steps are necessary. First, a process to group sonar readings to overcome their sparsity is performed. Then, the construction of the NDT grid is adapted to the noisy nature of sonar readings. Also, the readings in the scan to be matched are filtered. Finally, the trajectory followed by the robot while building the scans is corrected according to the scan matching estimate.

This paper is structured as follows. First, the notation used along the paper is presented in Section 1. Section 2 outlines the NDT approach. The main contributions of this paper are discussed from Section 3 onwards, which start by describing the scan generation and trajectory correction processes. Then, Section 4 focuses on the method to define a function representing one set of sonar readings. The process to filter scan readings is presented in Section 5. Experimental results validating our approach in terms of robustness, accuracy and convergence rate are presented in Section 6. Finally, Section 7 concludes this work and gives some insight for further research.

1. NOTATION

Let $S_{ref} = \{q_1, q_2, \ldots, q_n\}$ be a set of n points gathered at frame A, which is called the *reference scan*. Let $S_{cur} = \{p_1, p_2, \ldots, p_m\}$ be a set of m points gathered at frame B, which is called the *current scan*. Let x_B^A be the scan matching estimate of the relative position between the frames A and B. A common assumption is that the error in scan matching is Normal. Thus, we will represent the scan matching estimate as a Normal distribution $x_B^A = N(\hat{x}_B^A, P_B^A)$. To deal with processes corrupted by Gaussian noise, a common approach in stochastic mapping and SLAM is the use of the operators $\oplus$ (composition of transformations) and $\ominus$ (inverse of a transformation). A detailed description of both operators can be found in [13]. The problem of scan matching is the one of finding the robot motion x_B^A that maximizes the overlap between the parts of the environment represented by S_{ref} and S_{cur}.

2. THE NDT

Biber and Straßer presented the NDT approach in [3] as an *ad-hoc* method to register point clouds. The idea was refined and a probabilistic interpretation was provided in [2]. The key idea of the NDT is to build a grid of Normal distributions modeling S_{ref}. Then, the robot motion x_B^A that maximizes the overlap between the points in S_{cur} and the mentioned grid is computed. Next, the construction of the NDT grid and the optimization are described as they were originally conceived but using the same notation that will be used along this paper. This description is necessary to fully understand the sNDT approach.

2.1. Building the Grid

The NDT grid is built from $q_i \in S_{ref}$. It starts dividing the space containing S_{ref} in N cells of size $L \times L$ and searching the set of points q_i lying inside each cell. Depending on the position of the grid's origin in the range $[0, L) \times [0, L)$, the points of S_{ref} will be divided in different groups. Let us denote by α a particular position of the grid's origin. Then, for each cell j containing at least three points, the following steps are executed:

1. Let $\Omega_{\alpha,j}$ be the set of n points $q_i \in S_{ref}$ contained in the cell j. Compute the mean $\mu_{\alpha,j}$ and the covariance matrix $P_{\alpha,j}$ of $\Omega_{\alpha,j}$.
2. Model the probability of having a reading at point x contained in cell j by the bivariate normal distribution $N(\mu_{\alpha,j}, P_{\alpha,j})$. Let $f_{\alpha,j}(x)$ be the PDF of this Normal.

The function $f_\alpha(x)$ associated to a particular grid's origin can be piecewise built using the previously defined $f_{\alpha,j}(x)$. Now each cell in the space containing S_{ref} is modeled by a Normal distribution. This may be a good approximation locally, however, important discontinuities appear in the cell limits. To reduce the effects of the discontinuities, the NDT approach proposes the use of four overlapping grids, instead of a single one, so that each point falls into four cells. Thus, four different functions are defined, each of them considering a particular position of the grid's origin. Using this four functions, the NDT models S_{ref} by the function $f : \Re^2 \rightarrow \Re$, which is a sum of the mentioned four f_α.

2.2. Minimization

The displacement and rotation that maximizes the overlap between the two scans is defined as the one that minimizes the score function :

$$s(x) = -\sum_{p_i \in S_{cur}} f(x \oplus p_i) \tag{1}$$

This score function evaluates the goodness of a certain transformation x by transforming each point in S_{cur} using x, evaluating the function f on the transformed point and summing up the results. Then, the value is changed to negative so that optimization may be defined as a minimization problem instead of a maximization one. The Newton's algorithm is proposed to carry out the minimization. The initial estimate for this algorithm can be obtained from odometry. Details regarding to the use of the Newton's algorithm in this particular context can be found in [3]. The x minimizing Eq. (1) constitutes $\hat{x}_B^A$.

3. SONAR SCAN MATCHING

In order to successfully match two scans, scan matching algorithms require dense sets of range readings, each gathered from a single robot pose. To perform scan matching localization with sonar, two additional processes have to be executed: the *Scan Generation* and the *Trajectory Correction*. The Scan Generation is executed before the matching process. It is in charge of grouping the sonar readings gathered along short robot trajectories to produce a dense set of readings. During this process, the robot pose is estimated by other means, for instance, odometry. The Trajectory Correction, which is executed after the matching process, is in charge of correcting the whole set of pose estimates involved in the Scan Generation by using the roto-translation provided by the scan matching. The description of these processes is out of the scope of this paper. A detailed description of both steps can be found in [4].

4. THE sNDT GRID

The modified NDT approach to work with sonar data will be referred to as sNDT (*sonar Normal Distributions Transform*). The sNDT process consists, similarly to the NDT, of a *Grid Building* step and a *Minimization* step. The main difference lies in the Grid Building step, which is going to be described next. The sNDT minimization step is not described here as it is based on the same foundations than the NDT.

The NDT approximates the probability of having a reading at a certain position in a given cell j of a particular grid α by the Normal $N(\mu_{\alpha,j}, P_{\alpha,j})$. For clarity purposes, subindexes will be dropped along this section and the Normal distribution written as $N(\mu, P)$. Both μ and P are directly obtained from the sensor readings lying inside the cell. Computing them in such way may be problematic in the presence of outliers. If sonar sensors are used the number of outliers is not negligible. Moreover, the low angular resolution of these sensors tends to create dense regions of outliers, called *artifacts*.

Our goal is to define a method to compute the mean and the covariance of the Gaussian that better fits the inliers of a given set of sonar readings. A common approach, specially in the computer vision community, to find the model that best fits to inliers (and discards outliers) of a given set of data points is the RANSAC (*RANdom SAmple Consensus*).

4.1. Gaussian Model Fitting

RANSAC was introduced by Fischler and Bolles [8] as an iterative process for fitting a model to experimental data containing a significant percentage of outliers. The general RANSAC algorithm has been modified in this paper so that the final estimate is, among the ones it can provide in a finite number of iterations, the best of them. This is accomplished by means of an error function that evaluates the models. This function makes also possible the introduction of some constraints on the desired model.

The proposed RANSAC approach to fit a Gaussian to a set of sonar readings will be referred to as RANSAC/GD (*RANSAC/Gaussian Determination*). It is described in Figure 1. The probabilistic meaning of the model is exploited, in line 11, to decide whether a reading should be part of the model or not. Line 20 is in charge of deciding if a model is better than any previous model. The narrowness factor λ and the narrowness resolution δ, used in lines 29 to 33, are in charge of avoiding singular and near singular covariance

```
Input:
        Q: Set of sonar data points
numIterations: Number of iterations to perform
        n: Number of points to be randomly selected
        m: Minimum number of points to consider a model
        β: Desired confidence level to accept a data point
        λ: Narrowness factor
        δ: Narrowness resolution
Output:
        μ_best: The estimated mean
        P_best: The estimated covariance
        ε_best: The quality of the estimated model
        S_best: The set of inliers used to build the model
begin
    k ←0;
    ε_best ←0;
    while k < numIterations do
        S1 ← random selection of n data points from Q;
        S2 ← ∅;
        μ ←mean(S1);
        P ←covariance(S1);
        foreach x ∈ (Q − S1) do
            if (x − μ)^T P^{-1} (x − μ) < χ²_{2,β} then
                S2 ← S2 ∪ {x};
            end
        end
        if |S2| ≥ m then
            μ' ←mean(S1 ∪ S2);
            P' ←covariance(S1 ∪ S2);
            ε_max ←max(eigenvalues(P'));
            ε_min ←min(eigenvalues(P'));
            if ε_max/ε_min > ε_best then
                ε_best ← ε_max/ε_min;
                μ_best ← μ';
                P_best ← p';
                S_best = S1 ∪ S2;
            end
        end
    end
    if ε_best > 0 then
        repeat
            ε_max ←max(eigenvalues(P_best));
            ε_min ←min(eigenvalues(P_best));
            P_best ← P_best + δI
        until ε_min/ε_max < λ ;
    end
end
```

Figure 1. The RANSAC/GD algorithm.

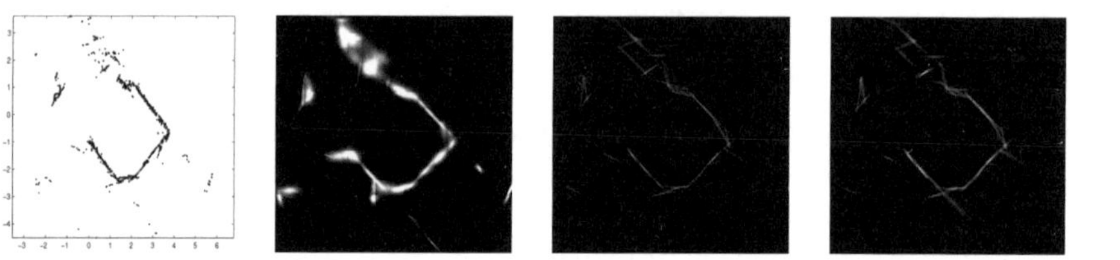

(a) (b) (c) (d)

Figure 2. The results of using RANSAC/GD to build the NDT grid. The effect of varying the narrowness factor λ can be observed. (a) Original set of sonar readings (b) The NDT grid (c) sNDT, $\lambda = 0.001$ (d) sNDT, $\lambda = 0.01$

matrices. Also, they are used to prevent the Normal distribution PDF to be too narrow in one direction. A detailed description of these parameters and how they should be tuned can be found in [5].

As an example, the sonar readings in Figure 2-a have been used to build different grids. Figure 2-b shows the NDT grid. Figures 2-c and 2-d show two sNDT grids using different values of λ, so that the effect of the narrowness factor can be appreciated. It can be observed that the use of RANSAC/GD produces a grid that fits better to the original data set and that the most part of spurious readings have been discarded.

5. SCAN FILTERING

The underlying probabilistic interpretation of the minimization described in Section 2.2 is that the points in S_{cur} can be seen as samples drawn from the Normal distributions in the grid. That is, the grid of Normal distributions defines a generative process for S_{cur}. As the process to build the sNDT grid discards outliers in S_{ref}, the outliers in S_{cur} should also be rejected. To accomplish this, we define the RANSAC/GF approach(*RANSAC/Gaussian Filtering*).

The RANSAC/GF approach is described now. First, the space containing S_{cur} is regularly divided in cells of size L. The points in S_{cur} will be divided in different groups

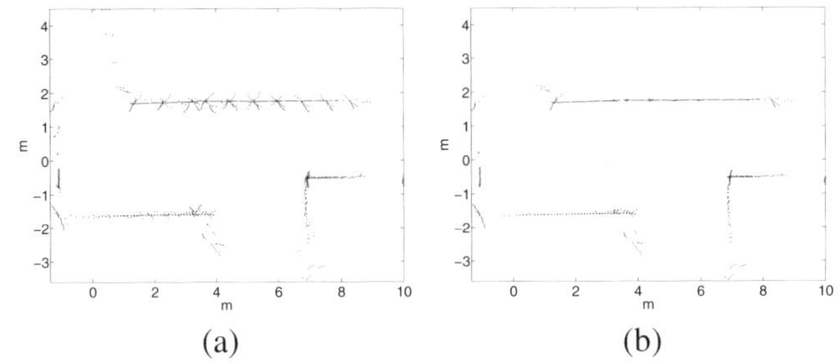

Figure 3. RANSAC filtering. (a) original sets of sonar readings. (b) filtered sets of sonar readings.

depending on the position of the grid's origin in the range $[0, L) \times [0, L)$, in the same sense it was described in Section 2.1. Let us denote by α a particular position of the grid's origin. Then, for each cell j containing at least five points, the set of accepted readings will be the set of inliers,S_{best}, provided by RANSAC/GD when applied to this cell. Let us denote this set as $\Omega'_{\alpha,j}$. Taking into account that four overlapping grids are used, thus, four different values for α, the set of accepted readings is built as follows:

$$\Omega' = \bigcup_{1 \leq \alpha \leq 4} \bigcup_{1 \leq j \leq N} \Omega'_{\alpha,j} \tag{2}$$

To ease notation, from now on, in the context of the sNDT, the term S_{cur} will refer to the set of accepted readings. In other words, $S_{cur} \leftarrow \Omega'$.

Figure 3 exemplifies the RANSAC/GF approach. It can be observed how isolated readings disappear after filtering. It can also be observed how spurious readings, specially those corresponding to artifacts, disappear after filtering. This fact suggests two possible benefits from using this technique. On the one hand, to select only the readings belonging to S_{cur} that make the generative process assumption valid and allow the use of such readings in the sNDT. On the other hand, to filter sonar readings so that clean sets could be used in other context, such as mapping or SLAM.

When the sNDT grid has been built as described and S_{cur} filtered, the score function $s(x)$ can be defined and the minimization can be performed. In the context of this paper, the scan matching estimate is a Normal of the form $x_B^A = N(\hat{x}_B^A, P_B^A)$. The x that minimizes the score function constitutes the mean $\hat{x}_B^A$. The computation of the covariance matrix P_B^A in the sNDT context is out of the scope of this paper. A description of how can this matrix be approximated can be found in [5].

6. EXPERIMENTAL RESULTS

The sNDT scan matching has been tested with real data obtained with a Pioneer 3-DX mobile robot endowed with 16 Polaroid ultrasonic range finders. The algorithm is compared to the original NDT as well as to three ICP-based algorithms, the sICP, the sIDC and the spIC [6]. The sICP and the sIDC are the sonar versions of the well known algorithms ICP and IDC. The spIC is a probabilistic approach to sonar scan matching which has proved to be more robust and accurate than sICP and sIDC.

The experiments discussed in this section are based on data sets gathered at four different environments of our university, which include structured and unstructured areas,

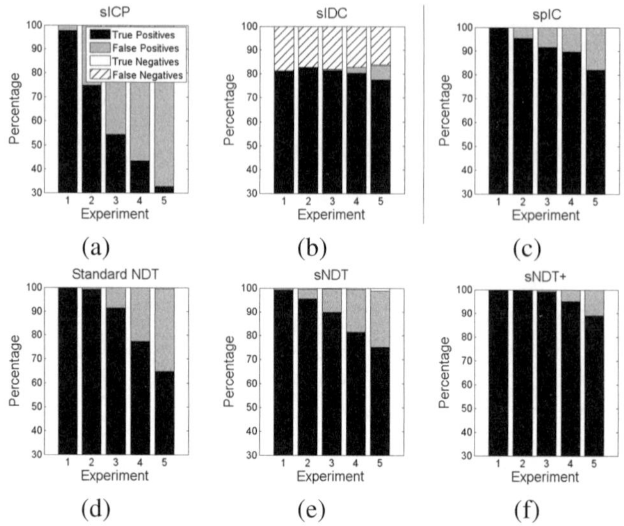

Figure 4. Robustness in front of initial errors. (a) sICP. (b) sIDC. (c) spIC. (d) NDT. (e) sNDT. (f) sNDT+Filtering.

posing difficulties to sonar sensing and odometry. Two scans have been gathered along the same robot trajectory at each environment. Although both scans have been gathered in similar conditions, they are not identical, constituting a realistic test bench for sonar scan matching algorithms. Moreover, as both scans have been obtained along the same robot trajectory, the ground truth $[0, 0, 0]^T$ is known. Five experiments have been carried out introducing different initial location errors. These errors correspond to the values assigned to the initial estimate, both in Newton's method and in ICP-based algorithms. These values have been randomly selected according to a uniform distribution between -0.05 m and 0.05 m in x and y, and between -9° and 9° in θ in Experiment 1. The amount of initial error increases with the experiment, up to random errors between -0.25 m and 0.25 m in x and y and between -45° and 45° in θ in Experiment 5. The procedure is repeated 1000 times per experiment and scan, which means a total of 20000 trials per method and a total of 100000 trials. In all the experiments, the sNDT has been parametrized with $\lambda = 0.5$. The convergence criteria for all the algorithms has been set to 0.00001.

6.1. Robustness

Figure 4 shows the results in terms of robustness. A trial is considered successful if the solution calculated by the algorithm is below 0.075m in x and y, and below 0.075 rad in θ. These thresholds have been used to distinguish between true and false positives and negatives. A true positive appears when the algorithm converges to the right solution. A false positive describes those situations where the algorithm converges to a wrong solution. True negatives appear when the algorithm do not converge, in a limited number of iterations, and the estimate generated in their last iteration was wrong. Finally, false negatives describe the situations where the algorithm do not converge, but their last estimation was correct.

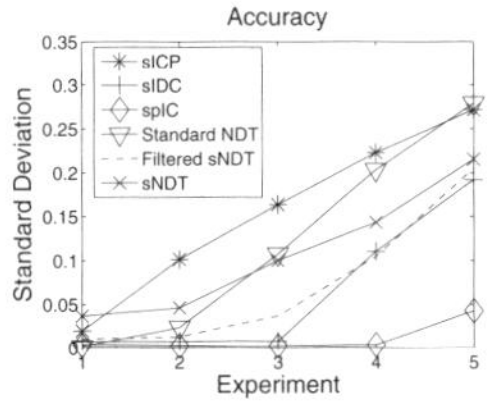

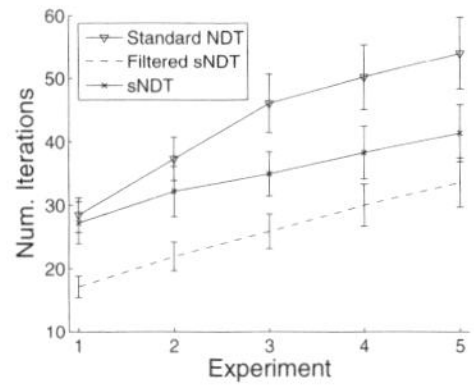

Figure 5. Evaluation of the sNDT (a) Accuracy. (b)Convergence.

Regarding to the NDT (Figure 4-d), the first thing to be noticed is that it is more robust than sICP (Figure 4-a), in terms of true and false positives. When comparing it with sIDC (Figure 4-b) it is clear that sIDC is better than NDT in presence of moderate and severe odometric noise and worse than NDT when only low or medium odometric noise is present.

The major problem of NDT appears when comparing it to the probabilistic ICP-based approach, the spIC (Figure 4-c). The spIC provides better results than NDT. Thus, although the NDT is better than sICP and sIDC to perform scan matching, it is significantly less robust than spIC.

Figure 4-e shows the robustness results for sNDT when the Scan Filtering (Section 5) is not used. Figure 4-f shows the robustness results when the Scan Filtering is applied. Thanks to the filtering approach, robustness increases, rendering the filtered sNDT approach remarkably better, in terms of robustness, than the non filtered sNDT and the Standard NDT. Moreover, robustness results with the filtered sNDT are even better than those provided by spIC. For instance, sNDT is able to achieve a 89.07% of true positives in Experiment 5, while spIC only finds the right solution in the 82.07% of the cases.

Two main conclusions can be obtained from the mentioned results. The first conclusion is that taking the range sensor behavior explicitly into account strongly increases the robustness of the scan matching. The spIC, which explicitly models the sonar behavior, is significantly better than the other ICP-based algorithms (and even the NDT). The sNDT, which explicitly considers the existence of outliers, is remarkably more robust than NDT (and even more robust than spIC). The second conclusion is that the filtered sNDT approach, subject of this paper, seems to be a better approach, in terms of robustness, to scan matching than any of the other tested methods.

6.2. Accuracy

In order to evaluate the accuracy of NDT and sNDT and compare it with sICP, sIDC and spIC, the standard deviation of the errors of the positive estimates is used. By focusing on positive estimates, the experiment evaluates the accuracy in real situations where no ground truth is available and the algorithm only can distinguish between positives and negatives. A low standard deviation represents that each trial has produced results close to the ground truth, which means that a good accuracy has been achieved. Higher standard deviations represent variations in the results obtained during the experiment, so that worse accuracy has been achieved.

Figure 5-a shows the mentioned standard deviation. In general terms, the spIC is the most accurate of the algorithms. It can be also observed how the accuracy of the

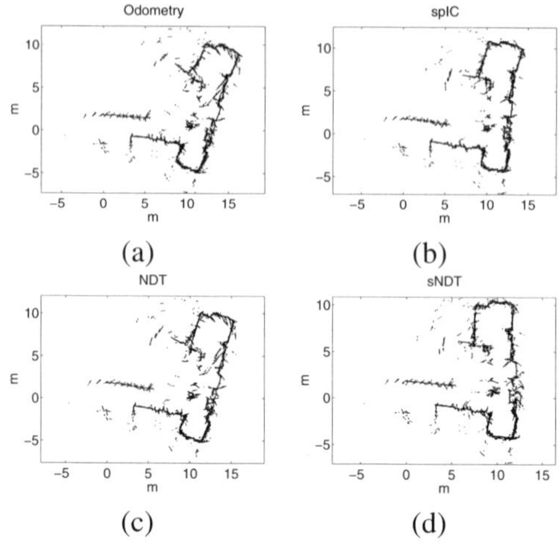

Figure 6. Sonar readings for visual inspection. (a) Odometry (b) spIC (c) NDT (d) sNDT

filtered sNDT is better than the ones of the Standard NDT and the non filtered sNDT. However, it is neither better than spIC nor sIDC. This is consistent with the fact that S_{ref} readings are not used directly by NDT and sNDT but approximated by a set of gaussians. Although this approximation solves the problem of exact point to point correspondence of ICP-based algorithms, it introduces additional uncertainty. Notice also that, as sNDT generates a more accurate representation of S_{ref} thanks to RANSAC/GD, it is able to achieve higher accuracy than NDT.

6.3. Convergence speed

Results regarding to convergence speed are shown for NDT and sNDT (both filtered and non filtered), but not for the ICP-based algorithms. An iteration in NDT and sNDT corresponds to one step in the Newton's algorithm. On the contrary, an iteration in an ICP-based algorithm corresponds to the minimization of one specific function. Thus, the number of iterations of NDT and sNDT can not be compared to those of ICP-based algorithms. Figure 5-b shows both the mean and the standard deviation of the number of iterations to convergence. The graphical representation of the standard deviation has been reduced to a 20% to make the image clear. It can be observed how the number of iterations required by the filtered sNDT is lower than the number of iterations required by non filtered sNDT and Standard NDT to achieve convergence in all the experiments.

6.4. Visual representation

Figure 6 depicts the results obtained when the robot is moving inside a building on our university campus. One more experiment has been conducted to evaluate the quality of our algorithm. As the ground truth is not available, the initial data with odometric error, as well as the results using spIC, NDT and filtered sNDT algorithms, are plotted for visual inspection. It can be observed how the results with all the scan matching techniques improve those obtained with raw odometry. Also, it can be observed how spIC and filtered sNDT results provide a map closer to the reality than the one provided by NDT.

7. CONCLUSION

In this paper, a new method to perform sonar scan matching has been presented. This method, the sNDT, is based on the construction of a piecewise function to model S_{ref}. The idea of building this kind of functions was first presented by Biber et al. [3] when describing the NDT approach to laser scan matching. The sNDT introduces new concepts to deal with the noisy and sparse sets of readings provided by sonar sensors. First, it deals with sparsity by grouping sonar readings to build a dense scan. Also, the whole trajectory involved in the scan building process is corrected. Second, it deals with noise and spurious readings by using the proposed RANSAC/GD algorithm to approximate sets of readings by gaussians.

The quality of sNDT has been evaluated in terms of robustness, accuracy and convergence speed, and compared to the NDT and to three ICP-based algorithms, the sICP, the sIDC and the spIC. Experimental results show that sNDT is the most robust of all the tested algorithms. They also show that, although sNDT is not as accurate as sIDC and spIC, it is more accurate than NDT and sICP. Finally, results regarding to convergence speed demonstrate that sNDT achieves convergence much faster then NDT.

In consequence, the final conclusion is that sNDT is the best choice in terms of robustness and convergence speed and that, although it is not as accurate as spIC, it is more than NDT and sICP.

References

[1] P. Besl and N. McKay. A method for registration of 3-d shapes. *IEEE Tran. on Pattern Analysis and Machine Int.*, 14(2):239–256, 1992.

[2] P. Biber, S. Fleck, and W. Straβer. A probabilistic framework for robust and accurate matching of point clouds. In *Proc. 26th Pattern Recognition Symposium*, 2004.

[3] P. Biber and W. Straβer. The normal distribution transform: a new approach to laser scan matching. In *Proc. IEEE/RSJ Int. Conf. on Intelligent Robots and Systems (IROS)*, volume 3, pages 2743 – 2748, October 2003.

[4] A. Burguera, Y. González, and G. Oliver. Probabilistic sonar scan matching for robust localization. In *Proc. of the Int. Conf. on Robotics and Automation (ICRA)*, Rome (Italy), April 2007.

[5] A. Burguera, Y. González, and G. Oliver. The sNDT: A grid-based likelihood field approach to robust and accurate sonar scan matching localization). Technical report, Dept. Matemàtiques i Informàtica, U.I.B., Palma de Mallorca (Spain), December 2007.

[6] A. Burguera, Y. González, and G. Oliver. A probabilistic framework for sonar scan matching localization (in press). *Advanced Robotics*, 2008.

[7] Young-Ho Choi and Se-Young Oh. Map building through pseudo dense scan matching using visual sonar data. *Autonomous Robots*, 23(4):293–304, November 2007.

[8] Martin A. Fischler and Robert C. Bolles. Random sample consensus: a paradigm for model fitting with applications to image analysis and automated cartography. *Commun. ACM*, 24(6):381–395, 1981.

[9] F. Lu and E. Milios. Robot pose estimation in unknown environments by matching 2D range scans. *Intelligent and Robotic Systems*, 18(3):249–275, March 1997.

[10] J. Minguez, L. Montesano, and F. Lamiraux. Metric-based iterative closest point scan matching for sensor displacement estimation. *IEEE Transactions on Robotics*, 22(5):1047–1054, October 2006.

[11] S.T. Pfister, K.L. Kreichbaum, S.I. Roumeliotis, and J.W. Burdick. Weighted range sensor matching algorithms for mobile robot displacement estimation. In *Proc. of the IEEE Int. Conf. on Robotics and Automation (ICRA)*, volume 2, pages 1667–1674, May 2002.

[12] S. Rusinkiewicz and M. Levoy. Efficient variants of the ICP algorithm. In *Proc. of the Int. Conf. on 3D Digital Imaging and Modeling (3DIM)*, 2001.

[13] J. D. Tardós, J. Neira, P. M. Newman, and J. J. Leonard. Robust mapping and localization in indoor environments using sonar data. *Int. Journal of Robotics Research*, 21(4):311–330, April 2002.

Intelligent Autonomous Systems 10
W. Burgard et al. (Eds.)
IOS Press, 2008

doi:10.3233/978-1-58603-887-8-74

Hardware and Software Architecture for Robust Autonomous Behavior of a Domestic Robot Assistant

Sven R. SCHMIDT-ROHR [a,1], Martin LÖSCH [a], Zhixing XUE [a] and Rüdiger DILLMANN [a]
[a] *University of Karlsruhe, Germany*

Abstract. We present a domestic robot assistant system designed around the objective to enable flexible and robust behavior in human centered environments. Emphasis is put on integration of different domains of abilities into a common framework to study techniques for robust and flexible human-robot interaction and general proactive decision making under uncertainty. The robotic system is designed to be fully autonomous and combines capabilities from the domains of navigation, object manipulation and multi-modal human-robot interaction. It is able to perform some typical missions in real world environments.

Keywords. Robot assistant, service robot, autonomous behavior, POMDP

Introduction

The requirements for robotic household assistants and companions span a wide range of ability domains, including autonomous navigation, grasping and manipulation of objects, multi-modal interaction between humans and robots as well as machine learning and abstract, proactive decision making. Real world domestic environments are characterized by being highly dynamic, not fully deterministically predictable and challenging to robot perception. Therefore, sophisticated algorithmic procedures are needed to enable autonomous, proactive behavior in complex mission scenarios. To evaluate such techniques in realistic settings, experimental robotic platforms are needed which integrate extensive domain specific capabilities with task and scenario level decision making in a common hard- and software framework.

In the following, this work presents such a robot assistant platform, integrating autonomous navigation, object manipulation and multi-modal human-robot interaction with mid-level natural task learning from observation and abstract, probabilistic decision making. In sec. 1, similar efforts to design autonomous robotic assistants are presented together with some references to probabilistic decision making. Sec. 2 gives an overview over the utilized hardware systems, while sec. 3 presents the overall software framework and sec. 4 describes the actual algorithmic parts. Experiments with the presented system are described in sec. 5.

[1]Corresponding Author: Sven R. Schmidt-Rohr, Institute of Computer Science and Engineering (CSE), University of Karlsruhe, Haid-und-Neu-Str. 7, 76131 Karlsruhe, Germany; E-mail: srsr@ira.uka.de

1. State of the Art

Concerning general architectures for control systems of autonomous robots, there has been a lot of investigation, so far. While there have been many quite different suggestions, a hierarchical approach for the design of reasoning and control systems for robots has proven to help coping with the complexity of task environments [1]. Three layer architectures are a very popular design [2]. The first layer performs low level processing of sensory data and reactive controlling of actuators. The second layer usually supervises an ongoing task on a symbolic level while the third layer handles the deliberative selection of abstract tasks. However, the supervisor and deliberative layer of most systems use classical planning which is not aimed at dealing with uncertainty.

Recently, several projects have undertaken the effort to design robot assistants, integrating autonomous navigation, object manipulation and human-robot interaction on a single robotic platform. The Stanford AI Robot (STAIR) [3] is such a project, as well as the humanoid domestic service robot ARMAR-III [4].

On the other hand, methods for probablistic reasoning and decision making have become popular in robotics to be able to cope with uncertainty in perception and dynamics of real world environments. Concerning high-level probabilistic decision making of rational agents (e.g. autonomous robots), a promising framework are partially observable Markov decision processes (POMDPs) [5], especially the class of discrete, model based POMDPs. A POMDP is an abstract environment model for reasoning under uncertainty [6], [7]. A POMDP models a flow of events in discrete states and discrete time.

As POMDPs handle partially observable environments, there exists only an indirect representation of the intrinsic state of the world. In POMDPs, the belief state, a discrete probability distribution over all states in a scenario model, forms this representation. At each time step, the belief state is updated by Bayesian forward-filtering. A decision about which action is most favorable for the agent when executed next, can be retrieved from a policy function which contains information about the most favorable action for any possible belief distribution.

The application of POMDPs in robotic domains has been stimulated by recent algorithmic advancements for calulating approximately optimal policies as *point based value iteration* (PBVI) [8] or HSVI2 [9]. A main focus of applications is robot navigation [10], where POMDPs can deal with sensor limitations, dynamic obstacles and low-level control glitches. Another, recent application is POMDP-controlled grasping of objects [11] and supervision based on visual clues in the caretaker domain [12].

To evaluate more high level POMDP decision making on an autonomous robotic assistant, a platform is needed, integrating the modalities of navigation, object manipulation and human-robot interaction with a central, abstract reasoning system. The following sections describe such a platform and framework.

2. Hardware systems

The robot Albert (see fig. 1) is mounted on a robust mobile platform with incremental encoders and a 180 degree SICK laser scanner for self-localisation. The platform also contains the batteries needed to power the robot, which last two hours when driving around continuously and moving the arm repeatedly.

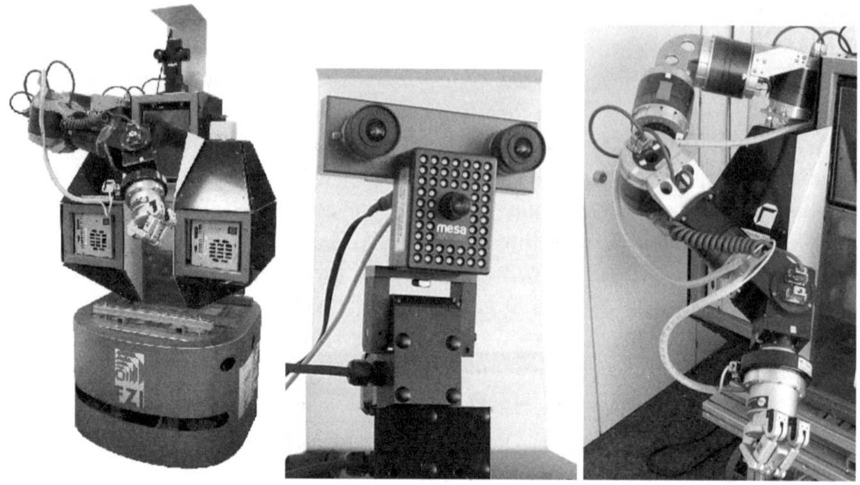

Figure 1. The service robot assistant Albert (left) and the vision system with stereo color camera and 3D–time-of-flight sensor (center) as well as the arm, force-torque-sensor and hand (right).

A robot arm with 7 degrees of freedom is mounted onto the shoulder of the torso, holding a 6 degree of freedom force-torque sensor with a 3 finger Barrett hand or alternatively a 4 fingered, anthropomorphic Schunk hand.

Mounted on a pan-tilt-unit (PTU), sits the vision system consisting of a stereo-color camera and a 3D-time-of-flight Swissranger camera with high image frequency (30 Hz). Combing color vision with 3D point cloud vision enables robust and powerful depth-field vision, concerning both object and human localization and tracking while the pan-tilt-unit allows to focus arbitrary regions of interest.

Off-the-shelf microphones and speakers enable spoken dialog while an embedded display may improve visual feedback. Finally, autonomous, onboard computation is provided by three computers, one in the platform and two in the torso, with two running Linux/RTAI and one running Linux without realtime extensions. One computer exclusively controls navigation and self-localization of the platform, one computer provides arm, hand, force-torque and pan-tilt control, while one computer processes vision, spoken dialog and high-level control. The onboard computers are connected via gigabit ethernet and can transmit information about the state of the system or individual algorithmic components via wireless ethernet.

3. Software framework

The software framework is designed along the practical consideration that manipulation control, vision and high-level coordination have quite different requirements, thus the system uses specialized frameworks for each domain which are integrated by hierarchical inter-framework communication.

The general system architecture is hierarchical, three layered, with independent low-level navigation, manipulation control and vision frameworks on the first level, an observation fusion and task coordination level in the middle and an abstract, POMDP decision making system on top (see fig. 2).

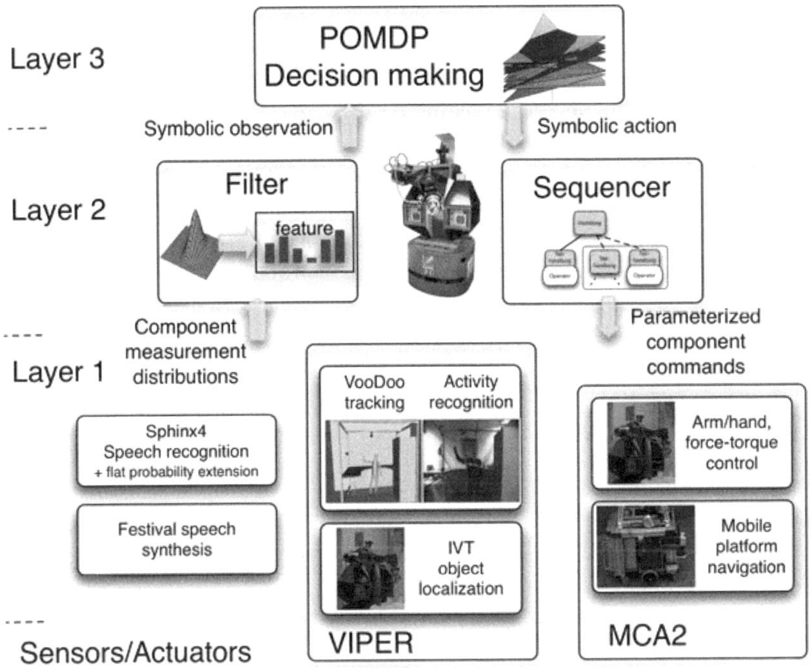

Figure 2. The general system architecture.

Mobile platform, arm, hand, force-torque-sensor and pan-tilt-unit are controlled by modules implemented and embedded into the Modular Controller Architecture (MCA2) framework [13]. MCA2 makes creating complex realtime control loop hierarchies, including many different hardware components, feasible. MCA2 targets at the design and development of low and medium level robot control software. It allows to easily reuse components and thus makes the process of getting new robot prototypes working faster. Individual controller modules in MCA2 are connected by edges, each transporting either sensory or actuator data. Modules can be aligned in hierarchical networks and groups, responsible for certain controller aspects. Network transparent edges are possible and used for data flow which is not realtime critical. To meet eventual realtime requirements, RTAI/LXRT is used to support realtime and non realtime threads within the same application.

Visual processing, including human body tracking and object localization (see also sec. 4) is done in the VIsual PERception framework (VIPER) [14] which is optimized to process streams of images from different sensors, e.g. stereo color images and 3D point clouds, in different algorithmic modules as e.g. body tracking and object localization. The modules in this framework can run distributed in a network with the idea to process large amounts of sensor data locally and provide abstract information for the synchronization of distributed modules (although it is also possible to process images distributed in a network). A further advantage of the framework is the option to "program" it graphically by simply selecting and connecting graphic representations for individual algorithmic modules to combine them into goal-directed data processing sequence.

The MCA2 controller framework, the VIPER vision framework, a speech recognition and a speech synthesis server have a network transparent interface to a central controller component, responsible for perception fusion, filtering and abstraction on the one hand and task sequencing and coordination on the other. The interface between MCA2 and the layer 2 central controller translates a control loop paradigm into an event-message paradigm also used by the other parts. On top, the POMDP decision making component has a network transparent interface to the layer 2 component, receiving symbolic observation distributions and sending symbolic action commands. The algorithmic procedures of these components are more closely described in the following section.

4. Algorithmic modules

4.1. Autonomous navigation

Autonomous navigation is possible on a known map in a structured indoor environment. Self-localization in enabled by laser-based detection of walls and other obstacles as well as incremental encoders measuring covered distance and turns. Fusing both perceptions with a common Kalman-filter based method results in information about the absolute position on the map, including uncertainty for x, y, θ in a covariance matrix. Navigation is possible either by driving on topological graphs between known nodes or by dynamically calculating new paths to dynamically inserted nodes by using an elastic band technique. Navigation is interfaced to the central control by delivering the current position including the covariance matrix and by receiving commands to drive to certain locations and flags indicating the belief to have reached a goal or being stuck.

4.2. Arm-Hand manipulation control

The arm with 7 degree of freedoms consists of 7 Amtec power modules. Closed-form analytic inverse kinematic is used. The additional redundancy angle at the elbow joint gives the arm anthropomorphic appearance, and can also avoid the collision of the elbow with obstacle. Besides position and velocity control, cartesian force control and cartesian impedance control are also implemented in MCA2. Force and torque values are measured by the force-torque sensor mounted at the wrist.

Grasp simulation system GraspIt! [15] is used to compute grasps for the Barrett hand. Grasps with high grasp quality are found offline and saved in a grasp database. During execution, the localized object position is given to the grasp module, which searches a good grasp in this grasp database, that can be performed while considering the reach ability of the arm. If a grasp could be found, the arm and hand are coordinated to grasp the object. The robot Albert can solve simple manipulation tasks such as pick-and-place, but also other more complex tasks in our daily life like opening the door.

4.3. Human body tracking and activity interpretation

The human body tracking is based on an Iterative Closest Points algorithm (ICP) and fits iteratively an articulated body model, built from degenerated cylinders as shown in figure 3, into a 3d point cloud measured using the SwissRanger depth camera [16]. The usage of the 3d point cloud instead of a more common approach using 2d cameras has the advantage of a better dynamic behavior concerning motion towards the camera, which is very hard to track using only 2d vision. The algorithm also allows the fusion with additional information like e.g. information from a 2d skin color tracking system to improve the tracking of the whole body configuration [17].

Figure 3. Articulated body model built from degenerated cylinders as it is used in *VooDoo* body tracking system (left) and two screenshots from actual tracking of a human holding an object (middle, right).

For the support of human-robot interaction, a human activity recognition approach based on this tracking has been developed. The recognition is organized in different steps as shown in fig. 4. First, different features are extracted from the body model like angles between limbs, velocities, statistical values of movements etc. A relevance analysis of the features is performed for each activity which shall be recognized, and only the most meaningful features for each activity are then used to train one classifier per activity. These steps are done off-line. Afterwards, the actual recognition can take place on-line, on the robot. The same features which were used to train the classifiers are extracted after each tracking step and fed to the recognition.

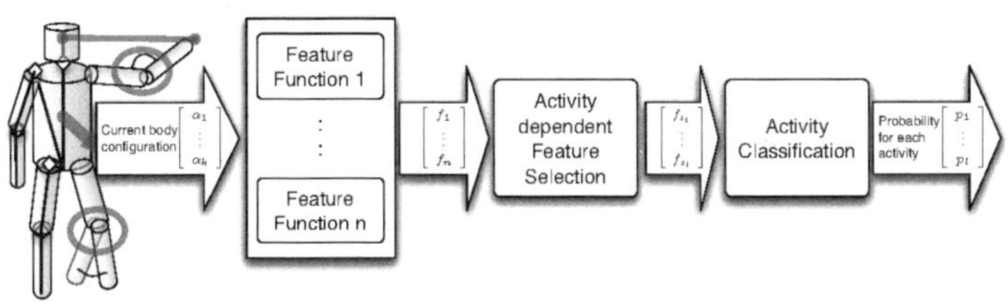

Figure 4. Steps in the process of activity recognition, from body model (left) to a probability distribution over activities.

The probability distribution resulting from the recognition step is not per se normalized to one (i.e. $\sum_i p(Act_i) = 1$), because the current set of activities can be varied independently and is not fixed. This allows to combine appropriate sets of activities, tailored for the current context and situation of the robot.

4.4. Object localization

The object localization module uses an approach which combines a global-appearance based approach and a model-based approach for the 6d recognition and localization of solid-colored objects in real-time [18]. So, this method is able to handle not only simple object geometries, while it can still identify the full 6d position of a found object.

4.5. Spoken dialog

Speech recognition is realized by using an onboard microphone and the Sphinx4 [19] speech recognition engine, modified to deliver discrete probability distributions over a set of human utterances: $\textit{percept human speech} = p(\textit{utter}_1), p(\textit{utter}_2), ..., p(\textit{utter}_n)$, $\sum_i^n p(\textit{utter}_i) = 1$, e.g. $p(\textit{"Bring me tea"}) = 0.2, p(\textit{"Bring a tissue"}) = 0.8$.

Natural speech synthesis simply receives a text string and converts it into an utterance by using the speech synthesis system Festival and speakers.

4.6. Central perception fusion and task sequencing

Perceptive information of the different perceptive systems has to be fused and component commands have to be coordinated to utilize the different capabilities in an integrated mission setting.

Except for the position of arm and hand which deliver very precise information, all the other perceptive systems deliver probability distributions to take into account information about the uncertainty of their measurements. However, measurement distributions are collected in different frequencies and in different representations, e.g. the position of the robot, a continuous distribution, is collected with 30 Hz, while a new, discrete speech recognition distribution is only delivered when the recognizer detected an utterance. Therefore, all perceptive distributions are stored with a timestamp in an observation storage module within the central controller. For every new single measurement update, each perceptive modalities is processed by a tailored Bayesian filter, considering passed time since the last measurement in that modality and being based on the last robot action for the predictive aspect. The resulting distributions which are still continuous are then discretized, e.g. self-localization by using a grid, and all modality-specific distributions mapped onto a common, discrete state-space by using an observation-state model. This resulting belief state distribution can then be used directly by POMDP decision making and is delivered to the decision making component. While this method needs quite some model knowledge, like the properties assigned to discrete states, it conserves uncertainty as perceived by the sensory components, usable for risk assessment in POMDP decision making.

While the filter in the central controller processes perceptive information to be used by POMDP decision making, the sequencer receives symbolic task commands, representing POMDP actions. It coordinates execution of these actions as sequential, hierarchical programs on the available actuators. The task is described as a hierarchical network of basic actions which is processed with a depth-first left-to-right search strategy and instantiated dynamically at runtime. A detailed description of the task description called *Flexible Programs* can be found in [20]. Typical tasks are *SpeakUtterance*, *GraspObject* or *DriveToLocation*. By decoupling the atomic sensor and actuator controlling from abstract reasoning, it is possible to reduce the POMDP decision state space to computationally reasonable dimensionality. The reasoning system decides on the global task to be carried out, while the sequencer performs the actual subtasks that reach the associated goal.

4.7. Natural learning of manipulation tasks

While medium-sized Flexible Programs can be easily assembled in a custom graphical editor, the intermediate sequencer is also used in the context of teaching manipulation tasks by natural human user demonstration. A system for Programming by Demonstration (PbD) exists [21], which can be used by a human to demonstrate typical household manipulation capabilities, as e.g. laying the table, in a PbD-Center equipped with sensors like cameras and datagloves. An automatic segmentation and interpretation process is then able to create Flexible Programs from natural user demonstrations. These Flexible Programs can be stored in a database and utilized when POMDP scenario level decision making decides, e.g. to place a cup onto a table.

4.8. POMDP decision making

At the most abstract level, POMDP decision making provides autonomy and the ability for proactive behavior. The main rational agent perceive – decide – act cycle runs with around 20 Hz, computing a new belief state b, retrieving a decision from the POMDP policy and executing a new, symbolic action in the sequencer, when applicable. The policy is queried using the new belief state and an action is retrieved to be executed next. It is actually executed in case the previous action has terminated in the sequencer.

The policy is a piecewise linear and convex (PWLC) maximum of a set of $|b|$-dimensional linear functions as calculated by an approximately optimal value iteration algorithm, e.g. PBVI [8] or a FRTDP/HSVI [9]. Each policy describes approximately optimal decision making in a predefined mission scenario and can be calculated online from POMDP mission models or offline and stored in a policy knowledge database.

5. Experiments

The presented system, as seen in fig. 5, can on the one hand be utilized to demonstrate autonomous behavior in a domestic setting in general, e.g. a butler mission, or on the other hand to evaluate different scenario-level decision making techniques against each other in realistic settings. Thus, it can serve as a platform for both pure demonstration as well as carefully devised science experiments.

A scenario, used for evaluation of proactive decision making, with common aspects of domestic service robot scenarios: verbal and non-verbal human-robot interaction, navigation and object delivery shall be outlined: The robot performs waiter duties by waiting for potentially interested persons, then tries to get into a dialog with them to finally offer to bring something. It constantly monitors if the human is facing the robot or not to have a better belief state about the dialog. When requested, it fetches a cup, then addresses the human again to ask to which destination shall be delivered, expecting an instruction with both gesture (pointing) and speech. If unsure, the robot may request a destination again. The cup is then delivered to the perceived destination and the robot returns to a suitable location to wait for further persons. Only onboard sensors are used, speech recognition is using an onboard microphone, activity recognition the onboard time-of-flight camera and navigation the onboard laser scanner. In this realistic setting, measurement uncertainties are high and human behavior and navigation glitches are preferably modeled as stochastic.

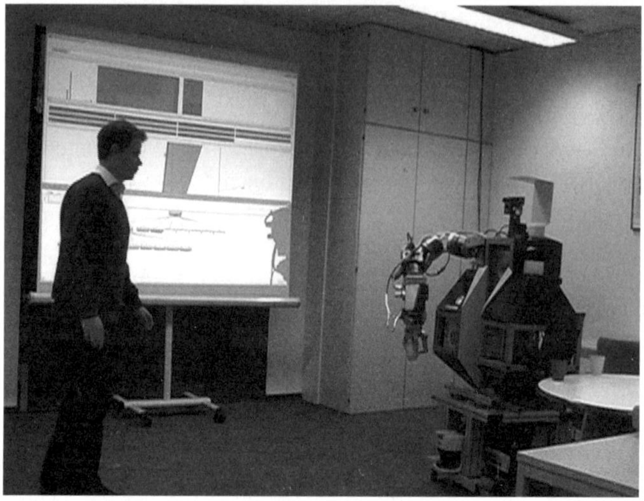

Figure 5. The robot during experiments with fully autonomous POMDP decision making. The live projection of POMDP belief state and Flexible Program processing, visible in the background, is based on data sent from the robot over WLAN.

For evaluation of different decision making techniques, the robot performed the presented waiter duty continuously for exactly half an hour, while the interacting human behaved stochastically, but on average according to the model. In one run, the robot was controlled by the presented POMDP system, while in the other, it was controlled by a Flexible Program in enhanced mode which can model a full finite state machine (FSM). A human supervisor recorded requests and actual robot behavior (correct or incorrect fetch/bring):

	Reassurance	Correct f./b.	Incorrect f. or b.
FSM	15	8	2
POMDP	5	16	2

The results in this case show that while the POMDP approach cannot avoid all errors, introduced by imperfect observations, it needs much less annoying reassurance questions, compared to the state machine. The POMDP utilizes inherent risk assessment, thus the robot knows when it is *sure enough* to take the risks to perform an action. Therefore the POMDP is less conservative than the state machine and can deliver more cups during the same scenario time. Accordingly, other methods, as e.g. MDPs, have been evaluated on this platform in realistic settings under full autonomy and with emphasis on different aspects of domestic scenarios.

6. Conclusion and outlook

This work has presented a hardware and software architecture of a domestic robot assistant, designed as an experimental platform for studying autonomous, robust and flexible behavior. The system integrates the domains of autonomous navigation, object manipulation and natural human-robot interaction in a single framework together with a centralized system for flexible, proactive decision making under uncertainty. The system has been used to evaluate different decision making mechanisms under realistic circumstances on the physical robot. Future enhancements will concentrate on learning and online acquisition of task and scenario knowledge. The concepts used in natural learning of manipulation tasks will be extended to online adaptation of models. With the presented platform, realistic and extensive online learning experiments are possible to evaluate such techniques.

References

[1] R. Simmons, R. Goodwin, K. Z. Haigh, S. Koenig, and J. O'Sullivan, "A layered architecture for office delivery robots," in *Agents'97*. ACM Press, 1997.
[2] E. Gat, "On three-layer architectures," *Artificial Intelligence and Mobile Robots. MIT/AAAI Press*, 1997.
[3] M. Quigley, E. Berger, and A. Y. Ng, "Stair: Hardware and software architecture," in *AAAI 2007 Robotics Workshop*, Vancouver, B.C, 2007.
[4] T. Asfour, K. Regenstein, P. Azad, J. Schröder, A. Bierbaum, N. Vahrenkamp, and R. Dillmann, "Armar-iii: An integrated humanoid platform for sensory-motor control," in *IEEE-RAS International Conference on Humanoid Robots*, 2006.
[5] K. J. Aström, "Optimal control of markov decision processes with incomplete state estimation," *Journal of Mathematical Analysis and Applications*, vol. 10, 1965.
[6] E. J. Sondik, "The optimal control of partially observable markov decision processes," Ph.D. dissertation, Stanford university, 1971.
[7] A. R. Cassandra, L. P. Kaelbling, and M. L. Littman, "Acting optimally in partially observable stochastic domains," in *In Proceedings of the Twelfth National Conference on Artificial Intelligence*, 1994.
[8] J. Pineau, G. Gordon, and S. Thrun, "Point-based value iteration: An anytime algorithm for POMDPs," in *International Joint Conference on Artificial Intelligence (IJCAI)*, August 2003, pp. 1025 – 1032.
[9] T. Smith and R. Simmons, "Focused real-time dynamic programming for mdps: Squeezing more out of a heuristic," in *Nat. Conf. on Artificial Intelligence (AAAI)*, 2006.
[10] A. Foka and P. Trahanias, "Real-time hierarchical pomdps for autonomous robot navigation," *Robot. Auton. Syst.*, vol. 55, no. 7, pp. 561–571, 2007.
[11] K. Hsiao, L. P. Kaelbling, and T. Lozano-Pérez, "Grasping pomdps," in *ICRA*, 2007, pp. 4685–4692.
[12] J. Hoey, A. von Bertoldi, P. Poupart, and A. Mihailidis, "Assisting persons with dementia during handwashing using a partially observable markov decision process," in *Proceedings of the International Conference on Vision Systems*, Bielefeld, Germany, 2007.
[13] K. Uhl and M. Ziegenmeyer, "Mca2 - an extensible modular framework for robot control applications," in *CLAWAR2007, 10th International Conference on Climbing and Walking Robots*, 2007.
[14] S. Vacek, G. Bocksch, and R. Dillmann, "A data-driven, component-based framework for visual perception," in *Robotik 2006*, 2006.
[15] A. Miller, S. Knoop, H. Christensen, and P. Allen, "Automatic grasp planning using shape primitives," in *ICRA*, vol. 2, 2003, pp. 1824–1829vol.2.
[16] S. Knoop, S. Vacek, and R. Dillmann, "Modeling joint constraints for an articulated 3d human body model with artificial correspondences in icp," in *Humanoid Robots, 2005 5th IEEE-RAS International Conference on*, Dec. 5, 2005, pp. 74–79.
[17] S. Knoop, S. Vacek, and Dillmann, "Sensor fusion for 3d human body tracking with an articulated 3d body model," in *ICRA*, Orlando, Florida, 2006.
[18] P. Azad, T. Asfour, and R. Dillmann, "Combining apperance-based and model-based methods for real-time object recognition and 6d localization," in *IROS*, Beijing, 2006.
[19] W. Walker, P. Lamere, P. Kwok, B. Raj, R. Singh, E. Gouvea, P. Wolf, and J. Woelfel, "Sphinx-4: A flexible open source framework for speech recognition," SUN Microsystems, Tech. Rep., 2004.
[20] S. Knoop, S. R. Schmidt-Rohr, and R. Dillmann, "A Flexible Task Knowledge Representation for Service Robots," in *IAS-9*, Tokyo, 2006.
[21] S. Knoop, M. Pardowitz, and R. Dillmann, "From abstract task knowledge to executable robot programs," *to appear in: Journal of Intelligent and Robotic Systems: Theory and Applications. Robotics, Virtual Reality, and Agents and their Body: A Special Issue in Memory of Marco Somalvico*, 2007.

Intelligent Autonomous Systems 10
W. Burgard et al. (Eds.)
IOS Press, 2008

doi:10.3233/978-1-58603-887-8-84

Topological Edge Cost Estimation through Spatio-Temporal Integration of Low-level Behaviour Assessments

Tim BRAUN [a] and Karsten BERNS [a]
[a] *University of Kaiserslautern, RR Lab, Germany*

Abstract. Edge cost measures for topological maps should be *consistent* with the actual difficulties experienced upon map edge traversal. This is a non-trivial problem for robots steered by a behaviour-based control subsystem, as little a priori information is available about the real trajectory emerging during motion. As solution, the paper proposes to learn consistent estimates of major cost factors for topological edges through a posteriori *observation* of situation assessments produced by the low-level control layer. These observations are then subjected to a spatial or temporal integration, yielding a multi-dimensional cost vector. Simulation results show that increasingly accurate cost estimates can indeed be derived using this strategy. This allows the use of the appealing abstract topological map representation for high-level navigation planning even in cluttered terrain, where consistent edge traversal cost estimates are indispensable for efficient path computation.

Keywords. Topological Maps, Behaviour Based Systems, Robot Navigation

1. Introduction

Efficient navigation in unstructured and cluttered terrain requires mobile robots to build a world model which records the traversability of possible navigational choices. This world model is commonly stored in form of either a metrical or a topological map.

Metrical approaches often arrange terrain traversability measures in form of a metric grid. After estimation of traversal costs for each grid cell in the working area, selection of an optimal path is easily accomplished using A*-type algorithms. Although accurate metrical maps allow optimal path planning, they must be kept consistent with new sensor input and therefore also require *accurate* robot localisation. Also, the large data structure imposes a high burden on the mobile robots scarce computational resources.

These drawbacks are somewhat circumvented by using a more abstracted *topological* world model. This technique does not explicitly model local terrain traversability but rather assumes that a subordinated obstacle avoidance layer copes with these spatially limited issues (often in a mainly reactive fashion). Pure topological maps only store 'reachability' information in an implicit form by representing reachable locations (nodes) and their connections (edges). After the traversal of an edge has been initiated by the path planner, the physical trajectory then emerges *in situ* through interaction of the obstacle avoidance layer and the terrain.

Unfortunately, this abstraction is a mixed blessing. On the positive side, the topological model allows more rigorous sensor data filtering and can therefore be more robust in cluttered environments. On the negative side, the lacking knowledge greatly complicates the computation of a cost measure actually reflecting the characteristics of the driven path. But without such a measure, cost-optimal path planning becomes effectively impossible.

In this paper, we propose a method that solves this problem and determines a physically plausible edge traversal cost estimate for navigation systems using a) a topological map for path planning and b) a behaviour-based subsystem for obstacle avoidance and vehicle control. As a result, we can now employ this appealing abstract representation for efficient high-level navigation planning in unstructured and cluttered *outdoor* terrain.

The main underlying idea of the presented method is to learn a consistent edge traversal cost measure *after* an edge has been traversed by *observing* the behaviour activities during movement. For this, we require that each behaviour generates a situation assessment or activity signal which *can* be observed. Many established behaviour-based systems do already provide such information, as such signals are also required for internal fusion of different behaviour outputs. The collection of the situation assessments during motion is then followed by a weighting and integration stage producing a single cost measure for the complete edge. We present two different integration schemes, a *temporal integration* for proprioceptive behaviours reacting to internal information, such as energy consumption, and a *spatial integration* for exteroceptive behaviours receiving stimuli from locations outside the robot such as obstacle detection behaviours.

A major benefit of the proposed approach is that by virtue of its design, it *guarantees* that the learned cost measure is consistent with the robot's actual behaviour guided by the behaviour-based subsystem. However, it is also computationally efficient, since the path planner does not need to evaluate any additional sensor information directly, but can effectively re-use the data interpretation already performed by the low-level layer.

The rest of the paper is structured as follows: Section 2 gives an overview of related work, section 3 introduces the observational learning method on the basis of a fictive, generic navigation system.In section 4, the proposed method is applied to an existing navigational system and simulation results are presented. In the last section, we give a conclusion and point out avenues for future research.

2. Related Work

While there are many existing approaches for large-scale navigation in complex outdoor environments that use metrical world models of varying complexity for path planning, such as grid maps [1]or 3D Point Clouds [2], the use of topological models is still rare. An existing approach is documented in [3]; but even here the focus is placed on topological place recognition, not on cost efficient navigation. In indoor environments, non-trivial edge cost models are more common and there exist models based on metrical node distance or multi-dimensional cost measures containing transition failure probability and energy consumption [4]. However, at the best of the authors knowledge, the use of a multi-dimensional cost estimate for topological outdoor navigation combined with a self-supervised, online adaption mechanism presents a novel contribution to the field.

Several researchers have recently proposed 'learn from observation' techniques that also integrate online experience in an unsupervised fashion. [5] proposes to 'learn from proprioception' and generates associations between the visual appearance of some surface patch and its traversability experienced later using accelerometers, bumpers and gyroscopes when the robot actually tries to pass over the patch. [6] essentially proposes the same, although here the traversability estimate is binary - surface patches are either marked traversable when the robot drives over them, or untraversable whenever a bumper hit occurs.

While the principal idea of using experience in an unsupervised learning process to optimize robot behaviour underlies both the approach proposed in this paper and the presented related work, they differ significantly in the level of abstraction. The approaches from literature try to learn a mapping from sensor data to traversability score in order to improve detection of unsuitable terrain. In contrast, our approach takes the result of sensor data interpretation, i.e. traversability scores, as *input*. These scores are then used to improve path selection. Therefore, both learning concepts can be used to complement each other; learning from proprioception can be used to improve obstacle discrimination over time, and topological cost adaption can be used to make high-level path planning more effective over time.

3. An Observation Framework for Behaviour Based Pilots

This section illustrates the proposed observation framework for behaviour-based local control systems by presenting its application to a fictive instance. The complete setup is shown in figure 1 and will now be discussed in detail.

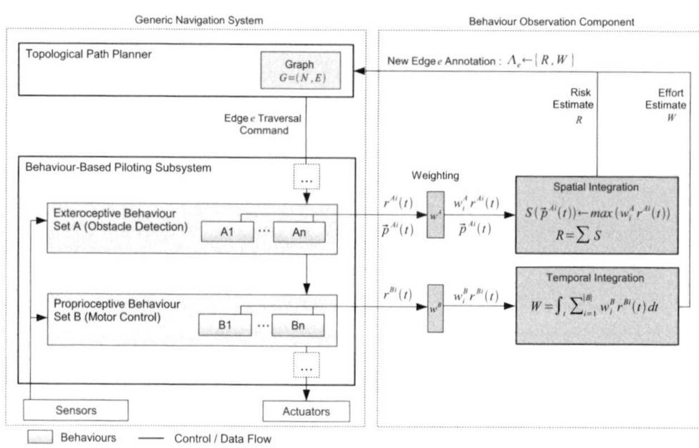

Figure 1. Schematic of a Fictive Topological / Behaviour-Based Robot Control System combined with the proposed Behaviour Observation Framework

3.1. Control System Layout

As can be seen in the figure, a robot's sensors and actors constitute the lowest level of the control hierarchy. The actors are controlled by the behaviour-based piloting subsystem, of which we single out two sets of behaviours A, B. We postulate that the 'Motor Control Behaviours' in set B are proprioceptive and that their situation assessment signal $r^{B_i}(t)$ reacts to the internal current fed into the actors at time t. We let r range from 0 if no current is applied up to a value of 1 at full current. The motor control behaviours are controlled by the set A of obstacle avoidance behaviours, which have the purpose of steering the robot safely around obstacles. In contrast to B, set A contains exteroceptive behaviours reacting to stimuli (obstacles) originating from a spatial location outside the robot. For these behaviours, we demand that the situation assessment $r^{A_i}(t)$ is 0 if behaviour A_i does not see any necessity to influence the robots trajectory and 1 if the behaviour wants to produce the maximal effect achievable in its scope. In addition, we require exteroceptive behaviours to export the spatial location $\vec{p}^{A_i}(t)$ of the stimulus responsible for the situation assessment. The topological path planner is placed at the uppermost part of this schematic. It keeps a topological map modeled as a directed graph structure $G = (N, E)$, with the nodes $N = \{n_1, \ldots\}$ specifying locations and edges $E = \{e_1, \ldots\}$ between nodes defining possible driving options.

3.2. Edge Cost Learning Component

In the following, we limit t to the execution time of a single edge traversal command e starting at time t_0 and ending at t_1 ($t_0 < t < t_1$). The edge cost learning component on the right side of figure 1 observes the situation assessments $r(t)$ of the two cost-relevant behaviour sets A, B. In this example, the assessments of set B are directly related to the motor currents, so it is reasonable to assume that their observation can yield some measure of how much energetical 'effort' the edge traversal requires. Similarly, the assessments of the obstacle avoidance set A contains a notion of how much trajectory deviation due to obstacles has occured. This should allow the estimation of a 'risk' cost factor. During robot motion, the scalar, normalized situation assessments $r^{A_i}(t)$ and $r^{B_i}(t)$ are first weighted by user-provided factors w_i^A and w_i^B. This allows specification of the relative cost impact of different behaviours.

Temporal Integration After weighting, the continual stream of assessments is integrated in order to generate one single cost estimate for the complete edge. For proprioceptive behaviours such as those in set B, this integration step amounts to a simple temporal integration of the weighted sum of the set's assessments:

$$W = \int_{t_0}^{t_1} \left(\sum_{i=1}^{\|B\|} w_i^B r^{B_i}(t) \right) dt \tag{1}$$

In our hypothetical robot control system, the result W is proportional to the total amount of energy spent by all motors during traversal of edge e.

Spatial Integration For exteroceptive behaviours, plain temporal integration leads to a dependency of the final cost estimate on the robot's motion speed, as a slower movement causes a longer exposition to the external stimulus. We argument that this is undesirable for the calculation of a risk estimate based on obstacle avoidance behaviours. It appears counter-intuitive that a slow robot judges a given obstacle situation more negatively than a faster model. As a solution, we advocate to use *spatial* integration of exteroceptive behaviours situation assessments. For this, the robot must provide a locally stable frame of reference *during edge traversal* onto which the source location $\vec{p}^{A_i}(t)$ of each behaviour assessment $r^{A_i}(t)$ is referenced. It is not required to provide an exact or globally stable coordinate frame, so accurate robot localization is still not needed. We introduce a *spa-*

tial map S modeled as a *grid map* with cell size equal to the lower accuracy limit of the sensor data or the reference frame. As notational conventions, $S(\vec{p})$ should be interpreted as an access to the grid cell of S which represents location $\vec{p}$. $\sum S$ denotes the sum of all grid cell values and is assumed to be zero initially.

Coming back to the spatial integration scheme, at each sampling time t, the $r^{A_i}(t)$ values of all behaviours are stored into S at their corresponding positions $\vec{p}^{A_i}(t)$ using a maximum update rule:

$$\forall i. S(\vec{p}^{A_i}(t)) := \max\left(S(\vec{p}^{A_i}(t)), w_i^A r^{A_i}(t)\right) \tag{2}$$

Once all assessments have been added to S, the integrated scalar risk cost estimate R for edge e can be derived through summation 'spatial integration' over S) using $R = \sum S$. After risk and effort estimates R, W have been generated for the topological edge e, they are attached to it as an *annotation* $\lambda = (R, W)$ and stored in the edge's *Annotation Set* Λ_e for later use in cost computation. This process is detailed in the next section.

3.3. Computing Traversal Costs from Annotations

Each time a topological edge is traversed, a new annotation with the experienced risk and effort cost factors is produced. In order to convert these annotations into a cost measure usable for path planning, we first combine all acquired annotations of an edge. Then, the two cost factors are merged into a single scalar value that can be used in standard graph path planning algorithms such as the dijkstra algorithm.

Let us assume that an edge e is annotated with the set $\Lambda_e = \{\lambda_1, \lambda_2, \ldots, \lambda_n\}$ where $\lambda_i = \left(R_e^i, W_e^i\right), 1 \le i \le n$. We expect that each annotation is a sample taken from an unspecified distribution centered at the true values of effort and risk. To find these values, we estimate the annotation set's *sample mean* and *unbiased sample variance*. Then, in order to include the variance information in the cost measure, we calculate a combined annotation $\overline{\lambda_e} = (\overline{R_e}, \overline{W_e})$ for e containing the estimated mean increased by an adjustable portion γ of the estimated standard deviation:

$$\mu_e = (\mu_e^R, \mu_e^W) = \frac{1}{n} \sum_{1 \le i \le n} (R_e^i, W_e^i) \tag{3}$$

$$\sigma_e = \sqrt{\frac{1}{n-1} \sum_{1 \le i \le n} \left((R_e^i - \mu_e^R)^2, (W_e^i - \mu_e^W)^2 \right)} \tag{4}$$

$$\overline{\lambda_e} = \mu_e + \gamma \sigma_e \tag{5}$$

The weighting factor γ can be an be utilized to produce exploratory behaviour, with γ determining the exploration / exploitation trade off. While high positive values of γ favor well established routes because edges having many annotations can be expected to have lower variance, using negative values for γ produces 'optimism under uncertainty' and actually favors connections that have been used rarely.

Once the two cost factors $\overline{R}_{e_i}$ and $\overline{W}_{e_i}$ are known for all edges e_i, these values must be brought together in a sensible fashion. This *multi-criteria path planning problem* has already received considerable attention in the literature [4]. Hierarchical approaches like

[4] propose to impose a hierarchy of constraints on each single cost criterion rather than weighting different cost factors against each other. While this avoid problems when cost aspects cannot be compared easily, the hierarchical structure does not allow smooth transitions of the relative importance of each factor. Since we deem this capability important, we propose to combine $\overline{R}_{e_i}$ and $\overline{W}_{e_i}$ to one single cost value C_{e_i} by plain weighted averaging. However, this requires a normalization in order to make the scales of the different cost aspects comparable. We normalize with the maximum cost factor value across all map edges, so the final cost factor combination is defined by:

$$C_{e_i} = \alpha \left(\frac{\overline{R}_{e_i}}{\max_j \overline{R}_{e_j}} \right) + \beta \left(\frac{\overline{W}_{e_i}}{\max_j \overline{W}_{e_j}} \right) \tag{6}$$

As can be seen, α and β specify the relative weights of each cost aspect. They can be interpreted as a *motivational state* of the global navigation system: α expresses the robots 'fear' level (high α values increase the cost of risky drive commands, making them less likely to be included in paths). Likewise, β denotes the robots 'impatience', penalizing arduous edges. In combination, α, β and γ allow an operator to influence the path planning strategy in an intuitive way, so that it can be set according to the current requirements of a mission.

4. Simulation Results

In order to validate its applicability, the cost learning method proposed in section 3 has been implemented as an extension of the navigation system of the mobile offroad robot RAVON (fig. 2). The RAVON project's overall goal is the development of both local

Figure 2. Offroad Robot Ravon

Set	Behaviour Name	w	Purpose
A	Slowdown	0.5	Reduces velocity near obstacles
A	Avoid Right	1	Turns away from obstacles on right
A	Avoid Left	1	Turns away from obstacles on left
A	Avoid Frontal	1	Decides to turn either right or left if obstacles on front
A	Halt	2	Stops robot just before obstacle contact
A	Traceback	2	Backward motion into free space if robot is blocked
B	Current Limit FR	1	Observes front right motor current, avoids overload
B	Current Limit FL	1	Observes front left motor current, avoids overload
B	Current Limit RR	1	Observes rear right motor current, avoids overload
B	Current Limit RL	1	Observes rear left motor current, avoids overload

Table 1. Observed Behaviours

and global navigation strategies able to cope with heavily vegetated and cluttered terrain in a robust fashion. The behaviour-based part of RAVON uses the iB2C framework [7] and contains a low level motor control part as well as a sophisticated obstacle avoidance component primarily based on two horizontal, planar laser-scanners combined with a tiltable 3D-scanner. All iB2C behaviours export a *target rating*, which is a measure of how 'unsatisfied' a behaviour is with the current situation. These ratings are taken as situation assessment values r in order to gather risk and effort cost factor estimates as detailed in section 3. Table 1 summarizes the selected behaviour sets A and B.

Figure 3 shows three risk estimation results obtained using the RAVON navigational system in the SimVis3D simulation framework [8], which simulates a three-dimensional

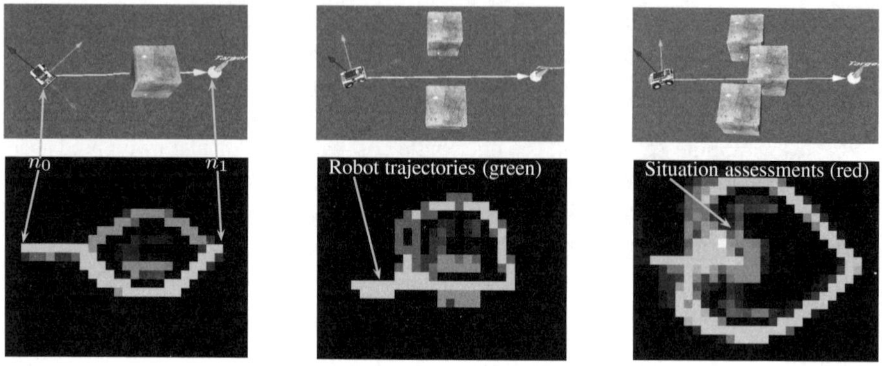

Contributor	Risk (μ_e, σ_e)
Total	**(1.8, 0.3)**
Avoid Left	75 %
Avoid Right	24 %

Contributor	Risk (μ_e, σ_e)
Total	**(5.6, 1.2)**
Avoid Right	53 %
Avoid Left	47 %

Contributor	Risk (μ_e, σ_e)
Total	**(11.4, 3.9)**
Avoid Left	35 %
Avoid Right	34 %
Avoid Front	26 %

Figure 3. Risk Estimates for different Obstacle Configurations

outdoor environment along with RAVON's sensor equipment. The top row shows a topological edge obstructed by three increasingly complex obstacle configurations. The middle row shows the corresponding spatial map S with a grid cell size of 0.75 m averaged over 50 edge traversals. Green pixels indicate locations passed by the robot, red pixels denote stored situation assessments $w_i^A r^{A_i}(t)$. In the bottom row, the resulting risk estimate is listed along with the three largest contributing behaviours. As can be seen, the increase in obstacle number and placement complexity corresponds well with the increase of the derived edge's risk estimate. The figure also illustrates the uncertainty of the actual trajectory driven by the robot, as minute differences in starting orientation can decide whether the obstacle is passed on the right or left side.

Figure 4 shows a larger scenario containing various obstacles such as trees or a ring of seats and an overlaid topological map. Again, each edge has been traversed multiple (≈ 10) times and the corresponding risk and effort cost values have been estimated. The resulting risk estimate has then been thresholded to give a qualitative impression where the behaviour observation approach has learned a high risk cost estimate. High risk edges are highlighted in white. It can be seen that these edges correlate well with the different obstacle situations.

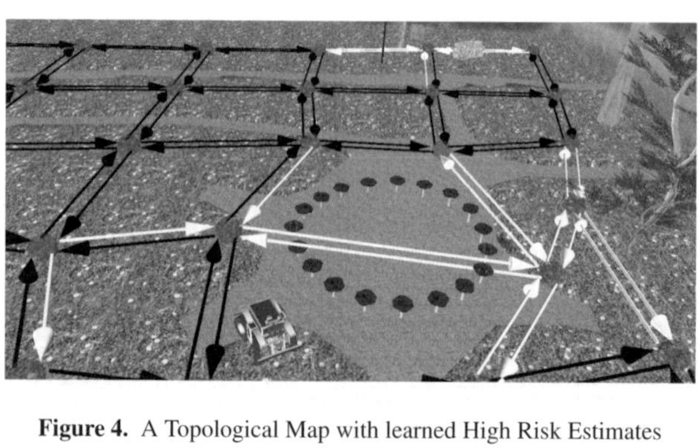

Figure 4. A Topological Map with learned High Risk Estimates

5. Conclusion

A generic method to learn consistent estimates of major cost factors for topological map edges through *a posteriori observation* of a low-level, behaviour-based control layer has been presented, and its applicability has been demonstrated for an existing navigation system. The observation strategy solves the problem of determining plausible cost measures for topological maps in difficult terrain, even though the concrete drive trajectories are not known in advance. With this concept, the use of topological maps in a high-level navigation system becomes feasible even if the terrain is very rich in variation.

Since the proposed learning scheme is only effective when a robot traverses edges multiple times, applications that involve a single passage of an area (such as exploration or the DARPA 2005 scenario) do not benefit from it. Instead, we envision the presented technique to be useful for patrolling applications like securing large corporate estates and repetitive transportation tasks, such as material transportation on construction sites. However, we want to point out that the learned costs also constitute a solid foundation for later cost extrapolation upon new map parts. Also, the general observational concept is not limited to navigation - it is a general approach for deriving semantically valuable information from a behaviour-based subsystem.

5.1. Future Work

Future work includes examination whether some basic assumptions on the metrical trajectory resulting from an edge execution can be made (or learned) before an edge is traversed. Based on this, the training of an unsupervised learning algorithm to link the derived cost estimates with corresponding camera images depicting the probably traversed terrain becomes possible. Then, new drive commands could be more accurately judged by combining the proposed cost extrapolation methodology with the learning algorithm's output. Also, longer tests on a larger map containing more different terrain and obstacle types will be performed.

References

[1] R. Manduchi, A. Castano, A. Talukder, and L. Matthies, "Obstacle detection and terrain classification for autonomous off-road navigation," in *Autonomous Robots*, 2005.

[2] C. Brennecke, "Ein scanbasierter ansatz zur exploratorischen navigation autonomer mobiler systeme in unstrukturierten outdoor-umgebungen," Ph.D. dissertation, University of Hannover, 2004.

[3] A. Vale, "Mobile robot navigation in outdoor environments: A topological approach," Ph.D. dissertation, Institute for Systems and Robotics, Lisboa, 2005.

[4] J. Fernández-Madrigal, J. González, L. Mandow, and J.-L. P. de-la Cruz, "Mobile robot path planning: A multicriteria approach," *Engineering Applications of Artificial Intelligence*, vol. 12, pp. 543–554, 1999.

[5] A. Howard, M. Turmon, L. Matthies, B. Tang, and A. Angelova, "Towards learned traversability for robot navigation: From underfoot to the far field," *Journal of Field Robotics*, vol. 23, 2006.

[6] D. Kim, J. Sun, S. M. Oh, J. M. Rehg, and A. F. Bobick, "Traversability classification using unsupervised on-line visual learning for outdoor robot navigation," in *ICRA*, 2006, pp. 518–525.

[7] M. Proetzsch, T. Luksch, and K. Berns, "The behaviour-based control architecture ib2c for complex robotic systems," in *30th Annual German Conference on Artificial Intelligence (KI)*, Osnabrück, Germany, September 10-13 2007, pp. 494–497.

[8] T. Braun, J. Wettach, and K. Berns, "A customizable, multi-host simulation and visualization framework for robot applications," in *13th International Conference on Advanced Robotics (ICAR07)*, Jeju, Korea, August 21-24 2007.

Intelligent Autonomous Systems 10
W. Burgard et al. (Eds.)
IOS Press, 2008

doi:10.3233/978-1-58603-887-8-92

Nonlinear Model Predictive Control of an Omnidirectional Mobile Robot

Xiang LI [a,1], Kiattisin KANJANAWANISHKUL [a] and Andreas ZELL [a]

[a] *Wilhelm-Schickard-Institute, Department of Computer Architecture, University of Tübingen, Sand 1, 72076 Tübingen, Germany*

Abstract. This paper focuses on motion control problems of an omnidirectional robot based on the Nonlinear Model Predictive Control (NMPC) method. The main contributions of this paper are not only to analyze and design NMPC controllers with guaranteed stability to nonlinear kinematic models, but also to show the feasibility of NMPC with a real fast moving omnidirectional robot.

Keywords. Nonlinear Model Predictive Control, Omnidirectional Robot, Path Following, Trajectory Tracking

Introduction

Recently, omnidirectional wheeled robots have received more attention in mobile robots applications, because they have full mobility in the plane, which means that they can move at each instant in any direction without any reorientation [4]. Path following and trajectory tracking are two fundamental motion control problems of mobile robots, where a robot is required to follow a geometric reference path and to track a time parameterized reference trajectory, respectively. According to the error kinematics of these two problems, many nonlinear controllers have been presented recently [1,2,5,6,8,14,15]. However, some methods rarely take the robot constraints into account, which are crucial for the robot performance, even destroying the control stability [10,16]. Moreover, only the current errors are considered in most control laws, which ignore the potential opportunity of improving the control performance by using more information of the given path or trajectory.

Motivated by the above considerations, we apply Nonlinear Model Predictive Control (NMPC) to solve the path following and trajectory tracking problem of an omnidirectional robot. As NMPC can easily take robot constraints into account, and utilize the future information to get current control inputs, it has been used in many robotics applications. However, many applications ignore considering stability [11,12]. Some methods can only guarantee local stability by linearizing the error kinematics [3,17]. Many researchers present detailed analysis of NMPC with mobile robots, but the applications are only in simulation [9,13]. The main contributions of this paper are to analyze and design NMPC controllers with respect to nonlinear kinematic models with guaranteed stability, and to show the feasibility of applying NMPC on a real omnidirectional robot.

[1]Corresponding Author: Xiang Li, Wilhelm-Schickard-Institute, Department of Computer Architecture, University of Tübingen, Sand 1, 72076 Tübingen, Germany; E-mail:xiang.li@uni-tuebingen.de.

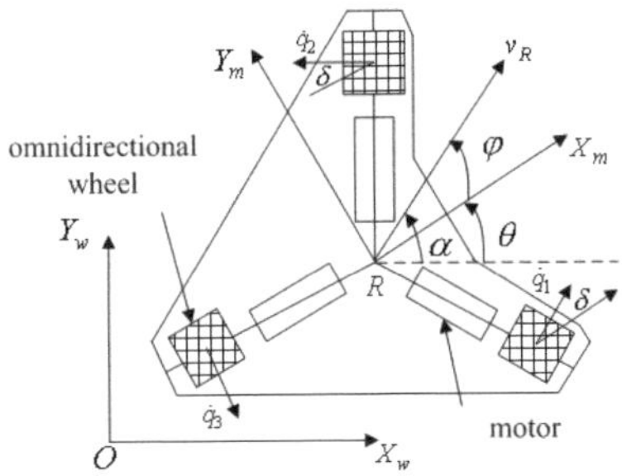

Figure 1. Kinematics diagram of the base of an omnidirectional robot

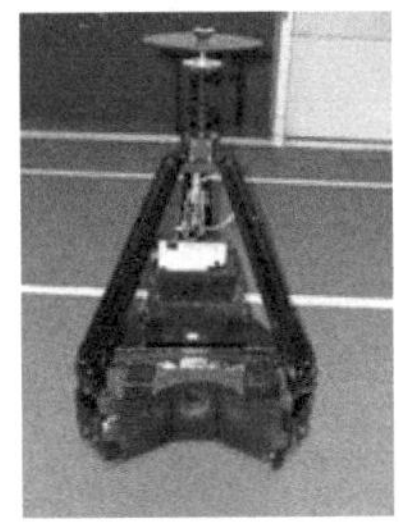

Figure 2. The real omnidirectional robot

1. Robot Kinematic Model

Figure 1 shows the base of our omnidirectional robot. Besides the fixed world coordinate system $\{W\}$, a moving robot coordinate system $\{M\}$ is defined. Angle θ between the axes X_m and X_w denotes the robot orientation. Angles α and φ denote the robot moving direction in the world and robot coordinate systems, respectively. Each wheel has the same distance L_w to the robot's center of mass R. δ refers to the wheel orientation in the robot coordinate system and is equal to 30 degrees. The kinematic model of the robot is as follows:

$$\dot{\mathbf{x}} = \begin{bmatrix} \dot{x} \\ \dot{y} \\ \dot{\theta} \end{bmatrix} = \begin{bmatrix} \cos\theta & -\sin\theta & 0 \\ \sin\theta & \cos\theta & 0 \\ 0 & 0 & 1 \end{bmatrix} \begin{bmatrix} u \\ v \\ \omega \end{bmatrix}, \tag{1}$$

where $\mathbf{x}$ is the robot state vector with respect to the world coordinate system, which is composed of the robot position x, y and the robot orientation θ; The inputs include the robot rotation velocity ω, the robot translation velocities u and v with respect to the axes X_m and Y_m. When we consider the wheel velocities, the lower level kinematic model of the robot can be deduced as:

$$\dot{\mathbf{q}} = \begin{bmatrix} \cos\delta & \sin\delta & L_w \\ -\cos\delta & \sin\delta & L_w \\ 0 & -1 & L_w \end{bmatrix} \begin{bmatrix} u \\ v \\ \omega \end{bmatrix}, \tag{2}$$

where $\dot{\mathbf{q}}$ is the vector of wheel velocities $[\dot{q}_1\ \dot{q}_2\ \dot{q}_3]^T$, and $\dot{q}_i (i = 1, 2, 3)$ denotes the i-th wheel velocity, which is equal to the wheel's radius multiplied by the wheel's angular velocity. As the motor's voltage and current are magnitude limited, the maximum wheel velocity is limited by $\dot{q}_m$, namely $|\dot{q}_i| \leq \dot{q}_m$. It is important to notice that the transformation matrices in the above kinematic models (1) and (2) are all full rank, which denotes that the translation and rotation of the omnidirectional robot are decoupled, and enables the separate control of these two movements.

2. Problem Formulation

Path following and trajectory tracking problems are illustrated in Figures 3 and 4, respectively. P and T denote the given path and trajectory, respectively. Point Q is the

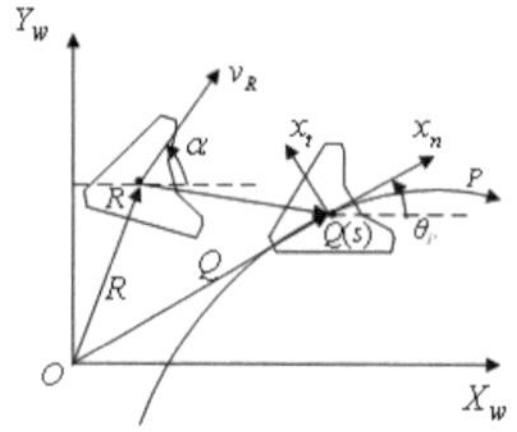

Figure 3. Path following problem with the desired position determined by parameter s.

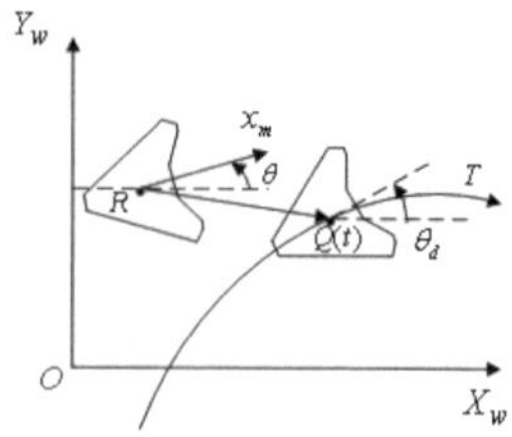

Figure 4. Trajectory tracking problem with the desired position determined by time t.

desired point of the robot. v_R refers to the robot translation velocity. The main difference between these two problems is point Q is determined by the parameter s and time t, respectively.

2.1. Path Following Problem

In the path following problem, a path coordinate system $\{P\}$ moving along the reference path P is introduced. The coordinate axes x_t and x_n direct the tangent and normal directions at point Q, respectively. θ_P denotes the path tangent direction at point Q. When we project the robot states in the path coordinate system, the error kinematic model can be deduced as,

$$\dot{\mathbf{x}}_e = \begin{bmatrix} \dot{x}_e \\ \dot{y}_e \\ \dot{\alpha}_e \end{bmatrix} = \begin{bmatrix} (y_e c(s) - 1)\dot{s} + v_R \cos\alpha_e \\ -x_e c(s)\dot{s} + v_R \sin\alpha_e \\ \dot{\alpha} - c(s)\dot{s} \end{bmatrix}, \tag{3}$$

where α_e presents the angular error as $\alpha_e = \alpha - \theta_p$, $c(s)$ denotes the path curvature at point Q. It is noticed from Eq. (3) that the controlling of v_R is decoupled from controlling of $\dot{s}$ and $\dot{\alpha}$, because the errors can stay at the equilibrium $(\mathbf{x}_e = 0)$ when $\dot{s}$ approaches v_R. Therefore, the path following problem is to find suitable values of $\dot{s}$ and $\dot{\alpha}$ driving errors x_e, y_e and α_e to zero with v_R to be steered to track any desired velocity.

2.2. Trajectory Tracking Problem

Unlike the path following problem, the robot has to track the pre-designed desired velocities with specified time in the trajectory tracking problem. If we take the robot coordinate system as the reference frame, the error kinematic model can be obtained with the desired robot kinematics,

$$\begin{bmatrix} \dot{x}_e \\ \dot{y}_e \\ \dot{\theta}_e \end{bmatrix} = \begin{bmatrix} \omega y_e - u + u_d \cos\theta_e - v_d \sin\theta_e \\ -\omega x_e - v + u_d \sin\theta_e + v_d \cos\theta_e \\ \omega_d - \omega \end{bmatrix}, \begin{bmatrix} \dot{x}_d \\ \dot{y}_d \\ \dot{\theta}_d \end{bmatrix} = \begin{bmatrix} \cos\theta_d & -\sin\theta_d & 0 \\ \sin\theta_d & \cos\theta_d & 0 \\ 0 & 0 & 1 \end{bmatrix} \begin{bmatrix} u_d \\ v_d \\ \omega_d \end{bmatrix}, \tag{4}$$

where the reference state $\mathbf{x}_d = [x_d, y_d, \theta_d]^T$ and corresponding reference input velocities u_d, v_d and ω_d are determined by time t, x_e and y_e denote the distance error, θ_e presents the orientation error between θ_d and θ. Therefore, the objective of trajectory tracking is to choose suitable robot inputs u, v and ω, such as the tracking error $\mathbf{x}_e$ converges to zero.

3. Nonlinear Model Predictive Control

As an attractive optimal control method, Nonlinear Model Predictive Control (NMPC) has been used in our control problems, because it can easily handle the system constraints and take future information into the controller design [7]. With respect to a normal nonlinear system described by the following differential equation:

$$\dot{\mathbf{x}}(t) = \mathbf{f}(\mathbf{x}(t), \mathbf{u}(t)), \text{ subject to } \mathbf{u}(t) \in U, \ \mathbf{x}(t) \in X, \ \forall t \geq 0, \tag{5}$$

where $\mathbf{x}(t) \subseteq R^n$ and $\mathbf{u}(t) \subseteq R^m$ are the n dimensional state vector and m dimensional input vector, respectively. Sets X and U include the feasible states and inputs, especially the system equilibrium ($\mathbf{x}(t) = 0$ and $\mathbf{u}(t) = 0$). The basic idea of NMPC is to execute the following iteration :

1. predict the system's future behavior over a prediction horizon T_p at time t based on the current system states;
2. find optimal inputs $\bar{u}(\cdot)\colon\ [t, t+T_p] \rightarrow U$ to minimize the value of the following objective function,

$$\mathbf{J}(t, \mathbf{x}(t), \bar{\mathbf{u}}(\cdot)) = \int_t^{t+T_p} \mathbf{F}(\bar{\mathbf{x}}(\tau), \bar{\mathbf{u}}(\tau))\, d\tau + +E(\bar{\mathbf{x}}(t+T_p)), \tag{6}$$

   subject to:

   - $\dot{\bar{\mathbf{x}}}(\tau) = \mathbf{f}(\bar{\mathbf{x}}(\tau), \mathbf{u}(\tau)), \ \mathbf{x}(0) = \mathbf{x}(0),$
   - $\bar{\mathbf{u}}(\tau) = \bar{\mathbf{u}}(\tau + T_c), \ \forall \tau \in [t + T_c, t + T_p]\,,$
   - $\bar{\mathbf{u}}(\tau) \in U, \ \ \bar{\mathbf{x}}(\tau) \in X, \ \ \forall t \in [t, t+T_p]\,, \ \ \bar{\mathbf{x}}(t+T_p) \in \mathbf{\Omega}$

   where T_c is the control horizon with $T_c \leq T_p$, $\mathbf{F}$ is the cost function specifying the desired control performance, the bar denotes the predicted values in the future, which are not same as the real values, the terminal penalty $E(\bar{\mathbf{x}}(t+T_p))$ and constraints $\bar{\mathbf{x}}(t+T_p) \in \mathbf{\Omega}$ are used to achieve close-loop stability;
3. take the first optimal input value $\bar{\mathbf{u}}(t)$ as the current input;
4. go back to the first step at time $t+1$.

As the prediction at each time is based on the current system states, NMPC is a feedback control law. To guarantee the stability, the terminal penalty and constraints can be selected according to the stability theorem presented in [7,9]: for a terminal penalty $E(\mathbf{x})$ with $E(\mathbf{0}) = 0$, and a closed region $\Omega \subseteq X$ including the origin, if there is a control law $\mathbf{k}(\mathbf{x}) \in U$ with $\mathbf{k}(\mathbf{0}) = 0$ such that

$$\dot{E}(\mathbf{x}) + \mathbf{F}(\mathbf{x}, \mathbf{k}(\mathbf{x})) \leq 0, \forall \mathbf{x} \in \Omega, \tag{7}$$

the former described NMPC method guarantees asymptotical stability of the closed-loop system.

To online solve the open-loop nonlinear optimization problem (6), sequential quadratic programming (SQP) is utilized in our project. We use the software *donlp2-intv-dyn* written by P. Spellucci, which is a general purpose nonlinear optimizer and can be found at *http://plato.la.asu.edu/donlp2.html*.

3.1. Path Following Control

As the translation and rotation of an omnidirectional robot are completely decoupled, the angular error α_e in Eq. (3) can be directly controlled. Therefore, the path following problem of an omnidirectional robot can be solved by finding suitable $\dot{s}$ and α_e. With the other freedom of controlling the robot orientation, we can drive the robot orientation θ to any desired orientation θ_d at the same time. Defining the orientation error θ_e, and a new vector $\mathbf{u}_e = [u_1, u_2, u_3]^T$, we get the following error kinematic model having the equilibrium at $\mathbf{x}_e = 0$ and $\mathbf{u}_e = 0$,

$$\begin{bmatrix} \dot{x}_e \\ \dot{y}_e \\ \dot{\theta}_e \end{bmatrix} = \begin{bmatrix} 0 & c(s)\dot{s} & 0 \\ -c(s)\dot{s} & 0 & 0 \\ 0 & 0 & 0 \end{bmatrix} \begin{bmatrix} x_e \\ y_e \\ \theta_e \end{bmatrix} + \begin{bmatrix} u_1 \\ u_2 \\ u_3 \end{bmatrix}, \quad \begin{bmatrix} u_1 \\ u_2 \\ u_3 \end{bmatrix} = \begin{bmatrix} -\dot{s} + v_R \cos\alpha_e \\ v_R \sin\alpha_e \\ \dot{\omega}_d - \omega \end{bmatrix}, \tag{8}$$

where $\theta_e = \theta_d - \theta$, ω_d denotes the desired rotation velocity.

We select the following cost function:

$$F(\mathbf{x}, \mathbf{u}) = \mathbf{x}_e^T \mathbf{Q} \mathbf{x}_e + \mathbf{u}_e^T \mathbf{R} \mathbf{u}_e, \tag{9}$$

with weight matrices $\mathbf{Q}$ and $\mathbf{R}$, whose elements are set to zero, except that the diagonal elements are positive constants. To guarantee the control stability, the following Lyapunov function can be selected as the terminal penalty:

$$E(\mathbf{x}_e(t + T_p)) = \frac{1}{2}\mathbf{x}_e(t + T_p)^T \mathbf{x}_e(t + T_p), \tag{10}$$

where $\mathbf{x}_e(t+T_P) = [x_{eT_p}, y_{eT_p}, \theta_{eT_p}]^T$ denotes the terminal state. When we choose the terminal feedback controller $\mathbf{u}^L = [u_1^L, u_2^L, u_3^L]^T$ as:

$$u_1^L = -\kappa x_{eT_p},\ u_2^L = -\beta y_{eT_p},\ u_3^L = -\gamma\theta_{eT_p} \tag{11}$$

with parameters $\kappa \geq 0$, $\beta \geq 0$, and $\gamma \geq 0$, the left side of stability condition (7) becomes

$$\begin{aligned} \dot{E}(x_e(t + T_p)) + F(t + T_p) =& x_{eT_p}^2(-\kappa + q_{11} + \kappa^2 r_{11}) + y_{eT_p}^2(-\beta + q_{22} + \beta^2 r_{22}) \\ &+ \theta_{eT_p}^2(-\gamma + q_{33} + \gamma^2 r_{33}). \end{aligned} \tag{12}$$

To satisfy the stability condition (7), the following requirements are necessary:

$$\kappa - q_{11} - \kappa^2 r_{11} \geq 0,\ \beta - q_{22} - \beta^2 r_{22} \geq 0,\ \gamma - q_{33} - \gamma^2 r_{33} \geq 0. \tag{13}$$

Furthermore, the outputs of feedback controllers (11) have to satisfy the system constraints, which are the bounded wheel velocities in our case. Combining (1), (8) and (11), respectively, we get the second part of terminal constraints as

$$-\begin{bmatrix} \dot{q}_m \\ \dot{q}_m \\ \dot{q}_m \end{bmatrix} \leq \begin{bmatrix} \cos\delta & \sin\delta & L_w \\ -\cos\delta & \sin\delta & L_w \\ 0 & -1 & L_w \end{bmatrix} \begin{bmatrix} -\kappa x_{eT} + \dot{s} \\ -\beta y_{eT} \\ -\gamma\theta_{eT} + \omega_d \end{bmatrix} \leq \begin{bmatrix} \dot{q}_m \\ \dot{q}_m \\ \dot{q}_m \end{bmatrix}. \tag{14}$$

3.2. Trajectory Tracking Control

With the same idea of solving the path following problem, we transfer the error kinematic model of the trajectory tracking problem (4) by introducing the following control variables u_1, u_2 and u_3:

$$\begin{bmatrix} \dot{x}_e \\ \dot{y}_e \\ \dot{\theta}_e \end{bmatrix} = \begin{bmatrix} 0 & \omega & 0 \\ -\omega & 0 & 0 \\ 0 & 0 & 0 \end{bmatrix} \begin{bmatrix} x_e \\ y_e \\ \theta_e \end{bmatrix} + \begin{bmatrix} u_1 \\ u_2 \\ u_3 \end{bmatrix}, \quad \begin{bmatrix} u_1 \\ u_2 \\ u_3 \end{bmatrix} = \begin{bmatrix} -u + u_R \cos\theta_e \\ -v + u_R \sin\theta_e \\ \omega_d - \omega \end{bmatrix}. \tag{15}$$

When we select the cost function, the terminal penalty and the feedback controller with the same forms as (9), (10) and (11), respectively, the similar terminal constraints can be deduced following the above process.

4. Experimental Results

The real-world experiments were made in our robot laboratory having a carpet covered field with a size of $5.1 \times 4.2\ m^2$. Figure 2 shows our middle-sized RoboCup robot, which is equipped with a Pentium-M 2GHz on-board PC with 1GB RAM. The wheels are driven by three $60W$ Maxon DC motors and have the maximum wheel velocity $\dot{q}_m = 1.9m/s$. Two kinds of experiments were made to test the above NMPC method: a path

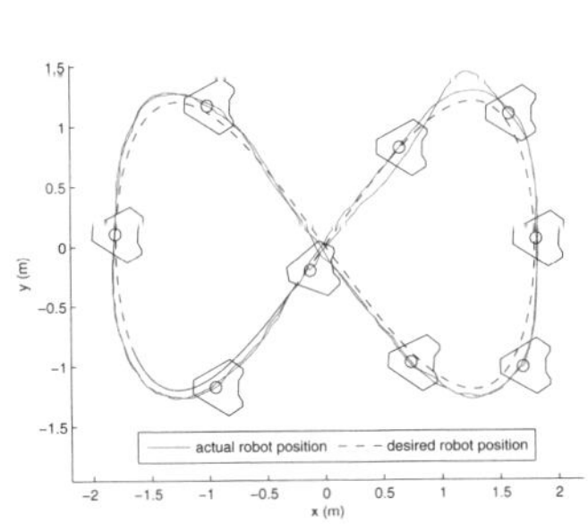

Figure 5. Reference trajectory and robot trajectory.

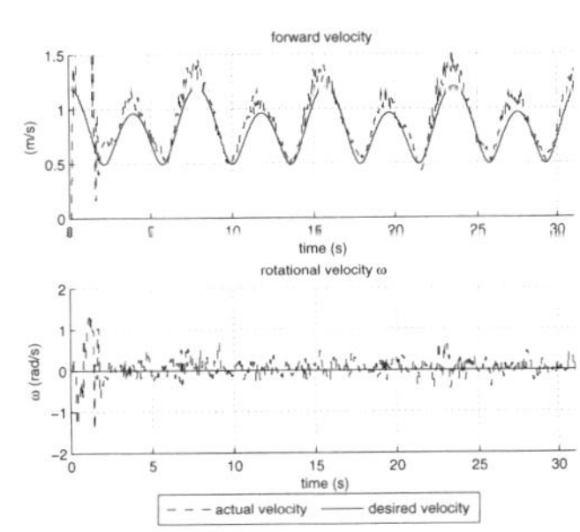

Figure 6. Velocities with respect to the robot frame.

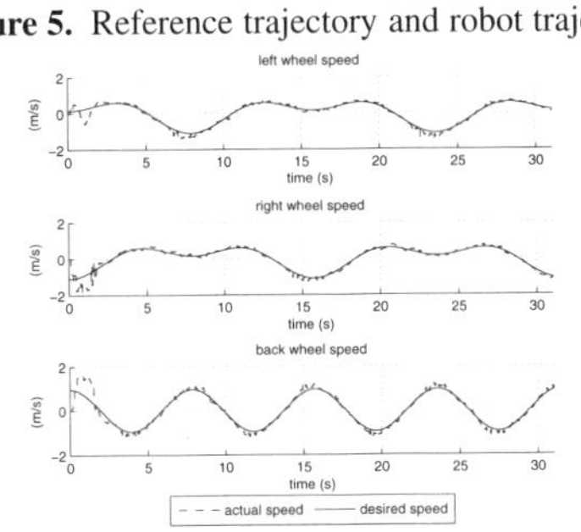

Figure 7. Wheel Velocities.

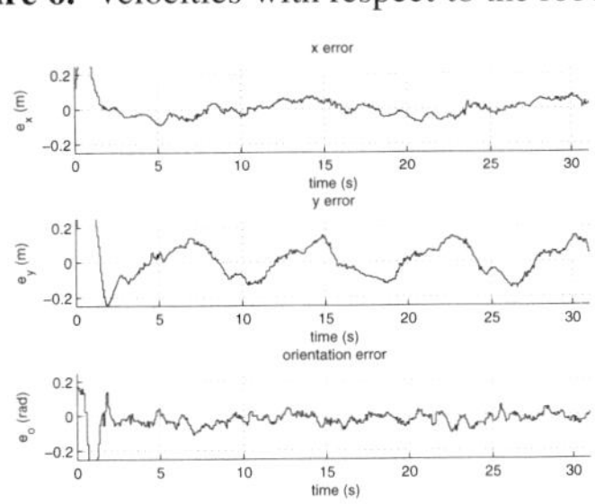

Figure 8. Tracking errors.

following control with a constant desired velocity of $v_R = 1.0m/s$ and a trajectory tracking control with changing desired velocity of v_R with time t. The desired robot orientations in two cases are $\theta_d = 0$. An eight-shaped curve is adopted as the reference

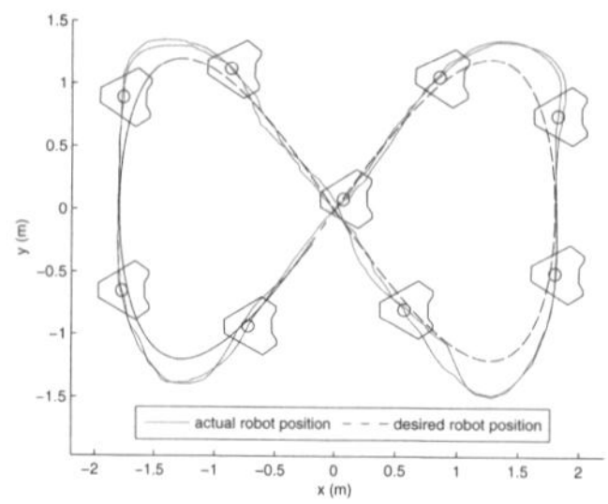

Figure 9. Reference path and robot path.

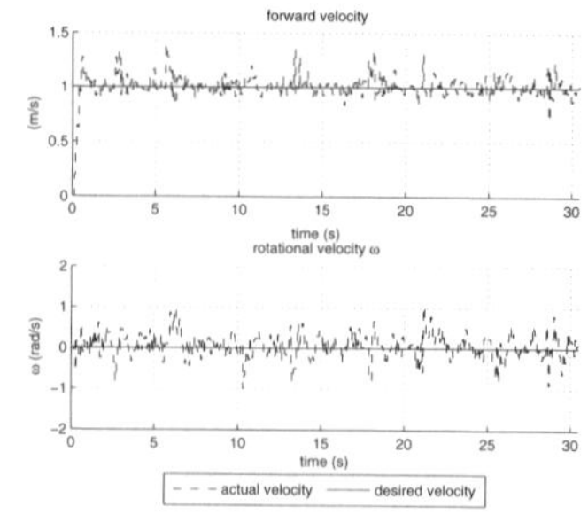

Figure 10. Velocities with respect to the robot frame.

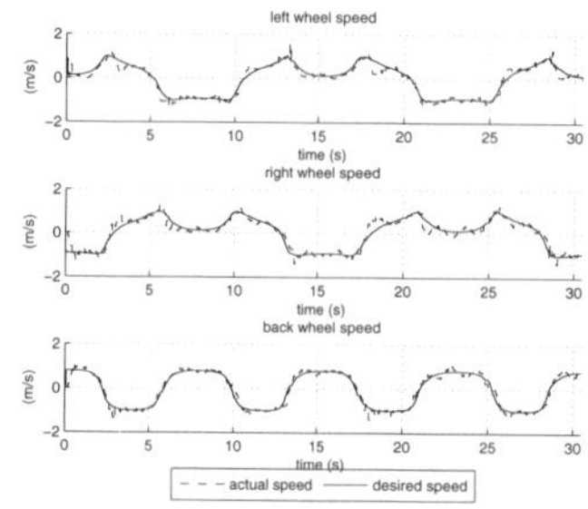

Figure 11. Wheel Velocities.

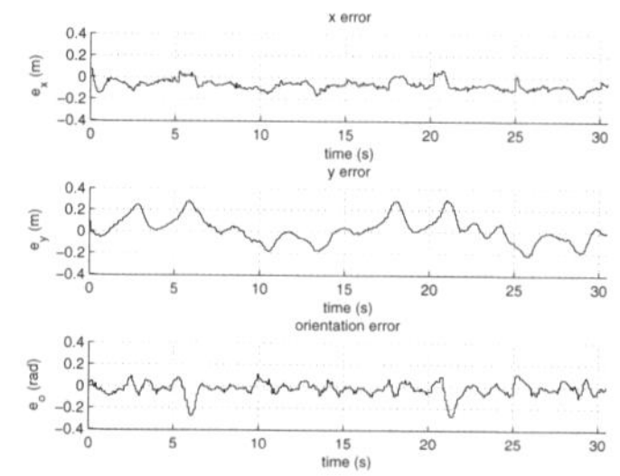

Figure 12. Following errors.

path and trajectory, because its geometrical symmetry and sharp changes in curvature make the test challenging. With respect to the world coordinate system, the coordinates of the reference path (x_P, y_P) and trajectory (x_T, y_T) are given by

$$\begin{cases} x_P(s) = 1.8\sin(s) \\ y_P(s) = 1.2\sin(2s) \end{cases} \qquad \begin{cases} x_T(t) = 1.8\sin(0.4t) \\ y_T(t) = 1.2\sin(0.8t). \end{cases} \tag{16}$$

In the path following and trajectory experiments, we chose the same parameters as: $\mathbf{Q} = [1.2\ 0\ 0; 0\ 1.4\ 0; 0\ 0\ 1.4]$, $\mathbf{R} = [0.08\ 0\ 0; 0\ 0.1\ 0; 0\ 0\ 0.1]$, $\kappa = \beta = \gamma = 2$, $\tau = 0.07s$, $T_P = T_c = 3\tau$.

Figures 5 and 9 illustrate the results of trajectory tracking and path following experiments. Figures 6 and 10 show that the desired translation velocity is changing according to the trajectory's curvature, but keeps constant in the path following problem, which increases the difficulty in following the sharp turning part of the given path. Figures 7 and 11 show that the controller guarantees the required wheel velocities under the boundaries. Figures 8, 12 show that the NMPC solved these real-time motion control problems with good performance, because at most time steps, the distance errors are less than 0.2m and 0.3m, the orientation errors are less than 0.1rad and 0.3rad in the trajectory tracking and path following tasks, respectively.

5. Conclusions

This paper focuses on motion control problems of an omnidirectional robot. Considering the constraints of robot systems and utilizing more known information to increase the robot performance, we use Nonlinear Model Predictive Control in this work. According to analyzing the path following and trajectory tracking problems, we formulate NMPC with respect to the corresponding error kinematic models, especially present details of choosing suitable terminal penalties and terminal constraints to guarantee the control stability. Experimental results with a real omnidirectional robot show that the Nonlinear Model Predictive Control not only can solve the path following and trajectory tracking problems with good performance, but also is feasible to be applied on a fast moving robot platform.

References

[1] A. P. Aguiar and J. ao Pedro Hespanha. Logic-based switching control for trajectory-tracking and path-following of underactuated autonomous vehicles with parametric modeling uncertainty. In *Proc. of the 2004 American Control Conference*, Boston Massachusetts, USA, 2004.

[2] A. P. Aguiar, J. P. Hespanha, and P. V. Kokotovic. Path-following for nonminimum phase systems removes performance limitations. *IEEE Transactions on Automatic Control*, 50(2):234–239, 2005.

[3] M. Bak, N. K. Poulsen, and O. Ravn. Path following mobile robot in the presence of velocity constraints. Technical report, Informatics and Mathematical Modelling, Technical University of Denmark, DTU, Richard Petersens Plads, Building 321, DK-2800 Kgs. Lyngby, 2001.

[4] G. Campion, G. Bastin, and B. D'Andréa-Novel. Structural properties and classification of kinematic and dynamic models of wheeled mobile robots. *IEEE Transaction on Robotics and Automation*, 12, 1996.

[5] F. D. del Río, G. Jiménez, J. L. Sevillano, S. Vicente, and A. Civit-Balcells. A generialization of path following for mobile robots. In *Proc. of the 1999 IEEE Int. Conf. on Robotics and Automation*, volume 1, pages 7–12, 1999.

[6] M. Egerstedt, X. Hu, and A. Stotsky. Control of mobile platforms using a virtual vehicle approach. *IEEE Transactions on Automatic Control*, 46(11):1777–1782, 2001.

[7] R. Findeisen and F. Allgower. An introduction to nonlinear model predictive control. In *21st Benelux Meeting on Systems and Control*, Veldhoven, Netherlands, 2002.

[8] R. Ghabcheloo, A. Pascoal, C. Silvestre, and I. Kaminer. Coordinated path following control of multiple wheeled robots with directed communication links. In *Proc. of the 44th IEEE Conf. on Decision and Control*, 2005.

[9] D. Gu and H. Hu. Receding horizon tracking control of wheel mobile robots. *IEEE Transaction on Control Systems Technology*, 14(4), july 2006.

[10] G. Indiveri, J. Paulus, and P. G. Plöger. Motion control of swedish wheeled mobile robots in the presence of actuator saturation. In *Proc. of the 10th annual RoboCup International Symposium*, 2006.

[11] T. Keviczky, P. Falcone, R. Borrelli, J. Asgari, and D.Hrovat. Predictive control approach to autonomous vehicle steering. In *Proc. of the 2006 American Control Conference*, 2006.

[12] B. Kim, D. Necsulescu, and J. Sasiadek. Model predictive control of an autonomous vehicle. In *Proc. of the 2001 IEEE/ASME Int. Conf. on Advanced Intelligent Mechatronics*, pages 1279–1284, 2001.

[13] F. Kühne, J. J.M.G. da Silva, and W. Lages. Mobile robot trajectory tracking using model predictive control. In *Proc. of the 2005 Latin-American Robotics Symposium*, 2005.

[14] X. Li and A. Zell. Motion control of an omnidirectional mobile robot. In *Proc. of the 4th Int. Conf. on Informatics in Control, Automation and Robotics*, Angers, France, 2007.

[15] K. Maček, I. Petrović, and R. Siegwart. A control method for stable and smooth path following of mobile robots. In *Proc. of the European Conference on Mobile Robots 2005*, 2005.

[16] A. Scolari Conceição, A. Moreira, and P. j. Costa. Trajectory tracking for omni-directional mobile robots based on restrictions of the motor's velocities. In *Proc. 8th Int. IFAC Symposium on Robot Control*, 2006.

[17] M. Seyr and S. Jakubek. Mobile robot predictive trajectory tracking. In *Proc. of 2005 ICINCO*, pages 112–119, 2005.

Intelligent Autonomous Systems 10
W. Burgard et al. (Eds.)
IOS Press, 2008

doi:10.3233/978-1-58603-887-8-100

A Self-Configuration Mechanism for Software Components Distributed in an Ecology of Robots

Marco GRITTI [a], Alessandro SAFFIOTTI [a]
[a] *AASS Mobile Robotics Laboratory, University of Örebro, Sweden*

Abstract.
Distributed heterogeneous robotic systems are often organized in component-based software architectures. The strong added value of these systems comes from their potential ability to dynamically self-configure the interactions of their components, in order to adapt to new tasks and unforeseen situations. However, no satisfactory solutions exist to the problem of automatic self-configuration. We propose a self-configuration mechanism where a special component generates, establishes and monitors the system configurations. We illustrate our approach on a distributed robotic system, and show an experiment in which the configuration component dynamically changes the configuration in response to a component failure.

Introduction

Robotic technologies begin to be employed in relatively inexpensive devices and consumer products, and they slowly start to appear in our homes, offices, and workshops. Many observers believe that these devices will be networked and will be able to share their information and coordinate their actions. This vision is becoming rather popular in robotics, and it has been spelled out under different names, like network robot systems [1], intelligent spaces [8], artificial ecosystem [16], and PEIS-Ecology [14]. In this paper, we generally refer to this vision as *ecology of robots*.

The software controlling these distributed robotic systems is usually constituted by components spread over the various devices. The strong added value of this approach comes from the potential of these systems to dynamically *self-configure* the component interactions, both within and across devices. This dramatically increases flexibility and adaptability, since the collaboration patterns are not fixed at design time but can change according to the task that should be performed, or to the current operating conditions. Much work has been done recently on the principles and techniques for automatic self-configuration in autonomous robotics [9,12] as well as in other areas of computer science like ambient intelligence [7], web services [13], or autonomic computing [17]. Despite this, no satisfactory solution exist to the problem of automatic self-configuration.

In this paper we propose a reactive approach to self-configuration, in which a special component dynamically creates, monitors and modifies configurations of a robot ecology. This component accepts task specifications in the form of *configuration templates*. This approach is inspired by ideas from the field of semantic web services and by the heritage of reactive architectures in robotics. The key elements of our approach are:

- *formal descriptions* of the components in the ecology;
- *configuration templates* expressing tasks in the form of desired components and desired component interactions; and
- mechanisms for *discovery* and *matching* of formal descriptions that allow to automatically find the components required by a given template.

In order to illustrate our approach, we apply it to a specific ecology of robots: the PEIS-Ecology. The approach, however, should be applicable to other distributed robotic platforms as well. The rest of this paper is organized as follows. In the next section we state the system models and requirements of our self-configuration mechanism. In Section 2 we give an overview of the constituents of this mechanism. In Section 3 we explain in detail our template based configuration component. In Section 4 we show an experiment in which our component generates a configuration for a task of cooperative navigation and dynamically changes such configuration in response to a failure.

1. Prerequisites of the System

The self-configuration mechanism proposed here ignores the internal logics of the software components of the system. In other words, all components are internally abstracted as *black boxes*. On the other hand, their external behavior must comply to the model of *asynchronous computation*: components accept the data that are set into their input ports, process them, and export the results by making them available on their output ports.

A uniform model of component interaction is also fixed: components are allowed to interact only through *connections* of input ports to output ports. Such input-output connections must be dynamic, in the sense that they are not to be decided at component design time, but it must be possible to establish or release them at runtime. According to this model, components exchange information in the form of *flows of data* on their input-output connections. This model of interaction is ubiquitous within robotic systems, e.g., in a sensor processing pipeline, or in a hierarchy of controllers. Such model also appears in other domains, e.g. multi-media content streaming through the Internet.

A convenient way to make all system components comply to the same model of interaction is to build them on top of a common *middleware*. In the present work, the PEIS-kernel [3] was adopted as middleware, but the achieved results do not depend on it. For the purposes of our self-configuration mechanism, it is sufficient that the middleware has an API that supports (1) asynchronous access to the component functions, and (2) dynamic connections of input ports to output ports. These two requirements are generally met by most of the existing middleware, being usually implemented through mechanisms of subscription to asynchronous events and of event notification. The self-configuration mechanism described below applies to any distributed system that satisfies the above prerequisites. The PEIS-Ecology [14] belongs to this class of systems.

2. Self-Configuration

The overall behavior of a distributed system is determined by the concurrent execution of the various components, which combine together their functionalities through specific interactions. We intend such concurrent executions organized in *system configurations*. A *system configuration* for a task t is a pair $< B, S >$, where:

- B is the set of running components whose functionalities contribute to t;
- S is the set of all the connections established between the components in B.

2.1. An Architecture Pattern for Self-Configuration

The problem of configuring a robot ecology consists in determining what components are eligible to execute in a common system-level task, and what are their suitable connections. In the terms of the above definition, this means to determine the two sets B and S opportune to solve a given task t. This computation can be done automatically, according to the architecture and interaction schemes of Figure 1 and Figure 2.

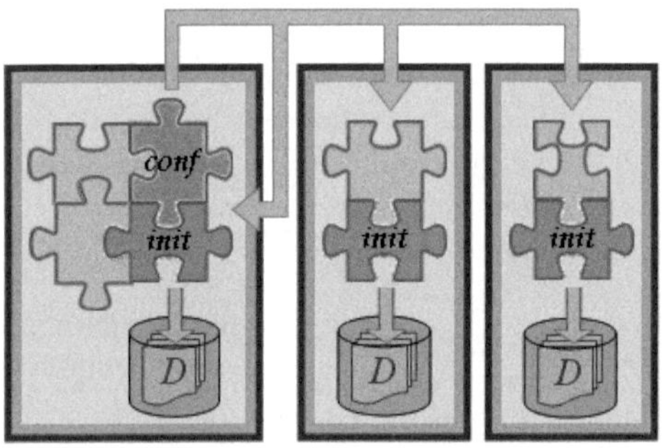

Figure 1. System architecture for self-configuration.

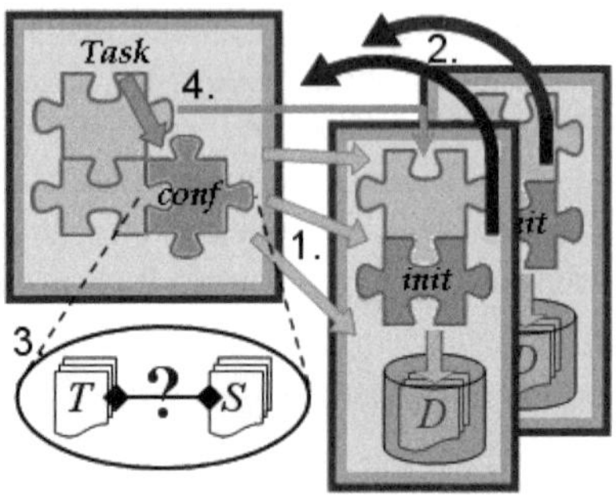

Figure 2. The self-configuration process.

As illustrated, repositories of component descriptions D should be deployed in the system. Special *init*-components should run on the same sites of the repositories, which fetch and propagate the descriptions. Other special *conf*-components (or *configurators*) should exist, capable of computing the configurations opportune to achieve the various system-level tasks using the information stored in the component descriptions.

2.2. Formal Component Descriptions

The formal component descriptions adopted here are called component *profiles*. These describe the components in both their *functional* and *non-functional* aspects.

The functional part of a profile lists the signals accepted in input and the signals produced in output. These are in one-to-one correspondence respectively with the component input and output ports. As an example, consider the advertisements in Figure 3 and in Figure 4. These publish the profiles of a navigation control component for mobile robots and of a motor drive component. The navigation control component has three input signals (named `sonar`, `position`, and `localize`) and produces one output signal (named `set.velocity`). The motor drive component accepts just one input signal and has no output signals. All signals are typed.

The non-functional part of a component profile includes the component type, and the list of component properties. Component types classify components according to the functionality they provide. Component properties declare additional features of the components, which allow to further specialize their description. For example, the type of the navigation component is `NavigationControl`; the type of the motor drive is `MotorDrive`. The motor drive has also a property, with type `Support` and value `ASTRID`. This property specifies the real object that carries the actuator to which such component is a bare interface. In this case such object is Astrid, a mobile robot.

Component profiles are stored into manifest files referred to as *advertisements*. We deploy the advertisements on the same sites of the component instances. This is a typical solution for allowing automatic component discovery in peer-to-peer frameworks.

```
<component>NavigationControl</component>
<compName>"thinkingcap"</compName>
<componentDescription>
  "A fuzzy logic behavior-based mobile
   robot control architecture."
</componentDescription>
<parameter>
  <paramName>"contr.b-goal"</paramName>
  <DataType>B-Goal</DataType>
</parameter>
<signalInput>
  <sigInName>"sonar"</sigInName>
  <DataType>DistanceArray</DataType>
</signalInput>
<signalInput>
  <sigInName>"position"</sigInName>
  <DataType>EncoderReading</DataType>
</signalInput>
<signalInput>
  <sigInName>"localize"</sigInName>
  <DataType>PlanarPose2D</DataType>
</signalInput>
<signalOutput>
  <sigOutName>"set.velocity"</sigOutName>
  <DataType>PlanarVel</DataType>
</signalOutput>
```

Figure 3. Advertisement of a navigation component.

```
<component>MotorDrive</component>
<compName>"driveastrid"</compName>
<componentDescription>
  "The drive controller of Astrid."
</componentDescription>
<property>
  <PropType>Support</PropType>
  <propValue>"ASTRID"</propValue>
</property>
<signalInput>
  <sigInName>"setpoint"</sigInName>
  <DataType>PlanarVel</DataType>
</signalInput>
```

Figure 4. Advertisement of the drive on Astrid.

```
<component>PositionSensor</component>
<compName>"persontracker"</compName>
<componentDescription>
  "A people detection and tracking
   system using stereo vision."
</componentDescription>
<signalOutput>
  <sigOutName>"person.pos"</sigOutName>
  <DataType>PlanarPose2D</DataType>
</signalOutput>
```

Figure 5. Advertisement of a tracker component.

2.3. *Use of Component Descriptions*

In our mechanism for automatic self-configuration, the component profiles are processed by the special *conf*-components for (1) selection of the components eligible to enter a configuration, and (2) check of mutual compatibility between components supposed to interact. These computations are done in compliance with the following rules:

Component selection rule: a component of profile p, is *eligible* to enter a system configuration if and only if this configuration needs a component b such that (i) the type of p matches the type of b, and (ii) for all properties r_i^p of p, if b has a property r_j^b such that the type of r_i^p matches the type of r_j^b, then their values are equal. In such case, it is also said that p is *compatible* with b.

Component mutual compatibility rule: given a set of components with profiles P, of which one, named p^{si}, represents a *sink* of data, and all the others, grouped in the set $P^{so} = P \setminus \{p^{si}\}$, represent *sources* of data, the sink profile is *mutually compatible* against all of its sources if and only if for each input i^{si} of the sink there is at least one output o^{so} among the outputs of the sources such that o^{so} matches i^{si}.

All the above matchings are purely *syntactic*: if the types of the compared quantities are equal, then the quantities match, otherwise they do not match.

2.4. *The Four Acts of Self-Configuration*

As depicted in Figure 2, having received a system-level task specification, the *conf*-component broadcasts queries (Act 1) that ask for components suitable for such task; then it collects relevant component descriptions received in reply from the various *init*-components (Act 2). These two acts constitute the *discovery* phase of the self-configuration process. After a query timeout, the *conf*-component evaluates the feasibility of the desired task given the descriptions it has received (Act 3). This is the *composition* phase.

In this phase the *conf*-component *searches* among all possible configurations featuring (some of) the discovered components for one that can solve the assigned task. If such a configuration is found, the *conf*-component deploys it into the system (Act 4). This is done by (i) activating all the components that have been selected, and (ii) establishing all the connections that have been computed.

3. The Reactive Configuration Component

The *conf*-components are the key elements of our self-configuration mechanism. We now describe the implementation of one sort of *conf*-component: the *reactive configurator*. Upon receiving in input a system-level task specification, the reactive configurator is able to configure the system in a way appropriate to achieve such task, and to re-configure it whenever a failure is declared by some component of the deployed configuration.

3.1. Template Configurations as System-Level Task Specifications

We specify a task by listing what *types* of components interact, what are the *properties* that those components must fulfill to be able to correctly perform their subtask, and finally what are the *interactions* according to which it is meaningful for these components to exchange data. Such kind of specification is a *template configuration*.

For example, consider the task of *playing the radio* in a given room. A template configuration for it could list as types of interacting components a radio tuner, and (at least) one speaker. The speaker component must comply with the property to have *place* in the desired room. The desired interaction is that the tuner should feed the speaker, and not *vice versa*. As another example, consider the task of *navigating a robot* in the free space. Its template configuration could list a position sensor (possibly a virtual sensor), a drive for the motors of the robot, and a navigation control component. It is necessary that the motor drive has the robot that should move as its *support*. If the position sensor is proprioceptive, it must have the same robot as its *support*. The desired interactions are the following: the position sensor should feed the navigation control component; the navigation control component, in turn, should feed the motor drive.

Figure 6. The template configurations for *playing the radio* (left) and for *navigating a robot* (right).

Figure 6 shows two possible template configurations for those tasks. The properties of the components become task-dependent *parameters* of the template configurations. For instance, the template of Figure 6 (right) can be parameterized by setting "*support* = `ASTRID`". Fully parameterized template configurations t are the input of the reactive configurator, whose functioning is illustrated in the following.

3.2. The Algorithm for Configuration Generation

The reactive configurator performs component discovery in a trivial way. Having received a system-level task specification in the form of a parameterized template configuration t, it broadcasts a generic query for discovering all the components of the system. The set P of all received component profiles identifies the set of all components currently available. At the beginning of the composition phase, the reactive configurator determines the set $F \subseteq P$ of profiles of the components which are eligible to enter a configuration for t. This is done applying the component selection rule of Section 2.3 to all the profiles in P, against all the components B^t listed in the template t. More precisely, the configurator computes a partition $F^1 \dots F^n$ of F such that for all $b_i \in B^t$, F^i contains all and only those profiles that are compatible with b_i.

Consider for example the task of *navigating* `ASTRID`. If the configurator receives the advertisement of a motor drive with "*support* = `PIPPI`", this will be immediately discarded applying the selection rule, because this component has a support property different from the one that is required. The selection rule will also make the configurator discard all the advertisements of the components whose types are not `PositionSensor`, `MotorDrive`, or `NavigationControl`. Discarding components because of type or property mismatch is a preliminary heuristic cut of the configuration search space.

Algorithm 1 CONSEARCH($t, F^1, \dots, F^n$)

Require: Template t, and n profile sets.
Ensure: A configuration $\tilde{c}$ or `FAIL`.

```
c̃₀ ← ∅, C ← {c̃₀}
while C ≠ ∅ do
  c̃ ← first(C)
  if |P^c̃| ≡ |B^t| then
    return c̃
  else
    X ← EXPAND(c̃, t, F¹, ..., Fⁿ)
    C ← rest(C) ∪ X
  end if
end while
return FAIL
```

Function 2 EXPAND($\tilde{c}, t, F^1, \dots, F^n$)

Require: Config. $\tilde{c}$; t, and F^i as before.
Ensure: All one-level expansions of $\tilde{c}$.

```
X ← ∅
U ← {b_i ∈ B^t | ∀p ∈ F^i is p ∉ P^c̃}
for all b_i ∈ U do
  for all p ∈ F^i do
    if local-compatibility(p, c̃, t) then
      X ← X ∪ {c̃ ⊕ p}
    end if
  end for
end for
return X
```

Given the profiles of the eligible components, partitioned in $F^1, \dots, F^n$, the configurator *searches* in the space of all their possible configurations for one that complies to the assigned template t. A *state* in this search problem is an internal representation $\tilde{c}$ of a *partial configuration*, where just some of the components in B^t are associated to some profiles of F, and just some of the interactions listed in t are associated to groups of input-output connections. The only available *action* of this search problem is to add to a partial configuration $\tilde{c}$ a new association between a profile $p \in F$ and a template component $b \in B^t$ that is compatible with it. This is denoted by the operation $\oplus$, defined among internal representations of configurations and component profiles. Algorithm 1 is the search algorithm implemented in the reactive configurator. It accepts in input a parameterized template configuration t, and the various $F^1, \dots, F^n$. Algorithm 1 implements a sort of uninformed search, where exploration is done expanding internal representations $\tilde{c}$ of partial configurations, grouped in a working set C.

Function 2 computes the set X of expansions of a partial configuration $\tilde{c}$. The function computes the set U of components of t for which no compatible profile is already in the set $P^{\tilde{c}}$ of the profiles already in $\tilde{c}$. For all the components $b_i \in U$, this function tries to expand $\tilde{c}$ with any of the profiles $p \in F^i$, i.e. the ones compatible with b_i. Each expanded configuration is obtained adding to $\tilde{c}$ a profile p through the operator $\oplus$. The expansion step is done in compliance with the following rule:

Local compatibility rule: a profile $p \in F^i$ can be associated to a component $b_i \in U$ if and only if (i) if all the components $b^f \in B^t$ which are *feeders of* b_i are already associated in $P^{\tilde{c}}$, then p must be mutually compatible against all the profiles that are associated to all the b^f; and (ii) for all the components $b^F \in B^t$ which are *fed by* b_i, if some b^F has all its other feeders already associated in $P^{\tilde{c}}$, then the profile associated to b^F must be mutually compatible against the set of profiles constituted by p plus the profiles associated to all the other feeders of b^F.

Mutual compatibility between component profiles is always tested according to the mutual compatibility rule of Section 2.3. When the mutual compatibility test fails, the local compatibility fails too, and the new profile is not added to the partial configuration. Otherwise, all the input-output connections between the tested profiles are added to the partial configuration. The information contained in the template configurations about what components have direct interactions with what other components permits another important heuristic cut of the search space. Without such information, the local compatibility test would explode in complexity, because at every expansion step the search algorithm would have to check the compatibility of the newly introduced profile against all the subsets of $P^{\tilde{c}}$, in all their possible interaction patterns.

3.3. Monitoring and Reconfiguration

Once a configuration has been generated and deployed, the configurator component starts to *monitor* the configuration execution. A configurator can monitor clean failures of the deployed components by connecting to a special output `FAIL` that all the components are supposed to export. Whenever a failure event is signaled by a component, the configurator erases its profile from the set of the discovered profiles and re-triggers the search algorithm to look for another system configuration that can solve the original task. This monitoring phase is intrinsically reactive.

4. Experimental Run

The experiment that follows was performed on a real PEIS-Ecology platform [3,14].

4.1. Setup of the Experiment

The experiment takes place in the PEIS-Home: a reconstruction of a small apartment of about 25 m^2. The living room of the PEIS-Home is under the field of view of a 3D stereo camera. A PeopleBot named Astrid (Figure 8) is present in the environment.

The assigned task is *moving* `Astrid` *to the* `bedroom`. This is specified through the parameterized template configuration of Figure 7. Five components are listed in the template. Their types are: `SonarArray`, `Encoder`, `PositionSensor`, `NavigationControl`, and `MotorDrive`. For each component, the template lists its actual parameters[1], the required properties, and the interactions in which it takes part. The navigation control component is given one parameter of type `B-Goal`. For all the

[1] In our complete framework, besides the input signals, components can also accept *operational parameters*, which are for setting or tuning their running mode. This increases the components flexibility of use.

```
                  *** components ***

("Comp;#SonarArray"       () (("Objects;#Support" "ASTRID"))    (0))
("Comp;#Encoder"          () (("Objects;#Support" "ASTRID"))    (1))
("Comp;#PositionSensor"   () (("Objects;#Support" "ASTRID"))    (2))
("Comp;#NavigationControl" (("Cpt;#B-Goal" "(AT BEDROOM)")) () (0 1 2 3))
("Comp;#MotorDrive"       () (("Objects;#Support" "ASTRID"))    (3))

                  *** interactions ***

(0 3) (1 3)    ; sonar->control    / encoder->control
(2 3) (3 4)    ; position->control / control->drive
```

Figure 7. Parameterized template configuration for the navigation task of the experiment.

other components the `Support` property is specified. If present in their advertisements, this property must be equal to `ASTRID`. The interactions meaningful in the task are listed at the end of the template: component number 3 (the `NavigationControl`) should be fed by all other components, and it should feed component number 4 (the `MotorDrive`). This template is an extension of the one in Figure 6 (right).

The sonar array, encoder, and drive are part of the equipment of Astrid. They are all grounded on the same software. This is based on Player [6]. The advertisement of the motor drive was shown in Figure 4. The other two are similar: all expose a `Support` property with value `ASTRID`. The advertisement of the navigation component was shown in Figure 3. The corresponding robot navigation software, called Thinking Cap [15], accepts high-level navigation goals in the form of logical propositions, called *behavioral goals* (or B-Goals). Two distinct localization systems are available for the role of position sensor. One of them is the person tracker advertised in Figure 5. The person tracker is a people detection and tracking system [10] that uses the 3D camera of the PEIS-Home. This system is also able to track Astrid. The other position sensor is a factitious component that just transforms the value of the odometry from the Player on Astrid into the absolute coordinates of the PEIS-Home. From the functional point of view the two position sensors are equivalent: they both issue a single output of type `PlanarPose2D`.

4.2. Execution of the Experiment

Configuration: After receiving the template of Figure 7, the reactive configurator queries the ecology, and receives the advertisements of all the components available at the moment, among which are the six components discussed above. These six components are the only eligible to enter the configuration. The robot sonars, encoder, and drive match the types and properties required by the template. The Thinking Cap is advertised as a `NavigationControl` component, precisely as needed. The person tracker and the absolute odometry are `PositionSensor`. They do not expose a `Support` property, hence this is not checked for them. During configuration search, all template components but the position sensor have just one compatible profile. In the mutual compatibility tests following the various expansions of the search algorithm, the drive component is verified to be mutually compatible with the Thinking Cap. In turn, the Thinking Cap is verified to be mutually compatible with the set of components constituted by the sonars, the encoders, and the person tracker. The generated configuration features the person tracker, which was the first position sensor to be tested.

Execution: The generated configuration is deployed. The snapshot of Figure 9 was taken with a system inspection tool just after the deployment was complete. Apart from

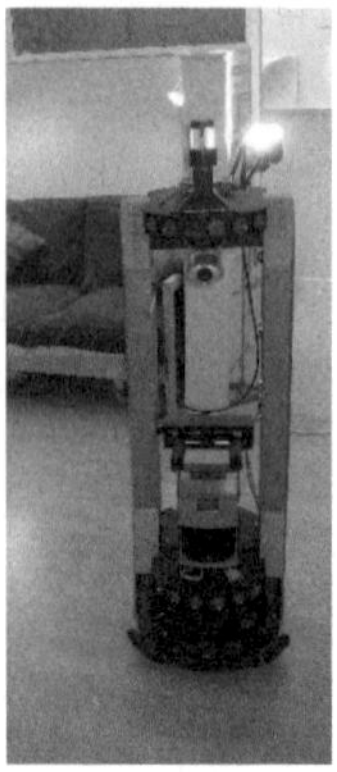

Figure 8. Astrid.

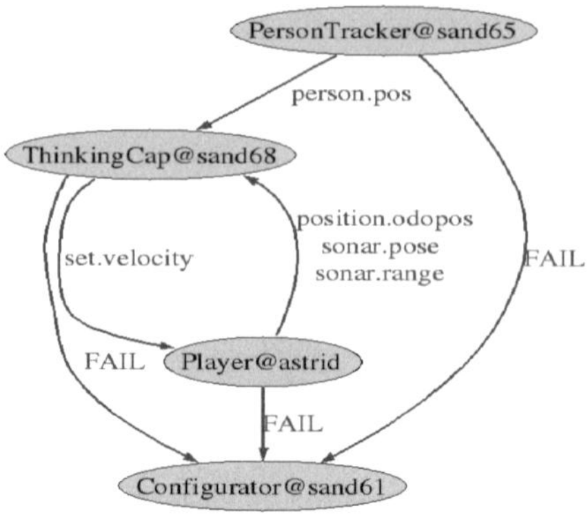

Figure 9. The deployed configuration.

Figure 10. Tracker view.

the connections between the task-related components, the snapshot shows also that the reactive configurator has connected itself to all the needed failure signals. After activating all the components and establishing all the connections, the configurator injects the B-Goal into the Thinking Cap. The Thinking Cap starts sending velocity setpoints to the drive of Astrid. Astrid starts moving. The person tracker extracts the visual signature of Astrid from the images received from the stereo camera (Figure 10), and transmits to the Thinking Cap the absolute position of Astrid in global coordinates. The navigation of Astrid proceeds smoothly until Astrid exits the field of view of the stereo camera. When that happens, the person tracker looses the robot, and raises its `FAIL` signal.

Reconfiguration: Upon receiving the failure signal, the configurator turns off the person tracker and removes its profile from the set of discovered components. Then it dismantles the deployed configuration. Astrid stops, because its drive is not receiving any more the velocity setpoint updates. The configurator starts from scratch another search for a new configuration. During the new configuration search, the configurator builds up the same associations as before, except for the position sensor, which is now associated to the absolute odometry component. As soon as the new subscriptions are deployed, Astrid resumes its course toward the bedroom.

5. Conclusions

The main contribution of this paper is to propose a task-driven, reactive approach to self-configuration for the software of distributed robot systems. We have shown the applicability of this approach to the PEIS-Ecology, but this approach should be applicable to any type of distributed robot system. The most notable benefit of our approach is to make the system more flexible and adaptive, by giving it the ability to automatically self-configure for a given task, and to automatically re-configure in case of failures.

Although other approaches have been proposed in literature for formal description and discovery of software components in distributed robotic systems [2,4,5,11], very few works exist that deal with the automatic, run-time composition of these components for a given task [9,12]. Contrary to the latter works, the approach presented in this paper is reactive. Two aspects of reactivity of our approach are the locality of configuration search in the neighborhood defined by the template, and the immediate return of the first solution found. As consequence, our approach can scale well with the size of the robot ecology, but it cannot guarantee optimality of the solution. Global planning approaches can overcome this problem, but have difficulties to scale-up. In the future, we intend to explore a hybrid global/local approach taking inspiration from distributed planning.

Our approach is inspired by ideas coming from the field of semantic web services, e.g. OWL-S. It should be emphasized, however, that one could not directly abstract components of a robot ecology as services of OWL-S. Web Services are similar to method invocations: they accept input parameters and issue a return value after a certain time. By contrast, typical robotic components are continuous processes that process and produce continuous flows of data. An important contribution of our work is to have devised a novel framework of formal descriptions tailored for components typical of robot ecologies and of distributed robotic systems of similar nature.

Acknowledgments

This work has been supported by the Swedish Research Council (Vetenskapsrådet), and by ETRI (Electronics and Telecommunications Research Institute, Korea). Our thanks to Mathias Broxvall for valuable discussions and technical help.

References

[1] T. Akimoto and N. Hagita. Introduction to a network robot system. In *Proceedings of the 2006 International Symposium on Intelligent Signal Processing and Communications*, Tottori, Japan, 2006.

[2] F. Amigoni and M. Arrigoni Neri. An application of ontology technologies to robotic agents. In *Proceedings of the International Conference on Intelligent Agent Technology*, Compiègne, France, 2005.

[3] M. Broxvall. A middleware for ecologies of robotic devices. In *Proceedings of the First International Conference on Robot Communication and Coordination*, Athens, Greece, 2007.

[4] L. Chaimowicz, A. Cowley, V. Sabella, and C. J. Taylor. ROCI: A distributed framework for multi-robot perception and control. In *Proceedings of IROS 2003*, Las Vegas, Nevada, USA, 2003.

[5] J. Fritsch, M. Kleinehagenbrock, A. Haasch, S. Wrede, and G. Sagerer. A flexible infrastructure for the development of a robot companion with extensible HRI-capabilities. In *ICRA'05*, Barcelona, 2005.

[6] B. Gerkey, R. Vaughan, K. Sty, A. Howard, G. Sukhatme, and M. Mataric. Most valuable player: A robot device server for distributed control. In *Proceedings of IROS 2001*, Wailea, Hawaii, USA, 2001.

[7] A. Kameas, I. Bellis, I. Mavrommati, K. Delaney, M. Colley, and A. Pounds-Cornish. An architecture that treats everyday objects as communicating tangible components. In *Proc. of PerCom'03*, 2003.

[8] J. Lee and H. Hashimoto. Intelligent space – concept and contents. *Advanced Robotics*, 16(3), 2002.

[9] R. Lundh, L. Karlsson, and A. Saffiotti. Plan-based configuration of an ecology of robots. In *Proceedings of the 2007 IEEE International Conference on Robotics and Automation*, Roma, Italy, 2007.

[10] R. Muñoz-Salinas, E. Aguirre, and M. García-Silvente. People detection and tracking using stereo vision and color. *Image and Vision Computing*, 25(6):995–1007, 2007.

[11] H. Noguchi, T. Mori, and T. Sato. Automatic generation and connection of program components based on rdf sensor description in network middleware. In *Proceedings of IROS 2006*, Beijing, China, 2006.

[12] L. Parker and F. Tang. Building multi-robot coalitions through automated task solution synthesis. *Proceedings of the IEEE*, 94(7):1289–1305, 2006.

[13] J. Rao and X. Su. A survey of automated web service composition methods. In *Proceedings of the ICWS'2004 International Workshop on SWSWPC*, San Diego, California, USA, 2004.

[14] A. Saffiotti and M. Broxvall. PEIS Ecologies: Ambient Intelligence meets Autonomous Robotics. In *Proceedings of Smart Objects & Ambient Intelligence (sOc-EUSAI 2005)*, Grenoble, France, 2005.

[15] A. Saffiotti, K. Konolige, and E. Ruspini. A multivalued-logic approach to integrating planning and control. *Artificial Intelligence*, 76(1-2):481–526, 1995.

[16] A. Sgorbissa and R. Zaccaria. The artificial ecosystem: a distributed approach to service robotics. In *Proceedings of ICRA'04*, New Orleans, Louisiana, USA, 2004.

[17] G. Tesauro, D. Chess, W. Walsh, R. Das, A. Segal, I. Whalley, J. Kephart, and S. White. A multi-agent systems approach to autonomic computing. In *Proceedings of AAMAS'04*, New York City, USA, 2004.

Intelligent Autonomous Systems 10
W. Burgard et al. (Eds.)
IOS Press, 2008

doi:10.3233/978-1-58603-887-8-110

Using the Body in Learning to Recognize Objects[1]

Mark EDGINGTON [a] José DE GEA [a] Jan Hendrik METZEN [b]
Yohannes KASSAHUN [a] Frank KIRCHNER [a,b]
[a] *Robotics Group, University of Bremen, Robert-Hooke-Str. 5, D-28359, Bremen, Germany*
[b] *German Research Center for Artificial Intelligence (DFKI), Robert-Hooke-Str. 5, D-28359, Bremen, Germany*

Abstract. Object recognition has traditionally been approached using primarily vision-based strategies. Recent research suggests, however, that intelligent agents use more than vision in order to comprehend and classify their environment. In this work we investigate an agent's ability to recognize objects on the basis of nonvisual proprioceptive information generated by its body. An experiment is presented in which an industrial robot collects and structures information about various objects in terms of its physical configuration. This information is then analyzed using a Bayesian model, which is used subsequently for classifying objects.

Keywords. Embodied cognition, Proprioception, Object categorization

1. Introduction

Through the years, the discipline of Artificial Intelligence (AI) has formed two major schools of thought in which the development of intelligent systems has been attempted. In the first school of thought, an approach is used in which cognitive processes are described as abstract information processing tasks [7]. In this approach, physical connections to the outside world are viewed as having a minimal role. The second, more recent, school of thought has focused on so-called "Embodied AI" approaches [13], where the features of an agent's body are given a major role in the process the agent goes through in forming concepts about its world. Researchers subscribing to the second school of thought claim that the representations used by an agent to describe its world must be based on its own abilities and experiences instead of on models provided by a human developer. An agent's attempt to recognize its environment is a typical scenario in which body-based representations can contribute to the agent's formation of concepts. We intend to explore this scenario in this paper.

The next section provides a review of works that relate to object and environment recognition. Following this, an experiment is described that investigates an agent's ability to recognize objects based on strictly proprioceptive information. The experimental results and repeatability of the methods in this experiment are presented. Finally, conclusions based on these results, and possible future directions to explore are discussed.

[1]This work is supported by the SFB/TR 8 DFG project, which is duly acknowledged.

2. Review of Related Work

Any agent in the real world needs the ability to distinguish between different types of objects. In other words, it must possess the ability to recognize objects. Object recognition is one of the most important, yet least understood, aspects of perception [16]. For many biological systems, the recognition and classification of objects is a spontaneous and natural activity, requiring virtually no conscious effort. In contrast, recognizing common objects is still a very difficult task for today's artificial recognition systems. Most of the research on object recognition has been conducted in the information processing framework under the umbrella term "machine vision". The intent of this section is to provide a general review of object recognition work done in the machine vision community, as well as a review of work done in the embodied cognition community.

There are various approaches to object recognition that have been proposed in the machine vision community. One can divide them, in general, into two groups: those using correspondence (feature matching), and those not based on correspondence (global matching). Global matching approaches involve finding a transformation that fits a model to an image without first determining the correspondence between individual parts or features of the model and the image data. Several of these approaches base themselves on parameters such as moments of inertia, eigenspace representations of an object's appearance, Fourier or other spatial frequency descriptions, tensor measures, etc. Representative examples include the works of Hu [5], Murase and Nayar [9], and Zahn and Roskies [17]. These methods are efficient, but are sensitive to occlusion, because they assume that an object of interest can be segmented from the rest of an image [3]. On the other hand, feature matching procedures try to find the correspondence between the local features of a model and the image data, and to then determine, for a given correspondence, the transformation between the model and the image data. The most commonly used local feature is the Scale-Invariant Feature Transform (SIFT) [8], which provides the distinctive local features of an object in an image. These methods are robust against occlusion, but they are less efficient than global matching procedures, because they have to solve the correspondence problem for every model that is assumed [3,16]. In general, machine vision approaches often suffer from the fact that they require restrictive conditions for illumination, backgrounds, and object pose. However, methods that involve the intentional physical interaction of an agent with its environment, where the information the agent receives is coordinated, are typically less susceptible to such restrictions.

Although learning to recognize objects is a difficult problem when viewed in terms of information processing, it is significantly simplified when viewed as a problem of sensorimotor coordination: embodied agents can structure their own sensory input, and thereby introduce regularities that significantly simplify learning [14]. For example, when a human grasps a cup, the cup can be easily brought into the range of the visual system due to the natural positions a human's arm can assume. Grasping can therefore be considered to be a process of sensorimotor coordination. Sensory stimulation generated in this way results in correlations within and between sensory modalities, which is a prerequisite for developing higher levels of cognition.

In recent years, the concept of sensorimotor coordination has gained increasing attention in artificial intelligence communities [11]. It has, among other things, been shown that sensorimotor coordination can be exploited to solve classification problems [1,2,6,10,15]. Classifying an object in this manner often involves the manipulation of objects in an environment. It is important to note that this process exploits not only the motor actions of the body, but also the intrinsic structure of the body itself. The bodily structure of a being or machine imposes certain constraints on the sorts of motions that can be performed. It provides natural configurations, in which energy consumption and stress are minimized, and thus is an important part of structuring the coordination of sensorimotor information. Seen from a different perspective, the constraints which a body's structure imposes are a form of *morphological computation* [12], in which the sensory data is preprocessed and, in a sense, "filtered" by the body's morphology.

3. Experiment

In order to understand the significance that an agent's body has in helping it learn about its environment, an experiment was designed in which an agent attempts to classify objects based strictly on the proprioceptive information it receives while interacting with them. The experiment described here demonstrates the feasibility of using proprioceptive information as a primary modality in a multi-modal object recognition system.

Figure 1. Equipment used in the experiment: (a) Mitsubishi PA-10 Industrial Robot, with (b) an internal view and (c) assembled view of the force sensor that we developed, which includes an integrated electromagnet.

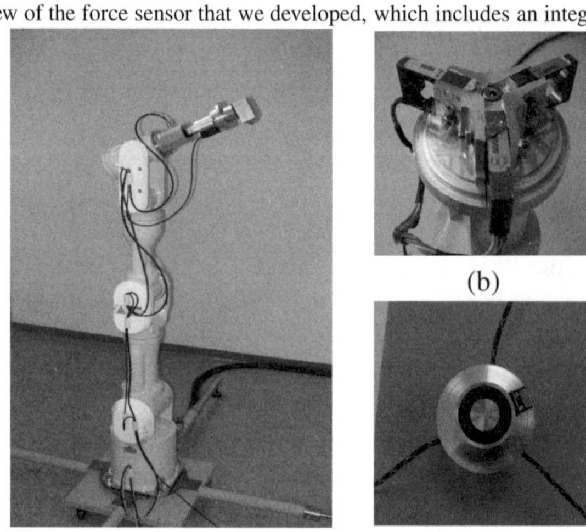

(a) (c)

3.1. Setup

The experiment makes use of a 7 degree-of-freedom Mitsubishi PA-10 industrial robotic arm, whose end-effector consists of a custom built 3-dimensional force-sensor connected with an electromagnet (both are pictured in Figure 1). The force-sensor is capable of measuring the forces exerted in all directions by an object. It consists of two interlocking pieces connected by three single-axis force sensors (the sensors are positioned 120° from each other). One piece is held fixed by the robot end-effector, while the other piece has an electromagnet attached to it. This integrated electromagnet is used by the agent to "grasp" one of several objects (the grasp-point of a particular object is fixed). Because

Figure 2. The objects used in the experiment. Objects (e) and (f) are alternate views of objects (b) and (c), respectively.

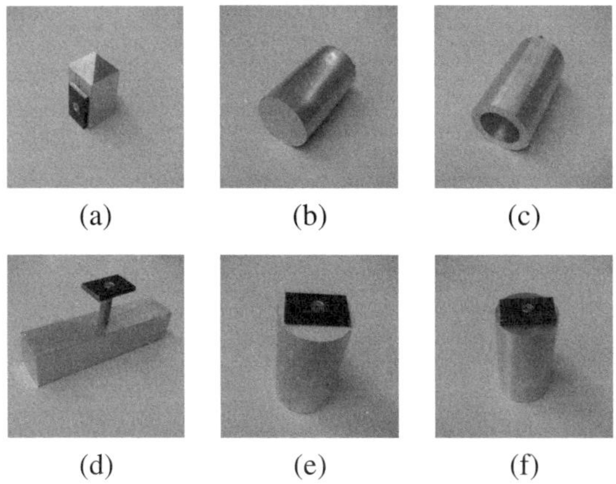

of the force sensor's geometric arrangement, each possible force (which does not exceed a predetermined maximum) that an object exerts (e. g. due to gravity and/or momentum) is uniquely mapped to the set of force-signal tuples that can be read from the single-axis force sensors. Due to the limited precision of its construction, the sensor provides inherently noisy data.

The grasped objects (see Figure 2) are of varying shapes and sizes. Some of the objects differ in shape, but have an identical mass and center-of-mass (Figure 2b and 2d), whereas other objects differ in mass, but appear identical from most vantage points (Figure 2e and 2f). In contrast to research done in the area of haptic sensing, where it is typical that only the static external properties of an object are sensed (for example, see [4]), the approach presented here senses the dynamic properties of objects.

3.2. Procedure

The experiment proceeds as follows: (1) The agent grasps an unknown object, and brings it to a predefined position in space. It then chooses one of three pre-programmed manipulation acts to perform on the object. A manipulation act is defined as a sequence of poses which the end-effector should assume over a given period of time, where pose is defined in terms of a (θ, ϕ) angle pair. These angles, as well as the different manipulation acts used in this experiment, are shown in Figure 3. (2) While a manipulation act is being performed, the pose of the end-effector and the force signals generated by the object being manipulated are recorded at various points along the action-path. The recorded force signals do not directly represent orthogonal forces in the Cartesian coordinate system, but instead form a basis of a space for which there is a bijective linear transformation to 3D Cartesian space which depends on the geometry of the force-sensor unit. (3) These pose-force data points are analyzed using a Bayesian model which generates and updates the conditional probabilities that the robot is manipulating a specific object, given these data points and a specific set of poses (a manipulation act) in the pose-space (see Section 3.3). This model is used to classify the object being manipulated.

Figure 3. Schematic depiction of manipulation acts: (a) moving the grasped object from $\theta = 0°$ to $\theta = 90°$, (b) the same, except with the object rotated by $\phi = 90°$, and (c) rotating the object 360° while $\theta = 90°$.

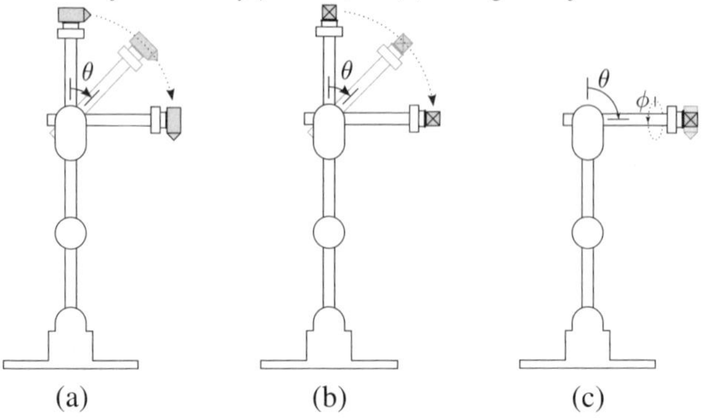

3.3. Data Analysis

We now present a formal description of the analysis technique used for classifying objects based on the measured proprioceptive data. Let $MA = \{ma_1, ma_2, \ldots, ma_N\}$ be a set of predefined *manipulation acts*, and let $O = \{o_1, o_2, \ldots, o_M\}$ be a set of objects to be recognized. Assuming that we do not have a preference of choosing a given manipulation act ma_i over the others for manipulating a given object, then the prior probability of choosing a specific manipulation act is $p(ma_i) = \frac{1}{N}, \forall\, ma_i \in MA$. Likewise, assuming there is no preference with regard to which object will be manipulated, one can write the prior probability of selecting a given object as $p(o_j) = \frac{1}{M}, \forall\, o_j \in O$. Since choosing a manipulation act and choosing an object to manipulate are independent of one another, the joint probability of choosing an object o_j and a manipulation act ma_i is given by

$$p(o_j, ma_i) = p(o_j)p(ma_i) = \frac{1}{NM}, \tag{1}$$

and the posterior probability of an object o_j given a manipulation act ma_i and a sensory data vector $\vec{x}$ (defined below) is given by

$$\begin{aligned} p(o_j|\vec{x}, ma_i) &= \frac{p(o_j, \vec{x}, ma_i)}{p(\vec{x}, ma_i)} = \frac{p(\vec{x}|o_j, ma_i)p(o_j, ma_i)}{\sum_{k=1}^{M} p(\vec{x}, o_k, ma_i)} \\ &= \frac{p(\vec{x}|o_j, ma_i)p(o_j, ma_i)}{\sum_{k=1}^{M} p(\vec{x}|o_k, ma_i)p(o_k, ma_i)}. \end{aligned} \tag{2}$$

Because $p(o_w, ma_z)$ is a constant $\frac{1}{NM}$ for all w, z (i. e., regardless of which object or manipulation act is chosen), (2) can be simplified as follows:

$$p(o_j|\vec{x}, ma_i) = \frac{p(\vec{x}|o_j, ma_i)\frac{1}{NM}}{\frac{1}{NM}\sum_{k=1}^{M} p(\vec{x}|o_k, ma_i)} = \frac{p(\vec{x}|o_j, ma_i)}{\sum_{k=1}^{M} p(\vec{x}|o_k, ma_i)}. \tag{3}$$

The sensory data vector $\vec{x}$ used in the above equation contains the readings of the three sensors at each of several poses along a manipulation path. Assuming that we have q poses at which data is recorded, one can write the sensory data vector in the form $\vec{x} = [x_{s_1}(1), \ldots, x_{s_1}(q), x_{s_2}(1), \ldots, x_{s_2}(q), x_{s_3}(1), \ldots, x_{s_3}(q)]^T$, where s_1, s_2 and s_3

refer to the three single-axis force sensors within the 3D force sensor. The likelihood $p(\vec{x}|o_j, ma_i)$ can be approximated using the multivariate normal density function in the form of

$$p(\vec{x}|o_j, ma_i) = \frac{1}{\sqrt{(2\pi\sigma_{ji}^2)}} \exp\left\{-\frac{1}{2\sigma_{ji}^2} \parallel \vec{x} - \vec{\mu}_{ji} \parallel^2\right\}. \tag{4}$$

This density function assumes that the covariance matrix contains no nonzero off-diagonal elements, and each of its diagonal elements is σ_{ji}^2. The elements of the mean vector $\vec{\mu}_{ji}$ and the variance σ_{ji}^2 *are determined from the sensory data collected during the training session.* If we represent the $\vec{x}$ data vectors taken during training with object o_j and manipulation act ma_i as the random variable $\vec{X}_{ji}$, then we can write

$$\vec{\mu}_{ji} = E[\vec{X}_{ji}]. \tag{5}$$

Likewise, we can write $\sigma_{ji,w}^2 = E[(\vec{X}_{ji}(w) - \vec{\mu}_{ji}(w))^2]$, where w indicates the index of one of $\vec{X}_{ji}$'s components. Finally, the value σ_{ji}^2 used in (4) can be written as

$$\sigma_{ji}^2 = \max_w \sigma_{ji,w}^2 . \tag{6}$$

Equation (3) gives the probability that we are manipulating an object with a given manipulation act *after we have observed the sensory data vector* $\vec{x}$. We use the maximum a posteriori probability (MAP) criterion in order to associate $\vec{x}$ with a specific object. A sensory data vector $\vec{x}$ for a given manipulation act ma_i is considered to have been produced by an object o_k if

$$k = \underset{j}{\operatorname{argmax}}\ p(o_j|\vec{x}, ma_i). \tag{7}$$

Using this criterion minimizes the likelihood of misclassification when assigning a sensory data vector $\vec{x}$ to an object o_j. Additionally, a discriminating power factor D is calculated as follows:

$$D(\vec{x}, ma_i) = \ln \frac{p(o_k|\vec{x}, ma_i)}{\underset{l \neq k}{\operatorname{argmax}}\ p(o_l|\vec{x}, ma_i)}, \tag{8}$$

where k is the value defined in (7). Intuitively, this factor is a measure of the distance that $\vec{x}$ is away from the decision boundary. It serves as a rough figure of merit estimating the confidence one can place in the classification results obtained with a given manipulation act.

4. Results

The experiment was performed a total of 60 times with each object. A stratified 10-fold cross-validation was performed on the resulting set of sample measurements, splitting them into training and validation sets. Each training set was used to generate the likelihood probability densities $p(\vec{x}|o_j, ma_i)$. These probability densities were used in determining the posterior probability of an object given the two variables that are known to the agent, namely the sensory data vector and the manipulation act. For each manipulation act, 5 poses were used to form the sensory data vectors.

4.1. Repeatability of Measurements

The repeatability of the measurements obtained from the force-sensor while manipulating an object influences the quality and resolution of the recognition. As the variance of the measurements from the mean decreases, the recognition quality and resolution increases. A lower measurement variance results in a minimal overlap of measurements for different objects and manipulation acts.

Figure 4 shows the repeatability of measurements taken over 30 trials for each combination of object and manipulation act. The average sensor data is plotted, along with its standard-deviation. As can be seen from this figure, the error bars for the different objects do not significantly overlap, and thus the repeatability is sufficient for recognizing the objects shown in Figure 2. The relatively small standard-deviations of these measurements may account for the cases in which there were no misclassifications in Table 1. Although the measurement error bars do overlap slightly for certain components of $\vec{x}$, for each pair of objects there is at least one component of $\vec{x}$ for manipulation act 3 in which they do not overlap. Because of this, it is expected that a learning method exploiting manipulation act 3 should be able to distinguish arbitrary pairs of objects from each other. For manipulation acts 1 and 2, objects B and D overlap significantly, and thus cannot be completely distinguished from each other unless data from manipulation act 3 is considered. This demonstrates the importance of using several different manipulation acts: where one fails to distinguish objects from each other, another will likely succeed.

4.2. Classification Rates

The validation data sets were used to calculate the posterior probabilities of each possible object, which in turn were used to identify which object was being manipulated. Table 1 shows the average correct-classification rate and discriminating power over 10 runs of a stratified 10-fold cross-validation (i. e. a total of 1000 classifications), for each combination of object and manipulation act. Manipulation act 3 yielded better results than acts 1 or 2, as expected due to the differences in the sensor measurements depicted in Figure 4. Particularly interesting is the fact that objects with identical shapes but differing weights (Figure 2e and 2f) are correctly distinguished from each other. If a strictly

Table 1. Mean classification rates (in percent) for each combination of manipulation-act and object. A stratified 10-fold cross-validation was performed 10 times. The mean discriminating power is shown in parentheses.

	Manipulation Act 1	Manipulation Act 2	Manipulation Act 3
Object A	100.0 (7.33)	92.8 (5.13)	100.0 (38.27)
Object B	63.3 (0.99)	55.0 (1.16)	99.8 (2.48)
Object C	100.0 (4.52)	89.2 (5.32)	100.0 (13.11)
Object D	81.0 (1.77)	97.5 (1.95)	100.0 (3.63)

Figure 4. Plots showing the mean and standard deviation of the sensory data vector $\vec{x}$ for different objects and manipulation acts. The standard deviation is depicted by the error bars.

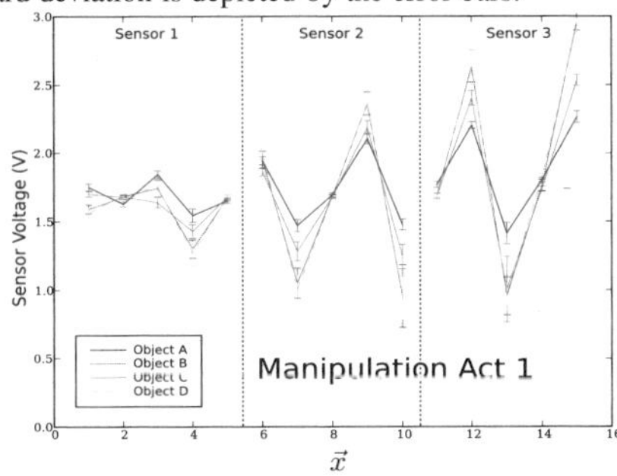

(a)

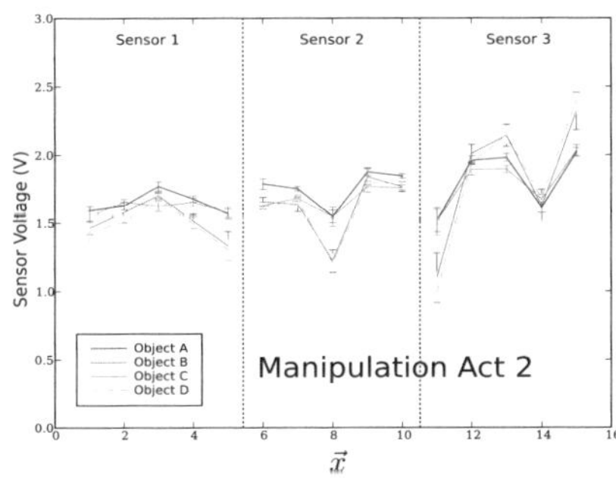

(b)

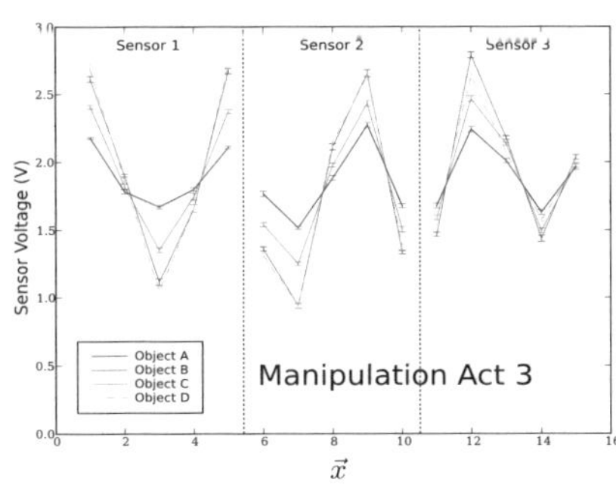

(c)

vision-based classification system were used for classifying objects, it might experience difficulty in distinguishing such objects from each other. On the other hand, one might mistakenly assume that proprioceptive data alone would not provide enough information to distinguish between objects with the same weight and center-of-mass (Figure 2b and 2d), but these are also correctly distinguished from each other. This can be accounted for by the fact that while performing a manipulation act on an object, the end effector doesn't necessarily record only static properties of the object, but is also able to capture the dynamic properties of the object, such as its moment of inertia.

5. Conclusions and Outlook

Based on the presented experimental results, one can see that an agent's knowledge of the state of its body with respect to additional sensory information it receives is central in helping the agent to recognize its environment, because it gives the agent a means of structuring the information in its environment. It therefore seems reasonable to conclude that information which has been structured according to proprioceptive information received from the body, when joined with other modes of sensory input, would serve to make existing single-modality object recognition systems more robust. In the future, this method can be extended to allow an agent to decide which combination of acts are optimal (i. e. requiring the least energy, or the fewest number of acts) for it to reach a high-confidence in correctly judging its environment. Additionally, it would be interesting to combine unsupervised learning approaches with the supervised approach we have presented.

References

[1] R. D. Beer. Towards the evolution of dynamical neural networks for minimally cognitive behavior. In *Proceedings of the 4th International Conference on Simulation of Adaptive Behavior*, pages 421–429, 1996.

[2] G. E. Edelman. *Neural Darwinism: The Theory of neuronal Group Selection.* Basic Books, New York, 1987.

[3] W. E. L. Grimson. *Object Recognition by the Computer.* MIT Press, Massachusetts, London, 1990.

[4] G. Heidemann and M. Schöpfer. Dynamic tactile sensing for object identification. In *Proceedings of the IEEE International Conference on Robotics and Automation (ICRA)*, pages 813–818, 2004.

[5] M. K. Hu. Visual pattern recognition by moment invariants. *IEEE Transactions on Information Theory*, 8:179–187, 1962.

[6] Y. Kassahun, M. Edgington, J. de Gea, and F. Kirchner. Exploiting sensorimotor coordination for learning to recognize objects. In *Proceedings of the 20th International Joint Conference on Artificial Intelligence (IJCAI-2007)*, pages 883–888, 2007.

[7] J. L. Lachman and E. C. Butterfield. *Cognitive Psychology and Information Processing.* Erlbaum, Hillsdalle, NJ, 1979.

[8] D. G. Lowe. Distinctive image features from scale-invariant keypoints. *International Journal of Computer Vision*, 60(2):91–110, 2004.

[9] H. Murase and S. Nayar. Visual learning and recognition of 3D objects from appearance. *International Journal of Computer Vision*, 14:5–24, 1995.

[10] S. Nolfi. Adaptation as a more powerful tool than decomposition and integration. In *Proceedings of the Workshop on Evolutionary Computing and Machine Learning*, 1996.

[11] S. Nolfi and D. Parisi. Exploiting the power of sensory-motor coordination. In F. M. D. Floreano, J-D. Nicoud, editor, *Advances in Artificial Life, Proceedings of the Fifth European Conference on Artificial Life (ECAL)*, pages 173–182. Springer Verlag, 1999.

[12] C. Paul. Morphology and computation. In *Proc. Int. Conf. on Simulation of Adaptive Behavior*, pages 33–38. MIT Press, 2004.

[13] R. Pfeifer and J. Bongard. *How the Body Shapes the Way We Think: A New View of Intelligence.* MIT Press, Massachusetts, London, 2006.

[14] R. Pfeifer and C. Scheier. *Understanding Intelligence.* MIT Press, Massachusetts, London, 1999.

[15] M. Takác. Categorization by sensory-motor interaction in artificial agents. In *Proceedings of the 7th International Conference on Cognitive Modeling*, pages 310–315, 2006.

[16] S. Ullman. *High-level Vision.* MIT Press, Massachusetts, London, 1996.

[17] C. Zahn and R. Roskies. Fourier descriptors for plane closed curves. *IEEE Transactions on Computers*, 21(3):269–281, 1972.

Intelligent Autonomous Systems 10
W. Burgard et al. (Eds.)
IOS Press, 2008

doi:10.3233/978-1-58603-887-8-119

Persistent Point Feature Histograms for 3D Point Clouds

Radu Bogdan Rusu, Zoltan Csaba Marton, Nico Blodow and Michael Beetz
Intelligent Autonomous Systems, Technische Universität München
Boltzmannstr. 3, Garching bei München, 85748

Abstract.
This paper proposes a novel way of characterizing the local geometry of 3D points, using persistent feature histograms. The relationships between the neighbors of a point are analyzed and the resulted values are stored in a 16-bin histogram. The histograms are pose and point cloud density invariant and cope well with noisy datasets. We show that geometric primitives have unique signatures in this feature space, preserved even in the presence of additive noise. To extract a compact subset of points which characterizes a point cloud dataset, we perform an in-depth analysis of all point feature histograms using different distance metrics. Preliminary results show that point clouds can be roughly segmented based on the uniqueness of geometric primitives feature histograms. We validate our approach on datasets acquired from laser sensors in indoor (kitchen) environments.

Keywords. persistent feature histograms, point clouds, geometric reasoning

1. Introduction

Understanding a scene represented by point clouds can not be done directly and solely on the points' 3D coordinates. In particular, geometrical reasoning techniques can profit from compact, more informative point features, that represent the dataset better.

Estimated surface curvatures and normals for a point [1] are two of the most widely used point features, and play an important role in applications such as registration [2], or segmentation [3]. Both of them are considered *local features*, as they characterize a point using the information provided by the k closest neighbors of the point. Their values however, are highly sensitive to sensor noise and the selection of the k neighbors (i.e. if the k-neighborhood includes outliers, the estimation of the features will become erroneous). Robust feature descriptors such as moment invariants [4], spherical harmonic invariants [5], and integral volume descriptors [6] have been proposed as point features and used for registering partial scans of a model [7,6]. All of them are invariant to translation and 3D rotations, but are still sensitive to noise.

In general it is not clear how one should select the optimal k support for a point when computing any of the above mentioned features. If the data is highly contaminated with noise, selecting a small k will lead to large errors in the feature estimation. However, if k is too big, small details will be suppressed. Recently, work has been done on automatic computation of good k values (i.e. *scale*) for normal estimation on 3D point cloud data [8,9] as well as principal curvatures [10,11,12] on multiple scales. Unfortunately, some of the above mentioned methods for computing an optimal scale require additional thresholds, such as $d1$ and $d2$ which are determined empirically in [8], and estimated using linear least-squares in [9] when knowledge about ground truth normal exists. In [10] the neighborhood is grown incrementally until a jump occurs in the variation-scale curve, but the method cannot be successfully applied to noisy point clouds, as the variations in the surface curvature are not modified smoothly with k. The selection of the T_c threshold in [11] is not intuitive, and the authors do not explain properly if the resulted persistent features are obtained using solely the intersection of features computed over different radii. The statistical estimation of curvatures in [12] uses a M-estimation framework to reject noise and outliers in the data and samples normal variations in an adaptively reweighted neighborhood, but it is unfortunately slow for large datasets, requiring approximately 20 minutes for about 10^6 points.

While the above mentioned descriptors can be considered good point features for some problems, they do not always represent enough information for characterizing a point, in the sense that they approximate a k-neighborhood of a point p with a single value. As a direct consequence, most scenes will contain many points with the same or very similar feature values, thus reducing their informative characteristics. Even if the feature estimation would be able to cope with noisy datasets, it can still be easily deducted that applications who rely on these 1D features will deal with multiple and false correspondences and will be prone to failure (e.g. registration). Alternatively, multiple-value point features such as curvature maps [13], or spin images [14], are some of the better local characterizations proposed for 3D meshes which got adopted for point cloud data. However, these representations require densely sampled data are not able to deal with the amount of noise usually present in 2½D scans.

The 3D object recognition community has developed different methods for computing multi-value features which describe complete models for classification: curvature based histograms [15], spin image signatures [16], or surflet-pair-relation histograms [17]. All of them are based on the local estimation of surface normals and curvatures and describe the relationships between them by binning similar values into a global histogram. A high number of histograms per object is required by [15], but the method can cope with up to 20% occlusions. The 4D geometrical features used in [17] and the spin image signatures in [16] need a single histogram and achieve recognition rates over 90% with synthetic and CAD model datasets, and over 80% with added uniformly distributed noise levels below 1% [17]. All of the above show promising results, but since they have only been tested against synthetic range images, it's still unclear how they will perform when used on noisier real-world datasets.

We extend the work presented in [17] by computing local point feature histograms in 16D for each point in the cloud. We make an in-depth analysis of the points' signatures for different geometric primitives (i.e. planes, sphere, cylinders, etc), and reduce the theoretical computational complexity of the algorithm by a factor of approximately 2. The uniqueness of a feature point is analyzed by discretely sampling over an interval of sphere radii (for k-neighborhood selection). We statistically analyze different distance metrics between each point's histogram signature and the mean histogram of the cloud (*μ-histogram*), and select the points outside the $\mu \pm \alpha \cdot \sigma$ interval as persistent features.

Furthermore, we show that: a) our point feature histograms are: (i) robust in the presence of outliers and noisy data; (ii) pose and scale invariant; (iii) consistent over different sampling densities in separate scans; and b) coupling them with persistence analysis yields accurate, and informative salient point features.

The remainder of the paper is organized as follows. Section 2 presents our implementation for computing point feature histograms, while Section 3 analyzes the histograms persistence over a spatial domain. We discuss experimental results in Section 4, and conclude with Section 5.

2. Point Feature Histograms

A problem that arises in point correspondence searches, is that the features usually used (e.g. surface curvature changes or integral volume descriptors) do not fully represent the underlying surface on which the point's neighborhood is positioned.

In order to efficiently obtain informative features, we propose the computation and usage of a histogram of values [18] which encodes the neighborhood's geometrical properties much better, and provides an overall point density and pose invariant multi-value feature. The histogram generalizes the mean surface curvature at a point p.

The input data consists of 3D $\{x, y, z\}$ point coordinates. For a given radius r, the algorithm will first estimate the surface normals at each point p by performing Principal Component Analysis (PCA) on the k-neighborhood defined by a sphere centered at p with radius r. Once the normals are obtained and consistently re-oriented(see [19] for a general algorithm for consistent normal orientation propagation), the histogram for p will be computed using the four geometric features as proposed in [17]. For every pair of points p_i and p_j ($i \neq j$, $j < i$) in the k-neighborhood of p, and their estimated normals n_i and n_j, we select a source p_s and target p_t point, the source being the one having the smaller angle between the associated normal and the line connecting the points:

$$\text{if } acos(\langle n_i, p_j - p_i\rangle) \leq acos(\langle n_j, p_i - p_j\rangle) \;\; p_s = p_i, p_t = p_j \text{ else } p_s = p_j, p_t = p_i$$

and then define the Darboux frame with the origin in the source point as: $u = n_s$, $v = (p_t - p_s) \times u$, and $w = u \times v$.

The four features are categorized using a 16-bin histogram, where each bin at index idx contains the percentage (of k) of the source points in the k-neighborhood which have their features in the interval defined by idx:

$$\left.\begin{aligned} f_1 &= v \cdot n_t \\ f_2 &= ||p_t - p_s|| \\ f_3 &= u \cdot (p_t - p_s)/f_2 \\ f_4 &= atan(w \cdot n_t, u \cdot n_t) \end{aligned}\right\} \Rightarrow idx = \sum_{i=1}^{i \leq 4} step(s_i, f_i) \cdot 2^{i-1}$$

where $step(s, f)$ is defined as 0 if $f < s$ and 1 otherwise. This means that by setting s_i to the center of the definition interval of f_i (i.e. 0 for features f_1, f_3, f_4 and r for f_2) the algorithm classifies each feature of a $\{p_i, p_j\}$ pair in p's vicinity in two categories, and save the percentage of pairs which have the same category for all features.

The four features are a measure of the angles between the points' normals and the distance vector between them. Because f_1 and f_3 are dot products between normalized vectors, they are in fact the cosine of the angles between the 3D vectors, thus their value is between ± 1, and 0 if they are perpendicular. Similarly, f_4 is the arctangent of the angle that n_t forms with w if projected on the plane defined by $u = n_t$ and w, so its value is between $\pm\pi/2$, and 0 if they are parallel.

The number of histogram bins that can be formed using these four geometric features is div^4, where div is the number of subdivisions of the features' value range. In our implementation, by dividing the feature values in two parts (f_i smaller or greater than s_i), we obtain a total of $2^4 = 16$ bins as the total number of combinations between the four features. Because the number of bins increases exponentially by the power of 4, using more than two subdivisions would result in a large number of extra dimensions for each point (e.g. $3^4 = 81$D), which makes the computational problem intractable.

Figure 1 illustrates the differences using our proposed 16D feature set between query points located on various geometric surfaces. The surfaces were synthetically generated to have similar scales, densities, and noise levels as our input real-world datasets. For each of the mentioned surfaces, a point was selected such that it lies: (i) on a plane, (ii) on the middle of an edge of a cube (normals pointing both inside and outside), (iii) on the lateral surface of a cylinder at half the height, and (iv) on a sphere. The 16D feature histogram was generated using all its neighbors inside a sphere with radius $r = 2cm$. The results show that the different geometrical properties of each surface produce unique signatures in the feature histograms space.

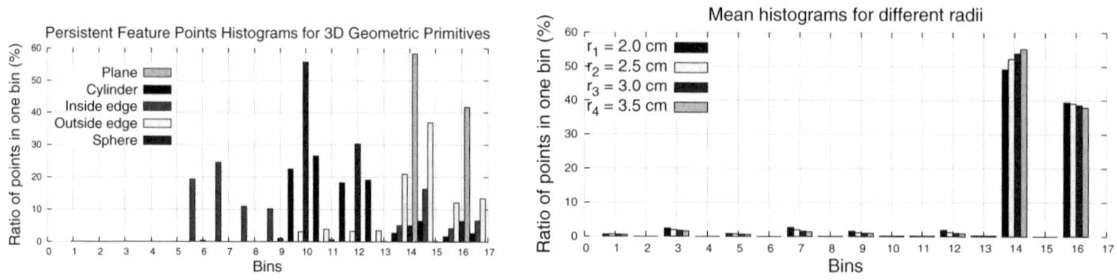

Figure 1. Feature Histograms for query points located on different geometric surfaces (left). Mean feature histograms over different radii for the kitchen scene (right).

Because of their properties, point feature histograms are promising to be more suitable candidates for problems like correspondence search while registering multiple scans under the same model. Figure 5 presents corresponding histogram features for similar points in two different overlapping point cloud datasets.

In the k-neighborhood around point p, for each pair of points a source is uniquely defined. Thus the computational complexity for each point changes from $O(k^2)$ to $O(k\cdot(k-1)/2)$, because of the above mentioned restrictions (i.e. $i \neq j$ and $j < i$).

3. Analyzing Feature Persistence

When characterizing a point cloud using point features, a compact subset of points P_f has to be found. The lesser the number of feature points and the better they approximate the data, the more efficient is the subsequent interpretation process. However, choosing the subset P_f is not easy, as it relies on a double dependency: both the number of neighbors k and the point cloud density φ. Our feature persistence analysis computes the subset of points P_f, that minimizes the number of points considered for further analysis from the input data set. Corresponding points in different point cloud views of a scene will be likely to be found as persistent features in both scans, which helps in registration but also for segmenting similar points into regions.

In order to select the best feature points for a given cloud, we analyze the neighborhood of each point p multiple times, by enclosing the point on a sphere S with radius r_i and the point p as its center. We vary r over an interval depending on the point cloud size and density, and compute the local point feature histograms for every point. We then compute the mean of the feature histograms of all the points in the cloud (*μ-histogram*). By comparing the feature histogram of each point against the *μ-histogram* using a distance metric (see below), and building a distribution of distances, we can perform a statistical analysis of each feature's uniqueness over multiple radii. More specifically, we select the set of points (P_{f_i}) whose feature distances are outside the interval $\mu \pm \alpha \cdot \sigma$, as unique features. We do this for every r and at the end, select the persistent features which are unique in both r_i and r_{i+1}, that is:

$$P_f = \bigcup_{i=1}^{n-1} [P_{f_i} \cap P_{f_{i+1}}]$$

For comparing the point feature histograms with the *μ-histogram* of the cloud, we have performed an in-depth analysis using various distance metrics from literature (see Table 1), similar to [17,15]. The symbols p_i^f and μ_i represent the point feature histogram at bin i and the mean histogram of the entire dataset at bin i respectively.

Two of the most used distances in Euclidean spaces are the Manhattan (L1) and Euclidean (L2) norms, particularizations of the Minkowski p-distance:

$$\text{Manhattan (L1)} = \textstyle\sum_{i=1}^{16} |p_i^f - \mu_i| \qquad \text{Euclidean (L2)} = \sqrt{\textstyle\sum_{i=1}^{16} (p_i^f - \mu_i)^2}$$

The Jeffries-Matusita (JM) metric (also known as Hellinger distance) is similar to the L2 (Euclidean) norm, but more sensitive to differences in smaller bins [20]:

$$\text{Jeffries-Matusita (JM)} = \sqrt{\textstyle\sum_{i=1}^{16} (\sqrt{p_i^f} - \sqrt{\mu_i})^2}$$

The Bhattacharyya distance is widely used in statistics to measure the statistical separability of spectral classes:

$$\text{Bhattacharyya (B)} = -ln \textstyle\sum_{i=1}^{16} \sqrt{p_i^f - \mu_i}$$

And finally two of the most popular measures for histogram matching in literature, the Chi-Square (χ^2) divergence and the Kullback-Leibler (KL) divergence:

$$\text{Chi-Square } (\chi^2) = \textstyle\sum_{i=1}^{16} \frac{(p_i^f - \mu_i)^2}{p_i^f + \mu_i} \qquad \text{KL divergence} = \textstyle\sum_{i=1}^{16} (p_i^f - \mu_i) \cdot ln \frac{p_i^f}{\mu_i}$$

The values of the r_i radii set are selected based on dimensionality of the features that need to be detected. Because small fine details are needed in our work at the expense of more data, (i.e. gaps between cupboard doors), we have selected $r_{min} = 2.0cm$ and $r_{max} = 3.5cm$. For our examples we fixed the value of α to 1, as only around $10 - 20\%$ of the points are outside the $\mu \pm \sigma$ interval for different radii (see Figure 2 for an example), thus selecting them as unique in the respective radius. By modifying the value of α one can roughly influence the number of persistent feature points resulting from the intersection and reunion operations. Figure 1 (right) shows the mean *μ-histograms* of the dataset for each radius. Notice how the resulting histograms are similar, and also very similar to the histogram of a point on a plane, telling us that most points in the dataset lie on planar surfaces. Because the deviation from the mean is small, selecting a small value of α is enough for identifying interesting points in the scan, as shown in Figure 5.

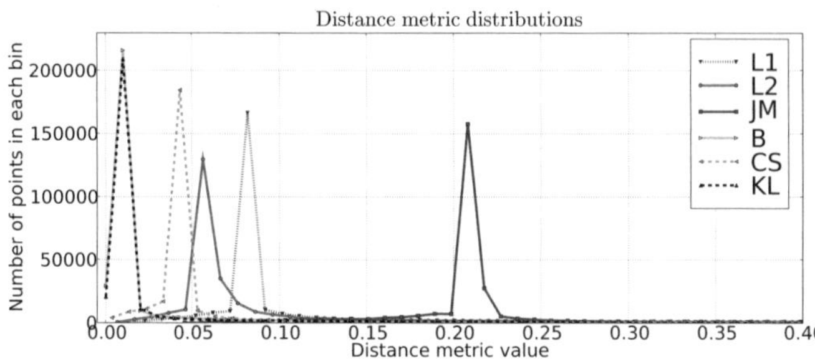

Figure 2. Distribution of feature histograms computed with different distance metrics for $r = 3cm$.

Table 1.: Analyzing the persistence of histogram features in a point cloud for 4 different radii (r_1 - red, r_2 - green, r_3 - blue, r_4 - black) on the left, and their appropriate distance graphs (for a narrow range of points for visualization purposes) on the right.

Metric	**Unique features for each r**	**Distances from *μ-histogram***
Manhattan (L1)		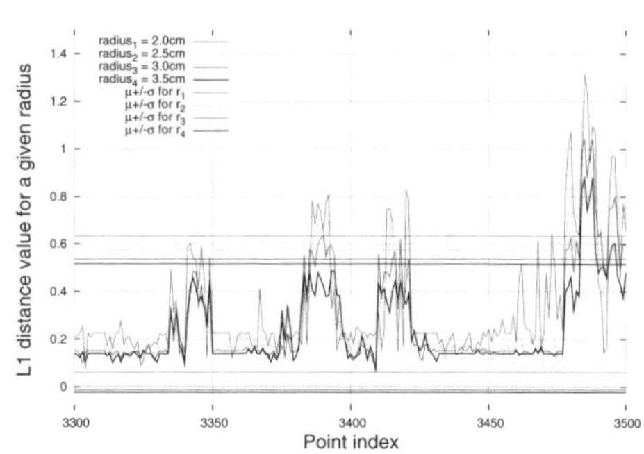
Euclidean (L2)	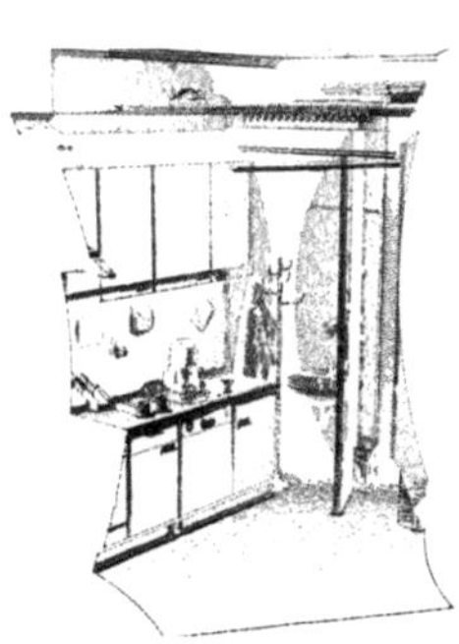	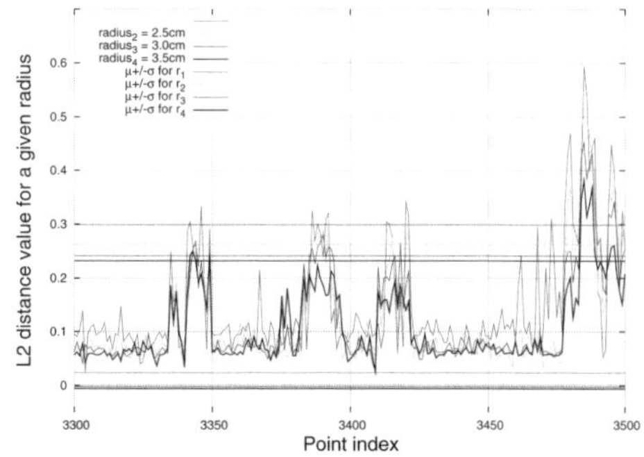

Continued on next page

Table 1 – continued from previous page

<table>
<tr><th>Metric</th><th>Unique features for each r</th><th>Distances from μ-histogram</th></tr>
<tr><td>Jeffries-Matusita (JM)</td><td>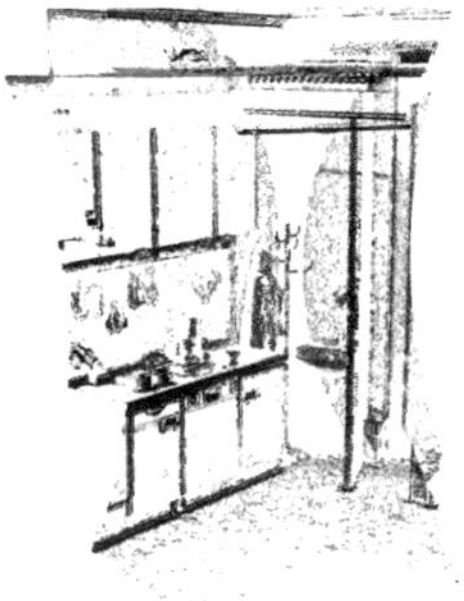</td><td>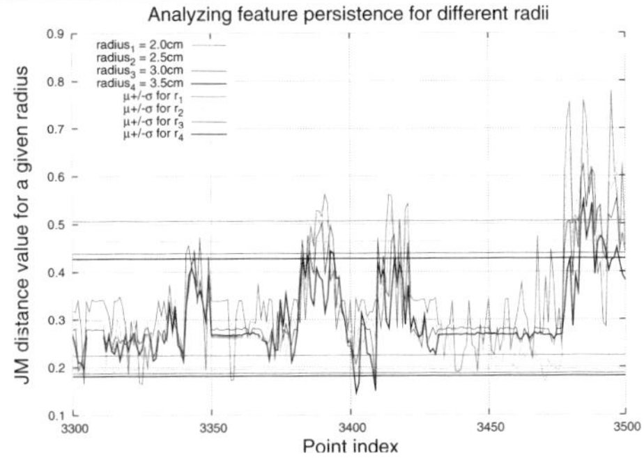
</td></tr>
<tr><td>Bhattacharyya</td><td></td><td>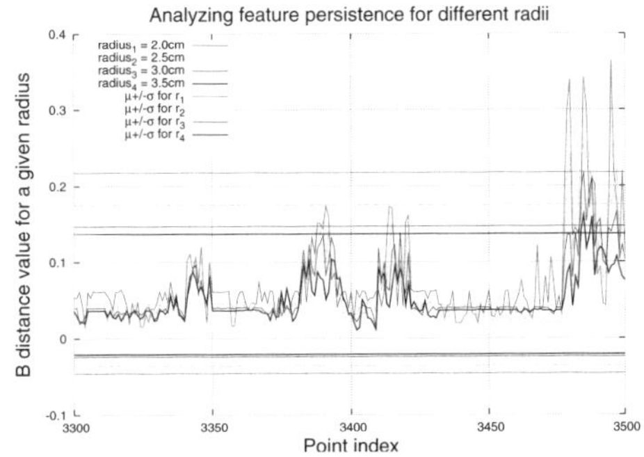
</td></tr>
<tr><td>Chi-Square</td><td>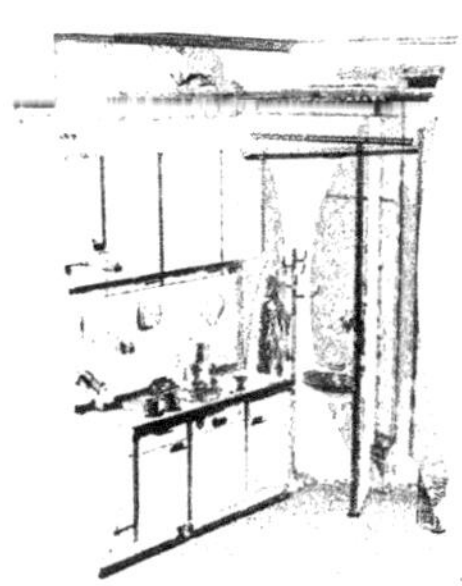</td><td>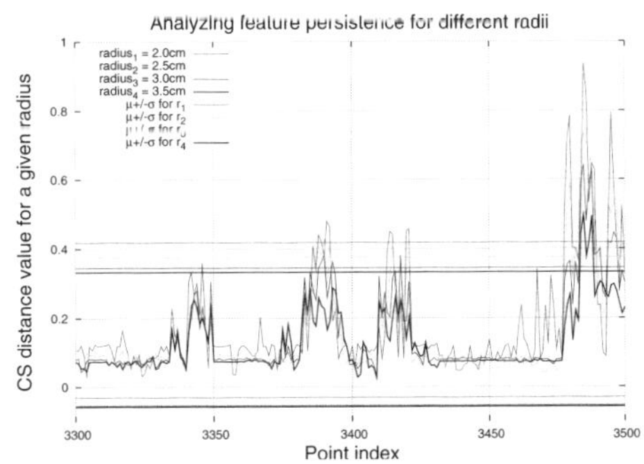
</td></tr>
<tr><td>Kullback-Leibler</td><td></td><td>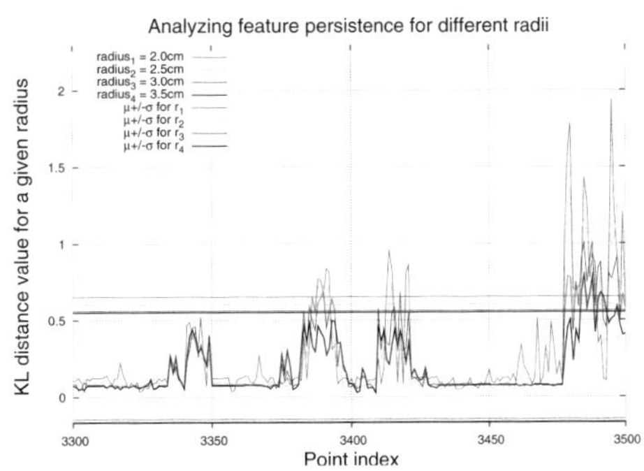
</td></tr>
</table>

4. Discussions and Experimental Results

Table 1 shows our persistence analysis on a point cloud using the presented distance metrics. We computed histograms for every point in the scan using 4 different sphere radii, and generated the mean histogram for each. For every radius, we used a different color to mark points in the scan whose distance to the respective mean histogram exceeded σ, and repeated this for the 6 distance metrics.

The s_1, s_3, and s_4 values are hard thresholds for angle values, thus they influence the decision of placing a value of $f_{1,3,4}$ in a certain bin. Since the computation of these values depends on the estimated surface normals, high degrees of noise can lead to small variations in the results. If the difference between two similar values is small, but they are on different sides of their respective threshold, the algorithm will place them in different bins. In our analysis pairs of points in a surface patch that lie on the same plane, or on perpendicular planes are particularly interesting. To ensure consistent histograms for planes, edges, and corners even under noisy conditions, we tolerate a deviation of $\pm 5°$ in those features by selecting $s_1 = s_3 = s_4 = -5° \approx -0.087\ radians$. The resulted histograms become more robust in the presence of additive noise for the mentioned surfaces, without influencing significantly the other types of surface primitives.

To illustrate the above, we show the computed feature histograms for a point located in the middle of a planar patch of $10cm \times 10cm$, first on synthetically generated data without noise, and then on data with added zero-mean Gaussian noise with $\sigma = 0.1$(see Figure 4). As shown in Figure 3, the estimated histograms are similar even under noisy conditions. Note that the histograms change only slightly as the noise level increases or the radius decreases, thus retaining enough similarity to the generic plane histogram.

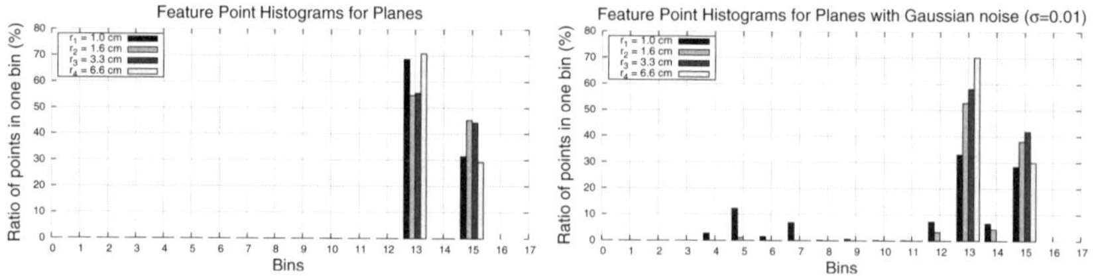

Figure 3. Feature Histograms over different radii for a point on a $10cm \times 10cm$ sized plane without noise – left; and with zero-mean Gaussian noise ($\sigma = 0.1$) – right.

Figure 4. Point (red) for which histograms were computed on a plane without noise – left; and with noise – right. Normals showed for $1/3$ of the points, computed for the smallest radius ($1cm$).

The validity of using feature histograms for registration is demonstrated in Figure 5. Points that are considered persistent features are marked in red. Note how the persistent feature analysis finds very similar points in the two scans. This speeds up applications like point cloud registration, since corresponding points are found easily and more robustly.

Another positive characteristic of the point feature histograms are their invariance to sampling density, due to the normalization of the histogram values with the number of points in each k-neighborhood. A rough classification of points based on their his-

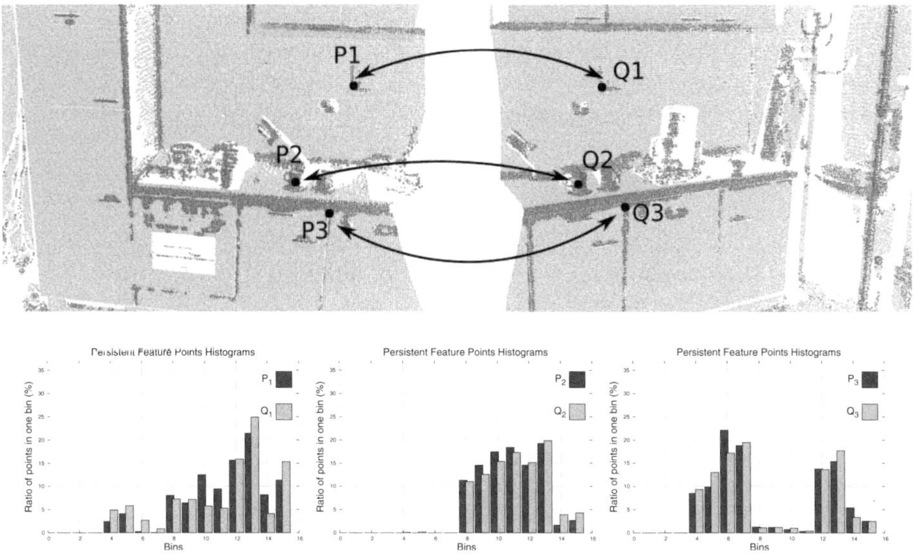

Figure 5. Feature Histograms for three pairs of corresponding points on different point cloud datasets.

tograms is shown in Figure 6, where the histograms of geometric primitive shapes (see Figure 1) are compared against all points. The test is performed using a simple distance metric, based on the comparison of the bin percentages at the location of the peaks in the geometric primitive shape histograms. This requires only few algebraic computations and comparisons for each point, i.e. $O(N)$.

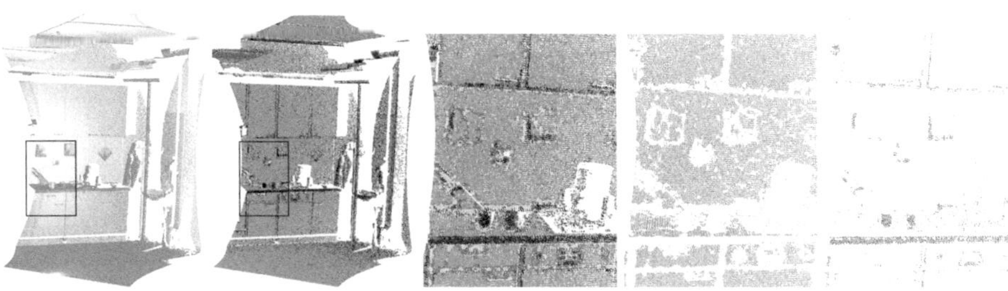

Figure 6. From left to right: partial scan of a kitchen; fast classification of each point's histogram (green: plane, red: edge, black: cylinder, blue: sphere, yellow: corner); detailed view of classification; points lying on planes and cylinders as determined by our method.

5. Conclusions

We have presented a method for computing feature histograms which approximate the local geometry and generalize the mean curvature at a given point p. The histograms are shown to be invariant to position, orientation, and point cloud density, and cope well with noisy datasets. The persistence of selected unique histograms is analyzed using different distance metrics over multiple scales, and a subset characterizing the input dataset is selected.

By creating unique signatures in this multi-dimensional space for points located on geometric primitives, the feature histograms show high potential for classifying and segmenting point cloud surfaces. Experimental results show that the proposed approach looks promising for solving the problem of correspondence search in applications such as point cloud registration.

References

[1] A. M. and A. A., "On normals and projection operators for surfaces defined by point sets," in *Proceedings of Symposium on Point-Based Graphics 04*, 2004, pp. 149–155.

[2] K.-H. Bae and D. D. Lichti, "Automated registration of unorganized point clouds from terrestrial laser scanners," in *International Archives of Photogrammetry and Remote Sensing (IAPRS)*, 2004, pp. 222–227.

[3] T. Rabbani, F. van den Heuvel, and G. Vosselmann, "Segmentation of point clouds using smoothness constraint," in *IEVM06*, 2006.

[4] F. Sadjadi and E. Hall, "Three-Dimensional Moment Invariants," *PAMI*, vol. 2, no. 2, pp. 127–136, 1980.

[5] G. Burel and H. Hénocq, "Three-dimensional invariants and their application to object recognition," *Signal Process.*, vol. 45, no. 1, pp. 1–22, 1995.

[6] N. Gelfand, N. J. Mitra, L. J. Guibas, and H. Pottmann, "Robust Global Registration," in *Proc. Symp. Geom. Processing*, 2005.

[7] G. Sharp, S. Lee, and D. Wehe, "ICP registration using invariant features," *IEEE Trans. on PAMI, 24(1):90–102, 2002.*, 2002.

[8] N. J. Mitra and A. Nguyen, "Estimating surface normals in noisy point cloud data," in *SCG '03: Proceedings of the nineteenth annual symposium on Computational geometry*, 2003, pp. 322–328.

[9] J.-F. Lalonde, R. Unnikrishnan, N. Vandapel, and M. Hebert, "Scale Selection for Classification of Point-sampled 3-D Surfaces," in *Fifth International Conference on 3-D Digital Imaging and Modeling (3DIM 2005)*, June 2005, pp. 285 – 292.

[10] M. Pauly, R. Keiser, and M. Gross, "Multi-scale feature extraction on point-sampled surfaces," pp. 281–289, 2003.

[11] Y.-L. Yang, Y.-K. Lai, S.-M. Hu, and H. Pottmann, "Robust principal curvatures on multiple scales," in *SGP '06: Proceedings of the fourth Eurographics symposium on Geometry processing*, 2006, pp. 223–226.

[12] E. Kalogerakis, P. Simari, D. Nowrouzezahrai, and K. Singh, "Robust statistical estimation of curvature on discretized surfaces," in *SGP '07: Proceedings of the fifth Eurographics symposium on Geometry processing*, 2007, pp. 13–22.

[13] T. Gatzke, C. Grimm, M. Garland, and S. Zelinka, "Curvature Maps for Local Shape Comparison," in *SMI '05: Proceedings of the International Conference on Shape Modeling and Applications 2005 (SMI' 05)*, 2005, pp. 246–255.

[14] A. Johnson and M. Hebert, "Using spin images for efficient object recognition in cluttered 3D scenes," *IEEE Transactions on Pattern Analysis and Machine Intelligence*, vol. 21, no. 5, pp. 433–449, May 1999.

[15] G. Hetzel, B. Leibe, P. Levi, and B. Schiele, "3D Object Recognition from Range Images using Local Feature Histograms," in *IEEE International Conference on Computer Vision and Pattern Recognition (CVPR'01)*, vol. 2, 2001, pp. 394–399.

[16] X. Li and I. Guskov, "3D object recognition from range images using pyramid matching," in *ICCV07*, 2007, pp. 1–6.

[17] E. Wahl, U. Hillenbrand, and G. Hirzinger, "Surflet-Pair-Relation Histograms: A Statistical 3D-Shape Representation for Rapid Classification," in *3DIM03*, 2003, pp. 474–481.

[18] R. B. Rusu, N. Blodow, Z. Marton, A. Soos, and M. Beetz, "Towards 3D Object Maps for Autonomous Household Robots," in *Proceedings of the 20th IEEE International Conference on Intelligent Robots and Systems (IROS), San Diego, CA, USA, Oct 29 - 2 Nov.*, 2007.

[19] H. Hoppe, T. DeRose, T. Duchamp, J. McDonald, and W. Stuetzle, "Surface reconstruction from unorganized points," in *SIGGRAPH '92: Proceedings of the 19th annual conference on Computer graphics and interactive techniques*, 1992, pp. 71–78.

[20] L. Green and K.-M. Xu, "Comparison of Histograms for Use in Cloud Observation and Modeling," 2005.

Intelligent Autonomous Systems 10
W. Burgard et al. (Eds.)
IOS Press, 2008

doi:10.3233/978-1-58603-887-8-129

The Robotic Busboy: Steps Towards Developing a Mobile Robotic Home Assistant

Siddhartha SRINIVASA [a], Dave FERGUSON [a], Mike VANDE WEGHE [b], Rosen DIANKOV [b], Dmitry BERENSON [b], Casey HELFRICH [a], and Hauke STRASDAT [c,1]
[a] *Intel Research Pittsburgh, Pittsburgh, USA*
[b] *Carnegie Mellon University, Pittsburgh, USA*
[c] *University of Freiburg, Freiburg, Germany*

Abstract. We present an autonomous multi-robot system that can collect objects from indoor environments and load them into a dishwasher rack. We discuss each component of the system in detail and highlight the perception, navigation, and manipulation algorithms employed. We present results from several public demonstrations, including one in which the system was run for several hours and interacted with several hundred people.

Keywords. Manipulation, Navigation, Grasping, Autonomous Systems

Introduction

One of the long-term goals of robotics is to develop a general purpose platform that can co-exist with and provide assistance to people. Substantial progress has been made toward creating the *physical* components of such an agent, resulting in a wide variety of both wheeled and humanoid robots that possess amazing potential for dexterity and finesse. However, the development of robots that can act autonomously in unstructured and inhabited environments is still an open problem, due to the inherent difficulty of the associated perception, navigation, and manipulation problems.

To usefully interact in human environments, a robot must be able to detect and recognize both the environment itself and common objects within it, as well as its own position within the environment. Robust and safe navigation approaches are required to effectively move through populated environments; a robot must differentiate between moving objects (e.g. people) and static objects, and know when it is appropriate to avoid or interact with people. Finally, complex manipulation techniques are required to interact with household objects in the environment; a robot must cope with the complexity of objects as well as errors in perception and execution. Most importantly, to function

[1]Corresponding Authors: Siddhartha Srinivasa and Dave Ferguson, Intel Research Pittsburgh, E-mail: {siddhartha.srinivasa, dave.ferguson}@intel.com

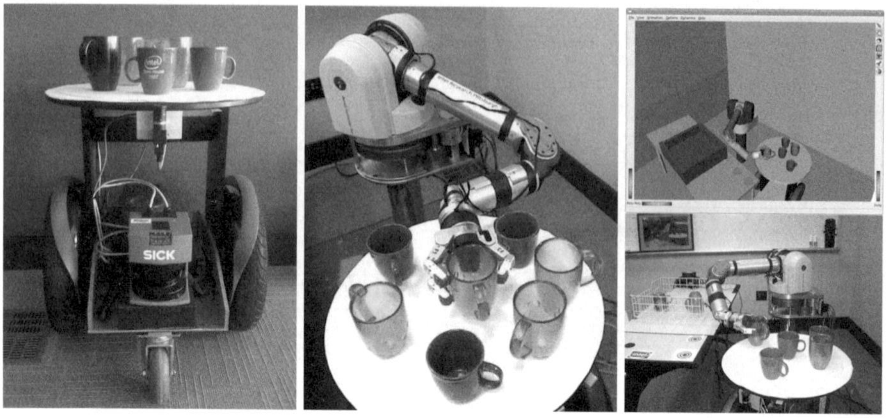

Figure 1. The Robotic Busboy composed of a Segway RMP200 and a Barrett WAM arm, with a snapshot from simulation and reality.

seamlessly in dynamic human environments, all of these actions must be planned and executed quickly, at human speeds.

Currently, although several researchers are working on robotic systems to satisfy some portion of these requirements for specific tasks, very few [1,2,3] have tried to tackle all of them in conjunction or achieved real autonomy. We focus our current work toward achieving this goal.

The Robotic Busboy is a project intended to encompass several of the challenges inherent in developing a useful robotic assistant while restricting the scope of the tasks performed. We concentrate on a dishwasher loading scenario in which people moving around an indoor environment place empty cups on a mobile robot, which then drives to a kitchen area to load the mugs into a dishwasher rack (Figure 1). This specialized task requires robust solutions to several of the challenges associated with general robotic assistants and we have found it to be a valuable domain for providing new research problems.

In this paper, we describe a multi-robot system to tackle this problem. We first introduce the architecture of the system and the interaction between the various components. We then discuss each component in detail and highlight the perception, navigation, and manipulation algorithms employed. We present results from several public demonstrations, including one in which the system was run for several hours and interacted with several hundred people, and provide a number of directions for future research.

1. System Architecture

In our approach to the Robotic Busboy task, we have a Segway mobile robot navigating through the environment collecting empty mugs from people. The Segway then transports the collected mugs to the vicinity of a Barrett WAM robotic arm which, in turn, detects and removes the cups from the mobile robot and loads them into a dishwasher rack. The overall system is divided into three main components: Segway navigation, vision-based mug and Segway detection, and grasp planning and arm motion planning.

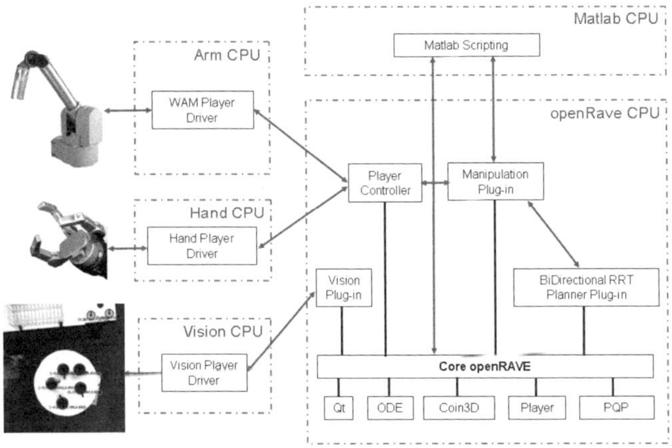

Figure 2. System Architecture Diagram.

We have developed a plugin-based architecture called OpenRAVE, released publicly on Sourceforge [4], that provides both basic services like collision detection and physics simulation, as well as a novel scripting environment that allows for the seamless interaction of many components to perform a combined task. Furthermore, the simulation environment provided by openRAVE allows for the testing of all subsystems via virtual controllers.

Figure 2 illustrates the high-level architecture of the manipulation portion of our system. Feedback and control of the arm and hand is achieved by a controller plugin which interfaces with Player/Stage [5] drivers for each hardware component. The vision plugin updates the scene with the pose of the Segway and the mugs. The manipulation plugin oversees the grasping and motion planning tasks and instructs the planner plugin to plan appropriate collision-free motions for the arm.

The following sections describe in detail the algorithms implemented in the individual subsystems as well as their interactions with each other.

2. Mobile Robot Navigation

To reliably navigate through highly populated indoor environments, the Segway needs to localize itself accurately within the environment and detect and avoid people and other obstacles that may be present. To facilitate this, the Segway is equipped with both a laser range finder and an upwards-pointing monocular camera.

For localization, we first build offline a 2D map of the (unpopulated) environment using the laser range finder (see Figure 4, left image for an example such map). This provides an accurate representation of the static elements in the environment and can be used effectively for determining the pose of the Segway during navigation in static environments (see, for instance, [6]). However, in highly populated environments such an approach is of limited accuracy, since the a priori 2D map doesn't accurately represent the dynamic elements encountered by the robot in the environment and these elements can significantly restrict the amount of the static environment that can be observed at any time.

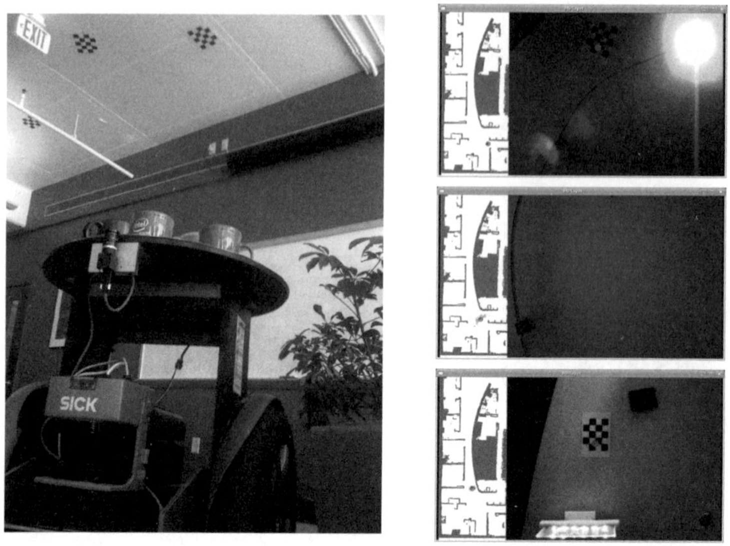

Figure 3. Camera-based indoor localization using ceiling markers. On the left is a view of the Segway with the upward-pointing camera and the ceiling markers. Images on the right show snapshots from localization performed using the ceiling markers. As the vehicle moves between the markers its position error accrues (particles from its particle filter-based localization are shown in red), but collapses upon reaching a marker.

To improve upon this accuracy, in populated environments we use a vision-based localization approach, exploiting an area of the environment that is never populated and thus never changing: the ceiling [7,8]. To do this, we place unique markers, in the form of checkerboard patterns, at various locations on the ceiling and store the pose of each marker with respect to the 2D map constructed offline[2]. We then use an upwards-pointing camera on the Segway to detect these markers and thus determine the pose of the Segway in the world (see Figure 3). This provides very good global pose information when in the vicinity of one of the markers, and in combination with odometry-based movement estimation provides a reliable method for determining the pose of the Segway in all areas of the environment.

For detecting and avoiding people and other objects that may not have existed in our prior 2D map, we use the laser range-finder to provide local obstacle information. This information is then combined with the prior 2D map and used when planning paths for the Segway. For planning we use the open-source Player implementations of a global wavefront planner and a local vector-field histogram planner [9]. The global planner is used to provide a nominal path to the goal, or value function, for the Segway and the local planner is used to track this nominal path by generating dynamically feasible, collision-free trajectories for the Segway to execute. Together, the global and local planners enable the Segway to reliably navigate through large populated indoor environments and find its way to the manipulator to deliver its collection of mugs.

---

[2]It is also possible to use ceiling-based localization without requiring markers, but for our application we found the marker-based approach to be easy to set up and very reliable.

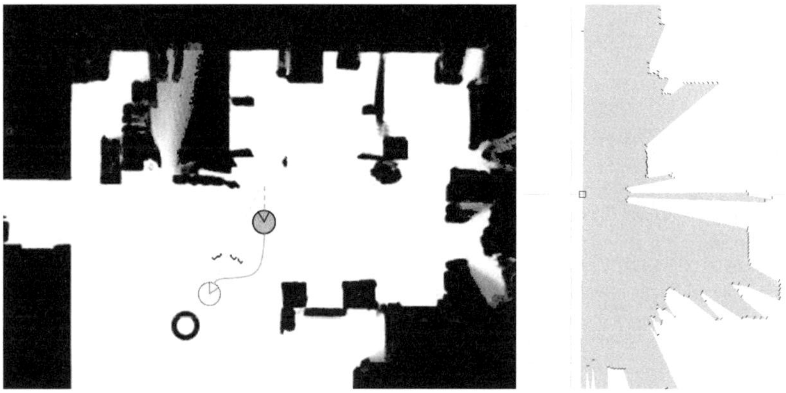

Figure 4. Laser-based obstacle avoidance. A person stands between the segway and its desired location. The person is detected by the Segway's laser range finder and is avoided.

3. Robotic Manipulation

After successfully navigating to the arm, the Segway and the mugs are registered by the robot using a camera mounted on the ceiling. Once registered, the arm unloads the mugs onto a dishwasher rack. The remainder of this section details our algorithms for detecting and manipulating the mugs and the tradeoffs between planning and execution speed.

The vision system tracks the 2D position and orientation of the Segway, mugs, and table in real-time while allowing users to add, remove or move mugs at any point during the demonstration. We used normalized cross-correlation template matching to detect the center of each mug and to find the orientation of the mugs we then search in an annulus around the center of each mug for a color peak corresponding to the mug's handle.

We have developed a grasping framework that has been successfully tested on several robot platforms [10]. Our goal for the grasp planner is to load the mugs with minimal planning delay. To achieve this, we perform most of the heavy computation offline by computing a grasp set comprising of hundreds of valid grasps, in the absence of any obstacles (Figure 5). Once a scene is registered with multiple mugs, we first select a candidate mug based on the proximity to the arm. The planner then efficiently sorts the mug's grasp set based on various feasibility criteria like reachability and collision avoidance, and tests the sorted grasps. Once a grasp is chosen, the arm plans to an empty position in the dish rack using a bi-directional RRT, and releases the mug. An illustration of the entire algorithm is shown in Figure 6.

Generated paths were smoothed and communicated across Player to the WAM. The grasp controller simply closed each finger until it exerted a minimum torque, securing the grasp. Once a cup is grasped, the configuration of the current hand is compared to the expected configuration for validation of a successful grasp. If the two configurations are substantially different, the arm replans the trajectory.

We added several heuristics specific to our problem. For aesthetic appeal, we prioritized loading the mugs face-down in the rack. The planner switched to the next best mug if it was unable to find a collision-free path within a given time. If the planner fails to repeatedly find grasps that put a mug face down in the rack, it would remove the face-down constraint and search for grasps regardless of final orientation. When the planner is no longer able to find any valid spot in the dish rack to place a mug, it requests for the dishwasher rack to be unloaded.

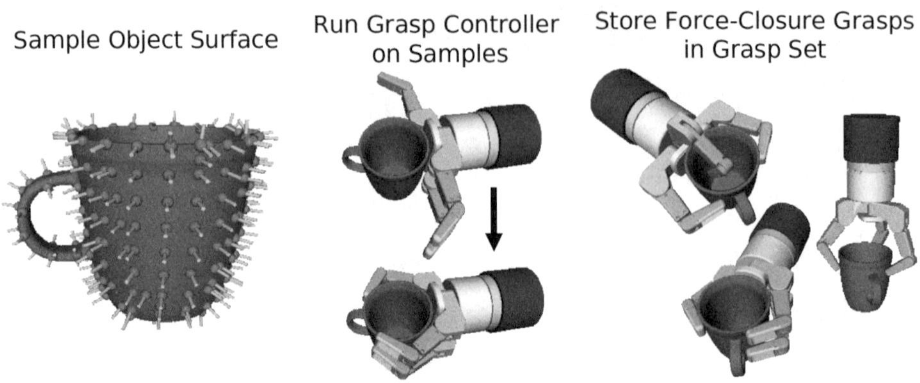

Figure 5. Offline grasp generation. Grasps are generated offline for a known object and stored for online evaluation.

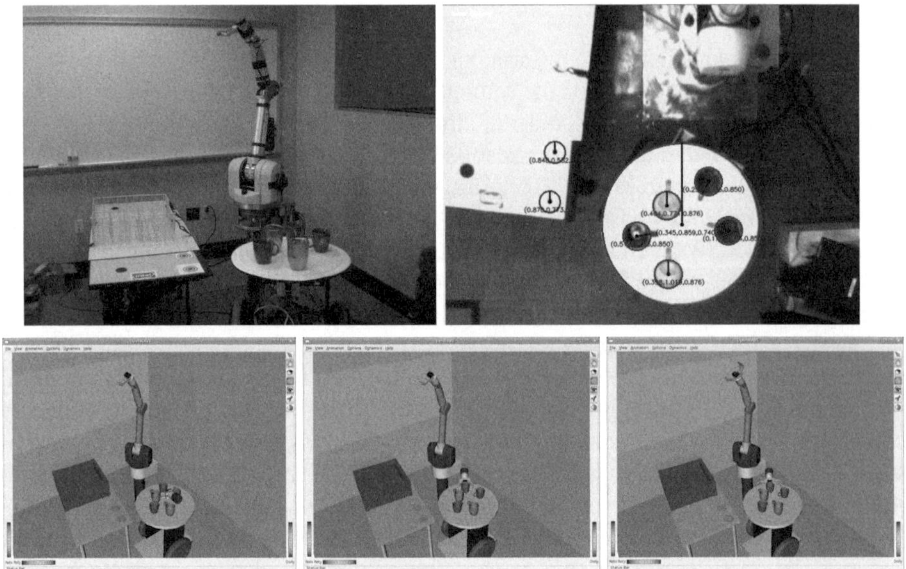

Figure 6. The real scene on the top left is registered by the ceiling camera on the top right and fed to OpenRAVE. The first grasp on the bottom left is invalidated by collision but the second one succeeds, triggering the arm planner which plans a collision-free path for the arm.

4. Results

The Robotic Busboy system has been developed and tested incrementally for more than 8 months. It has operated in populated environments dozens of times, often for several hours at a time. The largest demonstration, both in terms of duration and audience, was at the 2007 Intel Open House in October where the system operated continuously for four hours and interacted with hundreds of people. Figure 7 shows snapshots of the arm removing cups from the Segway and loading them into the dishwasher rack.

During the Open House demonstration the robot dropped or failed to pick up about 20 mugs and was successful about 200 times in picking up a mug and loading it into the

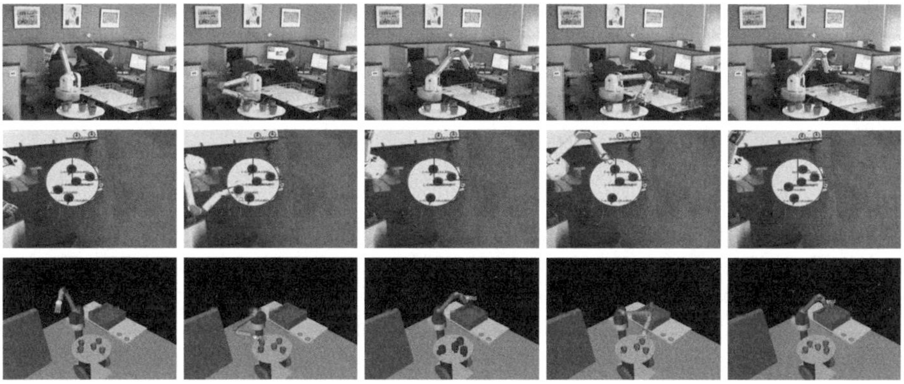

Figure 7. Removing cups from the Segway and loading them into a dishwasher rack during the Intel Open House Demonstration. The top images show the environment, the center images show the result from the ceiling camera vision system, and the bottom images show the corresponding OpenRAVE model. Note that while the system is running, users have added extra cups to the Segway and the system simply updates its model and continues.

dish rack. In most of the failures, the robot realized that it had not picked up the mug and indicated failure. In a more structured experiment with different placements of 4 mugs, we measured 19 out of 20 successes. We observed that failure to pick up a mug could be due to incorrect calibration of the arm, inaccuracies due to the loosening of cables that drive the arm, mis-registration of the mug handle by the vision system, and when someone moved the mugs when the arm was in motion to pick one up.

In our structured experiment, the average total time from giving the order to pickup a cup to the cup being released in the dishrack was measured to be 51 seconds, with a standard deviation of 4.9 seconds. We noted that execution times remained approximately consistent regardless of the number of mugs the arm had to consider. We believe that this is a testament to the dexterity of our arm and our grasp planner. We observed that the most challenging arrangement is with the mugs close together with the handles pointing outward.

As is the case for most complex autonomous systems, it was not until the entire system was functional and stress-tested that many of the key challenges to grasping and failure recovery were discovered. For example, if the arm fails to grasp a cup, it has to restart its grasping script gracefully without any human intervention. This was achieved by constantly comparing the hand's predicted encoder values vs the real encoder values. Using grasps that are robust against errors in the mug pose and arm configuration really helped increase the success rate. We picked these grasps by randomly moving the cups in simulation and testing whether force closure still persists. This ability to recover from error was very important for successful continuous operation, particularly in the presence of people endeavoring to test the system in challenging scenarios. The same technique is used to detect, and recover from, the case when a person has physically removed the cup from the hand of the robot.

5. Conclusions

We have presented an autonomous multi-robot system designed to play the role of a robotic busboy. In our system, a mobile robot navigates through a populated environment receiving empty cups from people, then brings these cups to a robotic manipulator on a fixed platform. The manipulator detects the cups, picks them up off the mobile robot, and loads them into a dishwasher rack. Although a specialized task, this problem requires robust solutions to several of the challenges associated with general robotic assistants and we have found it to be a valuable domain for providing new research problems and general purpose perception and planning algorithms. Our system is entirely autonomous and has interacted with people in dozens of public demonstrations, including one in which it was run for several hours and interacted with several hundred people.

We are currently working on adding compliance to the arm to enable smoother interaction with unexpected obstacles and to allow for humans to work in the arm's workspace and interact with the arm comfortably. While the Segway and the arm currently function as one coordinated robot system, we are also working on integrating them physically into a mobile manipulator and on migrating all of the sensing onboard. We believe that these improvements will enable us to perform complex mobile manipulation tasks like opening doors and cabinets, and interacting with people even more closely in unstructured human environments.

We believe that the ability to plan robustly under uncertainty is a compelling challenge. We are currently working on active vision for information gain, kinesthetic sensing on the hand and the arm for fine motion control while grasping, and on long-range people prediction and tracking for autonomous navigation in indoor spaces.

Acknowledgements

We are grateful for the contributions of James Kuffner, Ali Rahimi, and Chris Atkeson, and for the technical support of Barrett Technologies and Will Pong of Segway Inc.

References

[1] M. Quigley, E. Berger, and A. Y. Ng, "STAIR: Hardware and software architecture," in *AAAI Robotics Workshop*, 2007.
[2] H. Nguyen, C. Anderson, A. Trevor, A. Jain, Z. Xu, and C. Kemp, "El-e: An assistive robot that fetches objects from flat surfaces," in *Human Robot Interaction, The Robotics Helpers Workshop*, 2008.
[3] "Personal robotics program," www.willowgarage.com.
[4] R. Diankov *et al.*, "The OpenRAVE project," hosted at http://sourceforge.net/projects/openrave/.
[5] B. Gerkey *et al.*, "The Player/Stage project," hosted at http://playerstage.sourceforge.net.
[6] F. Dellaert, D. Fox, W. Burgard, and S. Thrun, "Monte carlo localization for mobile robots," in *Proceedings of the IEEE International Conference on Robotics and Automation (ICRA)*, 1999.
[7] W. Burgard, A. Cremers, D. Fox, D. Hähnel, G. Lakemeyer, D. Schulz, W. Steiner, and S. Thrun, "The interactive museum tour-guide robot," in *Proceedings of the National Conference on Artificial Intelligence (AAAI)*, 1998.
[8] S. Thrun *et al.*, "MINERVA: A second generation mobile tour-guide robot," in *Proceedings of the IEEE International Conference on Robotics and Automation (ICRA)*, 1999.
[9] I. Ulrich and J. Borenstein, "Vfh+: Reliable obstacle avoidance for fast mobile robots," in *Proceedings of the IEEE International Conference on Robotics and Automation (ICRA)*, 1998.
[10] D. Berenson, R. Diankov, K. Nishiwaki, S. Kagami, and J. Kuffner, "Grasp planning in complex scenes," in *Proceedings of IEEE-RAS International Conference on Humanoid Robots (Humanoids)*, 2007.

Intelligent Autonomous Systems 10
W. Burgard et al. (Eds.)
IOS Press, 2008

doi:10.3233/978-1-58603-887-8-137

Control of a Virtual Leg via EMG Signals from Four Thigh Muscles

Massimo SARTORI [a,1], Gaetano CHEMELLO [b] Monica REGGIANI [c] and Enrico PAGELLO [a,b]

[a] *Department of Information Engineering, University of Padua, Italy*
[b] *Institute of Biomedical Engineering, National Research Council, Padua, Italy*
[c] *Department of Management and Engineering, University of Padua, Italy*

Abstract. The work presented in this paper is our first step toward the development of an exoskeleton for human gait support. The device we foresee should be suitable for assisting walking in paralyzed subjects and should be based on myoelectrical muscular signals (EMGs) as a communication channel between the human and the machine. This paper concentrates on the design of a biomechanical model of the human lower extremity. The system predicts subject's intentions from the analysis of his/her electromyographical activity. Our model takes into account three main factors. Firstly, the main muscles spanning the knee articulation. Secondly, the gravity affecting the leg during its movement. Finally, it considers the limits within which the leg swings. Furthermore, it is capable of estimating several knee parameters such as joint moment, angular acceleration, angular velocity, and angular position. In order to have a visual feedback of the predicted movements we have implemented a three-dimensional graphical simulation of a human leg which moves in response to the commands computed by the model.

Keywords. EMG signals, human knee, exoskeleton, simulation

Introduction

Several research projects are currently focused on the development of devices to support human movements and although the results are promising none of the proposed solutions provide an effective control system for such machines [2,7,9,11,13]. This work represents our initial effort toward the realization of a wearable robotic device (exoskeleton) to assist people who are denoted by a limited control of their lower limbs during basic but pivotal motion tasks such as sitting on a chair, standing up, staying erect, starting and stopping walking. Although those tasks might seem quite simple, they are actually essential for an effective rehabilitation process as they provide a first level of autonomy to the patient. The device we want to develop will be endowed with a control system capable of understanding the subject's intentions through the analysis of myoelectrical signals (EMG signals) and defining a proper level of actuation to independently move the robotic orthosis for assisting injured patients.

[1]Corresponding author: Massimo Sartori, Department of Information Engineering, University of Padua, Via Gradenigo 6/B, 35131 Padova, Italy; E-mail: sartori1@dei.unipd.it

We decided to focus our efforts on the development of supporting devices for the human leg because of the lack of research regarding such machines compared with the advancements on upper extremity exoskeletons. While the latter have been studied for more than ten years, only recently particular attention has been put on lower extremity exoskeletons and human gait support despite the potentially large number of consumers for such machines [7,9,11,13]. As a matter of facts, every year a number of people are suffering neuropathologies affecting the lower limbs. Additionally, the amount of aged people is supposed to increase by the 2030 with a ratio estimated to be the 20% of the EU population [8].

From the beginning of our work we started a tight cooperation with the Department of Rehabilitation at Sant'Antonio Hospital in Padua, Italy. This allowed to clearly define patients' needs as well as to understand how individual thigh muscles are activated to generate joint moments and movements, and how they allow coordinated knee torques [12]. Only a proper model of the relations between EMG signals and the associated movements will yield to the design of an effective robotic exoskeleton suitable for supporting neurologically injured patients or aged people [2,6,7].

Hereafter, we will firstly motivate our choices regarding the knee features we decided to take into consideration (section 1). Then, we will describe the structure of a biomechanical model of the human leg (sections 2 and 3) with the purpose of recognizing the subject's intentions (we considered to be the number of flexo-extension movements) by performing an analysis of the electromyographical activity. Finally, we will present the implementation of a three-dimensional graphical simulation of a human lower extremity which moves in response to the prerecorded EMGs (section 4). The comparison between the number of varying knee flexo-extensions and the number of simulated ones allowed a validation of the model (sections 4 and 5).

1. Preliminary Remarks

We have chosen to concentrate on the knee as it plays a significant role in the human motion. We will only consider the knee torsion movement as it is the main task performed by the knee articulation. We neglected any modeling of the knee rotation as its contribution is not significant to the whole set of possible movements carried out by a subject [6]. In order to improve our previous work based on a simplified version of the virtual knee accounting for only two muscles [14], we now present a new model incorporating four thigh muscles. This extension is motivated by the partial unreliability of the previous model. The only extensor muscle we selected was, in fact, the rectus femoris which is mainly active during low force movements only. Therefore, whenever the subject performs higher force extensions the predicted movement might be imprecise [7]. The inclusion of only one flexor muscle could lead to rather accurate simulations when pure knee flexion motion tasks are studied. Nevertheless, for more complex movements the inclusion of additional flexor muscles is recommended [7]. The selected muscles with their *Physiological Cross-sectional Area* (PCA) [15] are the following ones: *rectus femoris* (8%), *vastus lateralis* (20%), *semitendinosus* (3%), and *biceps femoris* (10%). They cover a total of 41% of the cross-sectional area of all thigh muscles. The remaining area is occupied by the vastus medialis (15%), the semimembranosus (10%), the vastus intermedius (13%), the gastrocnemius (19%), the sartorius (1%), and the gracilis (1%) [7]. While the lat-

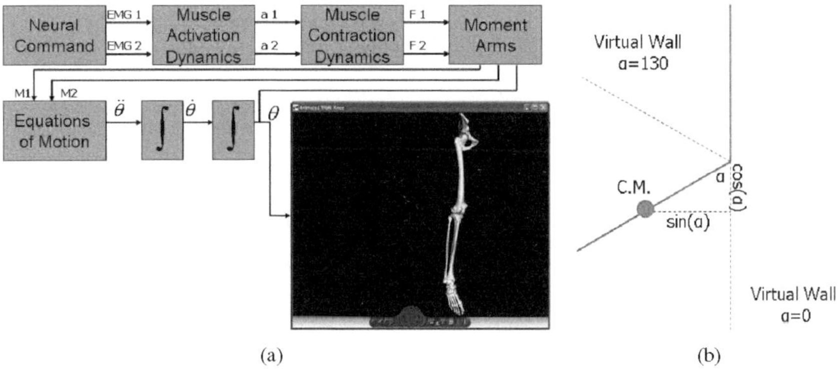

Figure 1. (a) Forward Dynamics Approach and Graphical Interface. (b) Virtual leg modeled as a rigid swinging body.

ter two are negligible due to their small force output, the gastrocnemius has not been recorded because it spans both knee and ankle articulations and it is not possible to relate its activity to the knee without considering also the ankle. The vastus intermedius is not recordable using surface electrodes [6]. Since we want to define the smallest set of muscles that assure correct simulations of the flexo-extension movement regardless of the amount of force involvement, we decided to leave out any contribution of the vastus medialis and semimembranosus and verify whether this simplification compromised the model accuracy (sections 4 and 5). Finally, we have used a forward dynamic approach in the study of the human movement (fig. 1a). This choice has been encouraged by the results in Buchanan *et al.* [3]. A detailed description of the phases involved in the control of the virtual knee will be provided in the following sections.

2. EMG Interpretation

2.1. Signal Acquisition and Muscle Activation

A raw EMG signal is a voltage that can be both positive and negative and changes as the neural command calls for increased or decreased muscular effort. It is quite difficult to compare the absolute magnitude of an EMG signal from different muscles because the magnitudes of the signals can vary depending on many factors (e.g. gain of the amplifiers, the types of electrodes, etc.) [3]. In order to use the EMG signals in a neuromusculoskeletal model, we first need to normalize them into a specific range (between 0 and 1) so that we can eventually compare them one with another. The signal resulting from the processing stage is a value that is called *muscle activation* and is meant to describe the process that makes the electrical activity to spread across the muscle causing its activation. Hereafter we will present the steps we adopted to perform this transformation which have been inspired by the Buchanan *et al.*'s study [3].

The acquisition stage comprises the sampling and the processing of the EMG signals while the subject executes flexions and extensions of his leg. The signals have been sampled at 1 kHz while the BIOPAC MP35 data acquisition unit was connected to a personal computer. During the acquisition stage the signals have been amplified (differential amplifier, gain of $1000\frac{V}{V}$) on both channels and successively bandpass filtered from 20 to 450 Hz [5]. Successively, the resulting signals have been full wave rectified and normalized via software to approximate the *activation* of the muscle, $a(t)$.

2.2. Muscle Force

We expressed the muscular force as a function of the muscular activation, $a(t)$, previously computed:

$$f_i(t) = \frac{1}{T}\int_{t-T}^{t} |a_i(t)|\, dt \tag{1}$$

where T is a temporal window specifying the dimension of the interval during which the calculation of every sample is executed. The index i has been introduced to identify which muscle, the force and the activation refer to. Equation (1) is a very rough approximation of the muscular force. However, we do not want to perform a careful clinical analysis of the muscles behavior. We just want to understand the time interval during which the muscles are active along with the intensity of contraction. Refer to section 4 to see how our approximations did not negatively affect the simulation phase.

2.3. Driving the Virtual Knee

Once all the muscle forces are calculated, we can compute the corresponding contribution to the joint moment. This requires knowledge of the muscle moment arms. According to the results of Herzog *et al* [10] they can be valuated as follow: $r_i(\theta) = b_0 + b_1 \cdot \theta + b_2 \cdot \theta^2 + b_3 \cdot \theta^3 + b_4 \cdot \theta^4$, where θ is the knee joint angle expressed in degrees, while b_0, b_1, b_2, b_3, b_4 are coefficients related to the i-th muscle. The corresponding joint moment M can now be estimated:

$$M(\theta, t) = \sum_{i=1}^{m}(r_i(\theta) \cdot f_i(t)). \tag{2}$$

The muscle force $f_i(t)$ is obtained from Equation 1. The joint moment, in turn, will cause the movements. The knee angular acceleration and the related command signal are calculated directly from the computed joint moment as explained in the following section.

3. The Biomechanical Model

The human lower extremity has been modeled as a rigid body swinging between 0° and 130° (fig. 1b). Our software simulates the action of the gravity affecting the rigid body during its movement, the action of the extensor and flexor muscle forces as well as some contact forces that limit the range within which the leg moves (see section 3.1). The anthropometric data for modeling the rigid body are defined in [4] and represent average values. We considered a center of mass of the rigid body placed at 21.65 cm from the knee joint and with a weigh of 3.8 Kg.

Figure 2a shows the structure of the biomechanical model which predicts the angular position from the subject electromyograms. In the first stage the muscular forces,

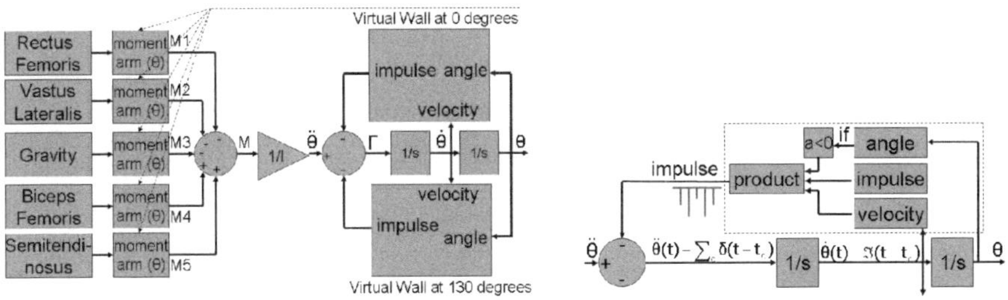

(a) (b)

Figure 2. (a) Biomechanical Model Structure. (b) Virtual Wall placed at α = 0° (in the outlined square).

along with the gravity, are coupled with their respective moment arms (according to the equation 2). The net knee joint moment, M, is then combined with the inertial coefficient and the resulting acceleration $\ddot{\vartheta}$ is composed with the action of the Virtual Walls (Section 3.1) placed at α = 0° and at α = 130°. The resulting signal is integrated twice to obtain the angular position ϑ used as a command signal for the virtual knee (see fig. 2a and 2b).

3.1. The Virtual Walls

The Virtual Walls generate impulsive forces for the purpose of stopping the motion of the leg before it reaches undesired positions. These forces are required only in the correspondence of the temporal instants in which α gets equal to 0° or to 130° (*critical instants*). These limits to the motion simulate: (1) the natural restriction affecting the knee torsion that prevents it to be further extended once the shank is aligned to the thigh (this is what happens in a healthy human knee), (2) the restriction of motion caused by an hypothetical exoskeleton worn by a subject that impedes the leg to be further flexed beyond a certain threshold (130° in our specific case). Hereafter we will concentrate on the description of the wall placed at α = 0° (fig. 2b) as the wall at α = 130° behaves in a similar manner. The virtual wall constantly checks the knee angle and as soon as it gets negative ($\alpha < 0°$) an unitary impulse is generated and multiplied by the scalar value of the velocity at which the leg swings (see fig. 2b). As this process always takes place when the leg reaches undesired positions, we obtain a sequence of impulses centered on the correspondence of critical instants. Each impulse has an area that is equal to the value of the velocity the leg assumed in each critical instant. This impulse train is the output of the virtual wall (see fig. 2b). In order to stop the motion of the leg we now need to set to zero the velocity that the leg assumes at every critical instant. To achieve this we first subtract the resulting impulse train from the acceleration signal, $\ddot{\vartheta}(t)$. Then, we integrate the resulting signal: $\Gamma(t) = \ddot{\vartheta}(t) - \sum_c \delta(t - t_c)$, where $\delta(t - t_c)$ is an impulse centered on the critical instant t_c. $\Gamma(t)$ is now integrated as follows: $\int_{t_{start}}^{t_{stop}} \Gamma(t)dt = \dot{\vartheta}(t) - \sum_c \Im(t - t_c)$ where t_{start} and t_{stop} indicate, respectively, the starting and stopping time of the simulation, while $\sum_c \Im(t - t_c)$ is a step train (resulting from the integration of the impulse train) whose steps are centered on the critical instants. The result is that the velocity signal $\dot{\vartheta}(t)$ is set to zero by the action of the integrated impulse train exactly in the correspondence of the critical instants (see fig. 3).

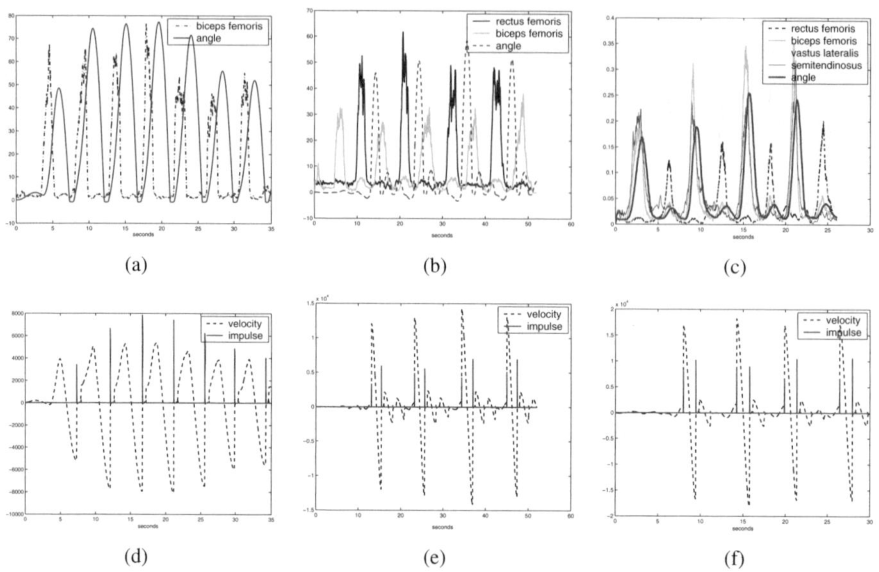

Figure 3. Graphs (a), (b) and (c) show the muscle force contractions overlying the computed control signals. Graphs (d),(e) and (f) show that the velocity is set to zero by impulsive forces at the critical instants.

4. Experimental Evaluation

4.1. Simulated Graphical Environment

To graphically see the behaviour of our model we implemented a virtual leg (fig. 1) driven by EMG signals. The original 3D image representing the lower extremity has been developed at the Department of Anatomy of the Université Libre de Bruxelles (ULB) and it is anatomically correct [1]. By using the *Data Manager* program, also developed at the ULB, we exported every single part of the 3D knee to the VRML format [1]. A VRML program has been written to integrate all the exported parts into a single virtual leg that can be controlled via EMG signals. We chose to adopt VRML for the rendering phase because it makes the communication between the MATLAB Simulink biomechanical model and the virtual knee easier to setup.

4.2. Experimental Results

To verify the accuracy of our simulator in predicting the human movement, we performed three tests. The testing phase is based on the gradual addition of muscles to the model for the purpose of observing changes in the system behavior depending on the number of included muscles. All the tests required the subject to stand up right and to flex and extend his leg several times. The computed command signal (angular position) and the number of knee torsions reproduced by the virtual knee have been compared with the actual subject's muscle contractions. The number of simulated torsions has been intended as the only assessing parameter. For the purpose of this work any validation of the computed angle has been neglected as we were only interested in recognizing the patient's intentions that in our case were expressed by the number of flexo-extensions.

In the first test we only considered the flexor muscle activity (biceps femoris). Figure 3a shows that after every muscle contraction (blue line) the biomechanical model generated an appropriate command signal (red line) which made the virtual knee to correctly reproduce all the seven torsions originally performed by the subject. In the second test we recorded the rectus femoris extensor muscle as well as the biceps femoris flexor muscle activity (fig. 3b, green and blue lines respectively). Figure 3b shows that after each couple of extensor and flexor muscle contractions, the biomechanical model generated an appropriate command signal (red line) that simulated all the four torsions originally performed by the subject. The third test was done by recording all the four muscles electrical activity. Four torsions have been performed by the subject and, as well as in the previous experiments, all the four flexo-extensions were correctly reproduced (see fig. 3c). However, likewise the second test, some undesired oscillations in between torsions had taken place due to *cross-talk* interferences between the recorded electromyograms. As the number of selected muscles increases, the amount of cross-talk interferences grows too. This behaviour can be adjusted by improving the processing stage of the EMG signals. Figures 3d, 3e and 3f show the behavior of the virtual walls previously described. More precisely, it is possible to see that, for each test, the angular velocity had been set to zero in the correspondence of every critical instant by the generation of impulsive forces.

5. Conclusions

This paper presented a study on the control of a virtual knee based on the analysis of biological signals. We developed a new *four-muscles-based model* that extends our previous one which was based on two muscles only [14]. This extension was motivated by two reasons. Firstly, to assure accurate predictions of flexo-extension movements regardless of the actual force involvement. Secondly, to allow a future study on more complex movements, such as sitting on a chair or standing up. Those movements would require a higher force involvement compared to the knee flexo-extension. Moreover, they could not be properly modeled by using EMG signals recorded from one extensor and one flexor muscle only [3,7]. The tests we carried out (section 4) were aimed at estimating the accuracy of our *four-muscles-based model* in recognising the flexo-extensions performed by the subject. Experimental results demonstrated two facts. First of all, our current model correctly recognised all subject's movements despite some cross-talk interferences. Second of all, a biomechanical model based on two extensors and two flexor muscles allows to carefully simulate the flexo-extension movement with no need of including the vastus medialis and the semimembranosus. This addition would eventually complicate the model without offering valuable improvements.

6. Future Work

Although the results derived from our experiments are quite satisfactory, the inclusion of a *geometry model* that takes into account the *force-length* relationship of muscles would significantly improve the accuracy of the model for future research on dynamic movements. Several studies show that the *EMG-to-force* relationship strongly depends on the muscle's fibers length indeed. A non-inclusion of it may lead to predictions off by 50% or more, depending on the joint angle [3,7]. We also intend to make our model both portable and able to run in real-time. Moreover, we would like to make use of angle sensors in order to properly compare the predicted angle with the one of the subject's knee.

Acknowledgments

We thank Dr. Christian Fleischer for his advices regarding the model design and Dr. Alessandro Giovannini (Sant'Antonio Hospital, Padua, Italy) for his guidance across the medical problems.

References

[1] Department of Anatomy of the University of Brussels. The ULB Virtual Human. http://homepages.ulb.ac.be/~anatemb/the_ulb_virtual_man.htm.

[2] A.H. Arieta, R. Katoh, H. Yokoi, and Y. Wenwei. Development of a multi-DOF electromyography prosthetic system using the adaptive joint mechanism. *Woodhead Publishing Ltd*, 3:1–10, 2006.

[3] T.S. Buchanan, D.G. Lloyd, K. Manal, and T.F. Besier. Neuromusculoskeletal Modeling: Estimation of Muscle Forces and Joint Moments and Movements from Measurements from Neural Command. *Journal of Applied Biomechanics*, 20:367–395, 2004.

[4] C.E. Clauser, J.T. McConville, and J.W. Young. Volume and Center of Mass of Segments of the Human Body. Ohio, August 1969. Aerospace Medical Research Laboratory, Aerospace Medical Division, Air Force System Command.

[5] S. Day. Important Factors in urface EMG Measurement. Bortec Biomedical Ltd. www.bortec.ca/Images/pdf/EMG%20measurement%20and%20recording.pdf.

[6] S.L. Delp, J.P. Loan, M.G. Hoy, F.E. Zajac, E.L. Topp, and J.M. Rosen. An Interactive Graphics-Based Model of the Lower Extremity to Study Orthopedic Surgical Procedures. *IEEE Transactions on Biomedical Engineering*, 32:757–767, 1990.

[7] Ch. Fleischer and G. Hommel. Calibration of an EMG-Based Body Model with Six Muscles to Control a Leg Exoskeleton. In *IEEE International Conference on Robotics and Automation (ICRA'07)*, Roma, Italy, April 10–14 2007.

[8] A. Giménez, A. Jardon, R. Cabas, and C. Balaguer. A portable light-weight climbing rogot for personal assistance application. In *Proceedings of the Clawar'05*, London, UK, 2005.

[9] T. Hayashi, H. Kawamoto, and Y. Sankai. Control Method of Robot Suit HAL working as Operator's Muscle using Biological and Dynamical Information. *IEEE International Conference on Systems, Man, and Cybernetics*, 2005.

[10] W. Herzog and L.J. Read. Lines of Action and Moment Arms of the Major Force-Carrying Structures Crossing the Human Knee Joint. *Journal of Anatomy*, 182:213–220, 1992.

[11] K.H. Low, X. Liu, and H. Yu. Development of NTU Wearable Exoskeleton System for Assistive Technologies. In *Proceedings of the IEEE International Conference on Mechatronics and Automation*, pages 1099–1106, Niagara Falls, Canada, 2005.

[12] E.N. Marieb and K.N. Hoehn. *Human Anatomy and Physiology (7th Edition)*. Benjamin Cummings, 2006.

[13] J.E. Pratt, S.H. Collins, B.T. Krupp, and K.J. Morse. The RoboKnee: An Exoskeleton for Enhancing Strength and Endurance During Walking. In *Proceedings of the 2004 IEEE International Conference on Robotics and Automation*, pages 2430–2435, New Orleans, LA, 2004.

[14] M. Sartori, G. Chemello, and E. Pagello. A 3D Virtual Model of the Knee Driven by EMG Signals. In *Proceedings of the 10th Congress of the Italian Association of Artificial Intelligence*, pages 591–601, Rome, 2007.

[15] G.T. Yamaguchi, A.G. Sawa, D.W. Moran, M.J. Fessler, and J.M. Winters. A Survey of Human Musculotendon Actuator Parameters. In J. Winters and S.L.-Y. Woo, editors, *Multiple Muscle Systems: Biomechanics and Movement Organization*, pages 717–773. Springer-Verlag, 1990.

Intelligent Autonomous Systems 10
W. Burgard et al. (Eds.)
IOS Press, 2008

doi:10.3233/978-1-58603-887-8-145

Autonomous Learning of Ball Passing by Four-legged Robots and Trial Reduction by Thinning-out and Surrogate Functions

Hayato KOBAYASHI [a,1], Kohei HATANO [b], Akira ISHINO [a], and Ayumi SHINOHARA [a]

[a] *Graduate School of Information Science, Tohoku University, Japan*
[b] *Department of Informatics, Kyushu University, Japan*

Abstract. This paper describes an autonomous learning method used with real robots in order to acquire ball passing skills in the RoboCup standard platform league. These skills involve precisely moving and stopping a ball to a certain objective area and are essential to realizing sophisticated cooperative strategy. Moreover, we propose a hybrid method using "thinning-out" and "surrogate functions" in order to reduce actual trials regarded as unnecessary or unpromising. We verify the performance of our method using the minimization problems of several test functions, and then we address the learning problem of ball passing skills on real robots, which is also the first application of thinning-out on real environments.

Keywords. Autonomous Learning, Four-legged Robot, RoboCup,

1. Introduction

For robots to function in the real world, they need the abilities to adapt to unknown environments, that is *learning* abilities. In particular, legged robots must acquire or learn many basic skills such as walking, running, pushing, pulling, kicking, and so on, in order to accomplish sophisticated behaviors such as playing soccer. In this paper, we address the learning of ball passing skills by four-legged robots as an instance of basic skills. Ball passing skills are used in RoboCup[2] and are essential to realizing sophisticated cooperative strategy.

We regard learning as optimization of an unknown score function $f : X \to \boldsymbol{R}$ on an n-dimensional search space $X \subseteq \boldsymbol{R}^n$, i.e. $\min_{x \in X} f(x)$ or $\max_{x \in X} f(x)$. In the case of robot learning, we must usually treat a noisy, high-dimensional, nonlinear function. Although such functions are surely difficult to optimize, it is more serious from a practical standpoint that each *trial*, which means evaluation of the function for a sampled point $x \in X$, brings the following various costs.

- Servo motors are damaged, and so robots themselves can be broken easily.

[1]Corresponding Author: Hayato Kobayashi; E-mail: kobayashi@shino.ecei.tohoku.ac.jp.

[2]RoboCup [1] is a competition for autonomous robots which play soccer and also an interesting and challenging research domain, because it has a noisy, incomplete, real-time, multi-agent environment.

- Human resources are often needed, e.g., we must restore a kicked ball for the next trial.
- It obviously takes more time compared to computations being done only on PCs.

There are two approaches in order to reduce those costs. One is an approach from experimental methodology, and the other is an approach from machine learning.

Approach from experimental methodology

Experiments in simulation environments are useful in the sense that each trial does not damage robots nor need human resources. Zagal and Solar [2] studied the learning of ball shooting skills by four-legged robots in a simulation environment. Kobayashi et al. [3] also studied the similar learning problem in another simulation environment and discovered sophisticated shooting motions. However, since simulation environments can not produce complete, real environments, we need to train robots in the real environment where basic skills heavily depend on complex physical interactions.

In experiments with real robots, *autonomous learning*, by which robots acquire some skills on their own without human intervention, is practically efficient, especially in environments that change frequently, such as RoboCup competitions. There have been many studies conducted on the autonomous learning of quadrupedal locomotion [4,5,6,7], which is the most basic skill for every movement. However, the skills used to control the ball are often coded by hand and have not been studied as much as quadrupedal locomotion. Fidelman and Stone [8] presented an autonomous learning method of ball acquisition skills that involves controlling a stopped ball by utilizing chest PSD sensors. Kobayashi et al. [9] studied the reinforcement learning to trap a moving ball autonomously. The goal of the learning was to acquire a good timing to initiate its trapping motion, depending on the distance and the speed of the ball, whose movement was restricted to one dimension.

Approach from machine learning

Memory-based learning is often utilized in robot learning, since the number of trials are highly restricted, which are up to and at most several hundred trials in many cases. Memory-based learning stores past evaluations in history and samples a new promising point based on the history. In memory-based learning, we can efficiently reduce the number of trials, even if computational complexity of the algorithm is too large.

Sano et al. [10] proposed Memory-based Fitness Evaluation GA (MFEGA) for noisy environments. They estimated more proper scores (or fitness values) by the weighted average of neighboring scores, so as to reduce the number of trials compared to multiple sampling methods, which evaluates a score function several times in each trial. Ratle [11] proposed an acceleration method of GA by utilizing *surrogate functions*, which are approximation models of original score functions. He chose kriging interpolation as surrogate functions and used it for evaluating some of the next generations. In our previous work [3], we took another approach for reducing the number of trials, based on the idea that it is able to theoretically determine whether or not selected points are worth evaluating if score functions are g-Lipschitz continuous described in Section 3.1. If a sampled point is unlikely to improve the results obtained so far, the trial for the point is not performed and just skipped over. We call this method *thinning-out* in search seedlings (or candidates), which contrasts to *pruning* in a search tree. One advantage of our method is that it is guaranteed to thin-out unnecessary candidates without any errors, if we know a proper Lipschitz function g. We can arbitrarily prepare the function g depending on the problems to be solved and also infer the function g for complex problems that are difficult for us to come up with it.

In this paper we propose an autonomous learning method of ball passing skills as an approach from experimental methodology as well as a hybrid method using thinning-out and surrogate functions as an approach from machine learning. This work is also important as the first application of thinning-out in the real world.

The remainder of this paper is organized as follows. In Section 2, we begin by illustrating the execution mechanism of passing motions and our training equipment for autonomous learning, and then formalizing our learning problem. In Section 3, we illustrate our proposed method and describe existing thinning-out and kriging interpolation utilized as a surrogate function. In Section 4, we apply the proposed method for the minimization of several mathematical test functions and learning of ball passing skills by four-legged robots. Finally, Section 5 presents our conclusions.

2. Preliminaries

2.1. Passing Motion

Ball passing is realized by accurate shooting motions that precisely move and stop a ball to an objective area, such as where a teammate is waiting. In this paper, we use AIBO as a legged robot, which is one of the robots permitted for use in the standard platform league in RoboCup. Shooting motions of AIBO are achieved by sending *frames*, consisting of the 15 joint angles for its head and legs, to OVirtualRobot every 8 ms. OVirtualRobot is a kind of proxy object that is defined in the software development kit OPEN-R for AIBO. In our framework, these frames are generated from *key-frames*, which are characteristic frames shaping the skeleton of each motion. For example, a kick motion needs at least two key-frames, since robots must pull and push its leg when executing it. We indicate the number of interpolations for each key-frame, so that whole frames can be generated by using a linear interpolation method. Thus, our motion takes $8n_i$ ms, where n_i is the total number of interpolations.

2.2. Training Equipment

Acquiring passing skills autonomously is usually difficult, because robots must be able to search for a kicked ball and then move the ball to an initial position. This requires sophisticated, low-level programs, such as an accurate, self-localization system, a ball acquiring skill, and a movement behavior with holding the ball. In order to avoid additional complications, we simplify the learning process a bit more. We can restrict ball passing to an anterior direction since our robot has rotational locomotion with holding the ball and can pass omni-directionally using only a forward passing motion. In other words, we can ideally treat our problem, which is to learn passing skills, one-dimensionally. In actuality though, the problem cannot be fully viewed in one-dimension because the ball might curve a little due to the grain of the grass.

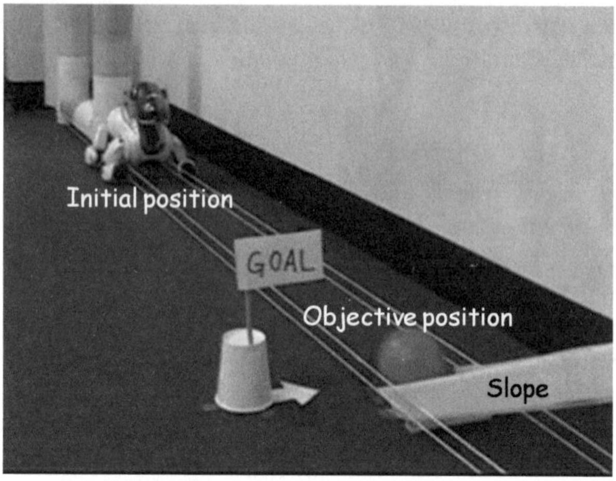

Figure 1. Training equipment for learning passing skills.

For autonomous learning of passing skills in one-dimension, we prepared almost the same training equipment as our previous work [9] as shown in Fig. 1. The equipment has rails of width nearly equal to an AIBO's shoulder-width. These rails are made of thin rope or string, and their purpose is to restrict the movement of the ball as well as the quadrupedal locomotion of the robot, to one-dimension. At the edge of these rails, there is a goal flag that shows the objective position of a passed ball. A slope just behind the flag can return the ball when passing is too strong, in the same way as an "automatic golf putting machine". Without the slope, the kicked ball would go beyond the objective position when the robot kicked it too strong, and it would take too much time for the robot to return the ball to the initial position where the robot first kicked it. Using the equipment, our robots can simply learn passing skills autonomously by kicking the ball straightforward and measuring the distance of the kicked ball on their own.

2.3. Problem Formulation

We directly utilize key-frames for learning of passing skills. It is possible to realize flexible search in the neighborhood of the skeleton without modeling the movement and setting extra-parametrization. All we do is create a sketchy motion as an initial motion, i.e., to indicate the key-frames for the motion. By adjusting the values of the key-frames of the initial motion, our robots learn an appropriate shooting motion, which is neither too strong nor too weak, so that the kicked ball will stop at an objective position.

We can apply the mirroring technique in the same way as Latzke et al. [12], since shooting motions are restricted to an anterior direction. In the mirrored situation, we can identify the angles of its right legs with those of its left legs and regard the horizontal "pan" angle of its neck as zero, although AIBO has 15 joint angles for its head and legs. As a consequence, the search space of the learning has $8n_k$-dimensions, where n_k means the number of key-frames.

Our learning is formalized as a maximization of a score function $f : X \rightarrow \boldsymbol{R}$ on an $8n_k$-dimensional search space $X \subseteq \boldsymbol{R}^{8n_k}$. The value $f(x)$ expresses the goodness of the shooting motion specified by motion parameters $x \in X$, that is, how closely the kicked ball stops at the objective position. The score is experimentally evaluated as follows:

First make our robot kick the ball by performing the motion, and then make the robot measure the distance d_{ball} from the robot to the point where the ball actually stopped. Since it obviously contains some noise in real environments, we treat the median $\hat{d}_{ball}$ of 5 executions of the trials for each x as the score $f(x) = \hat{d}_{ball}$. The score has the maximum value D_{goal}, which is the distance from the robot to the objective position shown by the goal flag in Fig. 1, because the ball will be returned back to the robot by the slope just behind the goal flag if it is kicked too strong. Moreover, for a constant D_{bounce}, we regard the score as zero if the returned ball is stopped within the distance D_{bounce}, because the ball might bounce off of its chest, and the estimated distance might be erroneous. In this paper, we set $D_{goal} = 800$ mm and $D_{bounce} = 300$ mm.

The distance 800 mm is used as the nearest distance between robots in the passing challenge, which was one of the technical challenges in the RoboCup four-legged league in 2007.

3. Learning Method

The score function in the previous section is obviously unknown, and so we utilize meta-heuristics for addressing maximization of the function. Meta-heuristics means heuristic algorithms that are independent of problems, such as Genetic Algorithm (GA), Simulated Annealing (SA), and Hill Climbing (HC). In this paper, we choose GA that was successfully utilized for the learning of shot motions by virtual four-legged robots in Kobayashi et al. [3].

For reducing the number of trials of robot learning, we utilize *thinning-out*, which will be described in Section 3.1. Thinning-out can judge whether or not a point sampled by meta-heuristics is promising according to the history of past trials. For the sake of discussion for now, we describe a point sampled by GA as a *candidate*. If a sampled candidate is regarded as unpromising, we can thin-out (or skip over) the candidate without evaluating it. In meta-heuristics with thinning-out only, when a sampled candidate is thinned-out a new candidate is simply resampled with random perturbation, i.e. mutation in GA, as shown by the left of Fig. 2. Therefore, samples with lower scores tend to be evaluated even in a later phase of learning, since samples in an unknown area where candidates have not been sampled yet can be regarded as promising. In order to make the resampling process more efficient, we propose a new method combined with a surrogate function $\hat{f}(x)$, which is an approximation model of $f(x)$. In this method, when a sampled candidate is thinned-out, the candidate is evaluated by the surrogate function instead of the original score function, and a new candidate is resampled by meta-heuristics, as shown by the right of Fig. 2. In this paper, we utilize kriging interpolation, which will be described in Section 3.2, as a surrogate function. We anticipate that the method has the advantage of accelerating the learning process, since meta-heuristics may be more efficient than random perturbation, and the disadvantage of increasing the possibility of convergence to a local optima, since the surrogate function might be wrongly approximated.

3.1. Thinning-out

In this section, we formalize our thinning-out technique for reducing unnecessary or unpromising trials. First, we assume that our score function is continuous and, to some

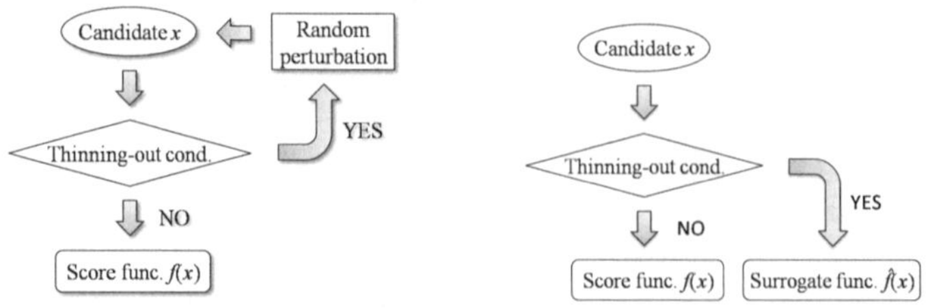

Figure 2. The left figure shows the concept of thinning-out only, and the right figure shows the concept of thinning-out with surrogate functions. When a sampled candidate is thinned-out, the former resamples a new candidate, while the latter evaluates a surrogate function instead of a score function.

extent, smooth over the search space. Our assumption seems to be reasonable, because each robot's movement is continuous, and thus small changes of parameters will not affect the score significantly. Based on this assumption, we can find out unpromising candidates theoretically by utilizing the degree of smoothness of the score function.

Now we define the local smoothness of the score function in terms of Lipschitz condition, which is found in standard textbooks on calculus. We use g-Lipschitz continuous for some function g, as a natural extension of c-Lipschitz continuous for some constant c in textbooks.

Definition 3.1 (Lipschitz condition) *Let $\boldsymbol{R}$ be the set of real numbers, X be a metric space with a metric d, and $f : X \to \boldsymbol{R}$ be a function on it. Given a function $g : \boldsymbol{R} \to \boldsymbol{R}$, f is said to be* g-Lipschitz continuous, *if it holds for any $x_1, x_2 \in X$ that*

$$|f(x_1) - f(x_2)| \leq g(d(x_1, x_2)).$$

The function g is called Lipschitz function.

From now on, we use d as the Euclidean metric. Supposing that a score function f is g-Lipschitz continuous, if a candidate satisfies the following thinning-out condition, it is guaranteed to safely thin-out the candidate, which will never become better than the current best score.

Definition 3.2 (Thinning-out condition) *Let x_b be the point with the current best score $f(x_b)$, x_c be a candidate point to try, and x_n be the nearest neighbor of the candidate. Given a Lipschitz function $g : \boldsymbol{R} \to \boldsymbol{R}$, x_c is said to* satisfy the thinning-out condition with respect to g, *if it holds that*

$$f(x_n) + g(d(x_c, x_n)) \leq f(x_b).$$

Since a Lipschitz function g is often not given in practical problems, we need infer the function g from the history of past trials. The naïve inferring method is Max Gradient (MG) method, which is defined by $g(x) = cx$ utilizing the maximum value c in the gradients between any two points in the history. We also proposed Gradient Differences (GD) inferring method. GD is intuitively a weighted average method of gradients of the score function using weights based on the inverse of the distance between any two points in the history of past evaluations. It will become a good approximation after enough evaluations, since a line connecting two close points can approximate the gradient of the function nearby the points.

3.2. Kriging Interpolation

Kriging [13] is one of the function interpolation or approximation methods of an unknown real function f, which is initially developed in geostatistics, and recently, many researchers successfully utilized kriging as a model of surrogate functions. Although there are several types of kriging, we choose *ordinary kriging* that is the most commonly used type of kriging. In ordinary kriging, its interpolator $\hat{f}$ at a point x^* is represented by a weighted linear combination as follows,

$$\hat{f}(x^*) = \sum_{i=1}^{n} w_i f(x_i),$$

where $f(x_1), ..., f(x_n)$ are the observed values of the function at some other points $x_1, ..., x_n$. In order to estimate the weights $w_1, ..., w_n$, we assumes that the observed values are the realization of a stochastic process with *second-order stationarity*, which means that the expected values are constants, i.e., $E[f(x_i) - f(x_j)] = 0$, and the covariances are dependent only on the distances, i.e., $Cov[f(x_i), f(x_j)] = C(d(x_i, x_j))$. The function C is called a *covariance function* and describes a correlation between any two points in terms of the distance. In this paper, we define $C(x) = \sigma^2 \exp(-\theta x^2)$ utilizing an isotropic Gaussian function, where $\sigma^2 (= C(0))$ is the variance $Var[f(x_i)]$, based on the heuristics that the closer points correlate outputs more positively.

The weights for x^* is chosen such that the error variance $V_e = Var(\hat{f}(x^*) - f(x^*))$ is minimized subject to $\sum_{i=1}^{n} w_i = 1$, which is given by the unbiased condition of the interpolator, i.e., $E[\hat{f}(x^*) - f(x)] = 0$, and the property of second-order stationarity. Utilizing a Lagrange multiplier λ, we can solve it by setting $V_e' = V_e + 2\lambda(\sum_{i=1}^{n} w_i - 1)$ and calculating $\frac{\partial}{\partial w_i} V_e' = 0$. Thus, the weights are given by the following $n+1$ equations.

$$\begin{cases} \sum_{i=1}^{n} w_i C(d(x_i, x_j)) + \lambda = C(d(x_i, x^*)) & \text{for } j = 1, ..., n, \\ \sum_{i=1}^{n} w_i = 1. \end{cases}$$

4. Experiments and Results

4.1. Minimization of Test Functions

For verifying the performance of our proposed method, we first address the minimization of the 6 mathematical test functions, Rastrigin, Schwefel, Griewank, Rosenbrock, Ridge, and Ackley. They have often been used for the performance evaluations of metaheuristics and are also utilized in Kobayashi et al. [3]. Rastrigin, Schwefel, Griewank, and Ackley have multiple peaks, although Griewank and Ackley have a single peak with a global view. Griewank, Rosenbrock and Ridge have the design variables' dependency.

In this section, SGA means a simple, real-coded GA with uniform crossover, whose rate is 0.3, Gaussian mutation with mean of 0 and variance of 10% of the domain size of each dimension, whose rate is 0.2, and roulette selection with elite strategy, and its

Table 1. This table shows the minimization results of 10-dimensional 6 test functions by SGA, GAT and GATS. Where, "min", "trial", and "error" represent the minimum scores in 100 trials, the trial rates in 100 candidates, and the error rates in 100 candidates, respectively. The trial rate means 100×#(tried candidates) / #(all candidates), and error rate means 100×#(wrongly thinned-out candidates) / #(all thinned-out candidates). Each value is the average over 100 experiments. As for GA, the trial rates and error rates are 100% and 0% (or undefined), respectively.

function	SGA	GAT			proposed GATS		
	min	min	trial (%)	error (%)	min	trial (%)	error (%)
Rastrigin	260	165	54.20	0.80	152	38.67	0.40
Schwefel	3583	1817	62.84	0.87	1305	42.63	0.17
Griewank	621	211	48.24	0.09	112	35.81	0.00
Rosenbrock	17472	3326	54.75	0.06	2265	39.34	0.00
Ridge	5.7e9	6.4e8	55.42	0.04	2.3e8	38.58	0.00
Ackley	21	21	60.37	0.92	21	43.26	0.05

population size is 20. GAT means SGA with ϵ*-thinning-out*, where $\epsilon = 0.01$, with MG inferring method. ϵ-thinning-out forcibly evaluates a candidate with probability ϵ, and otherwise, it skips over the candidate, so that it can hold out the possibility for evaluating candidates that are wrongly thinned-out once and avoid never halting by thinning-out all candidates. GATS means our proposed method, that is GAT with a surrogate function of ordinary kriging interpolation described in Section 3.2. The parameter θ of the covariance function is estimated by maximum likelihood estimation.

Table 1 shows the minimum scores, trial rates, and error rates of SGA, GAT, and GATS. The trial rate means the rate of tried candidates in all candidates, and the error rate means the rate of wrongly thinned-out candidates, whose scores (calculated by only for the error rate) were actually better than the current best score, in thinned-out candidates. Wrongly thinned-out candidates are examined by calculating thinned-out scores only for error rate. Note that even if candidates are thinned-out at random, the error rate becomes very small. All values of the minimum scores, trial rates, and error rates become better, as they become smaller. This section focuses on the comparison between GAT and GATS, since the comparison between SGA and GAT was already discussed in Kobayashi et al. [3]. As for GATS, we anticipated that the trial rates become higher, and the error rate become lower, because the efficient resampling method should choose promising candidates that can not be thinned-out. However, against our anticipation, the trial rate of GATS is lower than those of GAT, while, as expected, the error rates of GATS are lower than those of GAT. This was because evaluated points with good scores should make easy situations to thin-out new sampled candidates, although GATS surely tends to evaluate resampled candidates. Consequently, the fact that both trial rates and error rates are decreased leads to the good result that the minimum scores of GATS are better than those of GAT over all test functions.

4.2. Learning of Passing Skills

In this section, we address the learning of ball passing by four-legged robots, i.e., maximization of the score function formalized by Section 2.3. SGA, GAT, and GATS are almost the same in the previous section. The only following parameters are different. SGA utilizes discrete mutation, which randomly adds one from $\{-r, 0, +r\}$ in each dimension, where $r = \pi/36$ radians, and the population size is 10. GAT utilizes GD inferring method, which was effective for score functions over high dimensional spaces in our previous experiments [3].

The initial motion for learning is a motion of pushing a ball with its chest, which is actually used as a shooting motion in games in RoboCup. The motion is performed not only by moving the chest ahead, but also by moving the whole body so as to enhance the power of the shot. As the shooting motion involves the movement of almost all joint parts of the robot, it is not easy to even adjust the motion. Therefore, it is quite difficult for humans to design the shooting motion so that the ball goes to the objective position. The motion takes 482 ms because it has 54 total interpolations ($n_i = 54$), and can move the ball to approximately 1500 mm. Since the motion consists of 6 key-frames ($n_k = 6$), the search space of this learning problem is 48-dimensions.

Fig. 3 shows a comparison of the learning processes obtained by running SGA, GAT, and GATS for 50 trials. Unlike the previous experiment, the scores on these processes become better, as they become higher. Since the score of each trial is calculated by the median of 5 motion executions, our robot, in fact, needs 250 motion executions in total for 50 trials. Each experiment took more than 3 hours, and we must have exchanged more than 6 batteries since our robot needs one battery every 30 minutes. The figure indicates that GATS and GAT are obviously more efficient than SGA, and GATS seems to be slightly better than GAT. Unfortunately, we could not conduct any more experiments, since we realized that the learning of shooting motions damages or breaks servo motors of robots more frequently than expected. At least in this section, however, we could verify that our thinning-out method was successfully applied to the learning problem in the real world for the first time. Our robot finally acquired the best score 777 by utilizing GATS, which means an accuracy of 23 mm, since the target distance is 800 mm in this experiment. The acquired motion was not much different from the initial motion. The fact implies that the adjustment of passing motions is difficult for humans. All movies of the earlier and later phases of our experiments are available on-line [3].

5. Conclusions and Future Work

In this paper, we proposed an autonomous learning method of ball passing skills and a hybrid method using thinning-out and surrogate functions. The former realizes that robots learn ball passing skills without human intervention, except for replacing discharged batteries. The latter reduces the number of trials efficiently with few errors and improves the result of learning. By using the proposed two methods, our robot successfully acquired an appropriate passing motion with an accuracy of 23 mm on their own. Our experiments were also the first application of thinning-out in the real world.

Future work includes extending passing skills into two-dimensions and toward arbitrary objective distances. We can acquire more practical passing skills by replacing the score function with a new score function that refers the average (or median) and variance of the position of each kicked ball in two-dimensions. We can also acquire more flexible passing skills by utilizing the interpolation of some passing motions.

[3] http://www.shino.ecei.tohoku.ac.jp/~kobayashi/movies.html#passing

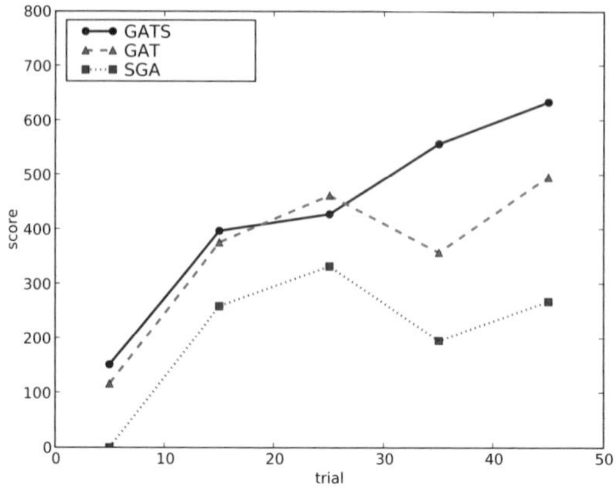

Figure 3. Learning processes of SGA, GAT, and GATS in terms of trials. The solid line, dashed line, and dotted line show the learning processes of GATS, GAT, and SGA, respectively. Each point means the average of each population, e.g. 10 trials.

References

[1] RoboCup Official Site. `http://www.robocup.org/`.

[2] Juan Cristóbal Zagal and Javier Ruiz del Solar. Learning to Kick the Ball Using Back to Reality. In *RoboCup 2004: Robot Soccer World Cup VIII*, volume 3276 of *LNAI*, pages 335–347. Springer-Verlag, 2005.

[3] Hayato Kobayashi, Kohei Hatano, Akira Ishino, and Ayumi Shinohara. Reducing Trials Using Thinning-out in Skill Discovery. In *Proceedings of the 10th International Conference on Discovery Science 2007 (DS2007)*, volume 4755 of *LNAI*, pages 127–138. Springer-Verlag, 2007.

[4] Gregory S. Hornby, Seichi Takamura, Takashi Yamamoto, and Masahiro Fujita. Autonomous Evolution of Dynamic Gaits With Two Quadruped Robots. *IEEE Transactions on Robotics*, 21(3):402–410, 2005.

[5] Min Sub Kim and William Uther. Automatic Gait Optimisation for Quadruped Robots. In *Proceedings of 2003 Australasian Conference on Robotics and Automation*, pages 1–9, 2003.

[6] Nate Kohl and Peter Stone. Machine Learning for Fast Quadrupedal Locomotion. In *The Nineteenth National Conference on Artificial Intelligence (AAAI2004)*, pages 611–616, 2004.

[7] Joel D. Weingarten, Gabriel A. D. Lopes, Martin Buehler, Richard E. Groff, and Daniel E. Koditschek. Automated Gait Adaptation for Legged Robots. In *IEEE International Conference on Robotics and Automation*, 2004.

[8] Peggy Fidelman and Peter Stone. Learning Ball Acquisition on a Physical Robot. In *2004 International Symposium on Robotics and Automation (ISRA)*, 2004.

[9] Hayato Kobayashi, Tsugutoyo Osaki, Eric Williams, Akira Ishino, and Ayumi Shinohara. Autonomous Learning of Ball Trapping in the Four-legged Robot League. In *RoboCup 2006: Robot Soccer World Cup X*, volume 4434 of *LNAI*, pages 86–97. Springer-Verlag, 2007.

[10] Yasuhito Sano, Hajime Kita, Ichikai Kamihira, and Masashi Yamaguchi. Online Optimization of an Engine Controller by means of a Genetic Algorithm using History of Search. In *Proceedings of the 3rd Asia-Pacific Conference on Simulated Evolution and Learning (SEAL)*, pages 2929–2934, 2000.

[11] Alain Ratle. Accelerating the Convergence of Evolutionary Algorithms by Fitness Landscape Approximation. In *Proceedings of the Fifth International Conference on Parallel Problem Solving from Nature (PPSN V)*, volume 1498 of *LNCS*, pages 87–96. Springer-Verlag, 1998.

[12] Tobias Latzke, Sven Behnke, and Maren Bennewitz. Imitative Reinforcement Learning for Soccer Playing Robots. In *RoboCup 2006: Robot Soccer World Cup X*, volume 4434 of *LNAI*, pages 47–58. Springer-Verlag, 2007.

[13] Georges Matheron. Principles of geostatistics. *Economic Geology*, 58(8):1246–1266, 1963.

Intelligent Autonomous Systems 10
W. Burgard et al. (Eds.)
IOS Press, 2008

doi:10.3233/978-1-58603-887-8-155

Robust Mission Execution for Autonomous Urban Driving

Christopher R. BAKER [a,1], David I. FERGUSON [b], and John M. DOLAN [a]
[a] *Carnegie Mellon University, Pittsburgh, PA, USA*
[b] *Intel Research Pittsburgh, Pittsburgh, PA, USA*

Abstract. We describe a multi-modal software system for executing navigation missions in an urban environment, focusing on the robust treatment of anomalous situations such as blocked roads, stalled vehicles and tight maneuvering. Various recovery mechanisms are described relative to the nominal mode of operation, and results are discussed from the system's deployment in the DARPA Urban Challenge.

Keywords. Error Recovery, Autonomous Vehicles, Urban Challenge, Tartan Racing, Boss

Introduction

The ability to cope reliably with unforeseen circumstances is a critical feature of any fully autonomous robotic system. This is especially true in the case of complex, highly structured environments such as urban road networks in the context of the Urban Challenge [6], an autonomous vehicle competition sponsored by the US Defense Advanced Research Projects Agency (DARPA). Contestant robots were required to autonomously execute a series of navigation missions in a simplified urban environment consisting of roads, intersections, and parking lots, while obeying road rules and interacting safely and correctly with other vehicles. Unlike the previous challenge [1,2], which focused on rough-terrain navigation, this competition combined the possibilities of local obstructions with the uncertainties of interacting with other traffic, forming a complex problem space that defied rigorous treatment *a priori*. As with other mission-oriented, fully autonomous robotic systems [3], robust mission execution required a system for selecting and executing various recovery maneuvers aimed at overcoming or circumventing anomalous situations. This paper describes the mission recovery mechanisms employed by Boss, Tartan Racing's winning entry to the Urban Challenge, shown in Figure 1(a). Section 1 provides a brief overview of the total software system to provide a context for discussion of the robot's motion planning capabilities and error recovery mechanisms in Sections 2 and 3. The performance of the system during the Urban Challenge Final Event is reviewed in Section 4, and we conclude with key lessons learned in Section 5.

[1]Corresponding Author: Christopher R. Baker, Robotics Institute, Carnegie Mellon University, 5000 Forbes Ave., Pittsburgh, PA 15213, USA; E-mail cbaker@andrew.cmu.edu

(a) Boss: Tartan Racing's Entry

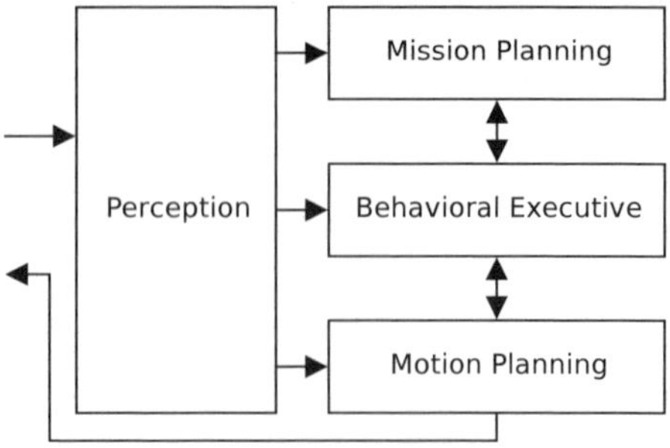

(b) Software Subsystem Architecture

Figure 1. Total system structure for the Urban Challenge

1. System Architecture

The software system that controls Boss is divided into four primary subsystems, shown in Figure 1(b). They communicate via message-passing according to the anonymous publish-subscribe [4] pattern and work in concert to ensure safe, reliable and timely mission execution. Their responsibilties and interactions, discussed briefly below, are presented more thoroughly in [9].

The *Perception* subsystem processes sensor data from the vehicle and produces a collection of semantically-rich data elements such as the current pose of the robot, the geometry of the road network, and the location and nature of various obstacles such as road blockages and other vehicles.

The *Mission Planning* subsystem computes the fastest route to reach the next checkpoint from all possible locations within the road network, encoded as an estimated time-to-goal from each waypoint in the network. This estimate incorporates knowledge of road blockages, speed limits, intersection complexity, and the nominal time required to make special maneuvers such as lane changes or U-turns.

The *Behavioral Executive* combines the global route information provided by the Mission Planner with local traffic and obstacle information provided by Perception to select a sequence of incremental goals for the Motion Planning subsystem to execute. Typical goals include driving to the end of the current lane or maneuvering to a particular parking spot, and their issuance is predicated on conditions such as precedence at an intersection or the detection of certain anomalous situations. These anomalous situations trigger the selection of recovery goals as described in Section 3.

The *Motion Planning* subsystem is responsible for the safe, timely execution of the incremental goals issued by the Behavioral Executive. The isolation of goal selection from goal execution promotes the development of powerful, highly general planning capabilities, which fall into two broad contexts: on-road driving and unstructured driving. A separate path-planning algorithm is used for each context, and the nature and capabilities of each planner have a strong influence on the overall capabilities of the system, including the nature of common failure scenarios and the options for attempting recovery maneuvers. The following section describes these planning elements in greater detail to provide a foundation for discussion of the rest of the recovery system.

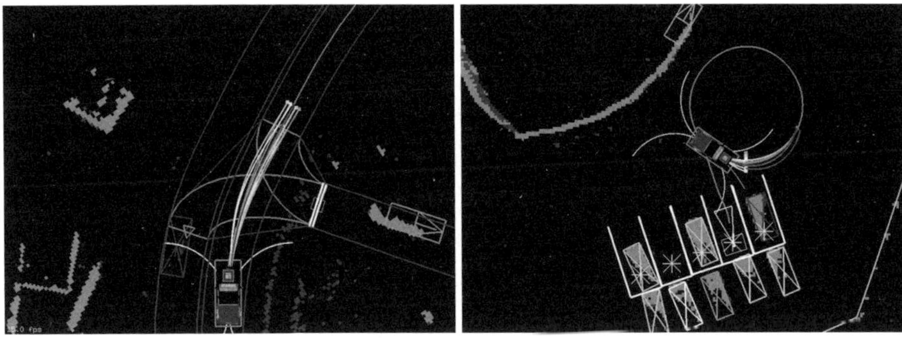

Figure 2. Left: Motion planning on roads. The centerline of the desired lane is shown (red) with the generated trajectories (various colors) extending along the lane. The obstacle map is overlaid (red cells are lethal; blue are curb estimates), and two detected vehicles are shown as rectangles. **Right:** Motion planning in unstructured areas. The vehicle is executing a path (red) to a parking spot between several other vehicles.

2. Motion Planning

Given a goal from the Behavioral Executive, the Motion Planner first creates a path towards the goal, then tracks it by generating a set of candidate trajectories following the path to varying degrees and selecting the best collision-free trajectory from this set. The algorithms used to generate this path are different for on-road and unstructured driving situations, as illustrated in Figure 2.

During on-road navigation, the goal from the Behavioral Executive describes a desired position along a lane. In this case, the Motion Planner constructs a curve along the centerline of the desired lane, representing the nominal path of the vehicle. A set of local goals are selected at a fixed longitudinal distance down this centerline path, varying in lateral offset to provide several options for the planner. Then, a model-based trajectory generation algorithm [7] is used to compute dynamically feasible trajectories to these local goals. The resulting trajectories are evaluated against their proximity to static and dynamic obstacles, their distance from the centerline path, their smoothness, and various other metrics, and the best trajectory is selected for execution by the vehicle. This technique is very fast and produces smooth, high-speed trajectories for the vehicle, but it can fail to produce feasible trajectories when confronted with aberrant scenarios, such as blocked lanes or extremely tight turns. In these cases, the recovery algorithms discussed in Section 3 call on the unstructured planner described below to extricate the system.

When driving in unstructured areas, the goal from the Behavioral Executive simply describes a desired pose for the vehicle, typically a parking spot or lot exit. To achieve this pose, a lattice planner searches over vehicle position (x, y), orientation (θ), and velocity (v) to generate a sequence of feasible maneuvers that are collision-free with respect to static and dynamic obstacles [8]. This planner is much more powerful and flexible than the on-road planner, but it also requires more computational resources and is limited to speeds below 15 mph. The recovery system leverages this flexibility to navigate congested intersections, perform difficult U-turns, and circumvent or overcome obstacles that block the progress of the vehicle. In these situations, the lattice planner is typically biased to avoid unsafe areas, such as oncoming traffic lanes, but this bias can be removed as the situation requires.

3. Behavioral Executive

During normal operation, the Behavioral Executive selects and issues goals to the Motion Planner according to three situational contexts:

- **Lane Driving**, in which the system traverses a road of one or more lanes while requiring maintenance of safe inter-vehicle separation and adherence to rules governing passing maneuvers and stop-and-go traffic;
- **Intersection Handling**, requiring the determination of precedence among stopped vehicles and safe merging into or across moving traffic at an intersection; and
- **Zone Maneuvering**, in which the system maneuvers through an unstructured obstacle or parking zone.

The active context is determined by the location of the system within the road network and by the best incremental action to take from that location to reach the next checkpoint. That action is issued as a goal to the Motion Planning subsystem, and its execution is monitored until it is either completed successfully or it is determined to have failed. Goal failures can either be directly reported by the Motion Planner or inferred by monitoring forward progress over some span of time. In both cases, the failure is treated equally and triggers the selection and issuance of a recovery goal.

A good recovery goal selection algorithm should be able to generate a non-repeating sequence of novel recovery goals in the face of repeated failures *ad infinitum*. It should be sensitive to the original driving context, as each calls out a specific underlying planner which has certain properties that must be considered when selecting recovery goals. In addition, each context is governed by a specific set of road rules that must be adhered to if possible, but disregarded if necessary in a recovery situation. The recovery system should also be minimally complex so as to be implementable in the available time.

Keeping these requirements in mind, recovery goals are selected as a direct function of the original failed goal, which encodes the driving context, and a notion of the current *recovery level*. When in normal operation, the recovery level is set to zero, and it is incremented whenever any goal, normal or recovery, fails. The intent is that smaller values for the recovery level yield low-risk, easy-to-execute maneuvers that are tuned to handle common and benign situations. If these initial recovery goals fail, increasing values for the recovery level yield higher-risk maneuvers, enabling more drastic measures in more convoluted situations. It is important to note that the recovery goal selection process does not explicitly incorporate any surrounding obstacle data, making it intrinsically robust to transient environmental effects or perception artifacts that may cause spurious failures in other subsystems.

All recovery goals are specified as a span of poses to achieve, leveraging the power of the underlying lattice planner to perform whatever actions are necessary to get the system back on-course. The successful completion of any recovery goal resets the system back to normal operation, requiring that all recovery goals terminate at valid locations in the world. This eliminates the possibility of complex multi-maneuver recovery schemes, but it was deemed that situations requiring a more complex recovery system were simply out of scope. Should the same, or a very similar, normal goal fail immediately after a seemingly successful recovery sequence, the previous recovery level is reinstated instead of simply incrementing from zero to one. This bypasses the recovery levels that presumably failed to get the system out of trouble and immediately selects goals at a higher level.

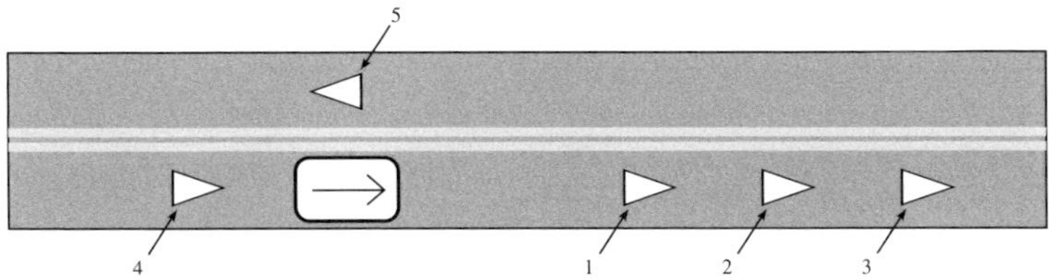

Figure 3. Example Lane Recovery Sequence

Each nominal driving context is backed by a unique recovery goal selection algorithm designed to complement the functionality of the underlying planner and observe as many situational rules as possible while still moving forward with the mission. The first and most commonly encountered recovery scenario is an error reported by the Motion Planner while driving down a lane. Typical causes for such a failure include:

- Small or transient obstacles, such as traffic cones, that do not block, but constrain the road such that the assumptions made by the lane planner are violated. In this case, a recovery goal specified a short distance ahead in the center of the lane will allow the lattice planner to quickly generate a path around the partial obstruction;
- Larger obstacles such as road barrels, or other cars that are detected too late for the distance-keeping behavior to come to a graceful stop, with the resultant pose requiring a back-up or other non-trivial maneuver to extricate the system;
- Low-hanging canopy that requires cautious and careful planning, perhaps including departure from the lane of travel, to guarantee the safety of the system;
- Lanes, especially virtual paths through an intersection, whose local shapes are kinematically infeasible, requiring a series of three-point turn maneuvers to re-align the system with the road.

Keeping these and other similar causes in mind, the algorithm for lane recovery goal selection, illustrated in Figure 3, selects an initial set of goals forward along the lane from the original failure position with the distance forward described by:

$$D_{\text{recovery}} = D_{\text{offset}} + \text{RecoveryLevel} * D_{\text{incremental}} \tag{1}$$

The values for D_{offset} and $D_{\text{incremental}}$ were tuned through testing to $20m$ and $10m$ respectively, providing good results in most situations. These forward goals (Goals 1,2,3 in Figure 3) are constrained to remain in the original lane of travel and are selected out to some maximum distance (roughly $50m$, corresponding to the system's high-fidelity sensor-range), after which a goal is selected a short distance behind the vehicle (Goal 4) with the intent of backing up and getting a different perspective on the situation.

After backing up, the sequence of forward goals is allowed to repeat once more with a $5m$ offset, after which continued failure triggers higher-risk maneuvers depending on the nature of the road. If there is a lane available in the opposing direction, the road is eventually marked as fully blocked, and the system issues a U-Turn goal (Goal 5) and attempts to follow an alternate path to goal. Otherwise, the system is trapped on a one-way road, so the forward goal selection process from Equation 1 is allowed to continue forward beyond the $50m$ limit. These farther-reaching goals remove the constraint of staying in the original lane, giving the lattice planner complete freedom to achieve the goals by any means possible. In either case, these maneuvers disregard all normal rules

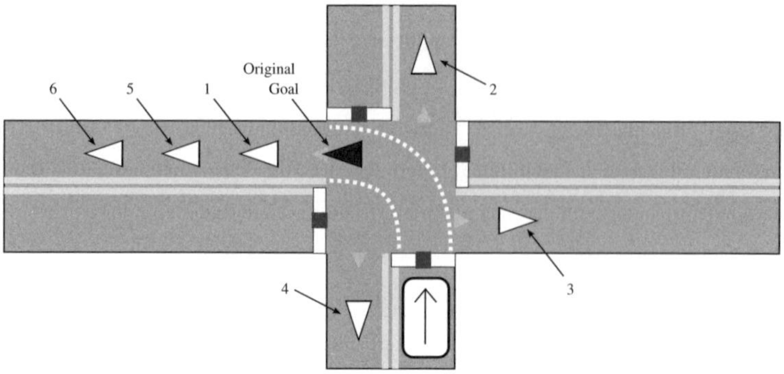

Figure 4. Example Intersection Recovery Sequence

governing interaction with other traffic on the road and thus are only issued at high recovery levels, typically greater than 20.

The second and perhaps most intricate scenario is recovering from failed intersection goals. The desired path through an intersection is a smooth interpolation between the two connected roads, as shown in Figure 4. It can often be narrow with high curvature and can intersect various obstacles such as guard rails and road signs, so the first and simplest recovery goal (Goal 1) is to try to achieve a pose slightly beyond the original goal using the lattice planner, giving the whole intersection as its workspace. This quickly recovers the system from small or spurious failures in intersections and compensates for intersections with turns that are beyond the one-pass kinematic capabilities of the robot.

If that first recovery goal fails, alternate routes out of the intersection (Goals 2,3,4) are attempted until all alternate routes to goal are exhausted. Thereafter, the system removes the constraint of staying in the intersection and selects pose goals incrementally deeper (Goals 5,6) into the lane or parking lot associated with the original failed goal. For lanes, this is almost identical to the selection of lane recovery goals, except that there is no initial offset. For parking lots, the goals are specified as incrementally deeper and larger spaces within the lot, as the specific terminal pose of the vehicle does not matter so long as it finds its way into the parking lot.

The case where specifics do matter, however, is the third recovery scenario, failures in parking lots. Because the lattice planner is always invoked when planning in parking lots, failure implies one of the following:

1. The path to the goal has been transiently blocked by another vehicle. While the rules guaranteed that parking spots will always be free, they made no such guarantees about traffic accumulations at lot exits, or about the number of vehicles that may be operating in the parking lot at any one time. In these cases, the system should try to move close to the original goal and allow the blockage some time to pass.
2. Due to some sensor artifact, the system believes there is no viable path to goal. In this case, selecting nearby goals that offer different perspectives on the original goal area may relieve the situation.
3. The path to the goal is genuinely fully blocked and the parking lot is effectively subdivided. This was considered to be beyond the scope and spirit of the competition and is very challenging to handle in the general case.

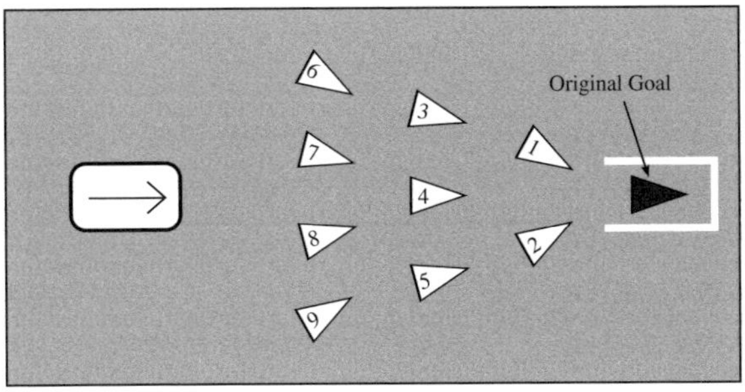

Figure 5. Example Parking Lot Recovery Sequence

The goal selection algorithm for failed parking lot goals selects a sequence of goals in a regular, triangular pattern, centered on the original goal as shown in Figure 5. If failures continue through the entire pattern, the next set of actions is determined by the exact nature of the original goal. For parking spot goals, the pattern is repeated with a small incremental angular offset *ad infinitum*, guided by the assumption that the parking spot is empty. Over time, all possible approaches to the parking spot may be attempted, but the system will not proceed on the mission until the parking spot has been attained[2]. Lot exits, on the other hand, will be marked as blocked and the system will attempt to re-route through alternate exits in a way similar to the alternate path selection in the intersection recovery algorithm above.

If all other exits are exhausted, the system issues goals extending outward along the road network from the exit according to a semi-random breadth-first search. In this fourth and last-ditch recovery strategy, increasing values of the recovery level call out farther paths, and the lattice planner is not constrained to stay on roads or in parking lots. These high-risk, loosely specified goals are meant to be invoked when each of the other recovery goal selection schemes has been exhausted. In addition to parking lot exits, these are selected for lane recovery goals that run off the end of the lane and similarly for intersection recovery goals when all other attempts to get out of an intersection have failed.

Through these four recovery algorithms, all possible paths forward can be explored from any single location, leaving only the possibility that the system is stuck due to a sensor artifact or software bug that requires a small local adjustment to escape, as opposed to the selection of a farther, riskier goal. To compensate for this possibility, a completely separate recovery mechanism monitors the vehicle's absolute motion, and if the vehicle does not move at least one meter every five minutes, it overrides the current goal with a randomized local maneuver. When that maneuver is completed, the entire goal selection system is re-initialized from the new location, clearing error recovery and execution state for another attempt. The goals selected by this recovery mechanism are meant to be kinematically feasible and can be either in front of or behind the vehicle's current pose. They are biased somewhat forward of the robot's position so there is statistically net-forward motion if the robot is forced to choose these goals repeatedly over time. This functionality *suppresses* the functionality of the other goal selection systems, and is analogous to the "Wander" behavior from [5].

[2]The parking spot could easily be bypassed after some number of attempts, but the penalty for skipping a checkpoint was unspecified and may have been interpreted as a failed mission.

	Mission Run Time	Recovery Events	Recovery Time Total	Recovery Time Average	Single-Level Recoveries	Multi-Level Recoveries
Mission 1	$89m, 41s$	10	$8m, 18s$	$49.8s$	8	2(9, 22)
Mission 2	$56m, 21s$	3	$39s$	$13s$	3	0
Mission 3	$85m, 53s$	4	$2m, 37s$	$39.2s$	3	1(23)
Cumulative	$231m, 55s$	17	$11m, 34s$	$40.8s$	14	3

Table 1. Recovery System Performance in the Urban Challenge Final Event

4. Event Analysis

The Urban Challenge Final Event (UCFE) consisted of a series of three missions, covering roughly 60 miles of urban roads and meant to be completed in less than 6 hours. Analysis of over 140 gigabytes of logs from the event shows that the recovery system played an important role in the vehicle's success, being called on to recover the system from over a dozen anomalous situations as summarized in Table 1.

Of the 17 recovery events during the UCFE, only three exceeded the first recovery level. These more complicated situations consumed $5m, 41s$, representing nearly half of the total time spent in recovery, and can be traced to small collections of software bugs or sensor malfunctions. The level-9 recovery in the first mission was incited by an *all-pause*[3] event that lasted roughly ten minutes. When the *all-pause* was lifted, recovery goals were issued along the road, but the planner was unable to find a collision-free action away from a roadside barrier due to GPS drift while the vehicle was paused. Eventually, the positioning system corrected itself and the robot was able to proceed. The primary contribution of the recovery system in this instance was the continual issuance of novel goals in the face of repeated failures, though no one particular goal resolved the situation.

The two higher-order recovery events, level 22 in mission one and level 23 in mission three, were due to a combination of perception and planning bugs that caused the system to believe that the path forward along the current lane was completely and persistently blocked by other traffic where there was, in fact, no traffic at all. After repeated attempts to move forward, the recovery system declared the lane fully blocked, commanded a U-Turn, and followed an alternate route to goal. This demonstrated the recovery system's ability to compensate for subtle but potentially lethal bugs in other components.

Discarding these more complex events, the remaining 14 were resolved in an average of 25 seconds and had a span of causes ranging from dust and transient sensor artifacts to vehicles proceeding out of turn at intersections. In all cases, the system recovered quickly and gracefully, demonstrating an appropriate emphasis on the selection of nearby, low-risk recovery maneuvers early in the recovery sequence.

5. Conclusions

The results from deployment in the UCFE show that the recovery system was well suited to the challenges posed by this competition. The decision against complex environmental reasoning in favor of comparatively simple, incremental goal selection algorithms led to

[3]Urban Challenge officials had the ability to temporarily disable all vehicles on the course to, for example, remove a disabled vehicle from the course or to resolve some dangerous situation.

a highly robust mechanism for overcoming a wide range of anomalous circumstances. The ability to quickly recover from simple failures also provided a safety net for all other elements of the system, as transient false-positives from obstacle detection systems or tracking failures from the Motion Planning subsystem were easily tolerated, often transparently so to an external observer. More complex scenarios often took longer than desirable to resolve, but the relative infrequency of these scenarios, when coupled with the fact that the system eventually recovered itself in every occurrence, demonstrate again the system's effectiveness in the scope of the Urban Challenge. That the Final Event did not exercise the deepest reaches of the various recovery mechanisms is a testament to the system's preparedness and to the importance of rigorous, clever and sometimes adversarial testing by a highly dedicated test team.

While many of the details of the recovery heuristics were tailored to the nature of the competition, several core concepts are generally applicable to autonomous mobile robots. Splitting the navigation problem into separate entities for goal selection and execution allows for the development of powerful, highly general motion planning capabilities which can then be leveraged by any goal selection mechanism to explore many possible resolutions to any one failure. Given a powerful underlying planner, the blind selection of incrementally higher-risk recovery goals is an effective means of quickly and safely resolving benign situations while retaining the ability to incrementally discard environmental rules regarding structure and interaction in favor of continuing onward. Most importantly, an effective recovery mechanism is a critical element of any fully autonomous robotic system, providing alternate paths forward from both internal faults and external obstacles and guaranteeing the timely completion of the system's mission.

Acknowledgments

This work would not have been possible without the dedicated efforts of the Tartan Racing team and the generous support of our sponsors including General Motors, Caterpillar, and Continental. This work was further supported by DARPA under contract HR0011-06-C-0142.

References

[1] Special Issue on the DARPA Grand Challenge, Part 1. *Journal of Field Robotics*, 23(8), 2006.

[2] Special Issue on the DARPA Grand Challenge, Part 2. *Journal of Field Robotics*, 23(9), 2006.

[3] Christopher Baker et al. A campaign in autonomous mine mapping. In *Proceedings of the IEEE Conference on Robotics and Automation (ICRA)*, volume 2, pages 2004 – 2009, April 2004.

[4] Len Bass, Paul Clements, and Rick Kazman. *Software Architecture in Practice*. Addison-Wesley, 2003.

[5] Rodney A Brooks. A robust layered control system for a mobile robot. *IEEE Journal of Robotics and Automation*, 2(1):14–23, March 1986.

[6] Defense Advanced Research Projects Agency (DARPA). Urban challenge website, July 2007. http://www.darpa.mil/grandchallenge.

[7] T. Howard and A. Kelly. Optimal rough terrain trajectory generation for wheeled mobile robots. *International Journal of Robotics Research*, 26(2):141–166, 2007.

[8] M. Likhachev and D. Ferguson. Planning Dynamically Feasible Long Range Maneuvers for Autonomous Vehicles. In *Proceedings of Robotics: Science and Systems (RSS)*, 2008.

[9] Chris Urmson et al. Autonomous Driving in Urban Environments: Boss and the DARPA Urban Challenge. *Accepted to Journal of Field Robotics*, 2008.

Intelligent Autonomous Systems 10
W. Burgard et al. (Eds.)
IOS Press, 2008

doi:10.3233/978-1-58603-887-8-164

Self-Organizing Stock Assignment for Large-scale Logistic Center

Masashi FURUKAWA [a,1], Yoshinobu KAJIURA [a], Akihiro MIZOE [b], Michiko WATANABE [c], Ikuo SUZUKI [a] and Masahito YAMAMOTO [a]
[a] *Dept. of Information Science and Technology, Hokkaido University, Japan*
[b] *System Labo Murata, Japan*
[c] *Dept. of Mechanical System Engineering, Kitami Institute of Technology, Japan*

Abstract. This Study focuses on efficient management for a large-scale logistic center and proposes a new method how to assign stocked products to racks in a warehouse so as to minimize a round-up time (the maximum make-span time) to fetch ordered products. Self-organizing map (SOM), one of artificial neural networks, is employed to determine stocked products' allocation to the racks. In applying SOM, a number of order forms for rounding up the products are treated as a set of input signals and rack positions are regarded as SOM topology. Numerical simulation proves that the proposed method allow us to determine an efficient assignment of the stocked products to the racks.

Keywords. self-organizing map, machine learning, stock assignment problem, logistic center, physical distribution system

Introduction

The more web-order and mail-order shopping becomes popular, the more large-scale logistic centers play an important role in a physical distribution system. In a world-wide scale, Amazon-com is famous for web-order book shopping and it has large scale logistic centers in the world. There are many web-order (on-line) and mail-order shops in many fields such as foods, cloths, shoes, writing materials, computers and so on in Japan as well. For instance, the web-order and mail-order shops for some writing materials prepare for about eight thousands products and receives about thirty thousands order forms for products (items) a day. Large-scale logistic centers for these shops must shipment ordered products by a due day in time. This requests quick and efficient products management for the logistic centers.

This management is regarded as a scheduling problem, in which workers with cart or automatically guided vehicles (AGVs) collects designated products according to the given product order forms by dropping in racks (storages) and picking up products at a warehouse as fast as they can. We have dealt with this problem as an agent problem. Namely, an agent must solve the following two problems, (1) how product order forms are assigned to the agent, and (2) how the agent makes the shortest path plan to collect

---

Corresponding Author: Professor, Information and Science Technology, Hokkaido University, Nishi 9 Kita 14 Sapporo, Japan; E-mail: mack@complex.eng.hokudai.ac.jp products designated by product order forms. The first problem belongs to one of assignment problems and the second one is considered as an n-traveling salesman problem. The first problem solution has a great effect with the second problem solution. Therefore, these problems can be thought as a quadratic assignment problem.

On the other hand, if we consider the problem as the agent problem, the agent activity must be affected by an environment. The problem that we have dealt with is no exception any longer. When different order forms are similar each other, it is better to allocate products designated by order forms to be stocked at closer racks. Therefore, products' allocation to racks is also an important environment problem against the agent problem.

This paper focuses on products' allocation to racks at a warehouse in a real large-scale logistic center and proposes a new method for allocating products to racks. For this purpose, products' characteristics extracted from real order forms are examined in detail and a self-organizing method, one of autonomous functions, is adopted for allocating products to racks.

This paper is composed as follows; related works are described in the second section, products characteristics are examined in the third section, a new method for allocating the products to the racks self-organizingly is proposed in the forth section, the proposed method is verified to give an efficient solution for the given problem through numerical experiments in the fifth section, and finally describe concluding remarks in the last section.

Related Works

There are three major decision-making that determine efficiency of warehouse scheduling: (1) how product order forms is assigned to an agent (a worker or AGV), (2) how an agent makes the shortest path plan to collect the products designated by the product order forms, (3) how products are assigned to racks in a warehouse in a logistic center. There are few researches on the first problem. The second problem is treated as an optimum routing problem or an n-traveling salesman problem (n-TSP). We have applied Heuristics or Meta heuristics to the second problem [1][2]. Many researches have been done for the third problem. C.M. Liu [3] proposed a clustering technique for stock location and order-picking. He formulated a problem as an assignment one and developed a primal-dual heuristic algorithm by rewriting the problem as the dual problem. However, this clustering algorithm is only applicable for a rectangular warehouse, where racks are arranged as columns and columns are allocated to be parallel each other (Figure 1). This allocation makes it possible to formulate the problem as an assignment problem. Since grouped columns are not allocated merely to be rectangular shape, this algorithm is not applicable for general stock allocation. M.B. Rosenwein [4] and J.M. Jarvis [5] were using a similar model as C.M. Liu. R.K. Koster, T. Le-Duc and K.J. Roodbergen [6] surveyed and reviewed many literatures for warehouse-picking. They illustrate "Class-based storage", "Zoning" and "Batching" and "Routing methods". However, a model used by them is a rectangular warehouse as seen in C.M. Liu. The actual warehouse is much more sophisticated than they used. L.C. Tang and E.P. Chew [7] proposed a batching algorithm for products based on the queuing theory and its result was applied for storage assignment. They also considered a picking system in a rectangular system. C.G. Peterson and R.W. Schmenner [8] evaluated routing and volume-based storage policies in an order picking operation.

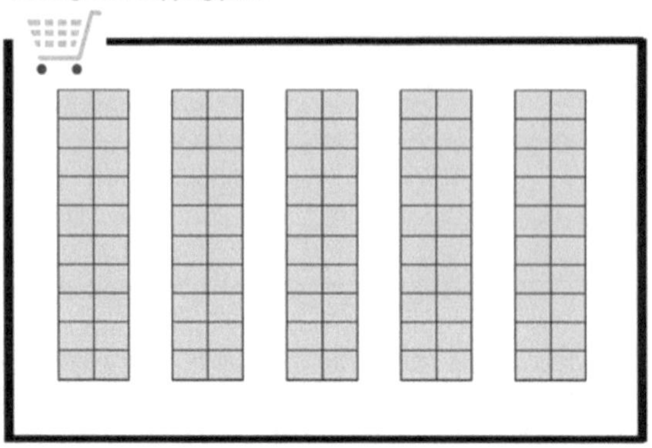

Figure 1 A rectangular warehouse

They indicated some geometric patterns for a warehouse to stock products based on ordered product volumes. They insisted that ordered products with the largest volume should be allocated close to a picking and dropping point (an exit point) in the warehouse. Their proposed patterns gradually make stripes from the picking and dropping point depending on volumes of ordered products. We pay attention their results because this rule is well known for warehouse engineers.

All related works assume that a warehouse adopts a rectangular system. This assumption makes it possible to formulate the problem mathematically. In general, racks in the warehouse are not allocated to be rectangular and a topology represented by connecting adjacent racks is so sophisticated. Therefore, no algorithm to the problem exists possible to apply to a real logistic center but heuristics. However, related works point out two directions to solve the problem: (1) making product clusters, and (2) making volume-based patterns in a warehouse.

Ordered Product Characteristics in a Large Scale Logistic Center

In this section, product characteristics stocked in a warehouse are examined based on real order forms at some company. This company is dealing with writing materials. The number of product kinds reaches almost seventy thousands and the number of racks for a warehouse is almost ninety thousands. One sheet of order forms contains three to ten kinds of product order statistically and the number of sheets a day, by which ordered products are shipped from the warehouse in a logistic center, is around a hundreds thousands.

At first, the number of each ordered product kind is counted as a product frequency. Figure 2 shows a product frequency, where the product is arranged and located along a horizontal axis from the largest frequency of ordered product to the smallest one and the frequency of ordered product is located in a vertical axis. As seen easily, the frequency follows a power law (a long tail distribution).

Then, the number of two ordered products in the same sheet of the order form is counted. If there are three ordered products (A, B, C) in the same sheet, it is counted that a pair of (A, B) and a pair of (B, C) are counted as one time, respectively. Figure 3 shows a frequency of a pair of product as arranged in the same way as Figure 2. This frequency follows the power law distribution as well.

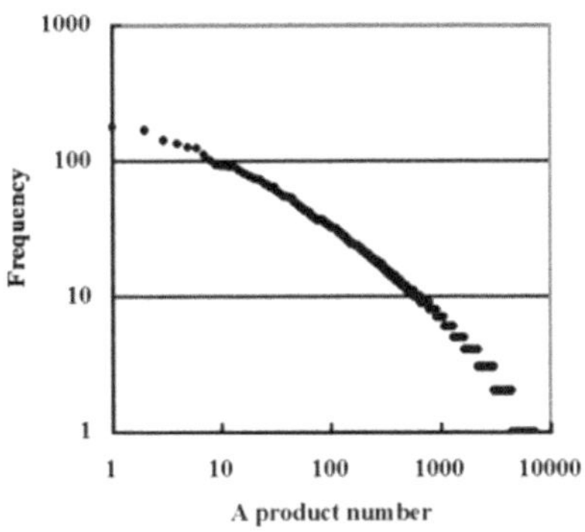

Figure 2 A product frequency

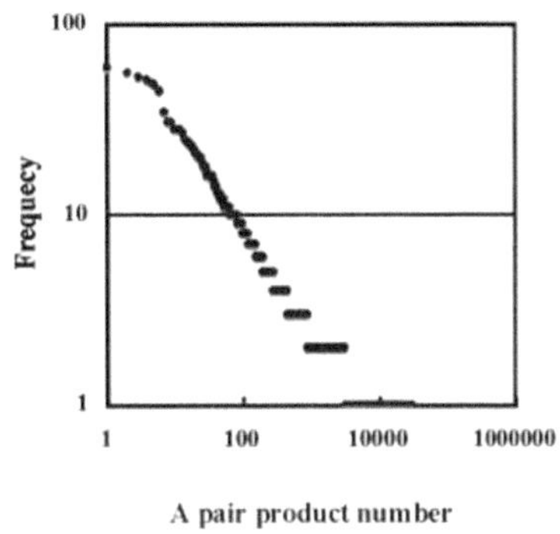

Figure 3 A co-ordered product frequency

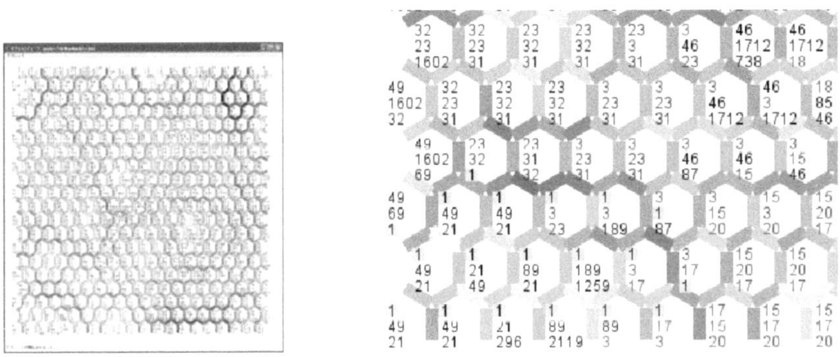

(a) SOM result

(b) A partially enlarged result

Figure 4 A Clustering data form obtained by SOM

After that, we examine whether any tendency (clusters) exist in order forms. For this purpose, SOM proposed by T. Kohonen [9] is adopted. In SOM procedure, a set of products contained in a sheet of order form is regarded as an input vector. As a SOM topology, hexagonal topology is employed. An obtained result is shown in Figure 4. In Figure 4, only three numbers are specified to show product kinds mapped on the SOM neuron. These products are the largest three product kind numbers in a sheet of order form. This map shows that there are some clusters in custom order.

These results tell us that a hundred kinds of product mostly occupies the order products and a hundreds pairs of product are mostly ordered together. For making picking time shorten, this means that a pair of products with a high frequency and ordered products making clusters should be stocked into closer racks each other.

Self-Organizing Stock Assignment

It is shown a frequency of ordered products follows a power law and there are clusters in the order forms. Therefore, products should be assigned to racks so as to make clusters according to the order forms and the products with high frequencies should be assigned to the racks close to a picking and dropping point.

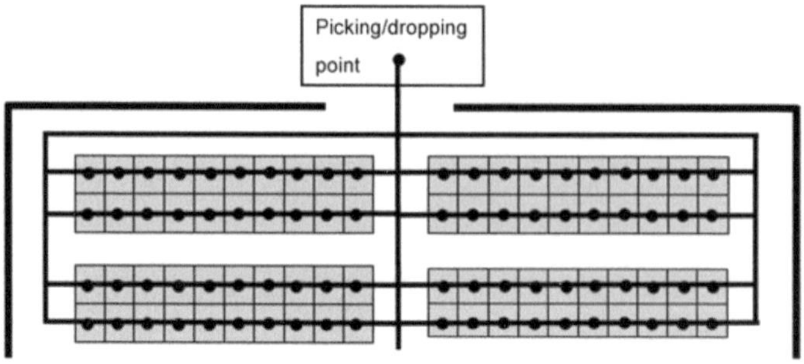

Figure 5 A warehouse topology drawn by rack relations

To achieve two purposes, SOM is employed for assigning products to racks, again. In order to apply SOM, we introduce necessary notations as follows.

$$\mathbf{V} = \{\mathbf{v}_i, i = 1,2,...,m\} \text{ where } \mathbf{v}_i = [v_{i0}, v_{i1}, v_{i2}, ..., v_{in}] \quad (1)$$

$$\mathbf{p} = [p_1, p_2, ..., p_n] \quad (2)$$

$$\mathbf{R} = (\mathbf{N}, \mathbf{E}) \text{ where } \mathbf{N} = \{N_j, j = 1,2,...,l\} \text{ and } \mathbf{E} = \{(N_j, N_k)\} \quad (3)$$

$$\mathbf{W} = \{\mathbf{w}_j, j = 1,2,...,l\} \text{ where } \mathbf{w}_j = [w_{j0}, w_{j1}, w_{j2}, ..., w_{jn}] \quad (4)$$

$\mathbf{V}$ is a set of order forms, m is the total number of order forms, the element vector $\mathbf{v}_i$ represents the i-th order form, v_{ih} is the number of h-th kind product, and n is the total number of different product kinds. v_{io} is defined later. $\mathbf{p}$ is a priority vector whose element p_u is the normalized number between [0,1] determined by a frequency of the u-th product kind. If p_u is set as 1, the u-th kind product has the highest frequency. If p_u is set as 0, the u-th kind product has the lowest frequency. v_{io} is defined by

$$v_{i0} = p_u \text{ where } p_u = \arg\max_u v_{iu}. \quad (5)$$

$\mathbf{R}$ represents a network relationship among racks. $\mathbf{N}$ is a set of network nodes where its element N_j corresponds to the j-th rack and l is the total number of racks. $\mathbf{E}$ represents a set of edges among $\mathbf{N}$. The edge (N_j, N_k) is defined in two cases. One is when the j-th rack is adjacent to the k-th rack. The other is when there is a reachable path from the j-th rack to the k-th one via an aisle as shown Figure 5. $\mathbf{W}$ is a set of synapse vectors whose element vector $\mathbf{w}_j$ is connected to the node N_j. w_{jo} is defined by

$$w_{j0} = \frac{d_{\max} - d(N_j, N_0)}{d_{\max} - d_{min}}, \quad (6)$$

where $d_{\max}$ is the longest distance among those from a picking and dropping point N_0 to the j-th rack via aisles, d_{min} is the shortest one, and $d(N_j, N_0)$ is the shortest distance via aisles between a picking and dropping point N_0 and the j-th rack.

Then, an algorithm for the stock self-organizing assignment is composed by two phases based on SOM as follows.

Phase 1

1) Set random numbers to the elements of vector $\mathbf{w}_j$ but w_{jo}.
2) Set t=0.
3) Select an input vector $\mathbf{v}_i$ at random.

4) Select a synapse vector $\mathbf{w}_j$ such as $\mathbf{w}_j = \underset{j}{\operatorname{argmin}} \left|\mathbf{v}_j - \mathbf{w}_j\right|$.

5) Update a synapse vector $\mathbf{w}^*_j \in \mathbf{W}^*$, where $\mathbf{W}^*$ is a set of synapse vectors reachable from the synapse vector $\mathbf{w}_j$ within the specified path length (in term of the graph theory), as follows;

$$\mathbf{w}^*_j = \mathbf{w}^*_j + \alpha(\mathbf{v}_j - \mathbf{w}^*_j). \tag{7}$$

6) Replace t with t+1. If t is equal to the terminated number, stop the algorithm; otherwise go back to the step 3).

Phase 1 assigns order forms to racks for making clusters on the network **R**. Phase 2 extracts a product from the order form assigned to the rack.

Phase 2

1) Select the j-th node N_j corresponding to the vector $\mathbf{w}_j$ with the maximum value w_{j0} among $\{w_{j0}, j = 1,2,...,l\}$.

2) Assign the h-th kind product to the j-th rack such as

$$w_{jh} = \underset{h}{\operatorname{argmax}}\ w_{jh}.$$

3) Remove $\mathbf{w}_j$ from a set of synapse vectors $\mathbf{W}$. If $\mathbf{W} = \phi$, stop the algorithm.

4) Replace the element w_{jh} of the synapse vector $\mathbf{w}_j \in \mathbf{W}$ with 0.

5) Go back to the step 1.

Numerical Experiments

Numerical experiments are performed to examine the efficiency of product assignment to racks. As a warehouse model, we used a real rack allocation shown in Figure 6. A picking and dropping point is located on the top of the warehouse. A network connecting racks is so sophisticated. Other conditions are as follows:

The number of order forms (m): 9000
The number of racks (l): 7174
The number of ordered product kinds (n): 7174
The number of picking carts (workers): 10
Cart velocity: 1m/sec
Picking time at a rack: 3sec
A distance between an input vector and synapse vector: Euclidian distance
Terminated learning times: 10,000

Then, a proposed method is applied to determine product assignment to racks. Figure 7 shows obtained products assignment. A hundred highest frequency products are grey-colored in Figure 7. The darkest color boxes correspond to the highest frequency product and the brightest boxes correspond to the lowest frequency product. The frequencies changing from the highest one to the lowest one are represented by gradually changing brightness from the dark color to bright one. Figure 8 and 9 shows random products assignment and product assignment designed by a skilled engineer (expert), respectively. It is observed that a concentric circle pattern appears in the obtained allocation from a picking and dropping point. This shows that the proposed method realizes a volume based allocation pointed out by C.G. Petersen [9]. between the obtained allocation and the random one. It is surmised that the skilled engineer might take consider in other conditions based on other evaluation factors.

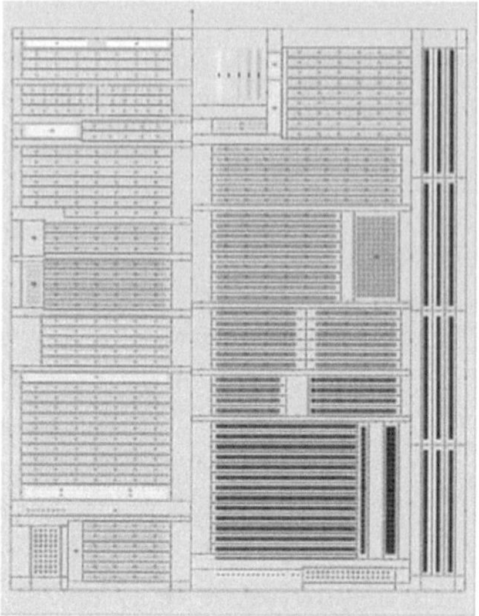

Figure 6 A rack allocation used for experiments

There is no pattern in the random allocation. It seems that the expert allocates the products Figure 7 shows the obtained products assignment. The highest a hundred frequency products are grey-colored in Figure 7. The darkest color box corresponds the highest frequency product and the brightest box corresponds to the lowest frequency product. The frequencies changing from the highest one to the lowest one are represented by gradually changing brightness from the dark color to bright one. Figure 8 and 9 shows the random products assignment and the product assignment designed by the expert, respectively. It is observed that a concentric circle pattern appears in the obtained allocation from a picking and dropping point. This shows that the proposed method realizes a volume based allocation pointed out by C.G. Petersen [9]. There is no pattern in the random allocation. It seems that the skilled engineer allocates the products between the obtained allocation and the random one. It is surmised that the skilled engineer might take consider in other conditions based on other evaluation factors.

In order to verify an efficiency of the obtained allocation for a warehouse, an order-picking simulator is applied to three allocation results. Figure 10 shows a bar chart on the maximum make-span time (MPST) of all carts required for ninety hundreds order forms in three allocations. It is clear that the obtained allocation brings the shortest MPST. The random allocation becomes the longest MPST.

Concluding Remarks

This paper analyzes a product distribution of a real large logistic center and propose the new allocation method which make it possible to assign the products to the racks in the warehouse based on self-organization. The numerical simulation clearly shows that the make-span time of the carts at the allocation achieved by the proposed method is shortest among other methods.

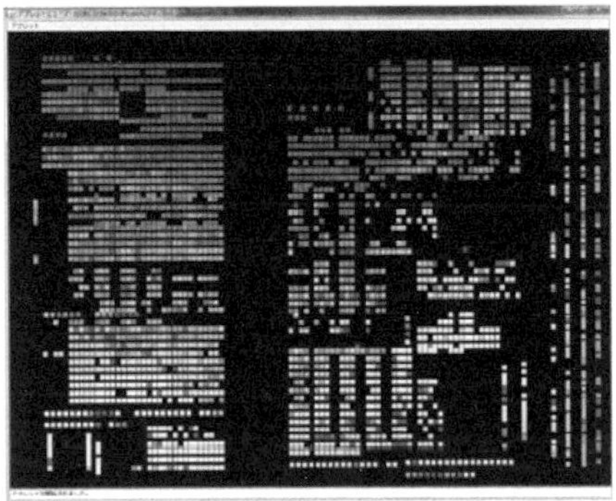

Figure 7 Products' assignment by a proposed method

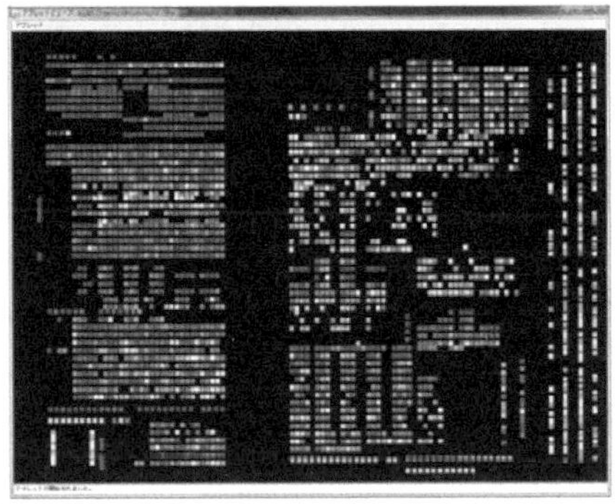

Figure 9 Products assignment by a skilled engineer

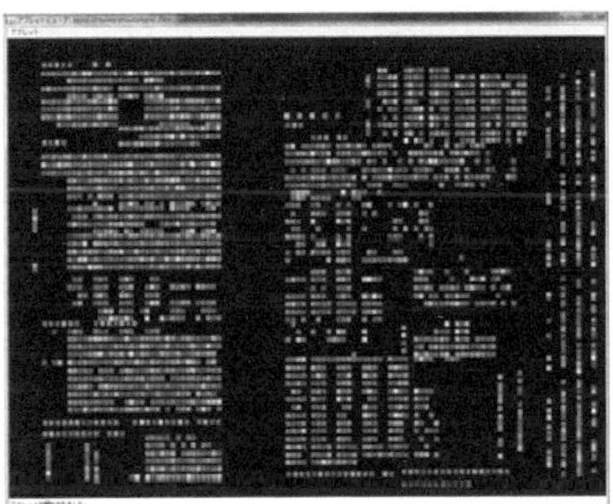

Figure 8 Products' random assignment

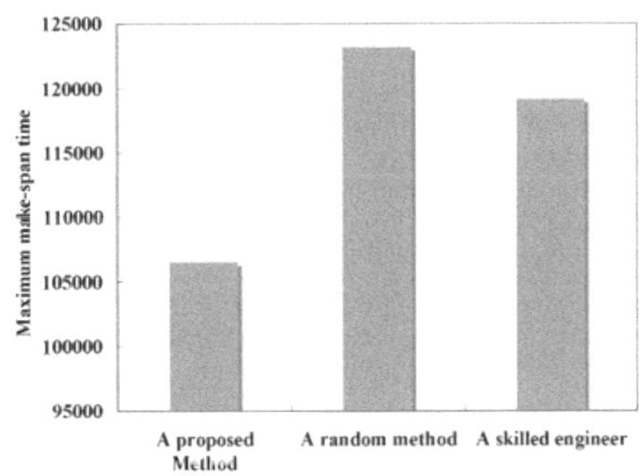

Figure 10 Comparison of three assignment method

This study is summarized as follows.

(1) A distribution of ordered product (item) numbers in the real logistic center follows a power law and a distribution of co-ordered products numbers follows the power law as well.

(2) By use of SOM, it is convinced that there are some clusters in the ordered forms.

(3) In order to satisfy two evaluations –making cluster and efficient order-picking- for the product assignment to the racks, the new method of self-organizing stock assignment is proposed based on SOM.

(4) In the proposed method, an input vector for SOM consists of two kinds of elements, namely the ordered products' elements and the priority element. The former is corresponding to the number of products in the ordered form and the later is corresponding to the maximum frequency of the ordered product in the order form, where frequencies of the ordered products are normalized from 0 to 1.

(5) The network with neighboring relation between two racks represents the racks' allocation and the racks are regarded as a set of nodes in SOM.

(6) The Synapse vector is defined at each node. This vector consists of two kinds of elements, namely the ordered products elements and the distance element. The former is corresponding to products being set initially at random and the later is corresponding to the distance between the rack and the picking and dropping point, where the distances are normalized from 0 to 1.

(7) The new method autonomously organizes the products' assignment to the racks so as to make clusters on the ordered forms and to allocate the products with high frequency to the racks with the shortest distance to the picking and dropping point.
(8) Numerical simulations for picking-order shows that the proposed method achieves the efficient make-span time scheduling for carts in order picking.

References

[1] M. Furukawa, M. Watanabe, Y. Tamayama and Y. Kakazu, Driving of Multiple AGVs for FMS by Use of SLA, Journal of Japan Society of Precision Engineering, vol. 62, 2 (1996) pp.260-264
[2] M. Furukawa, M. Watanabe, Y. Tamayama and Y. Kakazu, Multi-AGV Scheduling for FMS with One-way Driving Lane –Solving a Schedule problem by Use of GA-, Journal of Japan Society of Mechanical Engineers, 62, C (1996) pp.595-600
[3] M. K. Liu, Clustering Techniques for stock location and order-picking I a distribution center, Computers & Operation Research, vol. 26, (1999) pp.989-1002
[4] M. B. Rosenwein, AN APPLICATION OF CLUSTER ANALYSIS TO THE PROBLEM OF LOCATING ITEMS WITHIN A WAREHOUSE, IIE Transaction, vol. 26, No. 1 (1994) pp.101-103
[5] J.M. Jarvis, E.D. MacDowell, Optimum Product Layout in an Order picking Warehouse, vol. 23, No. 1 (1991) pp.93-102
[6] M.A. Hariga and P.L. Jackson, The warehouse scheduling problem: formulation and algorithm IIE Transactions, vol. 28 (1996)
[7] L.C. Tang and E.P Chew, Order Picking System: Batching and Storage Assignment Strategies, Computers ind. Engng, vol. 33, Nos3-4 (1997) pp.817-820
[8] C.G. Peterson II and R.W. Schmenner, An Evaluation of Routing and Volume-based Storage Policies in an Order Picking Operation, Decision Science, vol. 30, No. 2 (1999) pp.481-501
[9] T. Kohonen, Self-Organizing maps, Springer Verlag (1995)

Intelligent Autonomous Systems 10
W. Burgard et al. (Eds.)
IOS Press, 2008

doi:10.3233/978-1-58603-887-8-173

Double Container-handling Operation for an Efficient Seaport Terminal System

Satoshi HOSHINO [a,1] , Tomoharu FUJISAWA [b] , Shigehisa MARUYAMA [c] , Hisato HINO [d] , and Jun OTA [d]

[a] *Chemical Resources Lab., Tokyo Institute of Technology, Japan*
[b] *Ministry of Land, Infrastructure, Transport and Tourism, Japan*
[c] *Japan Association of Cargo-handling Machinery Systems, Japan*
[d] *Dept. of Precision Engineering, The University of Tokyo, Japan*

Abstract. For an efficient seaport terminal, we propose a novel operational model, namely, a double container-handling operation among operating machines, such as automated guided vehicles (AGVs), automated transfer cranes (ATCs), and quay container cranes (QCCs), in a seaport terminal system. In addition, a passing lane is provided in a container storage yard in order to activate the container-handling operation by the AGVs and ATCs. In this paper, the effect of the double container-handling operation and passing lane on the system utilization is examined. Finally, the effectiveness of the proposed operational model with a passing lane is discussed on the basis of the operating time and obtained number of operating machines for a given demand in consideration of a mega-container terminal.

Keywords. Double container handling, seaport terminal system, AGV, QCC, ATC

Introduction

In recent years, the volume of container trade at worldwide seaport terminals has increased significantly [1]. In this regard, it is required to promote the automation and improvement of a container-handling operation in a seaport terminal system. The authors have, so far, achieved the following design challenges in terms of the development of highly efficient and automated container-handling systems: for given demands, (I) combinations of the minimum number of operating machines were found [2]; (II) an efficient system layout was identified [3]; (III) operational models for machines were developed [4,5]; (IV) and (V) the machine performance and reliability were determined [6,7]. Moreover, we are currently developing efficient operational models for a case in which operating machines are in a maintenance mode in consideration of their reliability [8]. In addition to our research, many studies have focused on seaport terminal systems [9].

Previous studies have primarily taken into account import containers. Accordingly, the containers imported by a containership were transported to a terminal storage yard before being forwarded to other locations (i.e., containership → terminal system). How-

[1]Corresponding Author: Satoshi Hoshino, Tokyo Institute of Technology, 4259, Nagatsuta, Midori-ku, Yokohama, Kanagawa 226-8503, JAPAN; E-mail: hosino@pse.res.titech.ac.jp

ever, in an actual seaport terminal, export containers from a terminal to a containership also have to be included.

As for the present import and export container-handling operation at seaport terminals in Japan, first, import containers are all unloaded into a terminal system from a containership. After that, export containers in the terminal system are loaded onto the containership. Hence, a container transport vehicle cannot avoid traveling without containers on the outbound or homeward trek when making a round trip from a quay side to a storage yard in a terminal system. In this paper, we define the conventional operation as a single container-handling operational model. For this conventional operational model, in order to reduce the vehicle empty travel, we propose a novel operational model, namely, a double container-handling operation performed by operating machines in the terminal system. By using the proposed operational model, import and export containers are loaded and unloaded in the quay and storage yard at the same time. Furthermore, we provide a passing lane in the storage yard to activate the container-handling operation by container transport vehicles and movable container storage cranes.

Regarding the multi-item or container-handling operation, studies that deal with automated manufacturing systems, for example, that by Liu *et al.*, have used multi-load vehicles and addressed the task allocation problem [10]. On the other hand, usage of such a multi-load vehicle in a seaport container terminal has been considered by Grunow *et al.* only [11]. However, these studies have addressed the vehicle dispatching problem as a main topic, and, thus, the effect of a container-handling operational model on utilization of the system resources such as operating machines has not been investigated.

Therefore, in order to examine the effect of the double container-handling operational model on the system utilization, we compare the systems with the use of the (I) conventional and (II) proposed operational models on the basis of the operating time for a given task. Moreover, we impose a demand on the terminal system in consideration of the realization of a mega-container terminal in Japan. We then develop the systems and compare the obtained number of machines. Finally, we discuss the effectiveness of the proposed operational model.

1. Container-handling System in a Seaport Terminal

1.1. System Setting

Fig. 1 shows a horizontal container-handling system in a seaport terminal, which is the objective of this paper. The effectiveness of the system, compared to a vertical one, has been shown by controlling operating machines efficiently [3]. In a seaport terminal system, as shown in Fig. 1, quay container cranes (QCCs), automated guided vehicles (AGVs), and automated transfer cranes (ATCs) are used for container handling.

1.1.1. System layout

Fig. 1(a) shows one berth of the container-handling system in a seaport terminal. In this paper, the width of one berth is 350 [m] to respond to a cargo-carrying vessel. The depth of the berth is generally about 500 [m] in domestic terminals; however, the scale is variable according to the number of container storage locations in a yard.

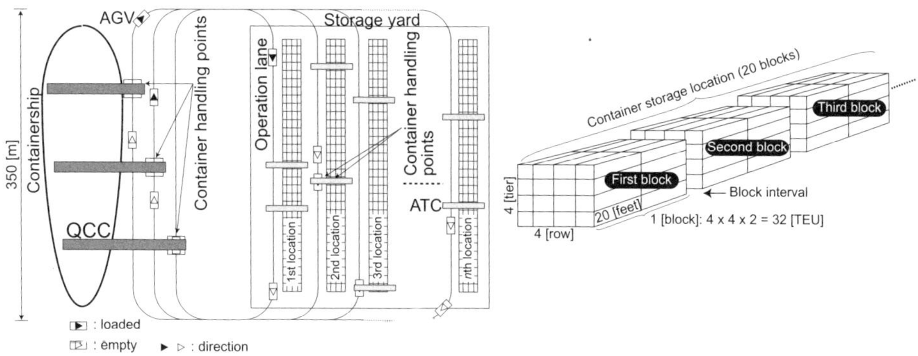

(a) Terminal system layout and containership (one berth, top view)

(b) Container storage location that consists of blocks, rows, and tiers

Figure 1. Horizontal Container-Handling System with AGVs, ATCs, and QCCs in a Seaport Terminal

1.1.2. Container storage location

A container storage location arranged in the yard is shown in Fig. 1(b). One location consists of blocks. One block has 32 (4 tiers $\times$ 4 rows $\times$ 2 (20-foot equivalent)) container storage spaces; in other words, the maximum storage height and width are those determined by the height and width of four containers. In this paper, we assume that one location consists of 20 blocks with a 3 [m] interval. This assumption is given considering general domestic seaport terminals.

1.2. Operating Machines and Machine Performance

As shown in Fig. 1(a), the QCCs for container loading and unloading at the quay side, the ATCs for container transfer, storage, and unloading in the storage yard, and the AGVs for container transport between the quay side and storage yard are the operating machines in this terminal system. The number of QCCs, AGVs, and ATCs are the input parameters for a container-handling simulation. Here, three QCCs are operating in the quay side because the scale of the berth is fixed (width = 350 [m]). As for the number of ATCs, two ATCs operate at one storage location in the yard.

The performance of the operating machines is as follows: for the AGV, the max. container transport and cornering velocities are 5.56 and 1.39 [m/s], respectively; the max. empty traveling and cornering velocities are 6.94 and 2.78 [m/s], respectively [12]; the acceleration and deceleration are 0.15 and 0.63 [m/s^2], respectively, regardless of the container-loaded or empty state. The max. ATC moving velocity is 2.5 [m/s]; the acceleration and deceleration are 0.1 and 0.4 [m/s^2], respectively [12]; the cycle time of one container transfer and storage or unloading and transfer operations is 135 $\sim$ 150 [s]. The cycle time of the QCC container loading or unloading operation is 90 [s]. These cycle times are given on the basis of average data of actual terminal systems in Japan: in an hour, 24 $\sim$ 30 containers are handled by an ATC and 90 containers are handled by a QCC.

1.3. Single Container-handling Operational Model

In general, import containers are first unloaded from a containership onto a storage yard; afterward, export containers are loaded from the yard onto the containership using single container handling, that is, the conventional operational model. The detailed operation procedures are described as follows:

1. QCCs unload import containers, which will be transported by AGVs, from a containership onto the AGVs. After all of the import containers are unloaded, the QCCs begin to load export containers from the AGVs onto the containership.
2. The AGVs transport the import containers from the quay side to target storage locations in the yard. For export containers, the AGVs travel from the quay side to locations without containers.
3. An AGV transfers an import container to an ATC that is available at a location. After all the import containers are transferred, an ATC transfers an export container from one location to an incoming AGV.
4. The AGV that has completed the container transfer goes back to a QCC (without a container). On the other hand, the AGV that has received the export container transports it to a QCC, that is, the containership.
5. ATCs unload the import containers from the AGVs and store them into the storage locations. On the other hand, the ATCs load the export containers from the locations onto the AGVs.

2. Double Container-handling Operational Model

2.1. Cycle of Operation

Fig. 2 illustrates a cycle of the double container-handling operational model between the quay (containership) and storage yard (location). The left side of the figure depicts the container loading and unloading operations in the quay, the right side depicts the container transfer, storage, and unloading operations in the yard, and the middle depicts the container transport operation by the AGV.

After an AGV arrives at a QCC, an export container that was transported from the storage yard is unloaded onto a containership by the QCC taking a constant amount of time. Right after the QCC finishes the unloading operation, it begins to load an import container from the containership onto the AGV taking a constant amount of time as well. The AGV transports the container to a designated area in the storage yard. As in the operation at the quay side, the container is transferred to an ATC from the AGV and then stored at a location by the ATC taking a constant amount of time. Afterward, an export container is unloaded and transferred to the AGV by the ATC taking a constant amount of time. The AGV begins to transport the container to the containership (a QCC). Thus, the AGVs are able to achieve the container transport without traveling with an empty load by handling containers twice in the quay and yard.

Regarding other operational models, we adopt the following models, which have been shown to be an efficient management strategy in a seaport terminal system. For import containers, the container-handling tasks are evenly divided among three QCCs; the containers are planned to be evenly transported to each storage site. For export as well

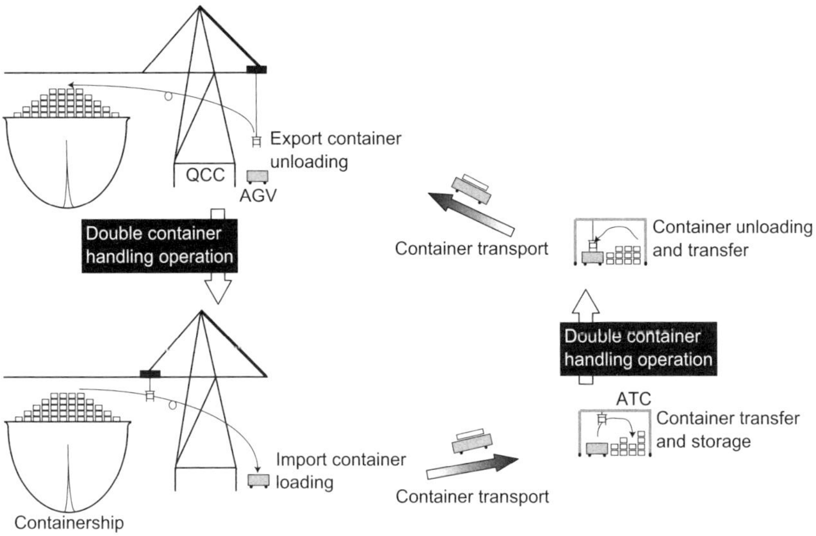

Figure 2. Cycle of the Double Container-handling Operation by a QCC, an AGV, and an ATC

as import containers, the tasks are evenly divided from each storage site. The containers are to be transported to three QCCs evenly. The execution sequence of the planned containers is scheduled so that the total distance of the ATCs for container handling is minimized. In the storage yard, an AGV calls out an ATC on the basis of the workspace-based ATC selection rule right after the AGV enters an operation lane. The detailed operational models are described in the literature [4,5].

2.2. *Single and Double Container-handling Operations*

As shown in Fig. 3, the single and double container-handling operational models have the following characteristics regarding the operational time among the machines and the empty travel of the AGV.

- Single container-handling operational model: All import containers are first unloaded from a containership into a terminal system; after that, export containers in a storage yard are loaded from the system onto the containership (see Fig. 3(a)).
  - ∗ The container-handling operations in the quay and storage yard among the AGVs, QCCs, and ATCs require "one-container" handling time.
  - ∗ The AGVs make the rounds in the system "n" times equal to the total number of import and export containers. Here, one half of them are the empty trips.
- Double container-handling operational model: Import and export containers are unloaded and loaded simultaneously in the quay and yard (see Fig. 3(b)).
  - ∗ The container-handling operations in the quay and storage yard among the AGVs, QCCs, and ATCs require "two-containers" handling time.
  - ∗ The AGVs make the rounds in the system "$\frac{n}{2}$" times. In other words, the AGV transports a container between the quay and storage yard on the outward and homeward trips.

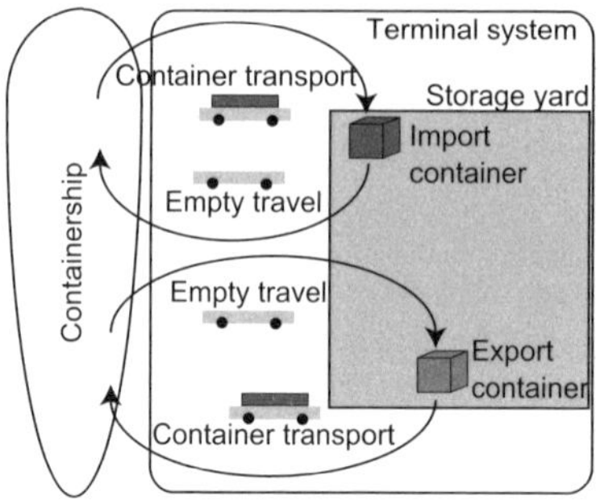

(a) Single container-handling operational model

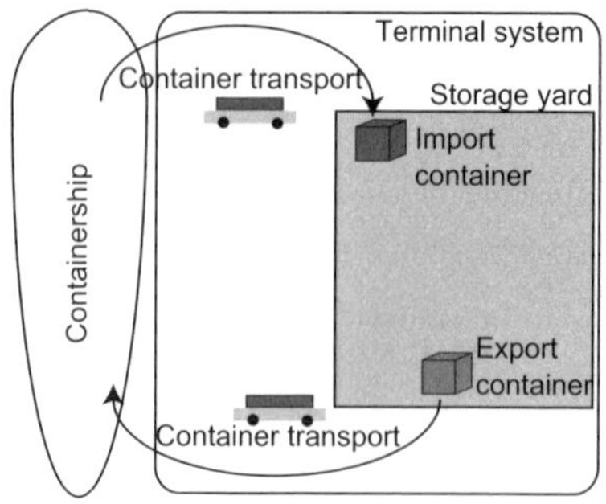

(b) Double container-handling operational model

Figure 3. Comparison of the Cycles of the Single and Double Container-handling Operation Models

2.3. Passing Lane

As described in 2.2, compared to the single container-handling operational model, although the double container-handling operational model halves the number of AGV trips with empty loads, the operational model doubles the container-handling time among the AGV, QCC, and ATC. In this paper, it is assumed that the number of QCCs at the quay is three, but the number of ATCs in the storage yard increases or decreases according to the number of locations. However, in a case in which an AGV stopped for container handling with an ATC impedes the container transport of other AGVs even if the number of locations (ATCs) is increased, the efficiency of the operation in the yard might be decreased.

To solve this problem, we additionally provide one adjacent passing lane for each operation lane, as shown in Fig. 4(a). The passing and operation lanes cross at the middle of the storage location. When an AGV goes into the storage yard, it selects either the passing or the operation lane according to its destination (upper or lower half of a location). An AGV that finished a container-handling operation at the upper half of the location goes to the junction; then, it changes lanes to the passing lane; finally, it leaves the yard to go to the quay. On the other hand, an AGV that operates with the ATC at the lower half of the location changes from the passing lane to the operation lane at the junction. In the vicinity of the junction, a control zone is installed, as shown in Fig. 4(b). Hence, in the control zone, the AGV that entered first has priority over other AGVs to go through the zone. Thus, collision avoidance for the AGVs at the junction is ensured.

3. Simulation Experiment

3.1. Simulation Scenario

Since the total number of import and export containers handled in the terminal system is the same even if the container flow between the terminal system and inland is considered, as described in [13,14], only the container flow between a containership and the termi-

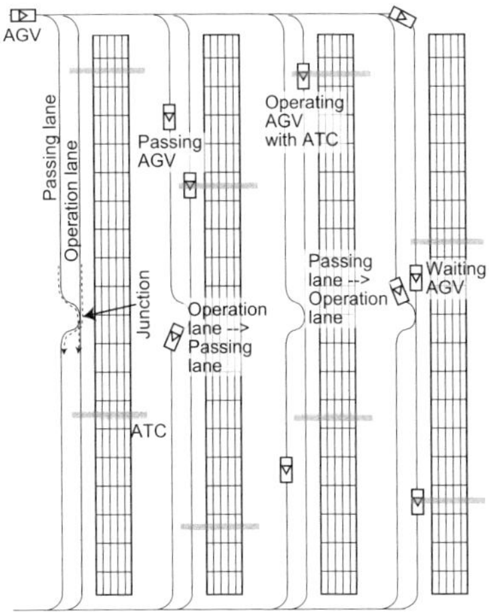

(a) Provided passing lanes adjacent to the operation lanes

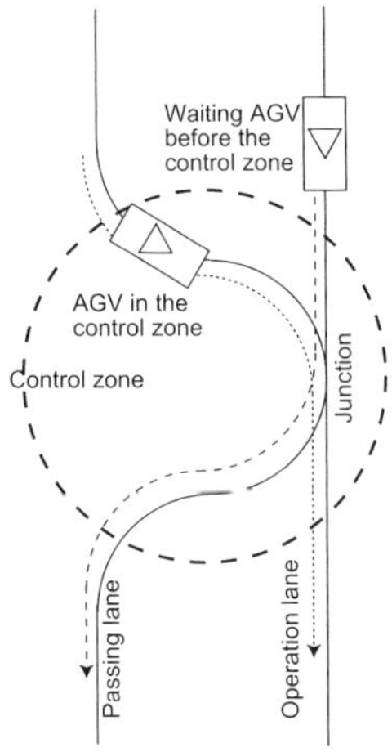

(b) Control zone around the junction

Figure 4. Behavior Model in the Container Storage Yard with the Provided Passing Lanes

nal system is considered in this simulation. As a simulation scenario, we assume that a cargo-carrying vessel that has 10,000 [TEU (Twenty-foot Equivalent Unit)] containers comes in. From our field surveys, we have found out that about 2,000 [TEU] containers in the vessel, at a maximum, are handled. Therefore, in the simulation experiment, the operating machines handle 2,000 containers. Half of the containers, i.e., 1,000, are import containers, and the other half, i.e., 1,000, are export containers. Here, as described in 1.1.2, since one location has container storage space of up to $32 \times 20 = 640$ [TEU] only, two locations, i.e., at least four ATCs, are used in the system. All combinations of three QCCs (fixed), $1 \sim 30$ AGVs, and $4 \sim 20$ ATCs represent the input parameter for the container-handling simulation.

Under the given scenario, the following three systems are prepared for the simulation to examine the effect of the double container-handling operation model on terminal system utilization. These three systems are then compared and evaluated on the basis of the operating time for 2,000 containers (tasks). The operation time between the AGV and ATC in the storage yard with the use of the single and double handling operation models is 135 [s] (single) and $150 \times 2 = 300$ [s] (double). This is because, in the system with the double container-handling model, it is not always true that an export container is located in the same block in which an import container will be stored.

1. Single handling with one operation lane (conventional system **1**)
2. Double handling with one operation lane (proposed system **2**)
3. Double handling with one operation and passing lanes (proposed system **3**)

3.2. Comparison Result of the Operating Time

Fig. 5 shows the operating time of three systems with QCCs, AGVs, and ATCs for 2,000 tasks. From the results of Fig. 5(a) $\sim$ Fig. 5(c), the operating time for all systems in-

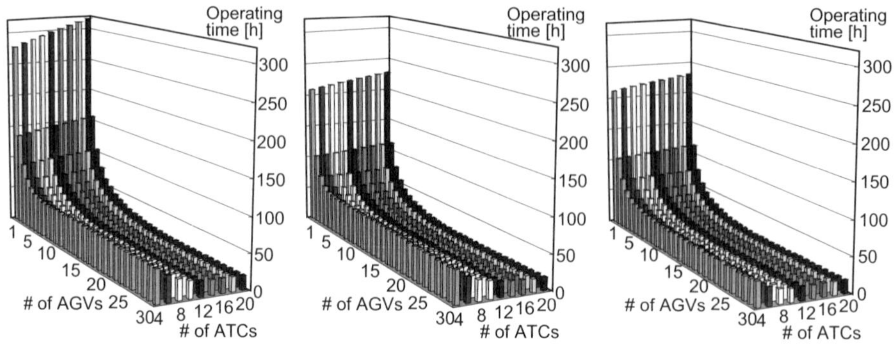

(a) Conventional system 1 (b) Proposed system 2 (c) Proposed system 3

Figure 5. Comparison Results of the Operating Time for 2,000 Container-handling Tasks

creases as the number of ATCs increases for a few AGVs. This is because the number of locations increases as the number of ATCs increases; eventually, the total container transport distance of the AGVs increases. From this result, it is noteworthy that a system with few AGVs and many ATCs might be inefficient.

From the comparison of Fig. 5(a) with Fig. 5(b), regardless of the number of ATCs, the operating time of the proposed system with few AGVs (e.g., $1 \sim 5$) decreased in all cases. The reason for this result is that the empty trips of the AGVs using the single container-handling model have more harmful effects on system utilization than the heavy workload caused by the AGVs using the double container-handling model. The difference in the operating time of those systems with few ATCs ($4 \sim 14$), however, decreases as the number of AGVs increases. This result indicates that the double container-handling operation with many AGVs causes a heavy workload in a case in which there are few locations; thus, the operation model eventually has harmful effects on system utilization. A comparison of system 2 in Fig. 5(b) with system 3 in Fig. 5(c), in which the passing lane is additionally provided, shows that the workload caused in system 2 is eased in system 3. As a result, the container-handling operation in the storage yard was fully activated; furthermore, the operating time of system 3 with $4 \sim 14$ ATCs decreased as the number of AGVs increased.

The average reduced operating time (the sum of the difference of the operating time / number of simulation runs) of system 2 is 8.01 [h] less than in system 1, and that of system 3 is 3.79 [h] less than in system 2. These results show the effect of the proposed double container-handling operation model and the passing lane on system utilization.

3.3. The Mega-Container Terminal

3.3.1. Problem description

Major seaport data in Japan in 2006 are described in Table 1. These seaports are identified as "super hub ports," according to the concept of a mega-container terminal proposed by the Ministry of Land, Infrastructure, Transport and Tourism of Japan. In the table, although the Yokohama Minami-Honmoku terminal consists of only two berths, its re-

Table 1. Major seaports in Japan aiming at a mega-container terminal

Port name	Number of berths	Annual volume [TEU]
Yokohama (Minami-Honmoku)	2	863,000
Tokyo (Ohi)	7	2,098,000
Nagoya (Tobishima)	1	266,000
Kobe (Port island)	6	1,061,000

Table 2. Combination of a number of AGVs and ATCs that meets the demand 68.49 [container/hour]

System number	# of AGVs	# of ATCs (locations)
1 (conventional system)	13	10 (5)
with single handling and one operation lane	12	12 (6)
2 (proposed system)	11	10 (5)
with double handling and one operation lane	10	12 (6)
	9	14 (7)
3 (proposed system)	11	6 (3)
with double handling and one operation	10	8 (4)
and passing lanes	9	10 (5)

cent annual container volume (in 2007) reached 1,000,000 [TEU], a remarkable number. Therefore, as in the Minami-Honmoku terminal, we aim at developing a terminal system that achieves a volume of 500,000 [TEU] annual containers per berth.

We assume that the automated container-handling system works 365 days a year and 24 hours a day. Generally, since a cargo-carrying vessel takes several hours to come in and sail out in addition to the maintenance time for the operating machines, in an actual seaport terminal, the daily working time of the system is 20 hours. Thus, we developed a system that meets a required throughput, $500,000/(365 \times 20) \simeq 68.49$ [container/hour], to achieve the handling of 500,000 containers per year per berth.

3.3.2. Comparison result of the obtained number of AGVs and ATCs

Combinations of the number of AGVs and ATCs that meet a required throughput, i.e., demand 68.49 [container/hour], are derived from Fig. 5 and shown in Table 2. In the table, note that a system with more AGVs and ATCs than mentioned meets the demand. Furthermore, in Table 2, combinations of the number of AGVs and ATCs are not mentioned, in which case the number of AGVs does not decrease for more ATCs. For instance, as for system 1, although a system with 12 AGVs and 14 ATCs meets the demand, the combination is not mentioned.

Regarding the results of systems 1 and 2, the number of AGVs obtained in system 2 with 10 and 12 ATCs was lower than that of system 1. Moreover, system 2, which consists of 9 AGVs and 14 ATCs, also met the demand. As for system 3, in which the passing lane was provided, even when the number of ATCs was 6 or 8, the system was able to meet the demand with 11 or 10 AGVs. Additionally, although the obtained number of AGVs in systems 1 and 2 with 10 ATCs was 13 and 11, respectively, the number of AGVs in system 3 was 9 and met the demand.

From the results, it is noticeable that the double container-handling operation model and the addition of the passing lane are also effective for high system utilization in terms of the realization of the mega-container terminal. Therefore, we confirm the effectiveness of the proposed operation model with the passing lane.

4. Conclusion and Future Work

In this paper, we proposed a novel operation model, which is double container handling among AGVs, QCCs, and ATCs for an efficient seaport terminal. Furthermore, in order to activate the operation among AGVs and ATCs in a container storage yard, we provided a passing lane in addition to the operation lane. Through simulation experiments, we demonstrated the effect of the double container-handling operation and the passing lane on system utilization. Furthermore, three systems were developed and considered for the realization of a mega-container terminal in Japan. From these results, we confirmed the effectiveness of the proposed operation model with the passing lane.

In future works, we will take into account the following issues which were the given assumptions in this paper: (I) variable operation time among the operating machines according to a container stacking condition and (II) trade-off analysis between the number of passing lanes and junctions (control zones) and the container storage yard space. Moreover, we will tackle a challenge, that is, the optimization of container-shuffling operations both on-ship and in-yard for more efficient seaport terminal systems.

References

[1] D. Steenken *et al.*, Container Terminal Operation and Operations Research - A Classification and Literature Review, *OR Spectrum*, **26** 1 (2004) 3–49.

[2] S. Hoshino *et al.*, Optimal Design Methodology for an AGV Transportation System by Using the Queuing Network Theory, *Distributed Autonomous Robotic Systems 6*, (2007) 411–420.

[3] S. Hoshino *et al.*, Hybrid Design Methodology and Cost-Effectiveness Evaluation of AGV Transportation Systems, *IEEE Transactions on Automation Science and Engineering*, **4** 3 (2007) 360–372.

[4] S. Hoshino *et al.*, Highly Efficient AGV Transportation System Management Using Agent Cooperation and Container Storage Planning, *IEEE/RSJ International Conference on Intelligent Robots and Systems*, (2005) 2330–2335.

[5] S. Hoshino *et al.*, Design of an AGV Transportation System by Considering Management Model in an ACT, *Intelligent Autonomous Systems 9*, (2006) 505–514.

[6] S. Hoshino *et al.*, Performance Design of Operating Robots in a Seaport Container-Handling System, *IEEE Conference on Automation Science and Engineering*, (2007) 692–697.

[7] S. Hoshino *et al.*, Design of an Automated Transportation System in a Seaport Container Terminal for the Reliability of Operating Robots, *IEEE/RSJ International Conference on Intelligent Robots and Systems*, (2007) 4259–4264.

[8] S. Hoshino *et al.*, Reactive Robot Control with Hybrid Operational Models in a Seaport Container Terminal Considering System Reliability, *IEEE/RSJ International Conference on Intelligent Robots and Systems*, (2008) submitted.

[9] H.-O. Günther *et al.*, Container Terminals and Automated Transport Systems, *Springer-Verlag*, (2005).

[10] F.-H. Liu *et al.*, Control Strategy for Dispatching Multi-load Automated Guided Vehicles in a Deadlock-Free Environment, *Journal of Mathematical Modelling and Algorithms*, **1** 2 (2002) 117–134.

[11] M. Grunow *et al.*, Dispatching Multi-load AGVs in Highly Automated Seaport Container Terminals, *OR Spectrum*, **26** 2 (2004) 211–235.

[12] MITSUBISHI HEAVY INDUSTRIES, LTD., Advanced Technology Cargo Handling Systems, Products Guide, (2004).

[13] J. Zhang *et al.*, Automated Container Transport System between Inland Port and Terminals, 83rd TRB Annual Meeting Interactive Program, (2004).

[14] W.C. Ng *et al.*, Yard Crane Scheduling in Port Container Terminals, *Applied Mathematical Modelling*, **29** 3 (2005) 263–276.

Intelligent Autonomous Systems 10
W. Burgard et al. (Eds.)
IOS Press, 2008

doi:10.3233/978-1-58603-887-8-183

A Convergent Dynamic Window Approach with Minimal Computational Requirements[1]

Javier ANTICH [2], Alberto ORTIZ
University of the Balearic Islands

Abstract. Many applications in mobile robotics require the safe execution of a real-time motion to a goal location through completely unknown environments. In this context, the *dynamic window approach* (DWA) is a well-known solution which is safe by construction —assuming reliable sensory information— and has shown to perform very efficiently in many experimental setups. Nevertheless, the approach is not free of shortcomings. Examples where DWA fails to attain the goal configuration due to the *local minima* problem can be easily found. This limitation, however, has been overcome by many researches following a common framework which essentially provides the strategy with a deliberative layer. Based on a model of the environment, the deliberative layer of these approaches computes the shortest collision-free path to the goal point being, afterwards, this path followed by DWA. In unknown environments, nevertheless, such a model is not initially available and has to be progressively built by means of the local information supplied by the robot sensors. Under these circumstances, the path obtained by the deliberative layer may repeatedly and radically change during navigation due to the model updates, which usually results in high-suboptimal final trajectories. This paper proposes an extension to DWA without the local minima problem that is able to produce reasonable good paths in unknown scenarios with a minimal computational cost. The convergence of the proposed strategy is proven from a geometric point of view.

Introduction

The problem of motion planning is a fundamental issue in mobile robotics. Responding to this interest, a huge variety of techniques has been developed along the past years. They all can be roughly classified into model-based/deliberative planners, sensor-based/reactive strategies, and hybrid systems, which are a proper combination of the two preceding approaches. Deliberative planners are generally characterized by computing, and later executing, the path to a given target based on a known map of the environment. Additionally, if the planner runs quickly enough, it can be applied in a feedback loop to continually replan whenever new sensor data update the environment model. Using abstract world models is, however, a time-consuming and error-prone process that reduces the potential correctness of a robot action in all but the most predictable scenarios [1]. To cope with this limitation, researchers put forward sensor-based/reactive approaches that control the movements of the robot by means of, exclusively, the sensory inputs. The sentence 'what you see is what you get' faithfully summarizes this idea. Typical members of this family of algorithms are: the artificial potential fields, the vector field histogram method (VFH, VFH+, and VFH*), the nearness diagram strategy, the curvature-velocity method, and the popular dynamic window approach (see [2] for an exhaustive survey of reactive techniques). The main advantage of these strategies lies in their high computational efficiency, which makes them well suited to navigate through unknown and dynamic scenarios. Nevertheless, they are only able to provide sub-optimal

[1]This study has been partially supported by project CICYTDPI2005-09001-C03-02 and FEDER funds. In addition, the authors of the paper would like to thank Javier Minguez for his valuable comments about the real advantages of the approach with respect to the current state of the art in the field
[2]Corresponding Author: Ctra. Valldemossa km. 7'5, 07122 - Palma de Mallorca, Spain; javi.antich@uib.es

results due to the locality of the information which is managed. In addition, robots implementing these techniques may not find a solution to the navigation task because of getting trapped into obstacles such as U-shaped canyons. As can be observed, neither deliberative nor reactive strategies are really free of problems. In the face of this situation, new approaches combining the strengths of the two aforementioned strategies and minimizing, at the same time, their corresponding weak points were proposed. In this kind of control systems, two components, one reactive and the other one deliberative, interact via an intermediate layer. The reactive part of the system takes time critical actions, while the deliberative one makes long-term decisions. Note that the work presented in this paper falls into the scope of such hybrid techniques.

Most approaches in the field of hybrid control systems share a common distribution of tasks among their different components/layers. To be precise, the deliberative layer of these strategies generally computes a global path to the given target using an up-to-date model of the environment. Afterwards, this path is securely followed by the reactive layer avoiding any unexpected obstacle in real time. There are many ways of computing such a reference path but the navigation function method has been frequently adopted by researchers due to the optimal trajectories obtained (see [3,4,5,6,7,8]). A navigation function (NF) is essentially an artificial potential field with an only minimum located at the target point or, in other words, NF is a function which maps every free space position to the length of the shortest collision-free path going from that position to the goal. Unfortunately, the construction of a navigation function is a demanding task requiring quite intensive computations. Many efforts have been done along the past years to reduce such a computational cost. Nowadays, according to [3,4], one of the simplest and more efficient ways of building a NF consists in applying a wave-propagation technique called NF1. This technique was shown in [4] to be fast enough to be executed inside the control loop of an XR4000 robot. However, there are no studies —to the authors' best knowledge— where NF1 has been computed in real time using low-cost/miniature robots equipped with a microcontroller. NF1, as will be seen later, is still a heavy technique when considering robots with the aforementioned limitations, which makes strategies such as [3,4,5,6,7,8] not especially suitable to be applied in this kind of problems. The authors of this paper are intended to fill this gap by proposing a lightweight strategy which, being well suited for low-cost robots, guarantees global convergence without constructing any navigation function.

Before going into details, it is important to note that the solution which is proposed in this paper moves away the idea of using sampling-based methods such as RRT (Rapidly-exploring Random Trees [9]) for obtaining/planning the global path, despite they significantly reduce the running time regarding NFs and other non-probabilistic approaches such as roadmaps and cell decompositions. Some of the reasons for such decision are: first of all, the weak form of completeness which is achieved; on the other hand, the lack of an upper bound for the length of the resultant path/s which may be, in consequence, highly sub-optimal; the difficulties which are typically found to generate samples inside narrow passages; and, finally, the inefficiency of these methods to become aware that the target is not reachable for the robot.

Now, being more precise, this paper puts forward a hybrid control system as an extension of the dynamic window approach (DWA) developed by Fox et al. in [10]. In short, DWA is modified in such a way that makes the robot behave like a blind person moving in an unknown environment as illustrated in fig. 1(a). As can be observed, the robot

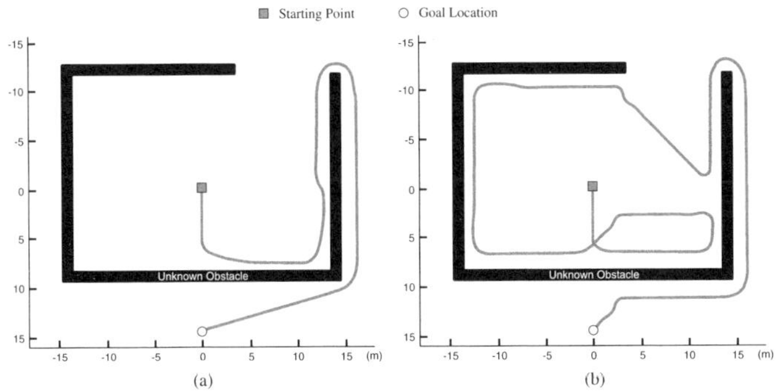

Figure 1. Navigation in an unknown environment. Comparing the path lengths of (a) the new proposal and (b) the global DWA [3].

Table 1. Some details about the runs of fig. 1. Regarding the global DWA, the execution time includes the time needed for recomputing the navigation function accordingly to the map updates. On the other hand, a 35m × 35m map implemented as a probabilistic occupancy grid was used with a resolution of 5cm (490000 cells)

Algorithm	Path Length (m)	Avg. Linear Velocity (cm / sec)	Execution Time (ms / cycle)		Memory Usage (floats, 1 float = 4 bytes)	
			Avg.	Max.	Avg.	Max.
Proposal	78.8	97.7	0.2	16.0	14649.6	15107
Global DWA [3]	146.2	98.7	243.6	1390.0	490000.0	490000

starts moving towards the given goal following a straight-line path. After the detection of an obstacle, on the other hand, the robot circumnavigates it by applying a contour following process. This process finishes when the robot heading is aligned with the current direction to the goal. By adding some extra details to this general behavior, the global convergence of the proposal can be proven from a geometric viewpoint.

To conclude the introduction, we bring forward some experimentation. Fig. 1(b) shows the trajectory generated by a competing algorithm in the same environment of fig. 1(a). The algorithm, described by Brock and Khatib in [3], is also a well-known extension of DWA where a deliberative component based on NF1 was integrated into the strategy. Note that, in the experiment, the U-shaped obstacle was unknown and, consequently, planning was based on a map progressively built from the information provided by the robot sensors, whose maximum range of detection was 3 meters (the dimensions of the scenario were, approximately, 35 meters × 35 meters). Table 1 compares the results of our proposal (see fig. 1(a) again) with the ones provided by [3] from the points of view of computational cost, memory usage, and path length performance. As can be seen, a better solution to the problem was found, using far less computational resources.

The rest of the document is organised as follows: section 1 revises the fundamentals of the dynamic window approach, while section 2 improves DWA but without guaranteeing global convergence, which is achieved in section 3; section 4 presents some experimental results using low-cost robots and, finally, section 5 concludes the paper.

1. Related Work: The Dynamic Window Approach

The *dynamic window approach* [10] (DWA) is a real-time obstacle-avoidance strategy that takes into account the kinodynamic constraints of a synchro-drive robotic platform.

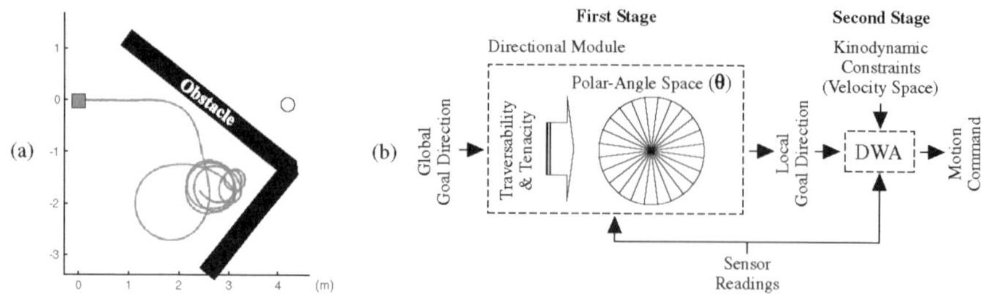

Figure 2. (a) limitations of the classic DWA; (b) block diagram for the DWA-based algorithm being proposed.

The basic scheme involves directly searching in the velocity space for the motion command which maximises a certain objective function. More precisely, the search space is the set of tuples (v, w) of translational v and rotational w velocities that are achievable by the robot within the next time interval given both its current velocity and its acceleration capabilities. This space is, afterwards, reduced by removing those velocity tuples that do not allow the robot to come to a stop before hitting an obstacle. Finally, an optimisation is performed over the resultant set of tuples to find the one that provides the highest utility according to the aforementioned objective function. This function, formally described by equation 1, includes in a weighted way a measure of progress towards a goal location, the forward velocity of the robot, and the distance to the next obstacle along the trajectory.

$$G(v,w) = \alpha_1 \cdot Heading(w) + \alpha_2 \cdot Vel(v) + \alpha_3 \cdot Dist(v,w) \tag{1}$$

DWA has been exhaustively tested showing in all cases consistent safe performance and high efficiency operating at high speed (above 1 m/sec). However, there are scenarios where the robot is not able to reach the target point. Fig. 2(a) shows the result of a real experiment where a Pioneer robot equipped with ultrasonic sensors got indefinitely stuck in front of an L-shaped canyon by executing the original dynamic window approach.

2. Extension of DWA to Avoid the *Local Minima* Problem

As was previously said, the dynamic window approach is a sensor-based strategy that allows the robot to get to a certain target location while avoiding in real time the obstacles found on its way. This general behavior stems from choosing, in each cycle of the algorithm, the velocity command that maximises the objective function defined in equation 1, which trades off safety, speed and goal-directness, where the last term favours the alignment of the robot with the target direction. We propose to change in an on-line way such preferred direction of motion to avoid the problem of *local minima* associated with the strategy. To this end, a new module is integrated into DWA which acts according to the so-called principles of *Traversability* and *Tenacity* —or T^2, in brief. These principles were introduced in [11] by the authors to allow a robot controlled by a behavior-based system [1] to successfully navigate in complex scenarios. Next, we show how to apply the T^2 principles into a completely different robot control scheme such as the dynamic window approach. Fig. 2(b) shows the two different stages in which the proposal has been divided. As can be observed, the first stage computes a preferred direction of motion while, in the second one, such preference is supplied to DWA to produce the velocity command that optimises navigation in the desired direction taking into account the kinematic and dynamic constraints of the robot.

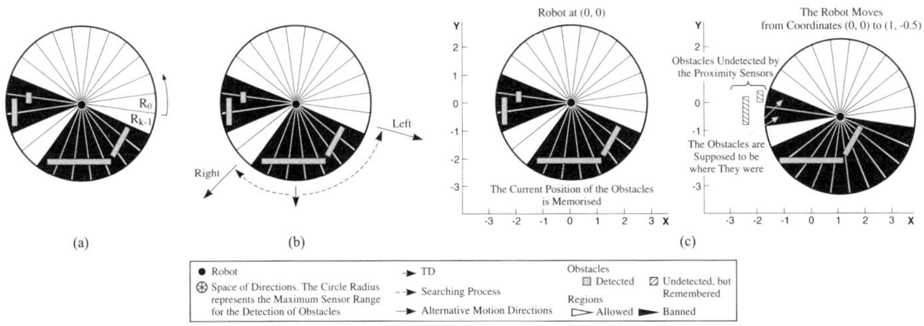

Figure 3. (a) division of the space of directions into K regions, labeling them as *allowed* or *banned*; (b) selection of two obstacle-free motion directions; (c) memorising the obstacle locations while navigating.

2.1. Computation of the Preferred Motion Direction

The calculation of the preferred motion direction is carried out according to the aforementioned *Traversability* and *Tenacity* principles. Note that concepts related with these principles can be separately found in some studies in the field of obstacle avoidance such as, for instance, in [12] and [13]. Such concepts, however, have never been combined. The proper definition and combination of the T^2 principles makes the robot behave in such a way that it is never trapped into local minima ([12,13] suffer from the local minima problem). In the following, the *Traversability* and *Tenacity* principles are sequentially applied in the given order taking as input the current direction to the target (*TD*, from now on). Both principles are explained next in a more detailed way.

2.1.1. The Traversability Principle

The application of the *traversability* principle requires the division of the space of directions around the robot into K identical angular regions as it is illustrated in fig. 3(a). Each region comprises a disjoint range of directions which can be classified as *allowed* or *banned*. Specifically, a region is said to be *allowed* when all the directions in its range are obstacle-free. In case this condition is not satisfied, the region is classified as *banned*.

Based on the previous information, this principle is intended to forbid the robot movement in directions where the presence of obstacles has recently been determined, avoiding thus unnecessary and unsuccessful displacements in the task of looking for a path towards the goal point. With this purpose, the viability of *TD* —by default, our preferred motion direction— is studied according to the above-mentioned premise. Changes are required only if such direction of motion lies in a *banned* region. In such a case, two alternative directions, generically labeled as *left* and *right*, are obtained as a result of a double searching process, clockwise and counterclockwise, for the first *allowed* region starting from the desired direction of motion (see fig. 3(b) for an example). The final decision about choosing one direction or another depends on the *tenacity* principle.

2.1.2. The Tenacity Principle

Remember that two alternative motion directions labeled as *left* and *right* are obtained when *TD* lies in a *banned* region. Under these circumstances, one of such directions has to be selected as the final preferred direction of motion. The *tenacity* principle is then applied by choosing *left* or *right* in coincidence with the last decision made.

2.1.3. Need for Remembering the Obstacles

The local information provided by the robot sensory equipment may not be enough to solve a given navigation task. Against this problem, the new module added to DWA keeps and uses past information regarding the location of the obstacles. By way of example, fig. 3(c) shows how the strategy remembers the presence of obstacles in directions where currently an obstacle-free space is being locally detected by the robot sensors. It is important to note that the position of the obstacles is only temporarily memorized, being removed when the robot is sure that they have been successfully overcome. Further details about this particular aspect of the control system will be given later.

2.2. Execution of the Dynamic Window Approach

The last stage of the strategy corresponds to the execution of the dynamic window approach. Two are the main differences regarding the classic algorithm. On the one hand, DWA now considers that the goal direction is the preferred motion direction (PMD) supplied by our T^2-based directional module. This involves some changes in the $Heading$ component of the objective function (see equation 2), which measures the expected alignment of the robot with respect to PMD assuming the execution of the command (v, w). On the other hand, we penalize in this stage the directions which have been previously banned by the directional method. It is done again in the $Heading$ function by returning a nil value under such a circumstance. In this way, the robot tends to navigate in directions where obstacles have not been detected, decreasing thus the risk of collision.

$$G(v, w, \text{PMD}) = \alpha_1 \cdot Heading(w, \text{PMD}) + \alpha_2 \cdot Vel(v) + \alpha_3 \cdot Dist(v, w) \qquad (2)$$

2.3. Resultant Emergent Global Behavior

The combined application of both the T^2 principles and the slightly altered version of the dynamic window approach results in an *emergent* global behavior which can be summarised as follows:

1. When the robot is navigating far from obstacles, it heads for the target point following a straight-line path.
2. After the detection of an obstacle, the vehicle follows its contour in a certain direction (left or right). The desired contour following direction is defined by the user before beginning the mission.
   On the other hand, the robot knows the detected obstacle has been overcome when the direction to the goal becomes free of obstacles, that is, not *banned.* At that moment, the obstacle information that was collected (see section 2.1.3) is removed since it is no longer necessary for the navigation task.

These stages are sequentially executed in the order specified so many times as obstacles the robot finds on its way towards the target (see fig. 1(a) for a simple example).

3. Pursuing Global Convergence

By adding a blind-person-type behavior to the classic dynamic window approach, we achieve a significant increase in the complexity of the missions which can be successfully solve. Several examples where the strategy reaches the given goal through unknown maze-like environments can be found in [14][3]. However, the proposal, as defined so far, does not guarantee convergence. Unfortunately, scenarios can be built where the robot is immersed in a cyclic behavior. The following provides a solution to the problem which consists in modifying the way how the robot decides to leave the contour of the obstacles.

3.1. Geometric Description of the Algorithm

First of all, a more formal (geometric) description of the global behavior of the strategy highlighted in section 2.3 is given in algorithm 1. The details and the notation used are examined next.

3.1.1. Notation

$S, T \in \mathbb{R}^2$ are, respectively, the starting and the target points of the mission. xy, $x, y \in \mathbb{R}^2$, represents the straight-line segment with endpoints x and y. On the other hand, O_i denotes a certain obstacle of the environment and ∂O_i its contour curve. $P_i^j \in \mathbb{R}^2$ is a point that belongs to the contour curve of O_i —$P_i^j \in \partial O_i$— and such that the line tangent to ∂O_i at P_i^j also contains T. Finally, given $x, y \in \mathbb{R}^2$ and γ a curve joining x and y, $\rho(\gamma, x, y) = \int_{p=x}^{y} \alpha'(p)\, dp$, where α is the angle of the tangent line of γ at p, and α' is the derivative.

3.1.2. Discussion

As can be observed in algorithm 1, our strategy exhibits two different behaviors: motion-to-goal and boundary-following. During the former, which is activated first, the robot moves towards the target (T) along a straight line. The boundary-following behavior, on the other hand, is invoked when the robot encounters an obstacle (O_i) on its way. The point where this obstacle is found is called hit point (H_k). Next, the robot follows the contour of the obstacle (∂O_i) to the left or right according to a user-definable parameter. During this contour-following process, a special situation may occur where the robot returns to H_k meaning that a loop around the obstacle boundary has been completed. In such a case, the target is inside the obstacle, not being thus achievable. More usual is, however, the situation where the robot heading agrees with the current direction to the target in a point away from the concave/inward parts of the obstacle boundary. This only happens in a subset of the points P_i^j which are filtered by means of the ρ function (see fig. 4(a) for an example). When the above-mentioned situation arises, a leave point (L_k) is defined and the motion-to-goal behavior is reinvoked.

3.2. A Convergent Geometric-based Algorithm

This section is intended to properly modify algorithm 1 in order to obtain a strategy which, not being computationally intensive[4], guarantees global convergence. Specifically, the changes made to algorithm 1 are highlighted in bold in algorithm 2 (the Euclidean distance is assumed for $d(A, B)$). As can be observed, the final strategy exhibits

[3]This report also shows that our strategy is much less sensitive to the tuning of the α-weights of the objective function —see equations 1 and 2— than the original DWA

[4]The proposal does not require the computation of a local minima-free navigation function

Algorithm 1

0) Initialise $k = 1$ and $L_0 = S$
1) Move along the straight-line segment $L_{k-1}T$ until one of the following occurs:
   a) T is reached. The algorithm stops
   b) An obstacle (O_i) is found. The hit point H_k is defined and step 2 is executed
2) Follow the boundary of the obstacle to the left/right taking into account the next three possible situations which can arise:
   a) T is reached. The algorithm stops
   b) The robot returns to H_k. The algorithm stops because the target is unreachable
   c) A point P_i^j is met satisfying that $\rho(\partial O_i, H_k, P_i^j) \leq 0$. In such a situation, define the leave point $L_k = P_i^j$, set $k = k + 1$ and, lastly, go to step 1

Algorithm 2

0) Initialise $k = 1$, $L_0 = S$, and $\boldsymbol{P} = \{\}$
1) Move along the straight-line segment $L_{k-1}T$ until one of the following occurs:
   a) T is reached. The algorithm stops
   b) An obstacle (O_i) is found. The hit point H_k is defined, $\boldsymbol{D = d(H_k, T)}$, and step 2 is executed
2) Follow the boundary of the obstacle to the left/right taking into account the next fourth possible situations which can arise:
   a) T is reached. The algorithm stops
   b) The robot returns to H_k. The algorithm stops because the target is unreachable
   c) A point P_i^j is met satisfying that $\rho(\partial O_i, H_k, P_i^j) \leq 0$. In such a situation, **if $\boldsymbol{P_i^j}$ is not found in set $\boldsymbol{P}$, then insert $\boldsymbol{P_i^j}$ into $\boldsymbol{P}$**, define the leave point $L_k = P_i^j$, set $k = k + 1$ and, lastly, go to step 1. Otherwise, continue in step 2
   d) **The straight-line segment $\boldsymbol{L_{k-1}T}$ is met at a point $\boldsymbol{Q}$ whose distance to $\boldsymbol{T}$ is less than $\boldsymbol{D}$. Under these circumstances, if the segment $\boldsymbol{QT}$ does not cross the obstacle $\boldsymbol{O_i}$ at point $\boldsymbol{Q}$, then define the leave point $\boldsymbol{L_k = Q}$, set $\boldsymbol{k = k+1}$ and, lastly, go to step 1. Otherwise, update $\boldsymbol{D (= d(Q, T))}$ and continue in step 2**

the same two behaviors as before: motion-to-goal and boundary-following. The main difference with respect to algorithm 1 is found in the condition which determines the transition from the boundary-following behavior to the motion-to-goal behavior. Such condition has been extended so that the robot now decides to leave the contour of the obstacles when one of the two following situations takes place: the first situation, which is described in point 2.c), agrees with the one of algorithm 1 with some additional details which will be discussed later when proving the algorithm's convergence. On the other hand, the second situation —see point 2.d)— allows the robot to abandon the contour following process when it reaches the line defined by the points L_{k-1} and T, that is, the line joining the last position where leaving occurred and the target point.

The formal proof regarding the convergence of algorithm 2 is too long to be included in the paper. However, we can expose the key ideas/lemmas on which the proof is based. Next, the term *leaving condition* is used to refer to one of the two aforementioned situations associated with the ending of the contour following process.

Lemma 1. *The leaving condition 2.c) can only be satisfied a limited number of times*. This is proved by assuming the workspace is bounded —that is, there is

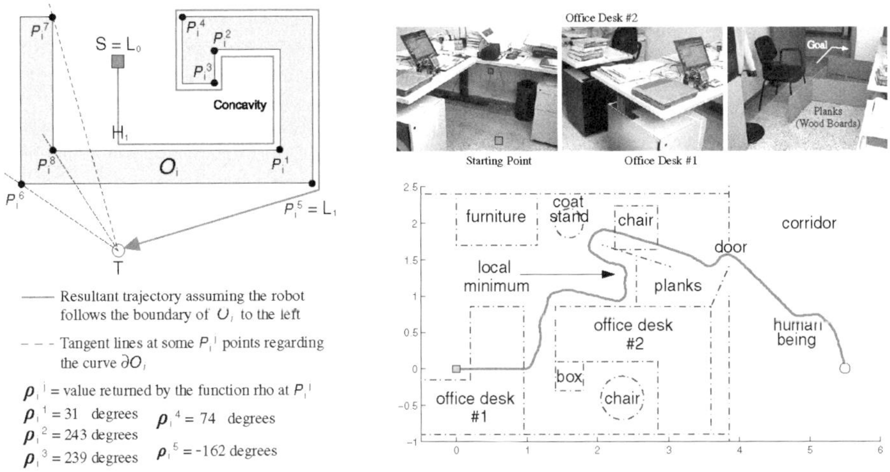

(a) (b)

Figure 4. (a) geometric solution of a mission according to algorithm 1; (b) results of a real test using the Soccerbot robot (the mission was completed in 45.09 seconds covering a distance of 9.02 meters).

a finite number of points P_i^j— and observing the robot is not allowed to leave more than once at the same P_i^j point.

Lemma 2. *The leaving condition 2.d) alone guarantees, whenever possible, reaching the target point.* Such a convergence is achieved by forcing the robot to abandon at points which are closer to the goal than the previous ones.

Proof. *Assuming the target point is reachable, algorithm 2 converges to it.* Once the leaving condition 2.c) is not more satisfied, the leaving condition 2.d) alone leads the robot to the goal according to lemma 2.

4. An Experiment with a Low-Cost Robot

A real test using a miniature robot called *Soccerbot* was carried out with a double purpose: on the one hand, to show that the computational cost of our convergent DWA-based algorithm is small enough to be run, in real time, on a microcontroller and, on the other hand, to highlight that good navigation results can be obtained even only having rough estimations of the obstacle locations provided by small and cheap range measurement sensors. In the experiment, only three infrared range sensors distributed to the left, right and at the front of the robot were used to measure the distances to the obstacles. The test was carried out in the office-like environment outlined in fig. 4(b), which was completely unknown for the robot. As can be observed, the mission consisted in going from one end of the office to the main corridor. The robot found many obstacles on its way to the goal point, being, some of them, artificially created by means of planks. The resultant robot trajectory, which was reconstructed with the odometry data, is superimposed on fig. 4(b).

At this moment, the authors are working on repeating the previous experiment but now applying the global dynamic window approach [3]. The preliminary results which have been obtained show, however, the high-computational requirements of the strategy. By way of example, note that, on average, the time needed to compute the navigation function —an improved version of a wave-front/NF1 planner— on the Soccerbot's microcontroller has been 13.43 seconds. This means the robot, when navigating in unknown environments, will get temporarily trapped into local minima ($\sim$ 13.43 sec for each, which is a very long time) because its reactive layer, that is DWA, is not able to escape from them.

5. Conclusions

This paper has presented an extension of the classic dynamic window approach [10] which guarantees global convergence requiring far less computational resources — processor power, memory, etc.— than the competing strategies ([3,4,5,6,7,8] only to name some of them). The extension has been progressively done in two sequential steps: the former made the robot adopt a simple but almost local minima-free behavior which consisted in imitating the actions that would be taken by a blind person moving with the only help of its walking stick, while the latter step incorporated into the previous behavior a geometric criterion for ensuring the achievement of the target whenever possible. Several experiments (see fig. 1 and 4, and table 1) have clearly shown the high-computational efficiency of the resultant strategy, which makes it particularly suitable for applications involving low-cost robots. This efficiency is essentially due to having removed the up-to-now common practice of building navigation functions (NFs) requiring quite intensive computations to prove convergence. Beyond the computational analysis, it is important to note that our approach is generally overcome, regarding the path lengths from start to goal, by the aforementioned NF-based strategies in both known and partially-known scenarios. However, when the environment is completely unknown, the path length performance of our proposal is similar and, on many occasions, better than the others (see fig. 1 for an example). Finally, as for the algorithm convergence, a formal proof has been given based on the geometry of the obstacles and under the assumption of a point-type robot as well as the correct parameter setting (for instance, the α-weights of the DWA's objective function described in equation 2). At present, the authors are working on the modification of the approach to generalize the proof to robots of arbitrary shape.

References

[1] R. Arkin. *Behavior-based robotics*. MIT Press, 1998.

[2] R. Siegwart and I. Nourbakhsh. *Autonomous Mobile Robots*. MIT Press, 2004.

[3] O. Brock and O. Khatib. High-speed navigation using the global dynamic window approach. In *Proceedings of ICRA*, 1999.

[4] J. Minguez, L. Montano, T. Simeon, and R. Alami. Global nearness diagram navigation. In *Proceedings of ICRA*, 2001.

[5] C. Stachniss and W. Burgard. An integrated approach to goal-directed obstacle avoidance under dynamic constraints for dynamic environments. In *Proceedings of IROS*, 2002.

[6] R. Philippsen. *Motion Planning and Obstacle Avoidance for Mobile Robots in Highly Cluttered Dynamic Environments*. PhD thesis, Ecole Polytechnique Federale de Lausanne, 2004.

[7] P. Ogren and N. Leonard. A convergent dynamic window approach to obstacle avoidance. *IEEE Transactions on Robotics and Automation*, 21(2):188–195, 2005.

[8] E. Demeester, M. Nuttin, D. Vanhooydonck, G. Vanacker, and H. Van Brussel. Global dynamic window approach for holonomic and non-holonomic mobile robots with arbitrary cross-section. In *Proceedings of IROS*, 2005.

[9] J. Kuffner and S. LaValle. RRT-connect: An efficient approach to single-query path planning. In *Proceedings of ICRA*, 2000.

[10] D. Fox, W. Burgard, and S. Thrun. The dynamic window approach to collision avoidance. *IEEE Robotics and Automation Magazine*, 4(1):23–33, 1997.

[11] J. Antich and A. Ortiz. Extending the potential fields approach to avoid trapping situations. In *Proceedings of IROS*, 2005.

[12] I. Ulrich and J. Borenstein. VFH+: Reliable obstacle avoidance for fast mobile robots. In *Proceedings of ICRA*, 1998.

[13] J. Minguez and L. Montano. Nearness diagram (ND) navigation: Collision avoidance in troublesome scenarios. *IEEE Transactions on Robotics and Automation*, 20(1):45–59, 2004.

[14] J. Antich and A. Ortiz. A dynamic window approach to navigate in complex scenarios using low-cost sensors for obstacle detection. Technical Report A-4-2007, University of the Balearic Islands, 2007.

Intelligent Autonomous Systems 10
W. Burgard et al. (Eds.)
IOS Press, 2008

doi:10.3233/978-1-58603-887-8-193

Multi-Robot Fence Patrol in Adversarial Domains

Noa Agmon, Sarit Kraus and Gal A Kaminka
Department of Computer Science
Bar Ilan University, Israel
{segaln,sarit,galk}@cs.biu.ac.il

Abstract. This paper considers the problem of multi-robot patrolling along an open polyline, for example a fence, in the presence of an adversary trying to penetrate through the fence. In this case, the robots' task is to maximize the probability of detecting penetrations. Previous work concerning multi-robot patrol in adversarial environments considered closed polygons. That situation is simpler to evaluate due to its symmetric nature. In contrast, if the robots patrol back and forth along a fence, then the frequency of their visits along the line is coherently non-uniform, making it easier to be exploited by an adversary. Moreover, previous work assumed perfect sensorial capabilities of the robots in the sense that if the adversary is in the sensorial range of the robot is will surely be detected. In this paper we address these two challenges. We first suggest a polynomial time algorithm for finding the probability of penetration detection in each point along the fence. We then show that by a small adjustment this algorithm can deal with the more realistic scenario, in which the robots have imperfect sensorial capabilities. Last, we demonstrate how the probability of penetration detection can be used as base for finding optimal patrol algorithms for the robots in both strong and weak adversarial environment.

Keywords. Multi-Robot systems, Adversarial/game domains, Multi-robot path planning

1. Introduction

The problem of multi-robot patrol has been investigated during the past few years [3,4,8]. The subject is of interest for a number of reasons, mainly its applicability in various domains, for example floor cleaning or lawn mowing [6]. In this problem, a team of robots is required to continuously visit the area, borders of a closed area or an open polyline, while monitoring it in order to detect change in state. Most of the studies in this area concentrated in assuring optimization of frequency criteria [4, 8]. However, several studies considered the problem of multi-robot patrol in adversarial environment, in which the robots are required to patrol along a given area in order to maximize the probability of detecting an adversary which attempts to penetrate through the patrol while trying to evade the robots. This problem is applicable in many security applications [1,2,11].

In this work we concentrate on multi-robot patrol in adversarial settings *along an open polyline*, for example a fence. In such environments, the k robots are required to travel back and forth along the line in order to maximize the probability of detecting penetrations.

Recent studies of multi-robot patrol in adversarial environment concentrated on patrolling around closed areas, i.e, a closed polyline (circle) [1]. The open polyline environment is more complicated to analyze, due to several reasons. The main reason lies in the fact that when a robot travels back and forth along a line, the frequency is inherently non-uniform. Therefore in adversarial environments the adversary can take advantage of this fact and manage to penetrate more easily. This forces the robots to adopt an asymmetric behavior in the sense that at each point along the line the best action of the robot might be different.

Another assumption made in previous work concerns the perfect sensorial capabilities of the robots with respect to detecting the penetration. In [1, 2] the authors assume that the robots detect penetrations with probability 1 once the adversary lies within the sensorial range of a robot. However, in reality this might not be the case. In many cases, the system can guarantee that the robot will detect the adversary up to some percentage, i.e., with probability p_d.

In this paper we tackle these two challenges. We first contrast the two environments - patrol along closed and open polylines, and show the consequence of the difference between the two with respect to the calculation of probability of penetration detection along the polyline.

We provide a polynomial time algorithm for finding the probability of penetration detection (ppd) in each segment along the open polyline, with respect to the location and direction of the robots. We then consider robot with imperfect detection, i.e., $p_d < 1$, and describe an algorithm that finds the probability of penetration detection in each segment with respect to this p_d value.

Finally, we show that after establishing the probabilities of penetration detection along the line, this information can be used as base for finding the *optimal* patrol algorithm in different adversarial environments. We demonstrate how the ppd functions are used in order to find the optimal patrol algorithm in a strong adversarial model, in which the adversary has full knowledge of the patrol scheme. We then show that if the robots are faced against a weak adversary that chooses its penetration spot at random, the optimal patrol that maximizes the expected ppd is no longer the deterministic algorithm as in the case of a closed polyline.

2. Related work

Systems of multiple robots engaged together in order to patrol in adversarial environments has been studied in various contexts. The problem is related to the multi-robot coverage (e.g. [5]), in which a team of robots is required to visit a target area *once*. However, in the area patrol problem (e.g. [3, 8]) and the fence patrol problem [9], the area is visited *continuously* by the team of robots while usually attempting to maintain some frequency criteria.

Agmon et. al. [1] introduced the multi-robot perimeter patrol in adversarial environments, along with the robotic model we base our work upon. In their work, they make worst-case assumptions regarding the capabilities of the robots, i.e., they assume the adversary has full knowledge of the robots' patrol algorithm. They provide a polynomial time algorithm for finding the probability of penetration detection throughout the perimeter, and describe a second algorithm for using these probability functions in order to find the patrol algorithm that maximizes the minimal probability of penetration detection along the perimeter. As mentioned previously, they assume that the detection of the adversary is perfect.

Agmon et. al. [2] further study the perimeter patrol problem in different adversarial environments. They study the case in which the adversary has no knowledge of the patrol algorithm, and chooses its penetration spot at random with uniform distribution, and show that the simple deterministic algorithm maximizes the expected ppd in this case. They examine the compatibility of different algorithms to different adversarial settings also empirically, and have shown that the algorithm intended for the worst case scenario failed miserably when working against a random adversary, and that the deterministic algorithm failed against a strong adversary. Also in this work Agmon et. al. do not refer to imperfect sensorial capabilities of the robots.

Other closely related work is the work by Paruchuri et. al. [11, 12], which consider the problem of placing security checkpoints in adversarial environments. They use policy randomization for the agents behavior in order to maximize their rewards. In their work, the adversary has full knowledge of the agents, therefore it uses it in order to minimize its probability of being caught in some checkpoint.

Elmaliach et. al [9] consider the fence patrol problem, however they aim to maximize optimization criteria using a deterministic algorithm, and do not assume the existence of an adversary. They provide a mathematical model that takes into consideration drawbacks of the system, for example odometry problems, when trying to find a patrol algorithm that maximizes the optimization criteria. Specifically, they show that overlapping the sections of responsibility given to each robot can be advantageous, and the optimality of the patrol algorithm is characterized by the extent of the overlap. In this paper, we show that in an open fence the deterministic algorithm is no longer optimal for both adversarial model, hence using this algorithm is not beneficial.

Theoretical work based on stochastic processes that is related to our work is the *cat and mouse* problem [7], also known as the *predator-prey* [10] or *pursuit evasion* [13]. In this problem, a cat is attempting to catch a mouse in a graph where both are mobile. The cat has no knowledge about the mouse's movement, therefore as far as the cat is concerned, the mouse travels similarly to a simple random walk on the graph. We, on the other hand, have different assumptions about the adversary, consider a *robotic* model, in which the movement is correlated to the movement of a robot, and in our model the robots travel along a fence, rather than in a graph or an area.

3. Background

3.1. Robotic and environmental model

In our work, we assume to be given a team of k homogenous robots that are required to patrol along an open polyline. The line is divided into N segments (not necessarily of the same length or orientation - see figure 1c.), where each robot travels through one segment per one time cycle. This allows us to consider heterogenous domains, i.e., domains in which the velocity constraints along the perimeter can change. Note that the polyline is not necessarily straight, but can be of any open shape. We then divide the N segments into k sections, where each section contains $d = N/k$ segments. We therefore focus our analysis of the system to analysis of a section, since the behavior in equivalent in all sections.

We assume the robots have directionality associated with their movement. In each time cycle, the robots have to decide where they should go. Since this is a line, they have the option to continue in their current course or turn around. Turning around may be costly, hence we model this cost by the number of time cycles it takes the system to stabilize after all robots turn around, denoted by τ.

The patrol algorithm of the robots is characterized by the probability p, i.e., the robots go straight forward with probability p, and turn around with probability $q = 1 - p$.

In our adversarial models, the adversary had to decide at time 0 through which segment to penetrate. We assume that the time it takes it to penetrate is not instantaneous, and lasts t time units.

3.2. Patrolling along a closed polyline vs. an open polyline

As stated previously, recent studies in multi-robot patrol in adversarial environments [1,2] assume the robots travel around a closed, circular, area. In the following, we describe why patrolling along an open polyline has a challenging nature compared to patrolling in cyclic environments (closed polyline).

The first reason lies in the fact that the robots are required to go back and forth along a part (or parts) of the open polyline, therefore the elapsed time between two visits of a robot in each point along this line can be almost twice as long as the elapsed time in a circular setting. In Figure 1, we are given two environments: one closed polyline (circle) (a) and an open polyline (b). Note that the open polylines b. and c. are equivalent in the sense that each robot travels through one segment per time step, regardless of the shape of the section. Both lines a. and b. are of the same total length l and with the same number of robots (4). In the circular environment, if it takes an adversary more than $l/4$ time units to penetrate - it will never be able to penetrate even if the robots simply travel with uniform distance between them continuously. However, if the robots travel along an open polyline (b), the maximal time duration between two visits of the robot — even in the best case, is $2l/4 - 2$ [9]. Therefore a weaker adversary that has penetration time which is almost twice as long as in the circular fence might still be able to penetrate.

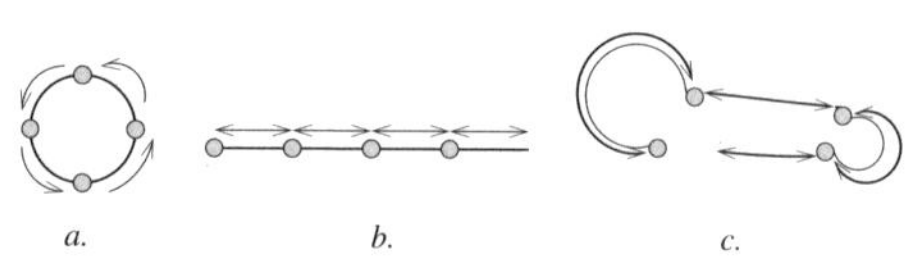

Figure 1. Illustration of the difference between patrolling along a line and patrolling along a circle, for different polylines

Another reason for the added complication in analyzing the probability of penetration detection in open polyline environments lies in the asymmetric nature of traveling in the segments along time. In a circular environment, if the robots are coordinated and switch directions in unison, then the placement of the robots is symmetric in each time unit. Therefore all segments in the same distance from some robot (with respect to its direction) have the same probability of penetration detection. Hence in order to calculate the optimal way of movement (in our case the probability p of turning around), it is enough to consider only one section of d segments, and the resulted p is equivalent throughout the execution [1]. In an open polyline environment this is not the case. The probability of penetration detection differs with respect to the current location and direction of the robot. Therefore the algorithm that finds the ppd for each segment, needs to calculate the ppd as a function of p for each segment s_i for each possible initial location of the robot inside the section. Therefore it results with a matrix of size $d \times d$ of the ppd functions (as opposed to a vector of d functions in the circular fence).

4. Patrol along an open polyline

The *best* patrol algorithm for a team of robots in adversarial environment requires to find the algorithm that maximizes some function of the probability of penetration detection (ppd). This function depends on the assumptions made on the capabilities of the adversary [2]. However, the basic step upon which other calculations are made is finding the probability of penetration detection in all segments along the fence. Therefore in this section we describe Procedure FindFencePPD that finds the ppd in the segments.

The ppd in a segment s_i is determined by the probability of the *first* arrival to s_i during t time units. This is due to the fact that we assume, at this stage, that if the robot arrives at a segment that is currently occupied by the adversary, it will detect it, hence it does not matter if the segment is visited more than once during t time units. For simplicity, we discuss the case in which $\tau = 1$.

Denote the ppd in segment s_i by ppd_i. We describe the system as a Markov chain (see Figure 2). Since the robots have directionality associated with their movement, we create two states for each segment: the first for traveling in a segment in the clockwise direction, and the second for traveling in the counterclockwise direction. The probability of turning around at the end of each section is 1, otherwise the robot is going straight with probability p, and turns around with probability $q = 1 - p$.

The algorithm is dynamic programming inspired, and uses the state transition rules describes in Figure 2 in order to fill in a matrix M gradually, until reaching the state of the system after t time units (last line). The algorithm works as follows. Each segment s_i is associated with a "phantom" variable f_i. Whenever an element goes through s_i's location in the matrix, it is multiplied by f_i. At the end, all visits to the segment are added, and all fantom variables are substituted with 1 *except* for f_i. Therefore the polynomial coefficients of f_i^t is the probability of the t'th visit to s_i, the coefficient of f_i^{t-1} represents the probability of the $t-1$ visit, and so on. Hence ppd_i in this case is the polynomial visit of f_i^1, i.e., the probability of first visit to s_i.

Recall that $d = N/k$. The returned value from Procedure FindFencePPD is a matrix of size $d \times d$, where each row i contains d functions representing the ppd in each segment, given that the robot is currently in segment i.

Time complexity: The time complexity of Procedure FindFencePPD is $\mathcal{O}(d^2 t)$, since we fill in the matrix M d times, where in each time it takes $d \cdot t$ time to fill it. Hence the total time complexity is $\mathcal{O}(d \cdot d \cdot t) = \mathcal{O}(d^2 t)$.

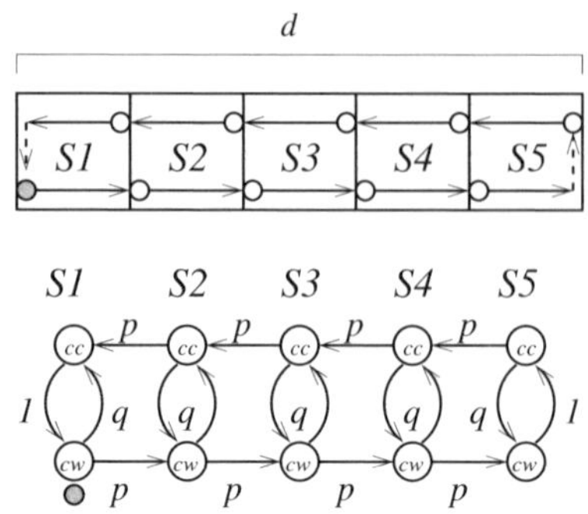

Figure 2. Description of the system as a Markov chain, as base for the FindFencePPD algorithm.

Algorithm 1 Procedure FindFencePPD(d, t)

---

for $loc \leftarrow 1$ to d **do**
  Create the matrix Res of size $d \times d$, initialized with 0s.
  Create the matrix M of size $2d \times (t+1)$ initialized with 0s.
  Set $M[1, s_{loc}^{cw}] \leftarrow f_{loc}$
  Fill all entries in M gradually using the following rules.
  for $r \leftarrow 2$ to t **do**
    Set $M[r, s_1^{cw}] \leftarrow f_1 \times M[r-1, s_1^{cc}]$
    Set $M[r, s_d^{cc}] \leftarrow f_1 \times M[r-1, s_d^{cw}]$
    for all other entries in row r **do**
      Set $M[r, s_i^{cw}] \leftarrow v_i \times \{p \cdot M[r-1, s_{i+1}^{cw}] + q \cdot M[r-1, s_i^{cc}]\}$.
      Set $M[r, s_i^{cc}] \leftarrow v_i \times \{p \cdot M[r-1, s_{i-1}^{cc}] + q \cdot M[r-1, s_i^{cw}]\}$.
  Create the vector V of size d
  for $j \leftarrow 1$ to d **do**
    $V[j] \leftarrow \sum_{i=1}^{t} M[i, s_j^{cw}] + M[i, s_j^{cc}]$
    For each entry i in V, substitute all f_k, $k \neq j$ with 1.
    Set $Res[loc, j] \leftarrow$ polynomial coefficient of f_j in $V[j]$.
  Set $Res_k \leftarrow VM$.
Return Res.

---

5. Imperfect detection

In many cases, even if the adversary passes through the sensorial range of the robot, it still does not necessarily detect it. Assume that the robot is currently monitoring some segment s_i. If the adversary penetrates through s_i while it is monitored, then the probability that the robot actually detects the adversary is p_d. The major difference when considering this case compared to the case in which $p_d = 1$, is that revisiting a segment can increase the probability of detecting the adversary. Therefore the probability of detection in a segment s_i (ppd_i) is *not* equivalent to the probability of *first* arriving at s_i, but the probability of detecting the adversary during some visit m to s_i, assuming it was not previously detected. Denote the probability of the y'th visit of some robot to segment s_i by w_i^y. Therefore ppd_i is defined as follows.

$$\mathsf{ppd}_i = w_i^1 p_d + w_i^1(1-p_d) \times \{w_i^2 p_d + w_i^2(1-p_d) \times \{\ldots\{w_i^t \times p_d\}\}\} \quad (1)$$

In other words, the probability of detecting the penetration is the probability that it was detected in the first visit ($w_i^1 \cdot p_d$) plus the probability that it was *not* detected then, but in later stages. This again is the probability that it was detected in the second visit ($w_i^2 \cdot p_d$) or in later stages, and so on.

Note that in t time units, $w_i^t = 0$ for all segments s_i currently unoccupied by some robot, and if a robot resides in s_i, then this value is exactly $p_d(1-p)^t$.

Algorithm 2 Procedure ComputeProbPPD(d, t)

Run Procedure FindFencePPD(d, t) while returning the entire polynomial, i.e., skipping line 16 of the procedure.
for $j \leftarrow 1$ to d **do**
  for $i \leftarrow 1$ to d **do**
    $w_i^y \leftarrow$ polynomial coefficient of f_i^y.
    $PRes[j, i] \leftarrow$ substitution of $w_i^y, \forall y$ in Equation 1.

Theorem 1. *For each segment s_i, Procedure* FindFencePPD *computes* ppd_i.

Proof. In order to prove the theorem, we need to show that the algorithm computes correctly the probability of the m'th arrival to s_i, for every $1 \leq m \leq t$, i.e., the coefficient of f^m is exactly w_i^m. Since each path is multiplied by f each time it passes through s_i, then necessarily if the path went through s_i m times, it is a coefficient of f^m. By adding all arrivals to s_0^{cw} and s_0^{cc}, we take into consideration all visits to s_i, hence all paths going through s_i are taken into consideration, and by using Equation 1, we generate exactly ppd_i. □

As in FindFencePPD, the returned value is a matrix of size $d \times d$, where each row i contains d functions representing the ppd in each segment, given that the robot is currently in segment i.

6. Applying knowledge of ppd functions in determining the patrol algorithm

Finding the ppd in all segments of the section is the first step in determining the best patrol algorithm for the robots. After obtaining this information, a suitable function can be executed in order to meet the desired criteria which depends on the adversarial model. The criteria could be maximizing the minimal ppd for a strong adversarial model, or maximizing the expected ppd for a weak one [2].

In the case of a fence, the ppd value depends on the current location of the robot, hence the optimal p value characterizing the patrol of the robots is different for each segment s_i, where $1 \leq i \leq d$. Note that there could be different optimal p values with respect to both location and *orientation* of the robot ($2d$ values). However, it is enough to calculate the ppd values only d times - only for one direction, as the other direction is a reflecting image of the first (see Figure 3).

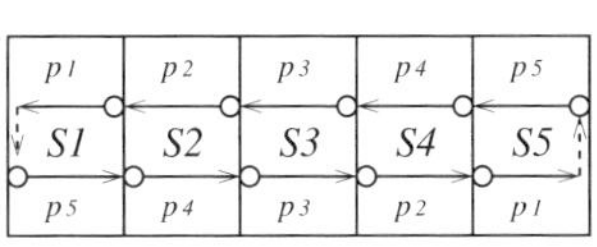

Figure 3. An illustration of the required output as a patrol algorithm, and where to use it.

6.1. Strong adversarial model - full knowledge

We demonstrate the use of Procedure FindFencePPD in the full knowledge adversarial model. In this model, when the adversary has to decide where to penetrate, it has full knowledge of the patrol algorithm and the current location of the roots. Therefore it will necessarily choose to penetrate through the segment in which the probability of penetration detection is minimal (the weak spot). Therefore in this case the robots should choose an algorithm that maximizes the minimal ppd along the section. For that we use algorithm FindP from [1] that finds the value p such that the minimal ppd is maximized. FindP computes this point by finding the maximal point in the integral intersection of all curves (ppd_i). The complete description of the algorithm is shown in Algorithm 3.

Algorithm 3 Procedure ComputeProbPPD(d, t)

$M \leftarrow$ FindFencePPD(d, t)
for $i \leftarrow 1$ to d **do**
  $OpP[i] \leftarrow$ FindP(d, t) [1] with given additional input $M[i]$ as vector of ppd functions.
Return OpP

6.2. Weak adversarial model - random penetration

We now turn to examine the case in which the adversary does not have full knowledge of the patrol algorithm. In this case, we assume the adversary chooses at random with uniform distribution its penetration spot from all unoccupied segments. Therefore we wish to maximize the *expected* ppd throughout the line.

In circular environments, it was proven that under this adversarial model, the optimal patrol algorithm is the deterministic algorithm. However, if the robot travels along a line - this is not the case any more. We show it by a counter example. The logic behind this fact is that in some segments, it is worthwhile to switch the direction of the robot before reaching the end of the section, and by that cover more segments.

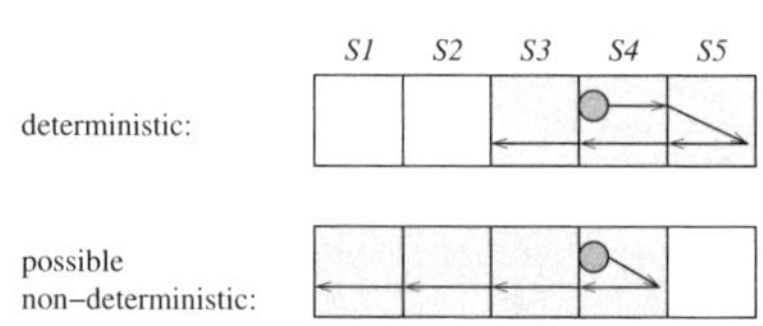

Figure 4. An illustration of a case in which the maximal expected ppd is obtained by a non deterministic algorithm. Each arrow represent a movement in one time cycle.

Consider the case in which $d = 5$ and $t = 5$. If the robot is placed in segment s_4, then the ppd values are as follows. $\mathsf{ppd}_1 = (1-p)p^4$, $\mathsf{ppd}_2 = (1-p)p^3$, $\mathsf{ppd}_3 = (1-p)p^2 + p^5 + (1-p)^3 p^2$, $\mathsf{ppd}_4 = 1$ and $\mathsf{ppd}_5 = p + (1-p)^2 p$, and the expected ppd is maximized for $p = 0.418$. The logic, as explained previously, is illustrated in Figure 4. It demonstrates that the robot can profit from turning around at s_4, since it can reach four segments, where if it travels deterministically through all segments, i.e., go straight to s_5, it reaches only three segments.

This strengthens our need to find the actual ppd values in all segments, since after obtaining these functions, it is possible to calculate the p value that maximizes the expected ppd throughout the line. This can be done using a procedure similar to ComputeProbPPD, while replacing the call to FindP to a function that calculates the p value that maximizes the expected ppd.

7. Conclusions and future work

This paper discusses the problem of multi-robot patrol along an open polyline in adversarial settings. We show that this case is more challenging than the problem of multi-robot perimeter patrol, which is symmetric by its nature. We provide a polynomial-time algorithm for finding the probability of penetration detection in each segment along the line. We then tackle another challenge posed by previous studies, and show that this algorithm can be altered into considering also more realistic cases in which the robots do not have perfect detection, i.e., the robot will detect the adversary if it is in its sensorial range with probability p_d. We then demonstrate how the probability of penetration detection functions can be used in order to find the optimal patrol algorithm in various adversarial models.

There are various points we wish to address as future work. First, we are interested in incorporating more realistic considerations in the model, such that it will better suit various practical systems, for example sensing ahead of the segment with changing probabilities. Moreover, we intend to check the implications of the overlapping of sections on the optimal patrol algorithm in various adversarial models. We are interested in using the ppd functions in order to find game-theoretic solutions to the the problem. Last, we would like to see how this algorithm can be adapted to patrol in other domains (area patrol, for example).

References

[1] N. Agmon, S. Kraus, and G. A. Kaminka. Multi-robot perimeter patrol in adversarial settings. In *ICRA*, 2008.

[2] N. Agmon, V. Sadov, S. Kraus, and G. A. Kaminka. The impact of adversarial knowledge on adversarial planning in perimeter patrol. In *AAMAS*, 2008.

[3] M. Ahmadi and P. Stone. A multi-robot system for continuous area sweeping tasks. In *ICRA*, 2006.

[4] A. Almeida, G. Ramalho, H. Santana, P. Tedesco, T. Menezes, V. Corruble, and Y. Chevaleyr. Recent advances on multi-agent patrolling. *Lecture Notes in Computer Science*, 3171:474–483, 2004.

[5] H. Choset. Coverage for robotics—a survey of recent results. *Annals of Mathematics and Artificial Intelligence*, 31:113–126, 2001.

[6] J. Colegrave and A. Branch. A case study of autonomous household vacuum cleaner. In *AIAA/NASA CIRFFSS*, 1994.

[7] D. Coppersmith, P. Doyle, P. Raghavan, and M. Snir. Random walks on weighted graphs and applications to on-line algorithms. *J. ACM*, 40(3), 1993.

[8] Y. Elmaliach, N. Agmon, and G. A. Kaminka. Multi-robot area patrol under frequency constraints. In *ICRA*, 2007.

[9] Y. Elmaliach, A. Shiloni, and G. A. Kaminka. A realistic model of frequency-based multi-robot fence patrolling. In *AAMAS*, 2008.

[10] T. Haynes and S. Sen. Evolving behavioral strategies in predators and prey. In *IJCAI-95 Workshop on Adaptation and Learning in Multiagent Systems*, pages 32–37, 1995.

[11] P. Paruchuri, J. P. Pearce, M. Tambe, F. Ordonez, and S. Kraus. An efficient heuristic approach for security against multiple adversaries. In *AAMAS*, 2007.

[12] P. Paruchuri, M. Tambe, F. Ordonez, and S. Kraus. Security in multiagent systems by policy randomization. In *AAMAS*, 2007.

[13] R. Vidal, O. Shakernia, H. J. Kim, D. H. Shim, and S. Sastry. Probabilistic pursuit-evasion games: theory, implementation, and experimental evaluation. *Robotics and Automation, IEEE Transactions on*, 18(5):662–669, 2002.

Intelligent Autonomous Systems 10
W. Burgard et al. (Eds.)
IOS Press, 2008

doi:10.3233/978-1-58603-887-8-202

Cooperative Behavior of Multiple Robots by Chain of Monolithic Policies for Two Robots

Keisuke KOBAYASHI [a,1], Ryuichi UEDA [a] and Tamio ARAI [a]
[a] *Univ. of Tokyo, Japan*

Abstract. We propose a novel method for cooperative behavior of multiple robots. To control more than two robots, the proposed method utilizes multiple policies that are created for not more than two robots. By chaining those policies redundantly, we can approximate the optimum policy for all robots, which is never obtained due to the curse of dimensionality. In simulations and experiments on the domain of RoboCup four-legged league, we verify that the proposed method can realize effective cooperation of robots.

Keywords. cooperative behavior, state-action map, dynamic programming, RoboCup

1. Introduction

When some robots work in an environment, one robot should consider the states of the others. The consideration avoids robots from colliding with each other at least. Moreover, some kinds of efficiency (e.g. time, energy,...) can be enhanced by cooperative behavior in some tasks.

Two well-known problems occur in such a multi-agent system. One is that some variables for decision cannot be obtained from sensors or network. The other one is the problem of computing complexity, which is called *the curse of dimensionality*. Even if all variables that relate to decision making are known, it never means that the robots can choose optimal actions for a task. To do so, search methods in large space are required for decision making. However, consideration of all variables is infeasible in almost all cases due to its computational cost.

In past studies, therefore, robots have mainly considered its own state variables [4,5,6,7]. Only when cooperation in required, some of other variables related to the cooperation are added to state variables. It means that another planner for cooperation, which may be composed of heuristics, is required. For example, GermanTeam uses an explicit teamwork framework in addition to decision rules. McMillen *et al.* also use a combination of roles. Each role (means a decision

[1]Corresponding Author: Keisuke Kobayashi, Intelligent Systems Division., Dept. of Precision Engineering, School of Engineering, Univ. of Tokyo, 7-3-1 Hongo, Bunkyo-ku, Tokyo 113-8656 Japan; E-mail: kobayashi@robot.t.u-tokyo.ac.jp

rule) considers state of one robot and a ball only. States of teammates are not considered in process of decision making.

On the other hand, Ueda *et al.* have created a monolithic policy for cooperation of two robots [2]. The policy, which is called a state-action map, is obtained by one week calculation of dynamic programming (DP) [1]. The creation of a state-action map has been done in computer simulation. In this case, cooperation of robots can be generated based only on a functional. Though this attempt is a brute force and suffers from some ill-effects by discretization, requirements of cooperation can be clearly explained. Of course this approach is not feasible now for more than two robots due to the amount of computation.

In this paper, we try controlling more than two robots by combinations of monolithic policies of *two robots*. This attempt is not only an extension for solving the problem of calculation amount, but also a different approach from the previous methods. A chain of robots connected by policies of two robots has a profound structure than any other structure that is composed of relations of each simplicial robot.

2. RoboCup Four-Legged League and Problem

As an example of multi-agent systems, we take up the RoboCup four-legged robot league [3]. In this league, the robots shown in Fig.2 play soccer on the field shown in Fig. 1. The robots work autonomously with wireless communication between colleague robots. They must decide its behavior in real-time in soccer games though information is incomplete and uncertain.

Figure 1. RoboCup Four-Legged Robot League

Figure 2. ERS-7 Made by Sony

For this game, we want to create a decision making policy that gives suitable behavior to every robot when the policy receives the state of the game in real-time. We should pay attention that cooperative behavior is not always required in RoboCup. Cooperation is not a purpose but a means of defense and offense. When heuristics are used for cooperation, in the ultimate sense, we cannot answer whether cooperative behavior is really required or not. Therefore, we want to avoid ad-hoc rules, and want to find a reasonable method that is based on the frame of optimal control problems. The trouble is that the curse of dimensionality must be considered. In this paper, a team of three robots and the position of the ball are only considered. It means that we do not care about the existence of opponent robots and the speed of the ball. Nonetheless, various problems will happen due to the curse of dimensionality when we want to obtain the policy that works in real-time.

3. Chain of Two Robot Policies

3.1. Decomposition of State Space

We assume that a state of a game can be identified by a set of variables: $\{x_1, x_2, \ldots, x_n\}$, are called state variables. The space spanned by the state variables is called state space, and written as $\mathcal{X}$ in this paper. In the case of the above RoboCup problem, 11 state variables are necessary at least. As shown in Fig.3, two variables are required for representing the position of the ball. As shown in the case of Fig.3, these variables are defined on a poler coordinate system from a robot. Three variables are then used for representing the position and the orientation of each robot on the field. In order to reduce the number of state variables, we neglect velocities of objects. Therefore actions that robots can execute are described as displacements of these variables.

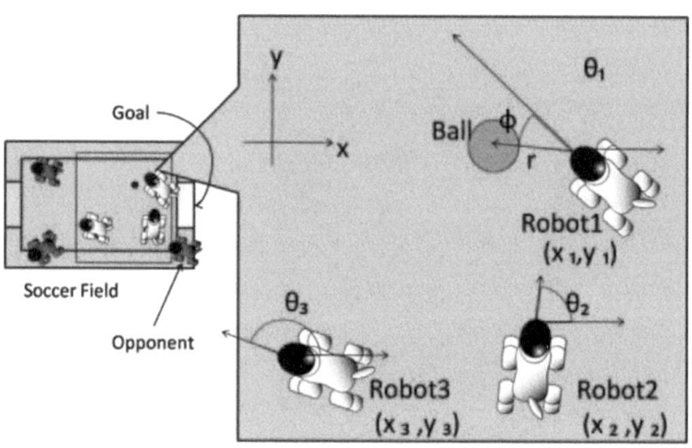

Figure 3. Coordinate System of Robots and the Ball

When a task is given to the robot team, behavior of the robots should do their best for the task. Put mathematically, a task is defined by an evaluation function, and the performance of the robots are measured by the integral of the function. Though the detail of the optimal control theory cannot be completely explained here, the following functions

$$\pi : \mathcal{X} \to \mathcal{A} \quad \text{and} \quad V^{\pi} : \mathcal{X} \to \Re \qquad (1)$$

must be at least defined for discussion. π is a mapping from a state in the state space to behavior of robots. The behavior is called *an action* hereafter. In this paper, an action means the total behavior of the three robots. A finite number of actions belong to the set $\mathcal{A}$. π is called a policy. When a policy is fixed, a state-value function V^π is defined in a time-invariant system. This function gives the value of the state in the task. π is also called a state-action map when it is obtained as a multi-dimensional look-up table. The state space is discretized by a multi-dimensional grid when a state-action map is created.

Even when the number of state variables are just eleven, it is difficult to solve a performable policy π and its state-value function V^π due to the costs of computation. The decomposition of the state space into subspace is frequently and implicitly used in those cases. (This problem is also defined in subspace because there is no consideration of opponent robots,) In many cases, moreover, the decomposition is done based on each robot as a unit.

In our case, subspace that contains two robots is also considered. This subspace takes an important role for elimination of heuristics.

3.2. State-Action Maps

In [2], a state-action map that contains eight state variables has been created. The eight variables represent state of two robots and the ball. In other words, this state-action map is obtained in the subspace with two robots, which is mentioned above. The state-action map contains 6×10^8 discrete states. This state-action map indicates an optimal policy for two robots. Performance of two robots using this map is cooperative and effective.

For the experiment in this paper, we have calculated a state-action map that contains a larger number of states again. Since the size of the field is enlarged, we need the calculation again. For the current field whose width and length are 4[m] and 6[m] respectively, a state-action map and its state-value function that contains 3×10^9 discrete states has been calculated in one month. By this map, each robot can choose 14 kinds of walking mode and four kinds of kicking mode. Since there are two robots, the number of actions in $\mathcal{A}$ becomes the square of the numbers of mode. In the state-action map that is used in this paper, however, the combination is limited to 74 due to the limitation of computing resources. We have now obtained a state-action map without this limitation. However, the limited one, which is called *the 8-dimensional (8D) offensive map* in this paper, is enough for explaining our method.

The following state-action maps have been then created.

- *The 5-dimensional (5D) offensive map*: a robot and the ball are considered. The task of the robots is to bring the ball to the opponent goal within a minimum number of steps.
- *The 3-dimensional (3D) defensive map*: a robot is considered. The task of the robot is to go to and to stay a fixed position near the goal of its team. The robot is not a goalkeeper.

The above two maps consider only one robot. They are used for evaluation accessorily.

3.3. Behavior Rule of Three Robots with State-Value Functions in Subspace

3.3.1. Behavior of Multiple Robots by Chain of Policies

The most important aspect of the proposed method is explained in Fig.4. The robots use the three kinds of maps with various combinations as described later. When all of the robots use the 8D offensive map, one of the robots belongs to both of the 8D space as shown in Fig.4(a). In this case, decision makings in each space by the 8D map conflicts with each other. We can resolve this conflict by the comparison of the state-value function. In the method, the values of the function are compared and the method chooses the action in the space whose value is larger than the other. This procedure is illustrated in Fig.4(b). In case where more than three robots exist, we can apply this method to them. First we decompose the state space into multiple subspaces, and we resolve conflicts by comparison of the state-value functions.

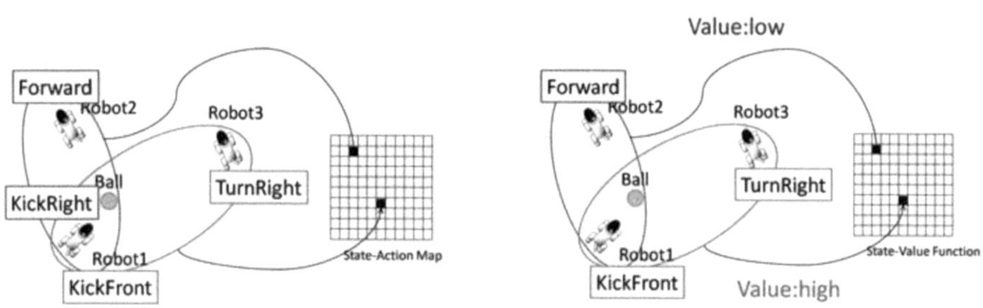

(a) Confliction of Actions Obtained from Two 8D Maps

(b) Comparison of Values for Action Choice

Figure 4. Confliction and Its Resolve

3.3.2. Use of Various Maps

When other maps are combined, the comparison of the values becomes difficult. Since the task or the scale of the state-value function in each subspace is different, the values cannot directly compared.

In this stage, we must design some heuristics. However, the basic strategy is still comparison of values. Each combination of maps is compared by weighted values.

When we assume that the closest robot to the ball chooses one of the offensive maps, the other robots choose the defensive map or one of the offensive maps respectively. In this case, the integral state value $V_{\text{integral}}(s)$, which considers three robots and a ball, is described as below:

$$V_{\text{integral}}(s) = \max \begin{cases} V_{\text{offense5D}}(s_1) + \alpha\{V_{\text{defense}}(s_2) + V_{\text{defense}}(s_3)\} \\ 2V_{\text{offense8D}}(s_{12}) + \alpha V_{\text{defense}}(s_3) \\ 2V_{\text{offense8D}}(s_{13}) + \alpha V_{\text{defense}}(s_2) \\ \max\{V_{\text{offense8D}}(s_{12}), V_{\text{offense8D}}(s_{13})\} \\ \qquad +V_{\text{offense8D}}(s_{12}) + V_{\text{offense8D}}(s_{13}) \end{cases} \quad (2)$$

We select the combination of maps which maximizes the above value.

s_1 denotes the discretized state of the ball and the robot that is the nearest one from the ball. For purposes of illustration, this robot is named Robot1. In the task, the owner robot of the name frequently changes. When a robot is temporarily named Robot1, other robots are called Robot2 and Robot3. s_{12} means the discretized state of the ball and the two robots in the 8D subspace. $V_{\text{offense8D}}$, $V_{\text{offense5D}}$, and V_{defense} mean the state-value function of each map respectively. Three maps are created independently therefore we cannot add three original state values. To add three state values, we introduce a parameter α. The parameter α is a positive value and decides the relative importance of defense and offense. In this paper, we determine α by varying it and observing behaviors of robots in preliminary simulation. When the combination is chosen, the behavior of each robot is chosen from the maps that are related to the combination.

4. Simulation and Experiments

4.1. Simulation

We verify the efficiency of our method by simulation. The environment of simulation is modeled as RoboCup four-legged league.

Our simulator is designed for evaluation of global decision making. Therefore it is different from the real environment as below.

- Complexity of robot body is not considered. All robots are described three variables which are position and angle.
- In this simulator, robots can get its correct positions and angles. Robots can also obtain state variables of teammates by communicating via wireless network. Any noise is not considered in perception and communication.
- In this simulator, robot can get state variables of opponent robots and a ball that exist in its sight area.
- In this simulator, robot can shot a ball when the distance from the robot to the ball is between 150[mm] and 250[mm] and the angle of the ball is between -95[deg] and 95[deg].

We simulate the game of RoboCup four-legged league. The task of three soccer robots is to score. We define the case where robots fail their task as below.

- The ball falls out of the soccer field.
- The Opponent team scores.

The teams of robots to be evaluated are as below.

- team A: three uncooperative robots,
- team B: three cooperative robots that use 8D map only, and
- team C: three cooperative robots that use three maps.

The opponent robots use only the 5D map. They all perform to score for itself. Therefore they do not consider each other. In simulation, collisions of robots are not considered. We execute 10000 trials and count the number of goals, which denotes achievement.

First, we show behavior of team A and team C in Figure 5 and in Figure 6. Robots in team A all tried to approach the ball and they were closely-spaced near the ball. Robots in team C used passes and three robots spaced each other. Moreover, by introducing defensive map, the left robot stayed in front of its goal after its pass. By the proposed method, three robots performed cooperatively.

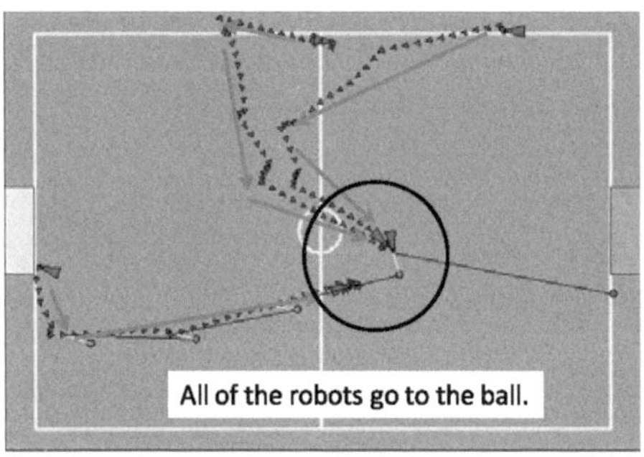

Figure 5. Behavior of team A

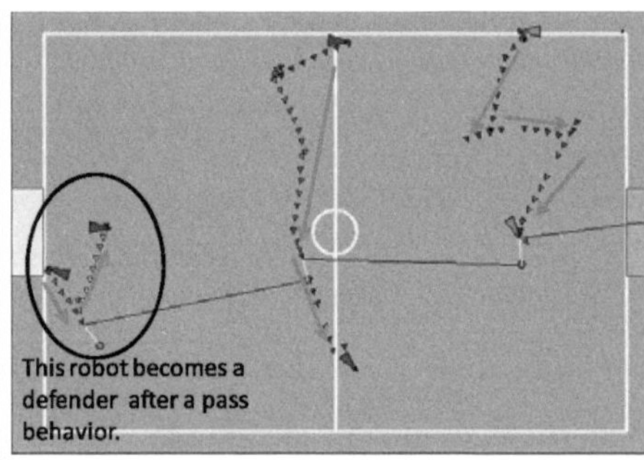

Figure 6. Behavior of team C

Result of evaluation is shown in Table 1. By introducing cooperation by 8D map, the number of goals increased by 5.5[%] from 5228(team A) to 5516(team B). By using 8D map, robots in team B succeeded their task before interception by opponents.

Moreover by using three maps, both offense and defense, the number increased by 7.4[%] from 5228(team A) to 5613(team C). Cooperation by the proposed method increased the number of achieving their task. However improvement by introducing three maps is small, which indicates the bad effectiveness of defensive map. We should modify the elemental maps.

Table 1. The result of simulation(without collisions)

team	goals
team A(uncooperative)	5228
team B(cooperative, only 8D map)	5516
team C(cooperative, three maps)	**5613**

4.2. Experiment

We implement our method to ERS-7. We compare performances of two teams. One team(team B) behave with the proposed method using only 8D offensive map.

The other team(team A) is uncooperative. They use only 5D offensive map. In our experiments, we do not use the opponent team and only observe the behavior of the robots.

The behavior of two teams are shown in Figure 7 and in Figure 8. All robots in team A tried to approach the ball and they conflicted. Robots of team B, which use 8D map, performed spacing each other. In Figure 8, when the robot in upper left approached the ball, other two robots stayed. After the robot in upper left kicked, the robot in lower left started to approach. Three robots in team B succeeded in performing cooperatively.

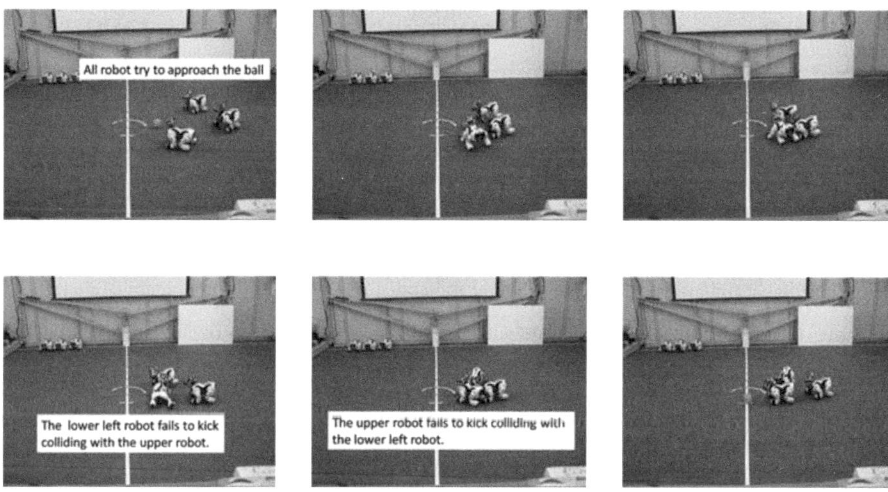

Figure 7. A performance of the uncooperative team(team A)

5. Conclusion

In this paper, we proposed the novel method to realize semi-optimum cooperation of three robots by combination of state-action maps. Before operation of robots, we created three optimum policies of robots by DP and applied them to three robots. To decide which policy the robots use, we used state-value functions. We evaluated combinations of policies by adding three state-value functions. In simulation modeled as RoboCup four-legged league, we verified that our method can realize cooperation and reduce time to achieve task of robots. In experiment, robots by the proposed method showed their cooperative performance spacing each other. For these reasons, we verified that our method can realize cooperation of three robots.

In future, we will introduce new state-action maps that consider state variables of opponent robots. By introducing new maps, we will try more complex combination of maps. Moreover we will propose an additional algorithm for online decision making to response opponents. We have proposed the off-line method

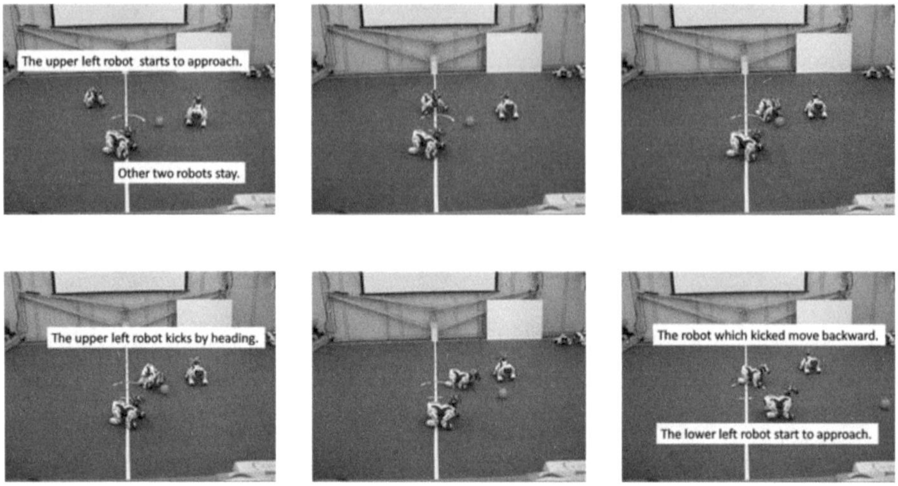

Figure 8. A performance of the cooperative team(team B)

in this paper, which is difficult to response opponent robots. An new algorithm will assist response to opponent robots.

Acknowledgements

This research is partially supported by the grant program of Tateishi Science and Technology Foundation.

References

[1] R.E. Bellman, *Dynamic Programming*, Princeton University Press, 1957.
[2] R. Ueda, K. Sakamoto, K. Takeshita and T. Arai, Dynamic Programming for Creating Cooperative Behavior of Two Soccer Robots –Part1: Computation of State-Action Map, Proc. of IEEE International Conference on Robotics and Automation (2007), 1–7.
[3] H. Kitano, M. Asada, Y. Kuniyoshi, I. Noda and E. Osawa, RoboCup: The Robot World Cup Initiative, Proc. of International Conference on Autonomous Agents (1997), 340–347.
[4] P. Stone and M. Veloso, Task Decomposition, Dynamic Role Assignment, and Low-Bandwidth Communication for Real-Time Strategic Teamwork, *Artificial Intelligence* **110** (1999), 241–273.
[5] M. Tambe, J. Adibi, Y. Al-Onaizan, A. Erdem, G.A. Kaminka, S.C. Marsella and I. Muslea, Building Agent Teams Using an Explicit Teamwork Model and Learning, *Artificial Intelligence* **110** (1999), 215–239.
[6] T. Röfer, R. Brunn, I. Dahm, M. Hebbel, J. Hoffmann, M. Jüngel, T. Lauel, M. Lötzsch, W. Nistico and M. Spranger, GermanTeam 2004 The German National Robocup Team, *RoboCup 2004: Robot Soccer World Cup VIII* (2004).
[7] C. McMillen and M. Veloso, Distributed, Play-Based Coordination for Robot Teams in Dynamic Environments, *RoboCup 2006: Robot Soccer World Cup X* (2007), 483–490.

Intelligent Autonomous Systems 10
W. Burgard et al. (Eds.)
IOS Press, 2008

doi:10.3233/978-1-58603-887-8-211

The SPICA Development Framework

Model-Driven Software Development for Autonomous Mobile Robots

Philipp A. BAER, Roland REICHLE and Kurt GEIHS
Distributed Systems Group, Kassel University, Germany

Abstract. The development of autonomous mobile robots is a challenge due to the complexity, interdisciplinarity and variety of requirements and components. It is thus quite difficult or even unrealistic to achieve an agreement on a standard development platform for robotic software. In this paper we present SPICA, a model-driven development framework for autonomous mobile robots. With SPICA the developer retains the freedom to choose the most appropriate programming language and underlying platform for the realisation of a specific functional component. Independently developed components are considered as service providers and consumers that interact in order to fulfil the desired functionality. For this purpose, SPICA provides modelling and transformation support to realise such an integration infrastructure for independently developed functional components and different robotic software systems. SPICA has been used successfully in the development of a team of soccer robots for the RoboCup competition.

Keywords. Robot Soccer, Networked Robots, Multi-Robot Systems, Design, Model-Driven Software Development, Modelling

1. Introduction

Looking at today's development of autonomous mobile robots it is quite obvious that there is no standard development strategy or platform. Neither a general agreement on how to develop software for robotic systems nor a general agreement on a software architecture as such have been achieved, although a number of different approaches were proposed in the past. This is not so much because these proposals lack any important features. The most important reason is that existing development and middleware frameworks build on specific standards for the underlying platform, the programming language, or the component interfaces. For autonomous mobile robots, however, such a restriction does not seem to be appropriate, as the development of autonomous mobile robots is a task with many facets and interdisciplinary requirements. Robots make use of hardware sensors and actuators along with appropriate control components, and typically require some form of information fusion, world modelling as well as AI planning strategies. Many of these aspects are often addressed through independently developed components, and for each of them a different programming language and underlying implementation framework may be most appropriate. For example, low-level hardware drivers are often implemented using C, an AI planning framework in Lisp, while Java may be used to implement user interfaces and visualisation tools. Encompassing all the different needs in a standardised development and middleware framework is a nearly impossible task. Thus, proprietary development frameworks and infrastructures are provided that reflect the particular importance of components and functions. In conclusion, heterogeneity issues are often observable even within a single robotic system. They are of course more predominant in heterogeneous groups that accomplish complex tasks.

In this paper we present SPICA, a development framework for autonomous mobile robots. With SPICA the developer captures component interaction and data management in reusable abstract models and retains the freedom to choose the most appropriate programming language and underlying platform for the realisation of a specific functional component. The resulting set of heterogeneous components is considered as a set of service providers and consumers (services), each of which integrates into the robotic software system in order to fulfil a desired functionality. The SPICA development framework facilitates the integration of the services by supporting the development of an integration infrastructure following the model-driven software development paradigm. In order to further boost interoperability and to deal with heterogeneity issues arising from cooperation of independently developed robots, an ontology serves as a common modelling basis. It is referenced by modelling elements and used in the transformation process from abstract models to source code.

The paper is organised as follows. Section 2 discusses our requirements. In section 3 the overall SPICA modelling approach is outlined, followed by a presentation of the transformation process in section 4. A real-world case study is shown in section 5. Section 6 discusses related work. Section 7 concludes the paper and points to future work.

2. Requirements

The development process of a robotic software system reveals a number of challenges. For the SPICA framework we have derived a set of requirements from the analysis of case studies and our own experiences gathered from several robotic projects. We do no claim completeness; our requirements are mostly related to the general development process and software architecture of robotic software as well as to communication and cooperation issues. Therefore, they also take into account aspects of software development for teams of autonomous robots.

Development process and structure of robotic software

- **Modularity**: Robots are self-contained entities composed of various functional components, ranging from hardware drivers to world modelling and behavioural decision making.
- **Diversity**: Each functional component may have different requirements to the platform, programming language, and provided development support.
- **Heterogeneity**: Heterogeneity issues are predominant in groups of autonomous robots. An integration infrastructure that bridges the gap between heterogeneous robotic systems is required, allowing seamless communication and collaboration.
- **Rapid Prototyping**: Robotic systems are often research subjects. This implies that the integration of new functional components for testing purposes must be possible in a simple and convenient manner.

Characteristics of Communication and Interaction

SPICA provides communication infrastructures that interconnect independently developed software components. Therefore, we have to take into account the requirements resulting from the interaction characteristics of autonomous robotic systems.

- **Event-orientation:** Robots observe their environment, changes in which trigger events. Freshness of information is most important here; lost events are less serious than delayed ones.
- **Asynchrony:** Information is predominantly generated in reaction to events and sent asynchronously.
- **Data aggregation:** Collaborative decision making depends on shared sensor data.
- **Efficiency:** Depending on the environmental constraints, wireless communication may be limited in bandwidth. Media access in general is further quite expensive.
- **Robustness:** Wireless communication may be unreliable, depending on the environmental constraints.

3. The SPICA Modelling Approach

Let us explain how these requirements are reflected in SPICA. Based on the modularity requirement, a robotic software infrastructure is assumed to be composed of components that interact with each other. There are many ways in which modularity can be supported in both the distributed and local case. Our scenario involves potentially loosely coupled, heterogeneous components. Thus, considering each component being a service, the generic approach of service-oriented architectures (SOA) appears adequate. Consumers and providers can be realised in a completely heterogeneous fashion involving different programming languages and utilising the most suitable development strategies. Then, the SPICA development framework provides modelling and transformation support to realise an integration infrastructure for these services. The model-driven development approach is very appropriate for SPICA as it explicitly aims at supporting different platforms through automatic code generation. Rapid Prototyping is explicitly addressed as well: existing or prototyped functional components may be integrated without having to be adjusted for specific software architectures. The integration code for components is generated from the abstract model specification.

The SPICA development framework (SpicaDF) along with its toolchain is aligned with the MDSD approach as shown in figure 1. Here, *SpicaML* is short for *SPICA modelling language*, covering domain-specific modelling languages such as *MDL* and *DFDL* (see section 3.1). An abstract SpicaML model builds the foundation for the communication infrastructure to be generated. It may be annotated semantically (based on an ontology) to help the model transformation tool with processing the model and generating the required conversion functions. Once parsed and converted into an intermediate representation (called AIR), the model is passed to the template engine (StringTemplate[1]). Here, the relevant target templates are loaded and the final code is generated. With the help of the utility library *Castor* and the service discovery engine *Geminga*, both developed

[1] `http://www.stringtemplate.org/`

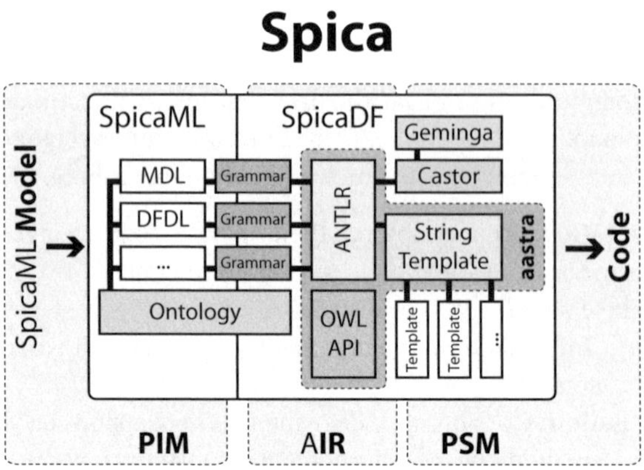

Figure 1. The SPICA development framework arranged with the MDSD classification.

by us for SPICA, service representatives are generated which dynamically establish the communication infrastructure as required.

Thus, SPICA provides a domain-specific, platform independent modelling language tailored to the domain of distributed autonomous robotics. The modelling elements assist with the specification of relevant robotic interaction traits as suggested above.

With regard to event-oriented data acquisition, a message-oriented communication model is most suitable. Compared to RPC, pure message-orientation involves less protocol interactions and is more reactive in unreliable environments. Missing a single message once in a while is acceptable in many typical application scenarios. Furthermore, message-oriented communication is more close to the freshness assumption as no message retransmission is involved. In terms of transmission efficiency, we consider group communication schemes most suitable.

Asynchronous communication decouples senders and receivers, but in turn introduces the requirement for the receiver to synchronise reception of incoming messages to the internal processing. Message buffering is one possible approach to overcome this issue. It allows receivers to decide when incoming messages are delivered by still retaining the decoupling property. The downside is an increase in transmission latencies, though.

SPICA also follows the message buffering approach. The transmission characteristics and buffering elements are covered by dedicated modelling elements of the SpicaML sub-languages presented in the following sections.

3.1. The Spica Modelling Language

The SPICA modelling language (SpicaML) comprises three different sub-languages, each of which was developed to address one or more of the requirements discussed above. The languages are only separated conceptually but tightly coupled in SpicaML.

Message Definition

The Message Description Language (MDL) of SPICA provides specifically tailored modelling means for hierarchically structured network messages, similar to but much more application domain oriented than ASN.1 or interface definition languages. It further provides a type concept with strong typing.

MDL basically consists of four modelling elements that model different aspects of network messages. An *enumeration* defines a mapping between names and constants. A *container* represents a construct for building complex data types, logically grouping together primitive or other complex types. SPICA containers may further be extended through single inheritance which is the basis for polymorphism. An *envelope*, the simplest form of which is a header, carries message control information. A *message* finally brings together envelopes and containers. Each message must be derived from an envelope and its payload has to be composed of zero or more containers. Messages thus represent data exchange entities in the SPICA infrastructure.

MDL has the ability to semantically annotate modelling elements. The semantic model follows the approach introduced in [10]: a semantic annotation is made up of a concept and a representation. The concept classifies an element or the data it contains. The representation defines the internal structure of an element. Semantic annotations are used in code transformation and by the Data Analysis Definition Language [8].

Components and Data Flow

Addressing the requirements discussed above, the *Data Flow Definition Language* (DFDL) provides modelling elements for the application domain of component interaction and multi-robot collaboration. It deals with data transmission, data management utilising *Data Management Containers* (DMC), service provisioning, and modules in terms of functional components.

DMCs implement a generic form of message buffer used mainly for message synchronisation but also for storing messages in general. Several modes of operation are supported. The most important one for realtime operation is when the messages are passed directly through. Besides, a DMC is capable of extracting container types from messages, taking this burden from developers.

In SPICA, links between modules are established dynamically based on the availability of resources. This is achieved through our automatic service discovery engine Geminga outlined in section 4. *Geminga channels* may operate in either of three modes. In static mode, a channel statically binds to both endpoints. In announce mode, the publisher only announces the availability of a service. In request mode, subscribers have to subscribe to an offered service.

Modules represent components and are mainly of conceptual interest. They act as communication endpoints for communication channels, one of which is created for each service type. A module definition comprises at most two blocks, one for outgoing and one for incoming data. Each block contains a list of messages and respective DMCs. Messages specified in the block for outgoing data are offered, the ones in the incoming block are requested by the module. Dynamic channel establishment is based on this information. In the SPICA development framework, services provide event-based message delivery or message streams; it does not restrict developers to only this option, though.

Figure 2 outlines the structure of a SPICA-generated communication infrastructure. It involves several modules (service representatives) and channels which may be established dynamically. Each SPICA module implements a four layer stack. The *routing layer* takes care of message routing and channel management. It utilises the Geminga service discovery which also handles service announcement. In the *filtering and mediation layer*, messages and message content may be further processed. The *data layer* is responsible for data management and data synchronisation. It utilises the DMC contain-

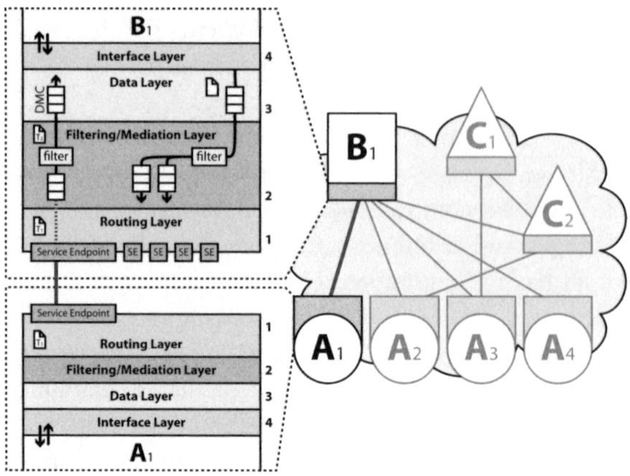

Figure 2. A SPICA-generated communication infrastructure with services A_x, B_x, and C_x

ers which are parameterised during modelling. The *interface layer* finally connects the representative to its service.

4. Transformation

The SPICA development approach employs the MDSD approach for achieving platform independence. The main challenge in MDSD is the specification of a model that is abstract enough but covers all the important characteristics of the application domain. With SpicaML we propose a modelling language for modular communication infrastructures.

SpicaML is specified in textual form. For a typical setup the specification will contain a set of container and message definitions. A single header is often sufficient. Then, depending on the complexity of the application, one or more module definitions with the respective message and DMC definitions have to be specified. The SPICA tool *aastra* performs the model transformation. It parses the model and generates an Abstract Syntax Tree (AST), i.e. the in-memory representation of the model. The parser component is generated by ANTLRv3 (ANother Tool for Language Recognition) [9], an LL(*)-based parser generator framework. aastra then transforms the AST into a SPICA-specific intermediate representation (Abstract Intermediate Representation; AIR) which is better suited for the final code transformation. It carries out the following steps:

- Resolve semantic shorthands
- Determine the hashing algorithm for type id generation
- Extract the namespace and enumerations
- Extract the MDL structures and resolve cross references; generate type id
- Extract the DFDL structures
- Pass the AIR to template engine and generate the implementation

MDSD comprises three model transformation stages, the Platform Independent Model (PIM), the Platform Specific Model (PSM), and the final target code. AST and AIR are both still platform independent, thus part of the PIM. The final code transformation is where the transition from PIM to PSM commences. The AIR tree is referenced directly from within the template so no additional step is required.

4.1. Target mapping

The template mechanism is a very flexible approach through which many target platforms can be supported. It nevertheless is a very complex task to write templates. We have thus tried to keep the template creation overhead low in SPICA. Castor provides the more complex techniques required for the communication infrastructure.

One important technique in Castor is Geminga, a light-weight distributed service discovery approach mentioned earlier. It cares about dynamic channel management and routing. Each generated module owns a Geminga instance which announces the locally subscribed and published services every few seconds. Remotely available services time out if no announcement was received within a given time interval. Explicit cancellation of published services is not implemented. A SPICA-based infrastructure together with Geminga guarantees operation in connected as well as disconnected state.

4.2. Implementation Notes

Containers, envelopes, and messages provide a clean interface with member accessors for each field. Data encoding is user-defined but a default scheme is provided for SPICA.

Modules are implemented as singletons with a private Geminga instance each. In order to integrate well into existing architectures, only data reception and data transmission interface operations are provided, i.e. no further configuration is required. The communication channels are setup automatically once the module instance is initialised.

The implementation generated from a SpicaML model is complete and does not require manual completion. When realising a functional component, a developer only has to use the interfaces of the generated service representatives and may concentrate on the service functionality.

5. Results

In this section we present an example that demonstrates the power of the SPICA development framework. During the last years we have developed a team of autonomous soccer robots that successfully participates in RoboCup competitions. The robot software architecture mainly consists of six independently developed components whose rather diverse functionality is outlined below. The component developers were allowed to freely choose a programming language and platform that appeared most suitable for them, as long as Linux support was available.

The *Vision* component is responsible for image processing with focus on feature detection. Since an omni-directional camera is the main sensor of a robot, the Vision constitutes the first and most important sensor module. For performance reasons the Vision is implemented in C++.

World modelling, object tracking, behaviour reasoning, and role assignment capabilities are concentrated in the *Base* component. Because of available code and experiences of the developers, the Base component is implemented in C#, adopting the Mono .NET framework[2].

[2] `http://www.mono-project.org/`

The *Motion* component interfaces to the motor controller and provides odometry feedback. A realisation in C was already available from a former robot project.

For communication with teammates and components located on remote systems we introduced the *Communication* component. It is implemented in C# but for no specific reason. All programming languages seemed to be more or less equally suited and there were no other requirements. The same applies to the *Joystick* component, which offers a joystick interface for debugging and testing purposes.

The *Controller/Viewer* serves visualization and remote control purposes. As Java provides libraries for graphical user interfaces in the standard development kit, we decided to implement this component in Java.

Vision, Base, Motion, and Communication modules are located on the robot, whereas Controller/Viewer and Joystick can be executed on arbitrary PCs. Setting up an infrastructure for these components manually is a quite time consuming task. For each component a developer has to provide appropriate IPC mechanisms, define and maintain messages, and take care of the data flow. The programming language and platform of the components have to be considered as well, resulting in a huge development overhead. SPICA reduces the effort substantially as follows:

- The developer defines the messages and data containers.
- The developer specifies the modules. This step involves defining the requested and offered message types, appropriate DMCs, and the transmission schemes.
- Elaborate filters are not required, so there is no need for a specification.
- The integration infrastructure is generated from an abstract model using the aastra model transformation tool. The service representatives provide an interface for data exchange with the service provider component itself.
- The developer integrates the representative into its service where it automatically connects to the SPICA-based infrastructure.

Compared to a manual implementation of the communication infrastructure, a much more consistent and easy to use development approach is provided by the SPICA development framework. Starting from a single, platform independent model, implementations on every supported target platform are generated with no further expense. Message structures are kept in sync, so almost no inconsistency is possible. The resulting systems not only proved to be realisable with low effort; they also proved to work as expected and with high performance during last year's RoboCup tournament German Open 2007.

6. Related Work

The presented SPICA development framework builds on our lessons learned in robotic software development and related work in this area. This section reviews some other approaches for robotic software development.

6.1. Modelling Approach

Research on robot software architectures in the past mostly focused on middleware frameworks. The SPICA development framework shifts the focus right to the development methodology. It is our intention to make the development process and the implementation more platform independent, enabling the developer to focus on the actual functionality rather than bothering with characteristics of the platform.

MIRO (Middleware for Robots) is an object-oriented framework for robotic applications [13]. It is CORBA-based utilising the ACE/TAO framework [11,12]. Abstraction layers relieve developers from dealing with software or hardware specifics.

CLARAty (Coupled Layer Architecture for Robotic Autonomy) [15] follows a similar approach. It is a two-layered object-oriented framework for robotic systems which focuses on reusability and integration of algorithms and components.

In contrast to the approaches outlined above, SPICA is not a middleware but a development framework aiming at platform independent specifications and automatic code generation. Middleware functionality is integrated into the generated code.

The Microsoft Robotic Studio [7] is a development environment for robotics that targets different robot platforms. It builds on the .NET framework and offers a runtime, a visual programming language, and a powerful simulation environment.

CoSMIC (Component Synthesis using Model-Integrated Computing) [4,1] is another development environment which follows the paradigm of MDSD. It is a collection of domain-specific modelling languages and generative tools for the development, configuration, deployment, and validation of component-based real-time systems.

The Microsoft robotics studio targets rapid development of robot control software but focuses more on prototyping than on efficient and domain-adapted solutions. CoSMIC is a model-driven approach that follows very similar goals. Our approach emphasises message-orientation and application in autonomous robotics. SPICA aims at rapid development for scientific application and ease of integration.

Regarding message-oriented communication, the open standard messaging middleware AMQP [14] (Advanced Message Queuing Protocol) implements a messaging behaviour similar to SPICA. Similar to the DMCs used in SPICA, AMQP provides queues to accomplish a store-and-forward semantics. Message routing and delivery is due to centralised message broker systems.

6.2. Automatic Configuration

Zeroconf [3] comprises of a set of technologies facilitating automatic network configuration for IPv4 as well as discovery of services and devices. A Zeroconf architecture builds on an extension of the Domain Name System (DNS) by embedding service description records (DNS-SD) [2].

The UPnP working group maintains similar specifications. The automatic network configuration approach is very close to the one of Zeroconf. Service discovery builds on the Simple Service Discovery Protocol (SSDP) [5] which itself relies on HTTP.

The Service Location Protocol (SLP) [6] is a service discovery protocol to find services in a local area network. Transmissions are mostly message-oriented, devices that join a network use multicast for initial service discovery.

The Geminga service discovery used in SPICA is tailored towards the integration in groups of distributed autonomous systems where network communication is unreliable and no centralised servers are possible. Geminga is furthermore targeted at the announcement of service types and addresses and at the automatic establishment of communication channels. Zeroconf and UPnP both reside in the context of automatic network configuration. The behaviour of SLP is very similar to Geminga but does not support the dynamic establishment of channels required by SPICA.

7. Conclusions

In this paper we presented SPICA, a model-driven development framework for autonomous mobile robots. The abstract and platform independent modelling language SpicaML, composed of domain-specific sub-languages, provides modelling means for communication and collaboration infrastructures as required for component interactions and team cooperation.

In contrast to existing robot middleware frameworks, SPICA targets a wide variety of platforms and programming languages. It does not dictate developers which language or which framework to use but integrates smoothly into existing applications and component-based architectures. It is further possible with only little effort to add further target languages or platforms by defining new transformation templates.

The SPICA approach has shown its viability and effectiveness during the last RoboCup tournaments and in internal matches and presentations of the Carpe Noctem Robotic Soccer team. It was demonstrated that the SPICA-based architecture allows for easy extendibility and exchange of components. To further improve the functionality and language coverage of SPICA, the Castor utility library will be extended and ported to other platforms. The set of predefined templates will be further extended as well.

References

[1] Krishnakumar Balasubramanian, Arvind S. Krishna, Emre Turkay, Jaiganesh Balasubramanian, Jeff Parsons, Aniruddha Gokhale, and Douglas C. Schmidt. Applying Model-Driven Development to Distributed Real-time and Embedded Avionics Systems. *International Journal of Embedded Systems, special issue on Design and Verification of Real-Time Embedded Software*, April 2005.

[2] Stuart Cheshire and Marc Krochmal. DNS-Based Service Discovery. IETF, Internet-Draft, 2006.

[3] Stuart Cheshire and Daniel Steinberg. *Zero Configuration Networking: The Definitive Guide*. O'Reilly Media, Inc., 2005.

[4] Aniruddha S. Gokhale, Douglas C. Schmidt, Tao Lu, Balachandran Natarajan, and Nanbor Wang. CoSMIC: An MDA Generative Tool for Distributed Real-time and Embedded Applications. In *Middleware Workshops*, pages 300–306. PUC-Rio, 2003.

[5] Yaron Y. Goland, Ting Cai, Paul Leach, and Ye Gu. Simple Service Discovery Protocol/1.0 – Operating without an Arbiter. IETF, Expired Internet Draft, 1999.

[6] E. Guttman, C. Perkins, J. Veizades, and M. Day. Service Location Protocol Version 2. RFC2608, 1999.

[7] J. Jackson. Microsoft robotics studio: A technical introduction. *Robotics & Automation Magazine, IEEE*, 14(4):82–87, 2007.

[8] Pedro Lima, editor. *Robotic Soccer*, chapter 1, pages 1–28. I-Tech Education and Publishing, Vienna, Austria, December 2007.

[9] Terence Parr. *The Definitive ANTLR Reference: Building Domain-Specific Languages*. The Pragmatic Bookshelf, Raleigh, 2007.

[10] R. Reichle, M. Wagner, M. U. Khan, K. Geihs, M. Valla, C. Fra, N. Paspallis, and G. A. Papadopoulos. A context query language for pervasive computing environments. In *5th IEEE Workshop on Context Modeling and Reasoning (CoMoRea)*. IEEE Computer Society Press, 2008.

[11] Douglas C. Schmidt. An architectural overview of the ace framework, 1998.

[12] Douglas C. Schmidt, Andy Gokhale, Tim Harrison, and Guru Parulkar. A high-performance endsystem architecture for real-time CORBA. *IEEE Communications Magazine*, 14(2), 1997.

[13] Hans Utz, Stefan Sablatnög, Stefan Enderle, and Gerhard K. Kraetzschmar. Miro–Middleware for Mobile Robot Applications. *IEEE Transactions on Robotics and Automation, Special Issue on Object-Oriented Distributed Control Architectures*, 18(4):493–497, August 2002.

[14] Steve Vinoski. Advanced message queuing protocol. *IEEE Internet Computing*, 10(6):87–89, 2006.

[15] R. Volpe, I. Nesnas, T. Estlin, D. Mutz, R. Petras, and H. Das. The CLARAty Architecture for Robotic Autonomy. In *Proceedings of the 2001 IEEE Aerospace Conference, Big Sky, Montana*, March 2001.

Intelligent Autonomous Systems 10
W. Burgard et al. (Eds.)
IOS Press, 2008

doi:10.3233/978-1-58603-887-8-221

Brain Control of a Smart Wheelchair[1]

Rossella BLATT [a], Simone CERIANI [a], Bernardo DAL SENO [a], Giulio FONTANA [a], Matteo MATTEUCCI [a] and Davide MIGLIORE [a]

[a] *Politecnico di Milano, Department of Electronics and Information – IIT Unit, Italy*

Abstract. This paper makes a contribution to the field of autonomous vehicles, especially for use on assistive wheelchairs. The aim of our work consists in the development of a low-cost autonomous wheelchair able to avoid obstacles, self-localize and explore closed environments in a safe way. In order to meet disabled people's requirements, we have designed our system in such a way that it can be simply modified and adapted to the users' needs. In particular, the user has the opportunity to choose among several autonomy levels and three different interfaces: a joystick, a touch-screen and a brain-computer interface (BCI). A BCI is a system that allows users to convey their intention by analyzing their brain signals.

Keywords. Assistive robotics, Interfaces, Localization.

Introduction

The possibility of moving in an autonomous way gives individuals a remarkable physical and psychological sense of well-being. Robotic wheelchairs are usually driven by a joystick and are addressed to those people that are not able to apply the necessary force to move a manual wheelchair. They often are people with low vision, visual field reduction, spasticity, tremors, or cognitive deficits. In order to give these people a higher degree of autonomy, and also to lighten the duties of those who assist them, a large number of solutions have been studied by researchers since the 1980s, by using technologies originally developed for mobile robots to create the so called "smart wheelchairs".

A smart wheelchair, or autonomous wheelchair, typically consists of either a standard powered wheelchair to which a computer and a collection of sensors has been added or a mobile robot base to which a seat has been attached. One of the first examples of autonomous wheelchairs was proposed by Madarasz et al. [1], who equipped a wheelchair with sonars and a vision system to identify landmarks and correct its trajectory in hallways. Another solution was by Levine et al. [2] with NavChair, an electric wheelchair provided with an obstacle avoidance algorithm and multiple task-behaviors to control the movements through doorways or to avoid collision with walls. A more sophisticated solution was Rolland III, proposed by Mandel et al. [3]: a semi-autonomous wheelchair, equipped with laser range finders, encoders and a camera, that is able to set the appropriate speed in the presence of obstacles and avoid them. Following the same topic, Monte-

[1]This work has been partially supported by the European Commission, Sixth Framework Programme, Information Society Technologies: Contract Number FP6-045144 (RAWSEEDS) and also by the Italian Institute of Technology (IIT).

sano et al. [4] presented an autonomous wheelchair for cognitive-disabled children, more robust in complex navigation situations such as narrow doors, and populated or cluttered scenarios.

The project LURCH (*Let Unleashed Robots Crawl the House*) fits in with this perspective: the aim of our work consists in the development of a affordable autonomous wheelchair able to avoid obstacles, self-localize and explore closed environments in a safe way. The wheelchair has been equipped with two embedded PCs (about 500 €), a video camera (less than 100 €), an inertial measurement unit (about 2500 €, not used in indoor environments) and two Hokuyo laser range finders (about 1500 €, used only for obstacle avoidance and replaceable with sonars); this equipment provides a self-localization capability and a safe navigation ability to the wheelchair. The smart wheelchair navigates, deals with all the low-level details, avoids obstacles and reacts to emergency situations, while the user decides where (and when) to go in a high-level fashion (e.g., "go to the kitchen"). In order to meet the variable requirements of disabled people, we have designed our system in such a way that it can be simply modified and adapted to users' needs. In particular, the user has the opportunity to choose among several autonomy levels and three different interfaces: a joystick, a touch-screen and a brain-computer interface.

The typical control system used by smart wheelchairs, based on the use of a joystick, is not suitable for totally paralyzed persons. Millions of people in the world suffer from several diseases (e.g., amyotrophic lateral sclerosis — ALS, multiple sclerosis, cerebral paralysis, spinal cord injury) that destroy the neuromuscular channels used by the brain to communicate and control body movements. This calls for the development of a flexible system, able to adapt also to the necessity of completely locked-in individuals. So challenging an objective can be achieved by a system able to determine user's intentions through the analysis of his/her cerebral signals and to transmit this information to an external device, such as a wheelchair. This system is called *brain-computer interface (BCI)* [5]. In general, a BCI is composed of a device that acquires the user's brain signals (e.g., through an electroencephalograph), and a computer that analyzes some predefined components of the signals and maps them into a particular command to be executed. There are several methods to acquire the cerebral signals; among them, we chose to consider the electroencephalogram (EEG) because of the advantages that it brings in terms of non invasivity, practicality, safety and low cost.

1. Wheelchair Design

The smart wheelchairs described in this paper has been designed to provide navigation assistance in a number of different ways, such as assuring collision-free travel, aiding in the performance of specific tasks (e.g., passing through doorways), and autonomously transporting the user between locations. Our aim is to reduce as much as possible the cost of the whole system (the total cost of the framework proposed for indoor environment, wheelchair not included, is less then five thousands of euros, which is cheap with respect to other works) and provide different kinds of interfaces (e.g., the BCI, see the next section for more details), in order to fulfill the needs of people with different disabilities, and to allow users to set the desired level of autonomy.

The LURCH system was designed to be easily adaptable to different kinds of electric wheelchairs. Figure 1 outlines a scheme of LURCH. As it is possible to notice from

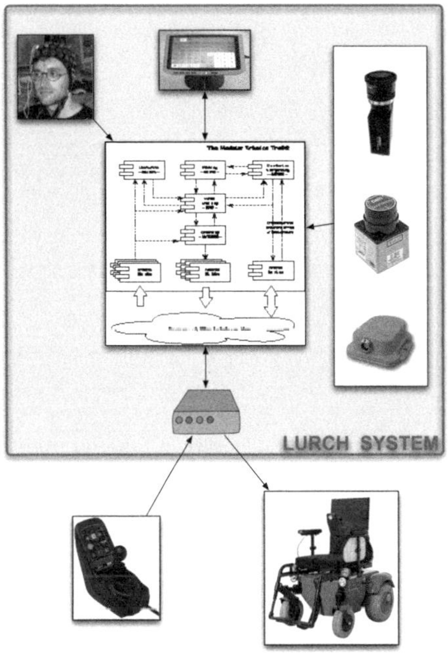

Figure 1. General scheme of LURCH system.

the image, our system is completely separated from the wheelchair, and the only gateway between LURCH and the vehicle is represented by an electronic board that intercepts the analog signals coming from the joystick potentiometers and generates new analog signals to simulate a real joystick and drive the joystick electronics. In other words, we do not integrate our system with the wheelchair at the digital control bus level, but instead we rely on the simulation of the signals from the joystick in the analogue domain. Though this choice could seem awkward, its motivations are twofold: first of all, it is often hard to obtain the proprietary communication protocols of the wheelchair controllers, or to understand how they exchange data with the motors and interfaces; second, this solution improves the portability, since it avoids a direct interaction with the internal communication bus of the wheelchair.

LURCH was designed by adopting a modular approach (as proposed in [6]):

- *localization module*: it estimates the robot pose with respect to a global reference frame from sensor data, using a map of the environment;
- *planning module*: using knowledge about the environment and the robot, this module selects the most appropriate actions to reach the given goals, while respecting task constraints;
- *controlling module*: it contains all the primitive actions, typically implemented as reactive behaviors that can be executed by the robot.

The localization algorithm operates using a video camera and some fiducial markers. These are placed on the ceiling of the environment, since this allows to avoid occlusions, and provide an accurate and robust pose estimation. Of course, this restricts the

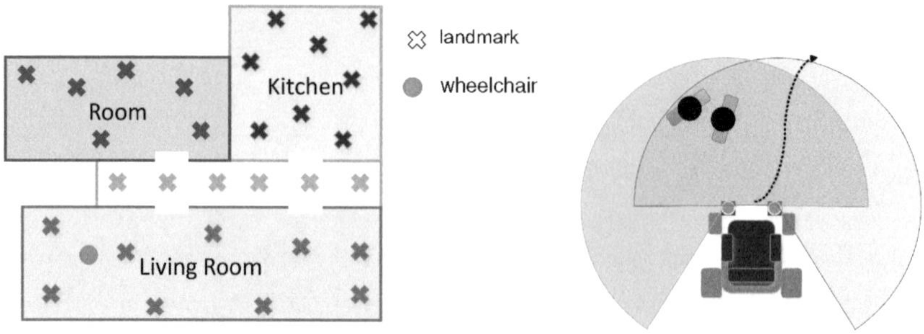

Figure 2. Example of localization and obstacle avoidance.

use of LURCH to indoor environments. Usually, a fiducial marker is a planar patch with a known shape that contains some encoded information. In this work we decided to use the ARToolKitPlus [7] system, where the markers are squares with a black border, and the information is encoded in a black and white image represented in the square. The marker identification process is defined by three steps: identification of possible markers in the image captured by the camera, rectification of the image, and comparison of the information represented in the markers with the database of known landmarks. If a marker is recognized, with the knowledge of its dimension, it is possible to estimate its 6 DoF position and orientation in the camera reference. Since the position and the orientation of the markers in the environment w.r.t. the absolute frame and also the position and the orientation of the camera w.r.t. the wheelchair are known, the system can estimate the pose of the wheelchair w.r.t. the world frame every time it is able to detect a marker. In fact, when a marker is detected, it is possible to estimate the matrix T_M^C that represents the rototranslation of the marker w.r.t. the camera frame. The position of the camera w.r.t. the marker can be easily estimated by calculating $T_C^M = (T_M^C)^{-1}$. Thus, by knowing the spatial relationship between the markers and the world, the position of the camera w.r.t. the absolute frame can be determined. In indoor environments, it is generally sufficient to know the 3 DoF pose of the wheelchair; thus, we decided to simplify the problem, improving in this way the robustness and the accuracy of the localization algorithm.

The trajectory planning is obtained by SPIKE (Spike Plans In Known Environments), a fast planner based on a geometrical representation of static and dynamic objects in an environment modeled as a 2D space (see [6] for more details). The wheelchair is considered as a point with no orientation, and static obstacles are described by using basic geometric primitives such as points, segments and circles. SPIKE exploits a multi-resolution grid over the environment representation to build a proper path, using an adapted A^* algorithm, from a starting position to the requested goal; this path is finally represented as a polyline that does not intersect obstacles. Moving objects in the environment can be easily introduced in the SPIKE representation of the environment as soon as they are detected, and they can be considered while planning. Finally, doors or small (w.r.t. the grid resolution) passages can be managed by the specification of links in the static description of the environment.

Our control module is MrBRIAN (Multilevel Ruling BRIAN) [8], a fuzzy behavior management system, where behaviors are implemented as a set of fuzzy rules whose antecedents match context predicates, and consequents define actions to be executed;

behavioral modules are activated according to the conditions defined for each of them as fuzzy predicates, and actions are proposed with a weight depending on the degree of matching (see [8] for more details).

2. Brain-Wheelchair Interface

There are several kinds of brain activities that can be used in a BCI. Some BCIs register the involuntary brain activity in response to stimuli associated with possible commands in order to infer the command intended by the user. Others analyze components of brain signals that can be controlled voluntary by the user; these BCIs may feel somewhat more natural to the user, as they do not need external stimulation, but users need some training.

In this study, we focus on the first approach; in order to detect user's commands, we used the event-related potential *P300*, while in order to detect and verify user's and autonomous control errors, we considered the *error potential* (ErrP).

2.1. The P300 Potential

The P300 is an event-related potential (ERP) which can be recorded via EEG as a positive deflection in voltage at a latency of roughly 300 ms in the EEG after a defined stimulus. It follows unexpected, rare, or particularly informative stimuli, and it is usually stronger in the parietal area. Stimuli can be visual, auditory, or even tactile.

A P300-based BCI presents the user with some choices, one at a time; when it detects a P300 potential, it selects the associated choice. The user is normally asked to count the number of times the choice of interest is presented, so as to remain concentrated on the task. As the P300 is an innate response, it does not require training on part of the user. Detecting a P300 in a single trial is very difficult; therefore, repeated stimuli are normally used to select the stimulus most likely to have generated a P300. The number of repetitions can be predetermined for each user to get the best trade-off between speed and accuracy.

Many techniques for detecting the P300 extract relevant features from the EEG signal and feed those features into a classifier. In these approaches, feature extraction becomes the key point, and doing it by hand can be at the same time cumbersome and sub-optimal. We have faced the issue of feature extraction by using a genetic algorithm able to retrieve the relevant aspects of the signal to be classified in an automatic fashion.

A P300-based BCI makes possible to build a very flexible system, which allows both communication and the control of different devices with a single graphical user interface. Such an interface renders all the possible choices in a menu on a screen, and highlights them one at a time. Different menus can be built for different tasks or environment, and menus can also be linked to each other. The use of P300 in the LURCH project makes the system versatile, as it is composed by context-sensitive menus that can have as many choices as they are needed, while the selection interface can be kept uniform.

2.2. The Error Potential

Real BCIs sometimes misclassify user intent, and much research is devoted to improving BCI classification ability in order to increase their performance. Another way to face the issue is, as previously mentioned, to repeat the process more than once until a sufficient

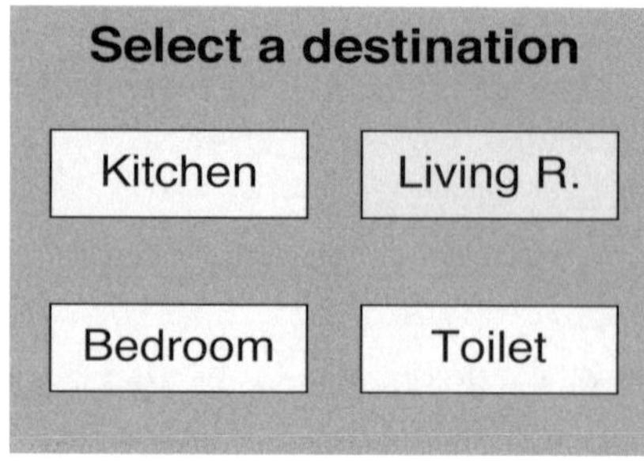

Figure 3. Menu for the selection of a room as destination

confidence is reached. Another possibility is, finally, to ask directly the user to confirm his/her choice, by adding another interactive interface.

In this work we adopted an alternative way to improve BCI performance, i.e., the early identification of errors and their automatic correction. It is known from the literature that user's and BCI errors elicit error potentials (ErrP), a particular kind of potentials that occurs in the EEG when a subject makes a mistake or, more relevant to BCI applications, when the machine the subject is interacting with does not behave as the user expects. Thus, the detection of ErrPs could be a viable way to improve BCI performance and to increase its reliability, without making the interaction with the user heavier.

3. Experimental Results

The LURCH data-acquisition and processing subsystem consists of two on-board computers (one to control the robot and acquire data from sensors, and another for the BCI), an industrial camera for the localization task, and two Hokuyo laser range finders for obstacle detection. An inertial measurement unit is currently used to estimate the wheelchair dynamic, though we are confident it can be replaced by rotation sensors on the chair wheels; work is in progress. The low-power computer runs the main software; in particular, the sensing modules acquire the sensor and input data (e.g., the joystick signals, the commands coming from the BCI system or from the touchscreen), the reasoning module elaborates the information and determines the wheelchair behavior, and the acting module generates the appropriate commands to control the wheelchair. As the computers rely on wheelchair batteries for power, special low-power custom designs were used.

We are developing a menu-based interface for the control panel of the wheelchair (see Figure 3 for an example) that presents the user with a choice of possible destinations, depending on the contest. The choices can be single rooms or particular positions within the room, e.g., near the table, near the window... When the wheelchair is located within a room, the available choices are the particular locations within the same room, plus all the other rooms of the house/environment. In this way, a hierarchy of location choices can be navigated while the wheelchair physically navigates the environment.

We already did experiments to detect P300s and ErrPs in EEG recordings [9]. Ten (healthy) volunteers participated in two kinds of experiments, where they were asked to interact with a computer while their EEG activity was recorded. In a setup, users operated a simple interface through a keyboard (like in [10]): the user is supposed to fill one of two bars displayed on the screen by pressing one of two buttons. Each bar is composed by ten parts, and every time the user press a button, the next part get filled, according to the button pressed; and when either bar is full, the trial ends. The user is free to choose either left or right, but the interface is programmed so that with a probability of 20% the wrong bar grows. An analysis of the EEG recordings showed that in three out of five cases an ErrP was detectable with good accuracy (80% accuracy for both error and correct cases).

In a second kind of experiment, subjects used a P300-based BCI, the *P3 speller* [11]. This BCI permits to spell words by selecting one letter at a time. A 6×6 grid of letters and symbols is presented to the user, and entire columns or rows are flashed one after the other in random order. Each one of the 6 rows and 6 columns is flashed exactly once in the first 12 stimulations; then another round of 12 stimulations is repeated, with flashing of rows and columns done in a new random order, and this procedure is repeated for a predefined number of times for each letter. After the fifth repetition, the P300 system detects the row and the column that are more likely to have elicited a P300, and selects the letter at their intersection. The selected letter is presented to the user in the rectangular frame, and it is also concatenated to the text being written. For P300 detection the genetic algorithm (GA) described in [12] was employed. Very briefly, the GA searches the features with which a given classifier perform best. We used features that can be written as the cross-correlation of the EEG signals with some simple templates, and a logistic classifier. This choices has permitted to combine feature extraction with classification, so the computational requirement for P300 detection is very low. The GA accuracy was sufficiently high to give the possibility to five out of seven subjects to make use of such an interface (accuracy in the range 50–90% for a 36-symbol matrix, at a rate of about 4 letter per minute). It is important to notice that by increasing the number of repetitions also the accuracy increases, and hence all subjects should be able to use the interface. Moreover, the experiment aimed at verifying that errors made by the BCI in recognizing the letter selected by the users induced an ErrP in them. The BCI selected the right letter with a probability of 80%, and the wrong one with a probability of 20%; the wrong selection elicited an ErrP in all users, and the automatic detection of ErrPs resulted to be possible in most of them (accuracy between 70% and 90% for both classes).

4. Conclusions and Future Work

We are integrating the wheelchair control system with a full six-degrees-of-freedom SLAM (Simultaneous Localization And Mapping) system based on a monocular frontal camera. SLAM permits to build and continuously update a 3-dimensional map of the surrounding environment. The use of three dimensions is important to avoid collision with low-hanging objects (and even tables, which appear to 2D sensors as just four legs), while the continuous update is useful to avoid moving obstacles.

The idea of integrating a BCI in a smart wheelchair has already appeared in the literature, but some differences with our work should be highlighted. For example, the Maia project is a European project which aims at developing an EEG-based BCI for controlling an autonomous wheelchair [13]. Behaviors like obstacle avoidance and wall following are built in the autonomous control of the wheelchair, while the user just gives simple commands like “go straight”, “go left”. Motor imagery [5] is used by the Maia BCI to drive the wheelchair, haptic feedback is given to the user, and the detection of error potentials in the user helps to lower the global error rate of the system. The main difference with our system is that the user must continuously drive the wheelchair, and no a-priori map of the environment is provided.

At the National University of Singapore a wheelchair controlled through a P300-based BCI is under development [14]. The wheelchair is autonomous, but its movements are constrained to predefined paths; the user selects a destination and the chair drives there. If an unexpected situation occurs, the wheelchair stops and waits until the user decides what to do. Classification of P300 events is done by a support vector machine fed with samples of the recorded EEG and its time derivative. This project has some resemblance to ours, but our focus is more on a versatile and reusable system than a single application. With respect to P300 detection, our methods are different and try to adapt automatically to the difficulty level of the task; regarding the wheelchair application, the integration with SLAM in our project should provide more flexibility.

References

[1] R. Madarasz, L. Heiny, R. Cromp, and N. Mazur., "The design of an autonomous vehicle for the disabled," *IEEE Journal of Robotics and Automation*, vol. RA-2, pp. 117–126, 1986.

[2] S. Levine, D. Bell, L. Jaros, R. Simpson, Y. Koren, and J. Borenstein, "The NavChair assistive wheelchair navigation system," *IEEE Transactions on Rehabilitation Engineering*, vol. 7, pp. 443–451, 1999.

[3] C. Mandel, K. Huebner, and T. Vierhuff, "Towards an autonomous wheelchair: Cognitive aspects in service robotics," in *Proceedings of Towards Autonomous Robotic Systems*, 2005.

[4] L. Montesano, J. Minguez, J. Alcubierre, and L. Montano, "Towards the adaptation of a robotic wheelchair for cognitive disabled children," in *IEEE/RSJ International Conference on Intelligent Robots and Systems*, 2006.

[5] J. R. Wolpaw, N. Birbaumer, D. J. McFarland, G. Pfurtscheller, and T. M. Vaughan, "Brain-computer interfaces for communication and control," *Clinical Neurophysiology*, vol. 113, pp. 767–791, 2002.

[6] A. Bonarini, M. Matteucci, and M. Restelli, "MRT: Robotics off-the-shelf with the modular robotic toolkit," in *Software Engineering for Experimental Robotics* (D. Brugali, ed.), vol. 30 of *Springer Tracts in Advanced Robotics*, Springer - Verlag, April 2007.

[7] D. Wagner and D. Schmalstieg., "ARToolKitPlus for pose tracking on mobile devices," *In Computer Vision Winter Workshop*, 2007.

[8] A. Bonarini, M. Matteucci, and M. Restelli, "A novel model to rule behavior interaction," *In Proceedings of the 8th Conference on Intelligent Autonomous Systems (IAS-8)*, pp. 199–206, 2004.

[9] B. Dal Seno, "Toward an integrated P300- and ErrP-based brain-computer interface," Tech. Rep. 2007.84, Politecnico di Milano – Department of Electronics and Information, December 2007.

[10] P. W. Ferrez and J. d. R. Millán, "You are wrong!—automatic detection of interaction errors from brain waves," in *Proceedings of the 19th International Joint Conference on Artificial Intelligence*, pp. 1413–1418, 2005.

[11] E. Donchin, K. M. Spencer, and R. Wijesinghe, "The mental prosthesis: Assessing the speed of a P300-based brain-computer interface," *IEEE Transactions on Rehabilitation Engineering*, vol. 8, pp. 174–179, June 2000.

[12] B. Dal Seno, M. Matteucci, and L. Mainardi, "A genetic algorithm for automatic feature extraction in P300 detection," in *IEEE World Congress on Computational Intelligence 2008*, 2008. Accepted.

[13] J. Philips, J. d. R. Millán, G. Vanacker, E. Lew, F. Galán, P. W. Ferrez, H. Van Brussel, and M. Nuttin, "Adaptive shared control of a brain-actuated simulated wheelchair," in *Proceedings of the 2007 IEEE 10th International Conference on Rehabilitation Robotics, June 12-15, Noordwijk, The Netherlands*, pp. 408–414, 2007.

[14] B. Rebsamen, E. Burdet, C. Guan, H. Zhang, C. L. Teo, Q. Zeng, M. H. Ang Jr., and C. Laugier, "A brain-controlled wheelchair based on P300 and path guidance," in *Biomedical Robotics and Biomechatronics, 2006. BioRob 2006. The First IEEE/RAS-EMBS International Conference on*, pp. 1101–1106, 2006.

Intelligent Autonomous Systems 10
W. Burgard et al. (Eds.)
IOS Press, 2008

doi:10.3233/978-1-58603-887-8-229

Simultaneous Learning and Recalling System for Wholebody Motion of a Humanoid with Soft Sensor Flesh

Tomoaki YOSHIKAI [a], Marika HAYASHI [a], Masayuki INABA [b]
[a] *Interfaculty Initiative in Information Studies, Graduate School of Interdisciplinary Information Studies, The University of Tokyo*
[b] *Dept. of Mechano-Informatics, The University of Tokyo*

Abstract. In order for robots to be taught new motions by human beings in any time they are directly touched, seamless ongoing learning system is necessary. In this paper, we focus on constructing ongoing learning system with simultaneous learning and recalling function for acquiring whole-body posture. Here, learning state monitor that checks a desirable state for learning and controlling when to learn continuously, a fast neural network model dealing with temporal correlative relationships, and a humanoid with soft sensor flesh that can enable itself to detect the touched place and to adapt to the humans, are key issues. Furthermore, proposed system is confirmed by a simultaneous learning and recalling experiment using an actual humanoid with soft sensor flesh.

Keywords. Neural network with time-delayed neuron sets, Learning state monitor, Humanoid with sensor flesh

1. Introduction

In order for robots to be taught new motions by human beings in any time they are directly touched, robots should prepare for learning in any time. It is desirable to change automatically from learning phase to recalling phase, or learning and recalling are conducted simultaneously.

Such seamless ongoing learning system for acquiring new motions are very important especially for a humanoid since it is very natural and easy for humans to teach motions by directly touching their arms or legs like they do against other humans. Also, by realizing anytime learning system, accumulation of new motions and slight modification to the existing motions becomes much easier and it can accelerate a trend of robot-human coexisting society.

For constructing a learning system to acquire whole-body motion of a humanoid, many studies have been tried so far. For example, R learning [1], one of reinforcement learning methods for continuous tasks, can be applied to the robot's ongoing task learning and Yoshikai *et al.* showed the continuous learning for musculo-skeletal humaonid's adapting behavior in pedaling a cycle[2]. But temporal correlative relationships cannot be acquired with their method since time related description is not included in their learning environment. For complex musculo-skeletal humanoids, Mizuuchi *et al.* proposed body information acquisition system, which collects the sensor data and always learns the relationships between sensor inputs and whole-body postures using the collected data in the layered type neural network[3]. However, this system also lacks the description of temporal correlation between sensor information. For acquiring temporal relationships within the motions, RNNPB , recurrent neural network with parametric bias, proposed by Tani *et al.*[4] is known to have capability to express several temporal relationships in one network structure, and Ito *et al.* showed an imitative motion learning of a small humanoid using RNNPB[5]. However, their method needs separate learning phase taking relatively long time and it is against our aim.

Therefore, we propose a new ongoing learning system fast enough to do simultaneous learning and recalling with a learning state monitor in this paper. By monitoring desirable state for learning and controlling when to learn continuously, ongoing learning system can be more efficient. Furthermore, proposed system is confirmed by a simultaneous learning and recalling experiment using a humanoid with soft sensor flesh, which can detect where it is touched through distributed tactile sensors.

2. Ongoing Learning System with Simultaneous Learning and Recalling for Acquiring Whole-body Motion

When a human teaches someone how to take a sequence of some postures, phases, such as 'Learning' and 'Recalling', are not always separated. Humans learn a posture and at the same time, they recall it and move themselves for confirming. Then, teacher can modify the posture again. Especially, such kind of 'ongoing learning with simultaneous learning and recalling' is important when a teacher tries to give every possible help with direct touch to others, as a mother does to her infant baby. As such direct teaching method, Dalla *et al.* proposed the intuitive teaching method based on tactile sensing information [6]. In their approach, 'Context information' is encoded from tactile, joint angle posture, body orientation, and velocity of the center of gravity. Using improved k-NN method, mappings from those context information to joint angle modifications are inferred in real-time from the examples that human programmer prepared in advance. Although they argue that their method can be learned online by interactively editing the database that gathers the past experienced examples, automatic seamless learning is our target and their method needs more extension including the temporal relationships among selected features. Therefore, we construct new ongoing learning system that can deal with temporal data relationships.

In order to construct, such ongoing learning system for whole-body motion, following points are important.

- Monitoring whether current state is desirable for learning, or not.
  Desirable situation for learning each tasks exists in the most of cases. For example, when a robot wants to learn how to walk, it should learn only when it is standing up and tries to walk ahead. Learning while it is falling down is meaningless for learning normal walking patterns. Therefore, the robot should monitor whether current state is desirable for learning, or not, and if it's not in the desirable learning state, it should move itself for recovering to the desirable learning state.
- Learning algorithm that is capable of dealing with temporal relationships.
  A sequence of posture taught to a robot can not be totally randomized. There is always some temporal relationships between each joint's movements. This is partly due to its embodiment. Body structure itself is a kind of constraint for making some relationships between joint movements. Therefore, learning algorithm adopted here should be able to deal with temporal relationships for learning a sequence of whole-body postures.
- Capability of the robot hardware itself for detecting where it is touched around whole body.
  Considering the occlusion problem, vision processing or torque sensors for each joints have disadvantages compared with distributed tactile sensors. Therefore, distributed tactile sensor system is used for detecting the touched place in this paper.

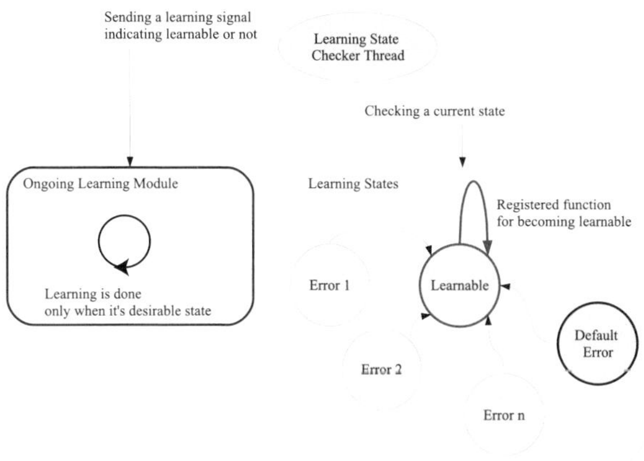

Figure 1. Monitoring system for learning states

In the following sections, detail implementation for these 3 points are described.

3. Monitoring Function for Ongoing Learning

In order to efficiently conduct ongoing learning, a robot should be able to judge when it is desirable for learning. For realizing such function, learning state monitor is constructed in this paper. The function of this monitor is 1) checking a current learning state, 2) calling registered function for becoming learnable, 3) sending a learning signal to a connected ongoing learning module. Schematics of the learning state monitor implemented here is shown in Figure.1.

Learning states consists of one learnable state, one default error state, and the other specific error states. Each states have their own checking rule functions. Learning state checker thread executes each checking functions and decides a current state. In each states, functions for becoming 'learnable' can be registered. They are executed if registered. By this function, error recovery to be learnable state is implemented. After the current state is decided, learning sate checker thread sends a message, whether it's learnable or not, to a connected ongoing learning module. Here, ongoing learning module is implemented as a thread that repeatedly executes learning only when it is informed that it is learnable.

4. Learning Algorithm for Forming Time Context: Neural Network with Time Delayed Neurons

For simultaneous learning and recalling of a sequence of whole-body posture online, fast learning method that can deal with temporal relationships between collected data is needed. Methods that strictly divide the phases into learning and recalling, and that needs enormous repetition times for learning are not appropriate for our aim. For such target, authors have developed a neural network model with time delayed neuron sets[7], [8]. Here, we apply the methods for constructing ongoing learning system with simultaneous learning and recalling functions.

4.1. Neuron Model with Constant Time Delay

To conduct ongoing learning on temporally related data of robots' joint angles, a new neuron model is proposed. First of all, the neuron itself is based on Mcculloch-Pitts model[9]. As shown in the Figure.2, each neurons have 2 layers. Upper layer

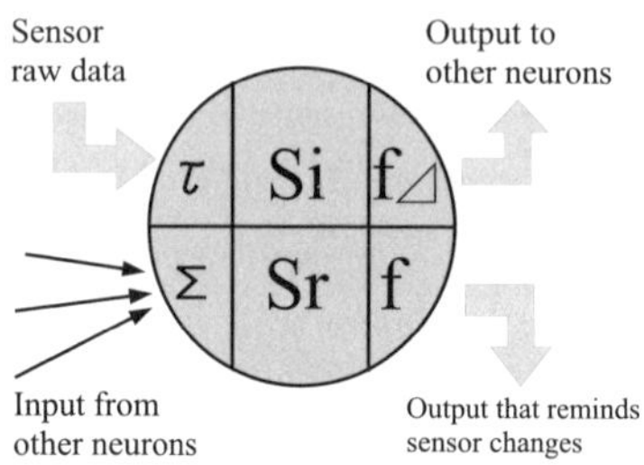

Figure 2. 2-layered neuron model with time delay

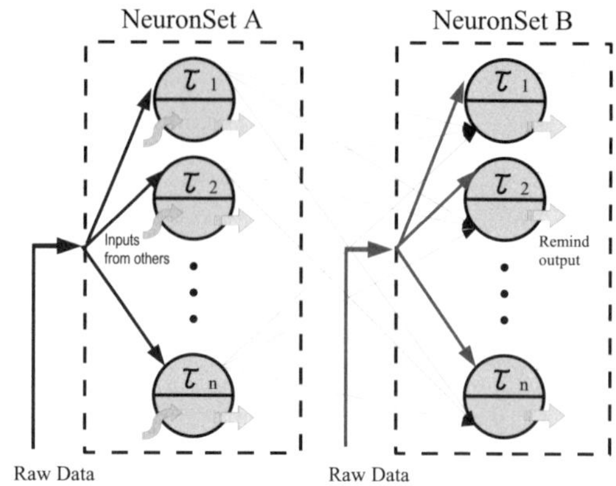

Figure 3. Connections between 2 kinds of neurons

receives one input I and outputs the temporal difference of the normalized input with time delay τ.

In upper layer, input I is nomarlized as $\hat{I}$ from -1 to 1. Using that normalized value, internal state for a sensor input is updated with time delay τ by Eq.1. Here, an output X_i for the upper layer is calculated as Eq.2, Eq.3, Eq.4. $F(x)$ is a nonlinear strictly increasing function defined as Eq.3. In Eq.3, c indicates some constant value. f_{adj} is a function for preventing the scale of the output from changing according to the time constant.

$$\Delta S_i = (\hat{I} - S_i)/\tau \tag{1}$$

$$X_i = F(f_{adj}(\Delta S_i, \tau)) \tag{2}$$

$$F(x) = \frac{2}{1 + e^{-cx}} - 1 \tag{3}$$

$$f_{adj}(\Delta S_i, \tau) = \tau \Delta S_i \tag{4}$$

Next, lower layer plays a roll for recalling the output according to other sensor's changes. In the lower layer, many inputs from other neurons' upper layer outputs are received. S_r, an internal state of this lower layer is calculated as Eq.5. In the Eq.5, k_r is a constant for reminding, and w_i is a connection weight for a neuron i. x_{si} expresses an output from neuron i's upper layer. Then, X_r, an output of the lower layer, is calculated using the same function of $F(x)$ in the Eq.3.

$$S_r = k_r \sum_i w_i x_{si} \tag{5}$$

$$X_r = F(S_r) \tag{6}$$

4.2. Network Structure and Learning Algorithm

Multiple neurons, each of which has different time delay constant, are prepared for one information source (each sensor or actuator information). Then, neurons receiving different information are fully connected to each other (Figure.3).Here, neurons are connected from one neuron's upper layer output to another neuron's lower layer input. This connections express temporal relationships between each

Time delay	1,10,50
Learning rate	0.05
Sigmoid constant	3.0
Forgetting rate	0.0001
Remind rate	1.4

Table 1. Network parameters

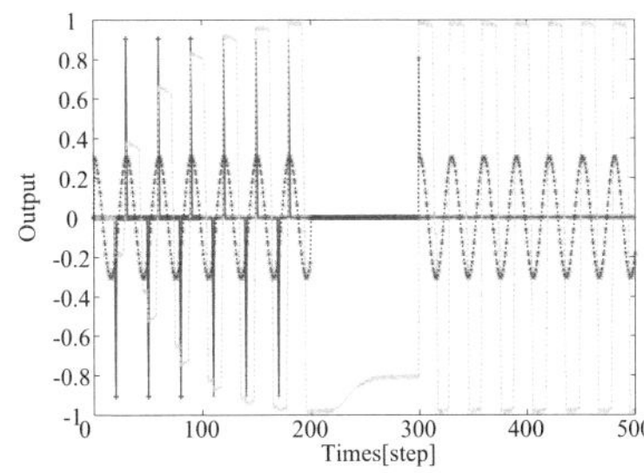

Figure 4. Simulation results: learning two correlative signals

sensor information, i.e., temporal information among different sensor information is expressed spatially in this network.

In this network, all connection weights between connected neurons are updated by Eq.7. In this equation, an amount of modification to a connection weight Δw_i is decided to be proportional to the product of two neurons' outputs, X_s and X_{st}, just like Hebb's law. Here, α is a learning constant and $beta$ is a forgetting rate constant. By adopting this simple updating calculation, fast learning process can be realized.

$$\Delta w_i = \alpha X_s X_{st} - \beta w_i \tag{7}$$

4.3. Behavior of the Constructed Network

Behavior of the Network is tested using two simulated signal (Eq.8, Eq.9). Although these signals have same frequency and they are correlative, the correlative relationships cannot be acquired by ordinary hopfield network. This is because timing when the signal becomes maxima or minima is not the same.

$$I_1 = \begin{cases} 1 \ (if\ 20 < (t\ mod\ 30)) \\ 0 \ (otherwise) \end{cases} \tag{8}$$

$$I_2 = sin(2\pi \frac{t}{30}) \tag{9}$$

At each time step, signals are input to the neurons and learning is conducted. At 200 [step], these two signal inputs are stopped. and after 300 [step] only the sine input restarts. Parameters are decided empirically and shown in Table.1. In the Figure.4, learning process is shown. Horizontal axis is time step, and vertical axis is an normalized output. Red solid line expresses the change of the internal state of the neuron for Eq.8 while blue dashed lines expresses the one for Eq.9. Green dashed line indicates the recalled output of the neuron for Eq.8. Here, outputs from neurons with least time delay are drawn in the figure. In the figure, amplitude of recalled output of the neuron for Eq.8 is getting larger as learning process proceeds. After 300[step], recalled output emerged again at the same timing of learning phase (0-200[step]) without any input to this neuron. This neuron can recall the learned changes for the pulse input triggering by the sine input.

5. A Humanoid with Soft Flesh Having Distributed Tactile Sensors

In order for robots to be touched by every position around its whole body, it has to be equipped with soft exterior allowing soft contacts at many points with humans. With a rigid body and exterior, many of existing robots have, it is not

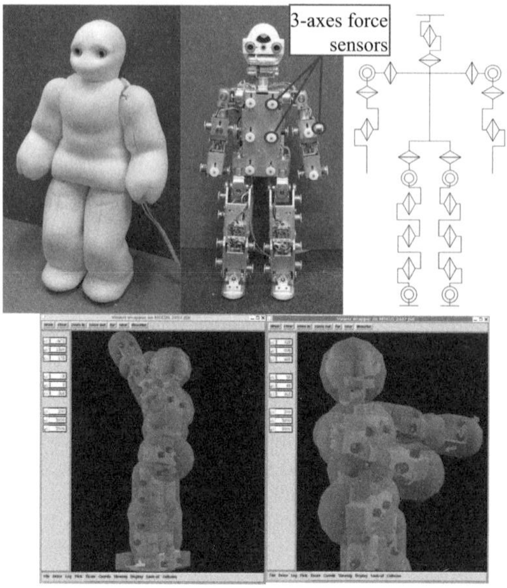

Figure 5. 'macra': a humanoid with soft sensor flesh (Upper left: whole-body image covered with soft flesh, Upper center: internal mechanical frame, Upper Right: an arrangement of DOFs, Lower left & right: robot's model with 3-axis force sensor outputs)

safe for humans to teach them a posture with direct touch to their body. Also, for being taught smoothly from humans, robots have to know where the teacher wants them to move and they have to be equipped with sensors which can detect touched place around their whole-body.

As a robot that meets above requirements, we have been developed a humanoid with soft sensor flesh, named 'macra' [10]. 'macra' (Figure.5) is a small humanoid which has soft urethane foam exterior and embedded tactile sensors. The height is about 700[mm], and the weight is about 7.5[kg]. Its internal mechanical frame has 22 DOFs. Some parts of the internal frame and circuit boards for driving actuators are built using HOAP-1 manufactured by Fujitsu Automation Ltd. As embedded tactile sensors, 49 small 3-axis force/torque sensors (PFS, manufactured by NITTA CORPORATION) are used. Since they are embedded in the soft urethane foam and can detect three dimensional force directions, deformation forces applied to the soft exterior can be calculated by the interpolation of the distributed force sensors. Therefore macra is possible to detect the whole-body contacts as a bundle of the flow of the applied forces. The lower row of the figure shows some examples of macra's sensing abilities. The sensor output vectors are illustrated as red arrows from corresponding sensors in the robot's geometric model. 3D force direction can be seen in the figure.

6. Simultaneous Learning and Recalling Experiment Using an Actual Humanoid

As an experiment for confirming feasibility of the proposed framework, direct teaching of a correlative swinging motion of both arms of macra is done. In the Figure.6, whole system for this experiment is illustrated. In this experiment, there are only two learning states, 'learnable' and 'default error'. Condition for being learnable in this situation is to be touched somewhere by the human. Therefore, checking function for 'learnable' state is judging macra is touched or not from distributed tactile sensing information. If somewhere is touched, and current state becomes 'learnable', registered function is executed. Registered function for 'learnable' state is an adapting behavior according to where it is touched. In the

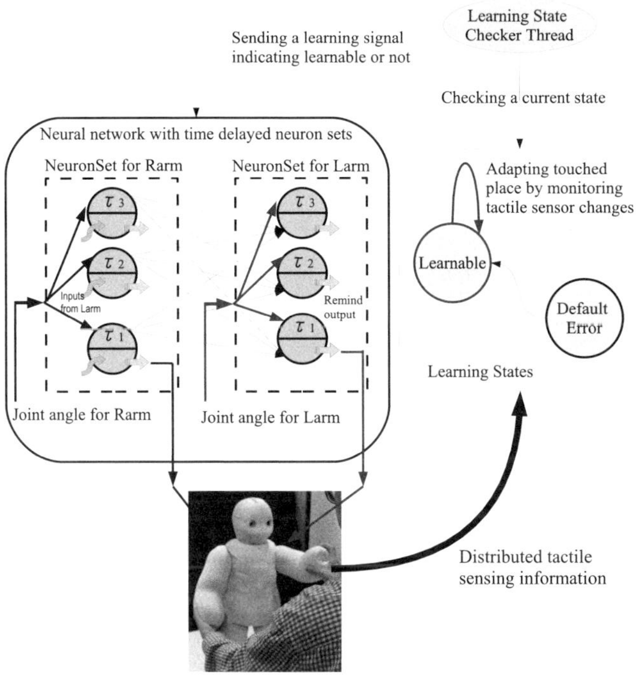

Figure 6. Experimental ongoing learning system with simultaneous learning and recalling functions for macra

touched limb, joints that are located inside from the touched sensors towards robot's torso are controlled to be 'current control mode' for adapting those joints. On the contrary, when the touched limb is released corresponding joints are controlled to be 'position control mode' again. By executing this function, macra can accept direct teaching from humans, and thus macra continuously becomes 'learnable'. When macra is not touched by any parts, learning state monitor judges that it is in 'default error' state. In this experiment, there is no registered function for this state since learning in this situation is passive and only human teacher can decide when it learns.

While macra is touched somewhere, it is learnable and learning state checker sends a signal to ongoing learning module, a neural network with time delayed neurons. In this experiment, 2 neuron sets, one for the right pitch axis joint of a shoulder, and the other for the left one, are prepared since target motion is macra's arm swinging motion. In each time step, joint angle sensor information is input to the neuron sets. Then their outputs from remind layer are multiplied by a converting rate and added to the current joint angles and are sent to the robot body. In this experiment, remind outputs from neurons with least time delays are sent to the robot. The connection weights are learned only when it receives a message that it is learnable.

In the Table.2, network parameters for this experiment are shown. Those parameters are decided empirically in this experiment. In the proposed method, basically the timing of the temporal changes among connected neurons are acquired, not the absolute value for the joint angle trajectory. Therefore, decision for the parameter 'converting rate from remind output to joint angle' is important for the robot to track the absolute trajectories of the taught motions. Although this parameter is decided arbitrary in this experiment, the automatic search for the converting rate by considering constraint condition for tracking absolute joint

Time delay	1.3,7,20
Learning rate	0.05
Sigmoid constant	3.0
Forgetting rate	0.0001
Remind rate	1.4
converting rate from remind output to joint angle	2.0

Table 2. Network parameters for acquiring swinging motion

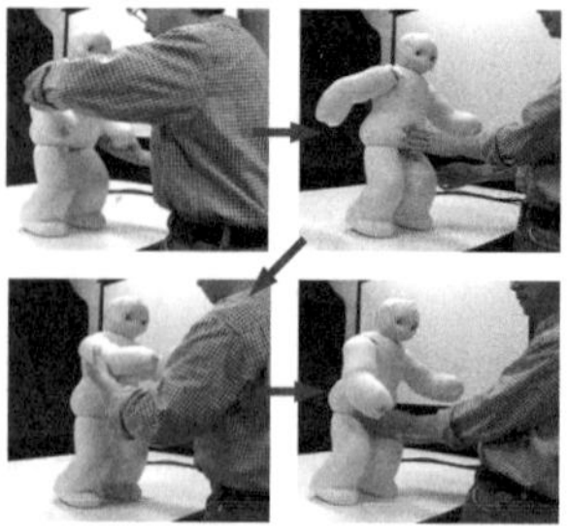

Figure 7. Simultaneous learning and recalling of arm swing motion

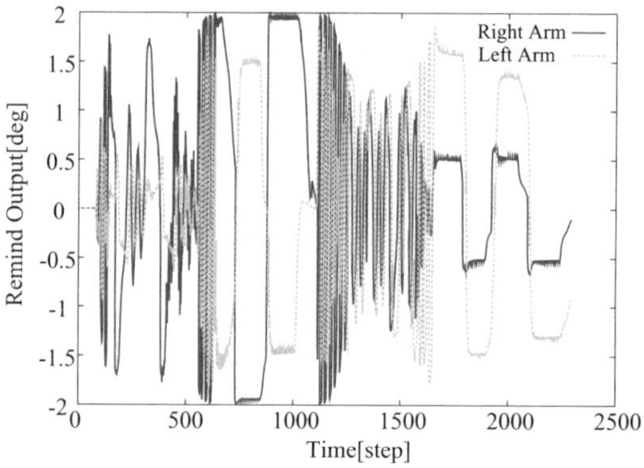

Figure 8. Transition of recalled output for each arm motions

trajectories is one of the important future works. Figure.7 shows actual behavior while learning and recalling simultaneously. At first, a human in front of macra teaches swing motion in which right and left arms of macra are moved forcedly by the human in the opposite phase. When confirming the acquired motion, the human just move one arm softly, then it triggers the opposite arm motion. Once the acquired motion is triggered, arm motions are repeatedly executed. After the first acquired swing motion is confirmed, second swing motion is taught continuously. In second motion, right and left arms move in the same phase. From the figure, the fact that second motions are acquired continuously after the first motion is confirmed.

Figure.8 shows a transition of recalled output for each arm motions. In this graph, horizontal axis expresses time step and vertical axis expresses remind output [1]. Red solid line indicates a remind output for the right arm while green dashed line indicates the one for the left arm. From 0 to about 600 steps in the graph, the human teaches first motion. After 600 until about 1200 steps, the human confirm that the first motion is acquired. Opposite phase motion of right and left arm can be seen from the graph. If the human moves one arm, it triggers another arm to move in opposite phase. From 1200 to about 1600 steps, second motion is taught by the human. Then, the second motion is confirmed after 1600 steps. Here, it is confirmed that the phase for both arm motion is almost the same and the second motion is acquired continuously. Thus, simultaneous learning and recalling ability for the timing of two temporally correlated information is realized by the fast online repetition of the learning cycle.

[1]Remind output for this network is a relative changes, not a absolute degree for the joint.

7. Conclusions

For constructing ongoing learning system with simultaneous learning and recalling function, 1) learning state monitor that checks whether a current state is learnable or not, 2) neural network with time delayed constants which can deal with temporal correlative relationships, and 3) a humanoid with soft sensor flesh and adapting mechanism using it, are developed in this paper. Furthermore, simultaneous learning and recalling system for the timing of temporally correlated sensors is confirmed by the experiment teaching several arm swing motions online.

Followings are listed as a part of interesting future works,

- Theoretically it is possible to learn more complex whole-body motion with the proposed method, the desirable network parameters, such as number of neurons in one neuron sets, or values for timed delayed constants, become hard to be decided empirically. Therefore, automatic acquisition of network parameters for the neural network with time delayed constants is needed.
- Learning state monitor system can be used for other ongoing learning systems. Also, the error recovery function of the monitor is useful for some tasks, such as learning of walking patterns, although it is not used in this paper.
- An application to ongoing learning system that forms a constraint of action by touching actively from robots against environments is another interesting future work.

References

[1] R.S.Sutton and A.G.Barto. *Reinforcement Learning:An Introduction(Adaptive Computation and Machine Learning)*. The MIT Press, 1998.

[2] T.Yoshikai, Y.Nakanish, I.Mizuuchi, and M.Inaba. Pedaling motion of a cycle by musculo-skeletal humanoid with adapting ability based on an evaluation of the muscle loads. In *Proceedings of The 9th International Conference on Intelligent Autonomous Systems (IAS-9)*, pp. 767–775, 2006.

[3] I.Mizuuchi, Y.Nakanishi, T.Yoshikai, M.Inaba, and H.Inoue. Body information acquisition system of redundant musculo-skeletal humanoid. In *Proceedings of The 9th International Symposium of Experimental Robotics*, p. paper164, 2004.

[4] J.Tani and M.Ito. Self-organization of behavioral primitives as multiple attractor dynamics: A robot experiment. *IEEE Transactions on Systems, Man, and Cybernetics Part A: Systems and Humans*, Vol. 33, No. 4, pp. 481–488, 2003.

[5] M.Ito and J.Tani. On-line imitative interaction with a humanoid robot using a dynamic neural network model of a mirror system. *Adaptive Behavior*, Vol. 12, No. 2, pp. 93–115, 2004.

[6] F.Dalla Libera, T.Minato, I.Fasel, H.Ishiguro, E.Menegatti, and E.Pagello. Teaching by touching: an intuitive method for development of humanoid robot motions. In *Proceedings of IEEE-RAS International Conference on Humanoid Robots (Humanoids 2007)*, 2007.

[7] K.Ikeya, K.Okada, M.Inaba, and H.Inoue. Generation of action triggered by environment using time context. In *Proceedings of The 22th Annual Conference on Robotics Society of Japan(in Japanese)*, p. 3B29, 2004.

[8] K.Ikeya, K.Okada, and M.Inaba. Teaching actions for a humanoid by reminding on a parallel with learning. In *Proceedings of The 22th Annual Conference on Robotics Society of Japan(in Japanese)*, p. 3L18, 2004.

[9] W.S.McCulloch and W.Pitts. A logical calculus of ideas immanentinnervous activity. *Bulletin of Mathematical Biophysics*, Vol. 5, , 1943.

[10] M.Hayashi, T.Sagisaka, Y.Ishizaka, T.Yoshikai, and M.Inaba. Development of functional whole-body flesh with distributed 3-axis force sensors for a humanoid to have close interaction. In *Proceedings of IEEE/RSJ 2007 International Conference on Intelligent Robots and Systems*, pp. 3610–3615, 2007.

Intelligent Autonomous Systems 10
W. Burgard et al. (Eds.)
IOS Press, 2008

doi:10.3233/978-1-58603-887-8-238

Learning to Extract Line Features: Beyond Split & Merge

Fulvio MASTROGIOVANNI, Antonio SGORBISSA and Renato ZACCARIA
DIST - University of Genova, Italy.

Abstract. The paper deals with the role of line features in self-localization, when an extended Kalman filter is adopted. First, a theoretical analysis is introduced, showing how the amount of range measurements contributing to lines extracted from 2D range data affects the localization accuracy. Second, a novel approach for line extraction, which takes the theoretical analysis into accountis considered. Experimental results are used to discuss the main properties of the system.

Keywords. Range Sensing, Feature Extraction, Self-Localization

Introduction

Self-localization is a central issue that – in its general formulation – has not been solved yet. Since dead-reckoning is not sufficient because of cumulative errors, the standard approach is to periodically compare observations collected from the environment with a map representation, and to consequently estimate the robot pose. This process is usually carried out adopting bayesian estimation. Among the most common approaches, Kalman filter based localization assumes that an unimodal probability distribution is adequate to represent the pose estimate [7], whereas the recent Markov localization [3] and particle filters algorithms [4] are able to represent multimodal probability distributions over the entire robot workspace. Within the same framework, observations can be used to update the map itself, thus simultaneously estimating both the robot and feature configurations: this is known as Simultaneous Localization And Mapping (SLAM in short) [9,5].

At the *core* of the *feature-based* localization algorithm two requirements are needed: (i) a noise model for sensors [7,11] and (ii) a feature model to be searched for in sensor data. Among geometric features, point and lines are the simplest ones [8]. In literature, a number of algorithms have been designed or adapted to extract line features from 2D range data [2,11]. Other authors focus on comparing these algorithms with respect to computational speed and quality of the extracted features [10]. Surprisingly enough, none of the works addressing line extraction investigates *the intimate relationships between feature characteristics and their subsequent use for estimating robot pose*. However, discussions about the roles among sensor data, features and statistical estimators are reported for SLAM frameworks, both for *dense* and *geometric* approaches [6].

The proposed approach relies on a laser rangefinder to correct the estimate provided by odometry through an extended Kalman filter (EKF in short), and assumes that a map of the environment is available. The main motivation is to investigate *how different line*

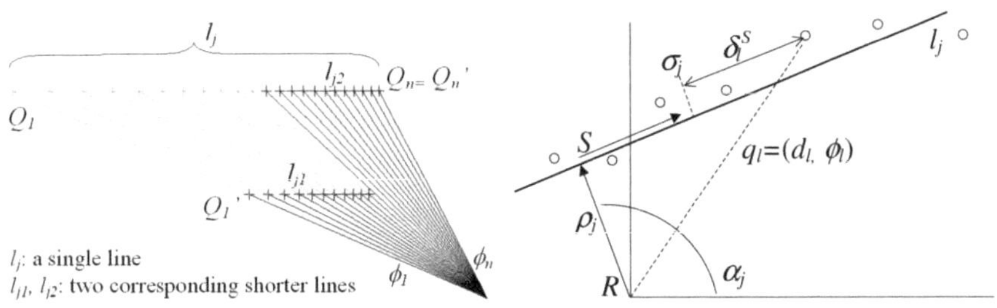

Figure 1. Fitting of range data.

features affect EKF-based self-localization. In Section 1 we discuss the relationships between line features and the localization accuracy when using an EKF estimator. Section 2 describes an algorithm taking into account these insights. Section 3 shows experimental results. Conclusions follow.

1. Line Features and Self-Localization

In this work, we assume a maximum likelihood approach $\mathcal{M}$ to extract line features l_j from a 2D range scan s_k, such that $\{l_j\} = \mathcal{M}(s)$, $j = 1, ..., |\mathcal{M}(s)|$ [11,10]. The $\rho\alpha$ model is adopted [8]: each line l_j is characterized by the distance ρ_j from the robot and the angle α_j with respect to its heading direction, i.e., $l_j = (\rho_j, \alpha_j)$.

Question 1 Each scan s is made up of $|s|$ range measurements. If $|l_j|$ measurements contributed to a line l_j used to correct the pose estimate, would the a posteriori covariance $\mathcal{P}$ of the estimated pose increase or decrease, if the same $|l_j|$ range measurements would instead produce two lines (l_{j_1}, l_{j_2} in Figure 1 on the left side) parallel to l_j?

Given an updating observation $z_j \equiv l_j$, characterized by the covariance matrix $\mathcal{R}_j$, the Kalman gain $\mathcal{K}$ inversely depends on $\mathcal{R}$, i.e., $\mathcal{K} \propto \mathcal{R}^{-1}$ holds. Therefore, the effect of z_j with smaller $\mathcal{R}_j$ can be compared with the joint effect of z_{j1} and z_{j2} with – respectively – higher covariances $\mathcal{R}_{j1}$ and $\mathcal{R}_{j2}$. By recursively writing the equations for $\mathcal{K}$ and $\mathcal{P}$ for two iterations, it can be inferred that two observations with covariance matrices $\mathcal{R}_{j1}$ and $\mathcal{R}_{j2}$ produce the same a posteriori $\mathcal{P}$ than a single observation with covariance matrix $\mathcal{R}_{j1,j2}$, where:

$$\mathcal{R}_{j1,j2} = \mathcal{R}_{j1}\big(\mathcal{R}_{j1} + \mathcal{R}_{j2}\big)^{-1}\mathcal{R}_{j2} \tag{1}$$

When $\mathcal{R}_j$, $\mathcal{R}_{j1}$, and $\mathcal{R}_{j2}$ are known, Equation 1 can be used to compare their effect on $\mathcal{P}$, by comparing $det(\mathcal{R}_j)$ and $det(\mathcal{R}_{j1,j2})$. Since $\mathcal{P}$ depends on the feature covariance matrices through the Riccati equations determining $\mathcal{K}$, it is necessary to investigate how the covariance $\mathcal{R}_j$ depends on the covariances of the range measurements which contributed to l_j. Therefore, the following question must be answered first.

Question 2. Given $|l_j|$ range measurements expressed as (d, ϕ) with – respectively – variance σ_d and σ_ϕ, is the "correcting power" higher when all of them contribute to a single line l_j, or when they contribute to two parallel lines l_{j1} and l_{j2} in the same scanning interval $(\phi_1, ..., \phi_{|l_j|})$ of l_j (see Figure 1 on the left side)?

We refer to [11] to describe the observation covariance $\mathcal{R}_j$ as a function of the covariance $\mathcal{Q}_l$ of each range measurement q_l, $l = 1, ..., |l_j|$. If we define $\mathcal{Q}_l^{RS}$ as the

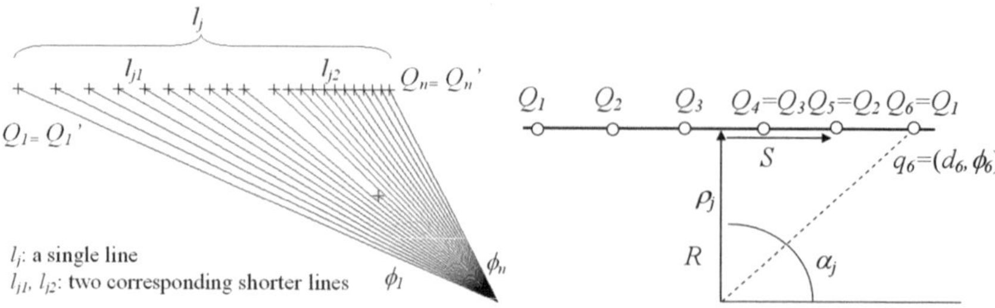

Figure 2. Left: Extracting segments. Right: The Pearson index tends to zero.

covariance matrix of q_l, represented with respect to a reference frame RS_j (see Figure 1 on the right side) associated with each observation z_j, then $\mathcal{R}_j$ is as follows:

$$\mathcal{R}_j = \begin{bmatrix} \mathcal{R}_{\rho\rho} \ \mathcal{R}_{\rho\alpha} \\ \mathcal{R}_{\alpha\rho} \ \mathcal{R}_{\alpha\alpha} \end{bmatrix} \tag{2}$$

where, defining $\delta_l^s = q_l^s - \sigma$, we obtain:

$$\mathcal{R}_{\rho\rho} = \frac{1}{\sum_{l=1}^{|l_j|}\left(\frac{1}{\mathcal{Q}_l^R}\right)} \quad \mathcal{R}_{\alpha\alpha} = \frac{1}{\sum_{l=1}^{|l_j|}\left(\frac{\left(\delta_l^S\right)^2}{\mathcal{Q}_l^R}\right)} \quad \mathcal{R}_{\rho\alpha}, \mathcal{R}_{\alpha\rho} = -\mathcal{R}_{\rho\rho}\mathcal{R}_{\alpha\alpha}\sum_{l=1}^{|l_j|}\left(\frac{\delta_l^S}{\mathcal{Q}_l^R}\right) \tag{3}$$

In other words, $\mathcal{R}_j$ measures the distance of the q_l scan point from the weighted mean σ of all the $|l_j|$ contributing points:

$$\sigma = \left(\sum_{l=1}^{|l_j|}\left(\frac{1}{\mathcal{Q}_l^S}\right)\right)^{-1} \sum_{l=1}^{|l_j|}\frac{q_l^S}{\mathcal{Q}_l^S} \tag{4}$$

If we temporarily ignore the α_j, $\mathcal{R}_j$ reduces to $\mathcal{R}_{\rho\rho}$. If $|l_j|$ is split into two subsets containing $|l_{j1}|$ and $|l_{j2}| = |l_j| - |l_{j1}|$ scan points, they can be used to extract two lines l_{j1} and l_{j2} characterized by $\mathcal{R}_{j1}$ and $\mathcal{R}_{j2}$. From Equation 3 we obtain:

$$\mathcal{R}_{\rho\rho1} = \frac{1}{\sum_{l=1}^{|l_{j1}|}\left(\frac{1}{\mathcal{Q}_l^R}\right)} \quad \mathcal{R}_{\rho\rho2} = \frac{1}{\sum_{l=|l_{j1}|+1}^{l_j}\left(\frac{1}{\mathcal{Q}_l^R}\right)} \tag{5}$$

and from Equation 1:

$$\mathcal{R}_{\rho\rho1,2} = \frac{1}{\sum_{l=1}^{|l_{j1}|}\left(\frac{1}{\mathcal{Q}_l^R}\right) + \sum_{l=|l_{j1}|+1}^{|l_j|}\left(\frac{1}{\mathcal{Q}_l^R}\right)} = \mathcal{R}_{\rho\rho} \tag{6}$$

If we *map* each point q_l in l_j with a corresponding point in l_{j1} or l_{j2}, characterized by the same covariance matrix $\mathcal{Q}_l$, and l_j, l_{j1} and l_{j2} are parallel, then $\mathcal{P}$ is identical after observing l_j or the couple l_{j1} and l_{j2}. $\mathcal{Q}_l$ itself varies with the range d_l associated to each scan angle ϕ_l: therefore, Equation 6 is not suitable to compare situations like in Figure 1 on the left. However, the result can deal with situations like in Figure 2 on the left: the covariance of the x- and y- state components does not change when an occluding object interrupts the sequence of collinear data.

Considering the general case in Figure 1 on the left, if σ_ϕ is small enough to be ignored, the covariance $\mathcal{Q}_l$ does not depend anymore on d_l, and $\mathcal{Q}'_l \approx \mathcal{Q}_l$ holds. Under these conditions, Equation 6 holds for any triplet of parallel lines l_j, l_{j1}, l_{j2} corresponding to the same scanning interval $(\phi_1, ..., \phi_{|l_j|})$. If we reintegrate α in the measurement, and assuming $\mathcal{R}_{\rho\alpha} = \mathcal{R}_{\alpha\rho} = 0$, Equation 1 can be applied to $\mathcal{R}_{\rho\rho}$ and $\mathcal{R}_{\alpha\alpha}$ independently. For $\mathcal{R}_{\alpha\alpha 1,2}$ a result similar to Equation 6 is obtained:

$$\mathcal{R}_{\alpha\alpha 1,2} = \frac{1}{\sum_{l=1}^{|l_{j1}|}\left(\frac{(\delta_l^S)^2}{\mathcal{Q}_l^R}\right) + \sum_{l=|l_{j1}|+1}^{|l_j|}\left(\frac{(\delta_l^S)^2}{\mathcal{Q}_l^R}\right)} \tag{7}$$

This suggests that $\mathcal{R}_{\alpha\alpha} = \mathcal{R}_{\alpha\alpha 1,2}$. However, the values of δ_l^S in Equation 7 are different than the corresponding values in Equation 3. The weighted mean σ must now be computed separately for scan points in l_{j1} and l_{j2}, thus yielding two different σ_{j1} and σ_{j2}. If parallel lines in the same scanning interval $(\phi_1, ..., \phi_{|l_j|})$ are considered, $\mathcal{R}_{\alpha\alpha 1,2}$ is necessarily bigger than $\mathcal{R}_{\alpha\alpha}$, since the distance δ_l^S of each scan point in l_{j1}, l_{j2} from the corresponding σ is smaller. This is true both for the situation in Figure 1 and – if σ_ϕ on each range measurement can be ignored – for the situation in Figure 2.

According to this theoretical analysis, the θ- component of $\mathcal{P}$ is expected to be lower, when range measurements are such that *longer* lines can be extracted: if $\mathcal{R}_{\rho\alpha} = \mathcal{R}_{\rho\alpha 1} = \mathcal{R}_{\rho\alpha 2} = \mathcal{R}_{\rho\alpha 1,2} = 0$, a single "more trustable" line l_j should be definitely preferred to the couple l_{j1} and l_{j2}. Unfortunately, ρ and α are correlated in most cases. Consider the Pearson correlation index:

$$corr_{\rho\alpha} = \frac{\mathcal{R}_{\rho\alpha}}{\sqrt{\mathcal{R}_{\rho\rho}\mathcal{R}_{\alpha\alpha}}} = -\frac{\sum_{l=1}^{|l_j|}\frac{\delta_l^S}{\mathcal{Q}_l^R}}{\sqrt{\sum_{l=1}^{|l_j|}\frac{(\delta_l^S)^2}{\mathcal{Q}_l^R}\sum_{l=1}^{|l_j|}\frac{1}{\mathcal{Q}_l^R}}} \tag{8}$$

If $\mathcal{R}_{\rho\rho}$ and $\mathcal{R}_{\alpha\alpha}$ are kept constant, a higher value for $corr_{\rho\alpha}$ means *less* uncertainty in the measurement. If $R_{\rho\alpha} = R_{\rho\alpha 1,2} = 0$ holds, it has been showed that the uncertainty R_j is always lower than the joint uncertainty $R_{j1,j2}$ of the couple l_{j1} and l_{j2}. Therefore, to answer *Question 1*, it would be finally required to demonstrate that the correlation $corr_{\rho\alpha}$ in $\mathcal{R}_j$ is always bigger than the correlation $corr_{\rho\alpha 1,2}$ in $\mathcal{R}_{j1,j2}$.

Unfortunately, a similar relationship cannot be found. According to Equation 8, $corr_{\rho\alpha} \to 0$ *iff* the numerator tends to zero. This could happen in two cases: (i) whenever $\mathcal{Q}_l^R$ and $\mathcal{Q}_l^S$ are constant for every ϕ_l, and (ii) when $\sigma_j = 0$ holds, as a consequence of range measurements which are symmetrically distributed along axis $\vec{S}$ on both sides of $\vec{R}$ (see Figure 2 on the right). In the first case, collecting $\mathcal{Q}_l^R$, we obtain:

$$corr_{\rho\alpha} = -\frac{\frac{1}{\mathcal{Q}_l^R}\sum_{l=1}^{|l_j|}\delta_l^S}{\sqrt{\frac{n}{(\mathcal{Q}_l^R)^2}\sum_{l=1}^{|l_j|}(\delta_l^S)^2}} = 0 \tag{9}$$

since $\sum_{l=1}^{|l_j|}\delta_l^S = 0$ when $\mathcal{Q}_l^S$ is constant. In practice, $\mathcal{Q}_l$ can be assumed to be "almost constant" when all scan points are "close enough" to each other, i.e., (d_l, ϕ_l) are almost the same for all the scan points q_l contributing to l_j. This happens with lines which scanning interval $(\phi_1, ..., \phi_{|l_j|})$ is small (e.g., *distant* or *short* lines). More interesting, when $\mathcal{Q}_l$ is almost constant for scan points in l_j, the same is obviously satisfied also for l_{j1} and l_{j2}, since each of them corresponds to a smaller scanning interval: i.e., when $corr_{\rho\alpha} \approx 0$, $corr_{\rho\alpha 1} \approx 0$ and $corr_{\rho\alpha 2} \approx 0$ as well, thus allowing to apply Equations 6 and 7. In the second case, each $\delta_l^S / \mathcal{Q}_l^R$ originates from a range point q_l^S that, when added to its symmetric $\delta^S_{|l_j|+1-l} / \mathcal{Q}^R_{|l_j|+1-l}$, sums up to zero. Unfortunately, $corr_{\rho\alpha} \to 0$ only for l_j, since – when $\sigma_j = 0$ – it is not possible that $\sigma_{j1} = 0$ or $\sigma_{j2} = 0$. Even if the correlation in $\mathcal{R}_j$ is very low, the correlation in $\mathcal{R}_{j1}$ and $\mathcal{R}_{j2}$ (and – consequently – in $\mathcal{R}_{j1,j2}$) is still high: the previous analysis is no more sufficient to univocally draw conclusions. However, it is still possible to claim that large sets of collinear range measurements should be preferred. In fact, it is always possible to purposely split the range measurements into two or more subsets (even when all of them are almost collinear), *if this is expected to produce an increment in the "correcting power" of the resulting observations.*

2. Split & Merge [& Split]

This Section introduces Split & Merge [& Split], a variant of Split & Merge taking into account insights from the previous theoretical analysis. Traditional Split & Merge outperforms other algorithms for line extraction from 2D range data [10]. At the end of Section 1, it is claimed that no theoretical result can be found supporting the idea of a *better shape* for line features. However, a line l_j originating from a great amount of scan points should be preferred in general, since either l_j is characterized by a small covariance matrix $\mathcal{R}_j$, or it can be split such that to originate two lines, namely l_{j1} and l_{j2}, which joint effect $\mathcal{R}_{j1,j2} < \mathcal{R}_j$ on the estimation process leads to an improved localization.

The key idea of Split & Merge [& Split] is to *learn* when the set of scan points contributing to l_j is amenable to be split into two parts. Split & Merge [& Split] is sketched in Algorithm 1. With respect to the original algorithm [10], Split & Merge [& Split] adds another step, called `split` in Algorithm 1: once collinear lines $\{l_j\}$ are merged together, `split` checks whether it is convenient to disregard some of the merged lines by using contributing points q_l to form two different lines instead.

In order to implement the `split` procedure the C4.5 algorithm [12] is run over a dataset obtained in simulation. For each run, l_j is compared to l_{j1} and l_{j2} which are parallel to l_j and span the same scanning interval $(\phi_1, ..., \phi_{|l_j|})$. Given the covariances $\mathcal{R}_j$, $\mathcal{R}_{j1}$ and $\mathcal{R}_{j2}$, $\mathcal{R}_{j1,j2}$ is computed according to Equation 1. Finally, $det(\mathcal{R}_j)$ and $det(\mathcal{R}_{j1,j2})$ are used to compute the ratio $\mathcal{R}_{j/j1,j2}$ between the areas of the two corresponding ellipsoids. When $\mathcal{R}_{j/j1,j2} < 1$, it is better to observe l_j than observing l_{j1} and l_{j2} in cascade. During simulation, $\mathcal{R}_{j/j1,j2}$ is computed for many different triplets (l_j, l_{j1}, l_{j2}), by varying the involved parameters: (i) the distance ρ_j between l_j and the robot; (ii) the lateral displacement σ_j of l_j; (iii) the length ϵ_j of l_j; (iv) the distance ρ_{j1} between l_{j1} and the robot, s.t. $\rho_{j1} < \rho_j$; (v) the ratios $|l_{j1}|/|l_j|$ and $|l_{j2}|/|l_j|$, such that $|l_j| = |l_{j1}| + |l_{j2}|$.

The result of the simulation is a dataset that, for each combination of line parameters, provides a boolean decision about the further splitting process. Finally, the dataset is then used to build a *decision tree* using C4.5: i.e., a collection of *if-then* rules abstracting actual data which are then embedded within the `split` procedure.

Algorithm 1 Split & Merge [& Split]

Require: A scan s. A stack $\mathcal{L}$. A counter j. A threshold τ.
Ensure: $\{l_j\} = \mathcal{M}(s), j = 1, ..., |\mathcal{M}(s)|$

$\mathcal{L}$ = push(s)
$j \leftarrow 1$
while $\mathcal{L} \neq \emptyset$ **do**
  $\mathcal{L}$ = pop(s_{top})
  $l_j \leftarrow$ fitting(s_{top})
  $q_k = \text{argmax}_q$ dist (l_j, q)
  if dist $(l_j, q_k) < \tau$ **then**
    $j \leftarrow j + 1$
    continue
  else
    $s_a \leftarrow$ sub(s_{top}, 1, k)
    $s_b \leftarrow$ sub(s_{top}, $k + 1$, $|s|$)
    $\mathcal{L}$ = push(s_a)
    $\mathcal{L}$ = push(s_b)
  end if
end while
$\{l_j\} \leftarrow$ merge($\{l_j\}$)
$\{l_j\} \leftarrow$ split($\{l_j\}$)

3. Experimental Results

3.1. Correcting Power of Line Features

Figure 3 on top shows how $\mathcal{R}_{j/j1,j2}$ varies when translating l_j either to the left or to the right over an infinite line with respect to the robot pose. Specifically, $-5m < \sigma_j < 5m$, $\rho_j = 7m$, $\epsilon_j = 4m$, $|l_{j1}| = 0.5|l_j|$ and $\alpha_j = 0$. Each curve corresponds to a different value such that ρ_{j1} between $3m < \rho_{j1} < \rho_j$. As expected, lower curves correspond to smaller values for ρ_{j1}: this is an obvious consequence of the presence of δ_l^S in Equation 7. In all the curves, $\mathcal{R}_{j/j1,j2}$ is maximum when $\sigma_j \approx 0$: this is because, as already discussed, $corr_{\rho\alpha} = 0$ for l_j, whereas it is not null for l_{j1} and l_{j2}. Moreover, the ratio $\mathcal{R}_{j/j1,j2}$ is minimum when $|\sigma|$ increases, since $\mathcal{Q}_l \rightarrow \hat{\mathcal{Q}}_l$ (i.e., a constant) over different measurements. When computing the mean over all the σ_j and ρ_{j1}, an average $Av(\mathcal{R}_{j/j1,j2}) = 0.17$ and a standard deviation $Std(\mathcal{R}_{j/j1,j2}) = 0.045$ are recorded. Figure 3 on the mid shows the same experiment when l_j is closer, and $1m < \rho_{j1} < 3m$. Notice that, when ρ_j decreases, $corr_{\rho\alpha}$ increases both for l_{j1} and l_{j2} (since now $\mathcal{Q}_l$ varies with l), whereas $corr_{\rho\alpha} \rightarrow 0$ for l_j when $\sigma_j \rightarrow 0$. $\mathcal{R}_{j/j1,j2}$ significantly increases up to a value bigger than 1. When averaging, this yields $Av(\mathcal{R}_{j/j1,j2}) = 0.28$ and $Std(\mathcal{R}_{j/j1,j2}) = 0.4$. Figure 3 on the bottom shows the case when $\rho_j = 3m$ is kept fixed, whereas ϵ_j is progressively reduced to $\epsilon_j = 2m$. When l_j is "shorter", $\mathcal{R}_{j/j1,j2}$ decreases for all σ_js (since $\mathcal{Q}_l$ varies less significantly), returning $Av(R_{j/j1,j2}) = 0.18$ and $Std(R_{j/j1,j2}) = 0.12$. Notice that the opposite is true as well: if, for example, when $\epsilon_j = 6m$, this results in a higher peak $max(\mathcal{R}_{j/j1,j2}) = 3.32$, with $Av(R_{j/j1,j2}) = 0.49$ and $Std(R_{j/j1,j2}) = 0.75$.

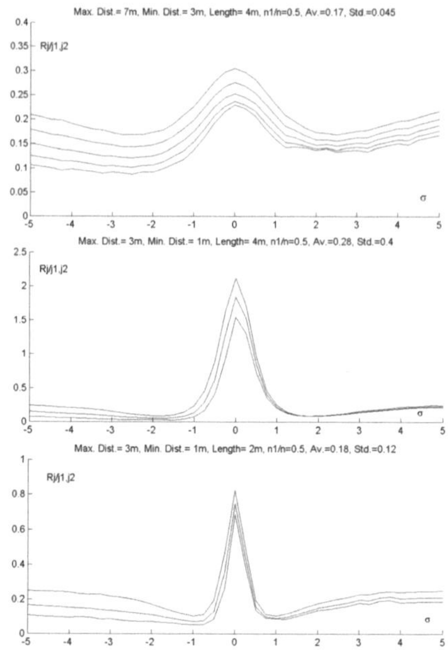

Figure 3. Plot of $\mathcal{R}_{j/j1,j2}$ versus σ_j.

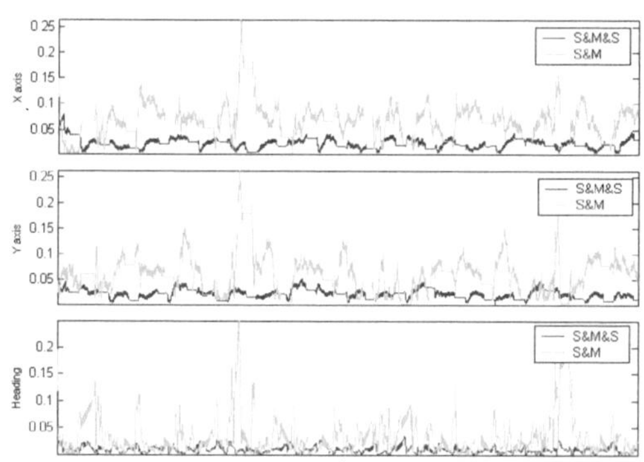

Figure 4. Experiments with Split & Merge [& Split].

Summarized results for $\mathcal{R}_{j/j1,j2}$ allow to claim that *longer* lines are expected to improve localization accuracy in a standard indoor environment with $1m < \rho_j, \rho_{j1}, \rho_{j2} < 10m$, and $2m < \epsilon_j < 10m$.

3.2. Localization accuracy with Split & Merge [& Split]

Our current `split` procedure is the result of a dataset containing 495.000 examples. Currently, we obtained a *decision tree* implementing more than 300 rules. Cross validation over the whole set of rules reported variable classification error rates, ranging from 4% to 13%, depending on the considered rule. However, current work is focused on adopting a learning framework with better performances.

Despite classification failures, experiments performed in simulation shows the superior performances of the introduced algorithm. Specifically, the system has been tested in different scenarios assuming *known* data association. Averaging over all the experimental runs, we noticed that Split & Merge [& Split] is characterized by a mean positioning error which is $0.02m$, whereas our current implementation of Split & Merge by $0.05m$. Inspecting the $x-$, $y-$ and $\theta-$ coordinates, it is evident how the major benefits are on robot heading, which is a fundamental achievement for self-localization.

4. Conclusions

The paper presented a discussion about the role of line features in mobile robot self-localization, when an extended Kalman filter is adopted for position tracking. First, a theoretical analysis has been introduced, investigating how different line features affect the localization accuracy. In particular, it is shown that when range measurements contribute to a smaller set of "longer" and "more trustable" features, the a posteriori covariance converges faster (on average), therefore increasing the estimate accuracy. Second, a novel approach to line extraction, taking into account the main findings of the theoretical analysis, is considered. Finally, experimental results have been used to validate the improved localization accuracy. Starting from these promising results, the system is currently being applied to SLAM.

References

[1] Y. Bar-Shalom, X. Rong Li, T. Kirubarajan. Estimation with Applications to Tracking and Navigation. Wiley Interscience, NY (USA), 2001.

[2] G.A. Borges, M.J. Aldon. Line Extraction in 2D Range Images for Mobile Robotics. In *Journal of Intelligent and Robotic Systems*, vol. 40, pp. 267 – 297, May 2004.

[3] D. Fox, W. Burgard, S. Thrun. Markov Localization for Mobile Robots in Dynamic Environments. In *Journal of Artificial Intelligence Research*, vol. 11, pp. 391 – 427, 1999.

[4] D. Fox, S. Thrun, W. Burgard, F. Dellaert. Particle Filters for Mobile Robot Localization. In A. Doucet, N. de Freitas (Eds.). *Sequential Monte Carlo Methods in Practice*, Springer-Verlag, New York, 2001.

[5] U. Frese. A Discussion of Simultaneous Localization and Mapping. In *Autonomous Robots*, vol. 20, pp. 25 – 42, January 2006.

[6] S. Huang and G. Dissanayake. Convergence and Consistency Analysis for Extended Kalman Filter Based SLAM. In *IEEE Transactions on Robotics*, vol. 23(5), pp. 1036 – 1049, October 2007.

[7] P. Jensfelt and H.I. Christensen. Pose Tracking Using Laser Scanning and Minimalistic Environmental Models. In *IEEE Transactions on Robotics and Automation*, vol. 17, no. 2, pp. 138–147, 2001.

[8] J. Folkesson, P. Jensfelf and H.I. Christensen. The M-Space Feature Representation for SLAM. In *IEEE Transactions on Robotics*, vol. 23(5), pp. 1024 – 1035, October 2007.

[9] M. Montemerlo, S. Thrun. FastSLAM: A Factored Solution to the SLAM Problem. Springer Verlag, July 2007.

[10] V. Nguyen, S. Gachter, A. Martinelli, N. Tomatis and R. Siegwart. A Comparison of Line Extraction Algorithms Using 2D Range Data for Indoor Mobile Robotics. In *Autonomous Robots*, vol. 23(2), pp. 97–111, August 2007.

[11] S. Pfister, S. Roumeliotis, J. Burdick. Weighted Line Fitting Algorithms for Mobile Robot Map Building and Efficient Data Representation. In *Proceedings of the 2003 IEEE International Conference on Robotics and Automation (ICRA'03)*, Taipei, Taiwan, September 2003.

[12] J.R. Quinlan. Improved Use of Continuous Attributes in C4.5. In *Journal of Artificial Intelligence Research*, vol. 4, pp. 77-90, 1996.

Intelligent Autonomous Systems 10
W. Burgard et al. (Eds.)
IOS Press, 2008

doi:10.3233/978-1-58603-887-8-246

Designing a Context-Aware Artificial System

Fulvio MASTROGIOVANNI [a,1], Antonello SCALMATO [a], Antonio SGORBISSA [a] and Renato ZACCARIA [a]

[a] *DIST - University of Genova, Italy*

Abstract. A framework is introduced that is aimed at integrating ontology- and logic-based approaches for context-awareness, suitable for use in Ambient Intelligence (AmI) scenarios. In particular, the context description language $\mathcal{CDL}$ is described, which allows to easily specify patterns of events which occurrences must be monitored by actual systems. As long as systems evolve, symbolic data originating from heterogeneous sources are first aggregated and then classified according to formulas described in $\mathcal{CDL}$. Experimental results performed in a Smart Home environment are presented and discussed.

Keywords. Context-Awareness, Cognitive Modeling

Introduction

During the past few years, many research efforts were undertaken to improve the quality of life of elderly and people with special needs [4]. Intelligent environments are considered a key technology to achieve these goals, since they are responsible for distributed data gathering, information extraction, context representation and interaction with users: they enable the effective use of such technologies as remote health care monitoring [3] or intelligent surveillance [7]. The semantic information associated with data flow is fundamental: intelligent user interfaces must be fed with ontology-grounded information able to capture the meaning of data being collected, whereas caregivers and system designers should be able to easily specify the overall system behavior, i.e., which contexts are considered relevant and how to face *anomalous* patterns of events. With respect to these requirements, a high-level symbolic representation for describing both the desired system behavior and the current context must be defined.

In literature, there is no widespread agreement about how to define and model contexts: (i) context structures must be effectively used to classify actual information; (ii) they can be composed with each other to generate complex hierarchical representations; (iii) they must grasp both relational and temporal information among events. The current debate is focused on issues related to formal description and reasoning capabilities, focusing on numerical, ontology- and logic-based approaches.

[1]Corresponding Author: Fulvio Mastrogiovanni, DIST - University of Genova, Via Opera Pia 13, 16139, Genova, Italy; E-mail: fulvio.mastrogiovanni@unige.it.

Numerical approaches are aimed at recognizing complex contexts from a limited amount of highly expressive information sources, mostly using probabilisitic state estimation techniques [10]. Associated drawbacks include high false classification rates of both measurements and data association, and the difficulty to provide numerical data with semantic labeling. *Ontology-based* approaches are characterized by a superior expressiveness in describing objects and relationships [12]. Generic structures are used to model concrete domain objects as well as abstract information (i.e., contexts and events). Ontologies have been widely used to model real-world high-level contexts and specific domains of interests [6], or location-based and topological models [11]. Some works [13,14] focus on modeling specific domains of interest, in order to design standard and reusable representations, whereas others (see [15] and the references therein) are aimed at formalizing context definition. Although these frameworks offer basic services for supporting reasoning, they lack in *generality* (as they focus on specific domains) and *extensibility* (since context models are hardly aggregated to build complex representations). *Logic-based* approaches manage contexts using facts which are stated or inferred from a given set of rules. In particular, two principles are identified [5]: (i) *locality*: reasoning must occur within a domain defined by a context; (ii) *cross-domains bridging*: relationships can occur between reasoning activities belonging to different contexts. Several logic-based architectures have successfully been designed and made effective [1,2]. In spite of the powerful reasoning capabilities offered by logic-based approaches, they are not suited to grasp the *general relationships* among entities in a compact way.

This paper proposes a framework aimed at integrating the benefits of ontology- and logic-based approaches for context-awareness in AmI applications. Based on the work described in [8], we assume the availability of a system architecture able to provide an ontology with information originating from distributed sources. In particular, we introduce a context description language, called $\mathcal{CDL}$, suitable to fill the gap between behavior specification and context-awareness algorithms. The paper is organized as follows: first, the proposed ontology and the $\mathcal{CDL}$ representation language are described; then, experimental results performed in a Smart Home set-up are reported. Conclusion follows.

1. $\mathcal{CDL}$: a Context Description Language

An intelligent environment is a distributed system where several entities cooperate to perform data gathering and context acquisition. In the following discussion we refer to [8], where a multi-agent system architecture for AmI applications has been described.

1.1. A Reference Scenario

Our reference environment (see Figure 1 on the left) comprises a `bedroom`, a `bathroom` and a living room, which is divided into two areas, namely `entrance` and `kitchen`. Several `ir` (i.e., infra red) sensors are placed in both doorways and passages, whereas `pir` sensors monitor interesting areas. Furthermore, we reasonably assume the availability of `temperature`, `bed` and `tap` sensors, and a number of intelligent appliances (e.g., `stove`, `fridge`, `shower` and `tv`). It is worth noticing that all the devices are either commercial products or they can be easily obtained (e.g., a *contact sensor* on a seat generate a *seat*

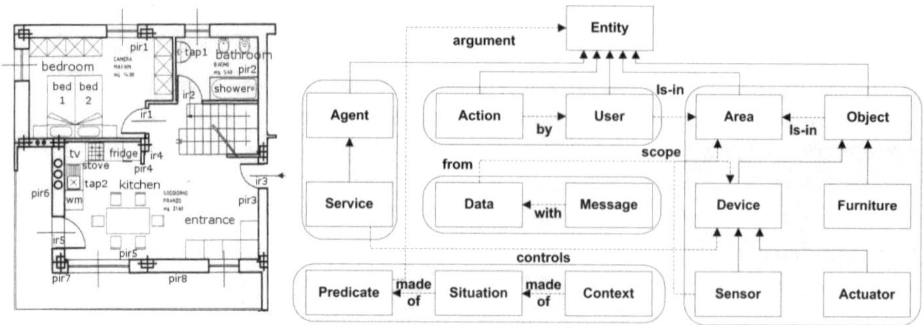

Figure 1. Left: a reference scenario. Right: part of the proposed ontology.

sensor). In the following discussion, we do not assume knowledge about `user` identity, which could be achieved in principle using RFIDs and wearable technology. For the sake of simplicity, we assume a single user scenario.

The system is designed to monitor the occurrence of specific contexts, i.e., collections of simultaneous events (i.e., *relational* contexts) or distributed over different periods of time (i.e., *temporal* contexts). For example, *taking lunch* could be modeled as a collection of *facts* including *using the stove*, *in the kitchen*, *near the table* and *on the chair*, which must occur either simultaneously or in sequence to infer that the user is taking lunch.

1.2. The Ontology Representation

The main goal of context assessment is to arrange information originating from distributed agents in templates relating numerical data to symbols. This process is handled by an *ontology* Σ that is represented using a Description Logic (DL for short), such that $\Sigma =$ `<TBox,ABox>` (see Figure 1 on the right). Reasoning capabilities are based on *subsumption*. Although a comprehensive description of Σ is beyond the scope of the paper, in the following paragraphs concepts and relationships which are relevant for the discussion are briefly outlined. The main purpose of Σ is twofold: (i) to manage a process of symbolic information fusion for assessing information [8]; (ii) to store instances of constants, variables, functions, predicates and terms for specifying the context definition language $\mathcal{CDL}$. With respect to Figure 1, the Σ can be divided into five main *domains*.

Entity layer. $\Sigma \models$ `Entity` is such that $\{$`Agent,Action,User,Area,Object`$\} \sqsubseteq$ `Entity`, which does not have instances: rather, it groups together shared descriptions for subsumed concepts.

Architecture and data modeling. This part represents definitions and instances of `Agent`, and `Data` which are exchanged within the system. An explicit representation is fundamental as it realizes both an *agent-directory* service and – with respect to instances of `Data` – a repository for *variables* used by $\mathcal{CDL}$. Examples of `Data` instances include $\Sigma \models$ `Data(pir-data)` or $\Sigma \models$ `Data(temp-data)`.

User modeling. $\Sigma \models$ `User` models users living within the Intelligent Environment. It is assumed that each user can be identified by matching, e.g., RFID tags with predefined labels. User identification is a research field in its own right. However, it is no longer discussed here.

Topology. It describes concepts, e.g., $\Sigma \models \{\texttt{Area}, \texttt{Sensor}, \texttt{Actuator}, \texttt{Furniture}\}$ used to describe the physical space. Relevant objects include areas, e.g., `Area(bedroom)` or `Area(bathroom)`, several devices, e.g., `PIR(pir`$_1$`)`, `Tap(tap`$_1$`)` or `Seat(seat`$_2$`)`, or *smart* furniture, such as `Stove(stove)`, `Fridge(fridge)` or `Bed(bed`$_1$`)`. With respect to $\mathcal{CDL}$, instances subsumed by concepts belonging to this class constitute useful *constants*.

Contexts descriptions. A $\Sigma \models$ `Predicate` describes *something* about the state of an `Entity`, which semantics depend on an interpretation $\mathcal{I}$. As long as a_i, such that $\{a_i|\Sigma \models \texttt{Data}(a_i), i = 1, ..., |\texttt{Data}|\}$ are updated with new sensor information (i.e., a variable assignment α over the `ABox` is defined), then corresponding p_j, such that $\{p_j|\Sigma \models \texttt{Predicate}(p_j), j = 1, ..., |\texttt{Predicate}|\}$ are updated as well. As a consequence, their truth values change accordingly. Useful instances `Predicate(pir-active)`, `Predicate(tap-open)` and `Predicate(seat-pressed)`.

$\Sigma \models$ `Situation` relates an `Entity` to one or more `Predicate`s, whereas, according to a multi-context perspective [5], a broader – yet partial – view of what is happening within the Intelligent Environment can be inferred by considering the union of relevant `Situation` instances. A `Context` is such an aggregate of one or more `Situation` concepts. `Context`s $\Sigma_c \in \Sigma$ are formulas which are described using $\mathcal{CDL}$, possibly involving the status of many entities.

In general, given an interpretation $\mathcal{I}$, $\texttt{Context}^{\mathcal{I}}$ is formally given by the conjunction of the semantics of the various constituent $\texttt{Situation}^{\mathcal{I}}$ and $\texttt{Predicate}^{\mathcal{I}}$. However, in real scenarios, more can be *afforded* from the mere conjunction of `Predicate` instances. For example, consider `Predicate(stove-active)`, `Predicate(pir`$_4$`-active)` (assuming that `pir`$_4$ is located in the `kitchen`) and `seat`$_1$`-pressed`: when considered as a whole, it is likely that someone is going to *take lunch*.

The corresponding `ABox` maintains the status of the system at *run time*. Specifically, for each time instant t, a different variable assignment α_t is used over `Predicate` instances. A properly defined individual, i.e., $\texttt{history} \sqsubseteq \top$, explicitly stores these instances occurred within a predefined temporal window of lenght n. As long as the system evolves, it happens that $(\Sigma, \mathcal{I}, \alpha_t) \vdash_i \Sigma_c^i(\texttt{history})$, i.e., `history` *can be subsumed* by a number of `Context`s Σ_c^i.

1.3. A Language for Describing Contexts

Herewith, we introduce the main aspects of $\mathcal{CDL}$, a language for context description. $\mathcal{CDL}$ is defined in terms of variables $\mathcal{CDL}_v$, constants $\mathcal{CDL}_c$, predicative symbols $\mathcal{CDL}_p$, functions $\mathcal{CDL}_f$, connecting symbols $\mathcal{CDL}_{cs}$ and words $\mathcal{CDL}_w$, which are divided into terms and formulas. $\mathcal{CDL}$ is a two-way representation between Σ and system designers. From the Σ side, $\mathcal{CDL}$ is grounded with respect to concepts, roles and instances in Σ. Furthermore, as concept instances are either created or updated with Σ, $\mathcal{CDL}$ specifications are updated and made available for system designers. From the designer side, as contexts are described using $\mathcal{CDL}$, new concepts instances are created within Σ and immediately used to check subsumption with respect to `history`. Hereafter, we assume the availability of the operator ξ: given a concept $\mathcal{C}$, $\xi\mathcal{C} = \{c_i|\Sigma \models \mathcal{C}(c_i), i = 1, ...|\mathcal{C}|\}$, i.e., all the individuals subsumed by $\mathcal{C}$.

Variables. $\mathcal{CDL}_v$ is a possibly infinite enumeration of symbols which, within the `ABox`, change over time as a consequence of a variable assignment α_t. Specifically, these symbols are mapped to ξ`Data` and ξ`Period`. Examples of ξ`Data` include either *boolean* (e.g., $\texttt{pir}_i\texttt{-data}$, $i = 1, ..., 8$, smoke-data or stove-data), or *ordered* (e.g., temp-data) domains. Each Predicate is satisfied by (Σ, α_t) for a given Period.

Constants. $\mathcal{CDL}_c$ is a possibly infinite enumeration of symbols which are closed with respect to α_t. These symbols are mapped to $\xi_{\texttt{Device}}$ and to thresholds used to discretize Device values originating in ordered domains. Examples of $\xi_{\texttt{Device}}$ include $\texttt{pir}_i$, $i = 1, ..., 8$, smoke, stove and $\texttt{ir}_i$, $i = 1, ..., 5$. Thresholds commonly used include, e.g., those for temperature or light sensor data.

Predicative symbols. $\mathcal{CDL}_p$ is a possibly infinite enumeration of symbols which are explicitly interpreted according to a semantics $\mathcal{I}$. With respect to Σ, they are either directly mapped to $\xi_{\texttt{Predicate}}$ or to immediately inferred facts. For example, interesting predicates include: $\texttt{pir}_i\texttt{-active}$ (and similar for smoke, ir or stove devices), or temp-cold (and similar for other ordered sensors, such as, e.g., light). Temporal duration of predicates is explicitly modeled using the *lenght* λ of a Predicate, i.e., a Predicate which is satisfied whenever the Period associated with the monitored Predicate exceeds a threshold τ. For example, we refer to the Predicate monitoring the length of smoke-active using $\lambda\texttt{smoke-active}$. Another particularly interesting abstraction is the *derivative* δ of a Predicate, i.e., a Predicate which is satisfied when a given monitored Predicate changes its status from *true* to *false* (or *viceversa*) at the time instant t. For example, we refer to the Predicate monitoring the activation of $\texttt{pir}_i$ using $\delta\texttt{pir}_i\texttt{-active}^{f \rightarrow t}$.

Functions. $\mathcal{CDL}_f$ is a possibly infinite enumeration of symbols which explicitly model the structured relationships among $\xi_{\texttt{Predicate}}$, $\xi_{\texttt{Situation}}$ and $\xi_{\texttt{Context}}$. Therefore, symbols belonging to $\mathcal{CDL}_f$ corresponds are mapped to roles within Σ. For example, recall the *take lunch* Context previously introduced: the constituent predicates stove-active, $\texttt{pir}_4\texttt{-active}$ and $\texttt{seat}_1\texttt{-pressed}$ are *functionally* related to a corresponding Context through the roles doing, where and pose, i.e., $\xi_{\texttt{TakeLunch}} \sqsubseteq \texttt{Context} \sqcap \exists \texttt{doing.stove.active} \sqcap \exists \texttt{where.pir}_4\texttt{-active} \sqcap \exists \texttt{pose.seat}_1\texttt{-pressed}$. From this example, it is evident how the generation of complex words, such as terms and formulas, heavily relies on functions.

Connecting symbols. $\mathcal{CDL}_{cs}$ is a finite enumeration of propositional connectors, which include:

- *Logic connectors*: $\neg$ (i.e., negation), $\sqcap$ (i.e., DL-like conjunction and $\sqcup$ (i.e., DL-like disjunction).
- *Inference connectors*: $\sqsubseteq$ (i.e., subsumption) and $\times$ (i.e., role filling). Subsumption is the main reasoning scheme provided by DLs. Given two concepts, namely C_1 and C_2, we say that C_1 is subsumed by C_2 (and we write $C_1 \sqsubseteq C_2$) iff C_2 is more general than or equivalent to C_1. Role filling identifies associations between functions and constants or variables.
- *Temporal connectors*: λ (i.e., length of a Predicate), δ (i.e., derivative of a Predicate) and $\overleftarrow{\sqcap} \sqsubseteq \sqcap$ (i.e., *and before*). Specifically, $\overleftarrow{\sqcap}$ is used to concatenate sequences of predicative symbols which temporal relationship is important. For example, given C_1 and C_2, we say that $C_1 \overleftarrow{\sqcap} C_2$ is true iff C_2 occurs before than C_1. It is worth noting that, semantically speaking, the relation $\overleftarrow{\sqcap}^{\mathcal{I}} \sqsubseteq \sqcap^{\mathcal{I}}$ holds, i.e., *and before* is a special case of *and*.

Words. $\mathcal{CDL}_w$ comprises terms and formulas. With respect to Σ, terms represent partial descriptions of concepts, whereas formulas are properly defined concepts. In or-

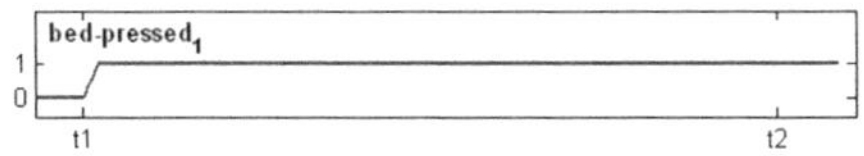

Figure 2. Behavior of $\texttt{bed}_1\texttt{-pressed}$ in Experiment 2.

der to specify $\xi_{\texttt{Context}}$, a designer is requested to write formulas, according to the specifications of $\mathcal{CDL}$. A term is defined as follows: (i) a generic symbol $x \in \mathcal{CDL}_v$ is a term; (ii) a generic symbol $a \in \mathcal{CDL}_c$ is a term; (iii) if $f \in \mathcal{CDL}_f$ is a function, and $t_1, ..., t_2 \in \mathcal{CDL}_w$ are terms, then $f \times t_1 \times ... \times t_n$ is a term. Terms are useful in order to build more complex formulas. In particular, an *atomic* formula is a word in the form $p(t_1, ..., t_n)$, where $p \in \mathcal{CDL}_p$ and $t_1, ..., t_n \in \mathcal{CDL}_w$. Therefore, a formula is defined as follows: (i) atomic formulas $p \in \mathcal{CDL}_w$ are formulas; (ii) if $A \in \mathcal{CDL}_w$ is a formula, then $\neg A \in \mathcal{CDL}_w$ is a formula; (iii) if $A, B \in \mathcal{CDL}_w$ are formulas, then $A \sqcap B \in \mathcal{CDL}_w$ and $A \sqcup B \in \mathcal{CDL}_w$ are formulas (please notice that, as $\overleftarrow{\sqcap} \sqsubseteq \sqcap$ holds, also $A \overleftarrow{\sqcap} B \in \mathcal{CDL}_w$ is a formula); (iv) if $A \in \mathcal{CDL}_w$ is a formula, then $\lambda A \in \mathcal{CDL}_w$ and $\delta A \in \mathcal{CDL}_w$ are formulas.

2. Experimental Results and Discussion

The designed experiments are aimed at validating the general behavior of the proposed architecture. Experiments have been performed in the Know-House@DIST setup, described in [8]. In the following paragraphs, we introduce and discuss a number of key examples.

Experiment 1. In bed while the shower is active. $\mathcal{C} \sqsubseteq \texttt{Context} \sqcap (\texttt{bed}_1\texttt{-pressed} \sqcup \texttt{bed}_2\texttt{-pressed}) \sqcup \texttt{shower-active}$. This is an anomalous context because, when the shower is opened, *one is supposed to use it*, not to stay in bed. The interesting thing is that, although $\mathcal{C}$ is a relational context, it subsumes more complex *temporal* patterns in $\texttt{history}$, such as, e.g., $\texttt{bed}_1\texttt{-pressed} \overleftarrow{\sqcap} \texttt{ir}_1\texttt{-active} \overleftarrow{\sqcap} \texttt{ir}_2\texttt{-active} \sqcap \texttt{shower-active}$: i.e., the user left the $\texttt{bathroom}$ and went to the bed forgetting the $\texttt{shower}$ opened (referring to Figure 1 to track user movements, remember that $\overleftarrow{\sqcap}$ is meant to be considered from *right* to *left*). In other words, $\mathcal{C}$ subsumes all the descriptions subsumed by the constituent formulas (i.e., instances of $\texttt{Predicate}$): no matter what else happens, the anomaly is detected.

Experiment 2. In bed more than usual: $\mathcal{C} \sqsubseteq \lambda\texttt{bed}_1\texttt{-pressed} \overleftarrow{\sqcap} \delta\texttt{bed}_1\texttt{-pressed}^{f \to t}$. This is a temporal $\texttt{Context}$, as it relates predicates which mutual occurrence is constrained. In order to $(\Sigma, \mathcal{I}, \alpha_t) \vdash \mathcal{C}(\texttt{history})$, a change in the truth value of $\texttt{bed}_1\texttt{-pressed}$ must be detected. Then (remember to consider the $\overleftarrow{\sqcap}$ from *right* to *left*) the *length* of $\texttt{bed}_1\texttt{-pressed}$ is checked: if it exceeds the threshold τ, $\lambda\texttt{bed}_1\texttt{-pressed}$ is satisfied, and then $\mathcal{C}$ is satisfied as well. Here, only one $\texttt{Predicate}$ in Σ is really involved in practice, i.e., $\texttt{bed}_1\texttt{-pressed}$ (see Figure 2): in t_1, the user goes to bed. As the time passes, $\texttt{bed}_1\texttt{-pressed}$ does not change: when reaching t_2, the relationship $t_2 - t_1 > \tau$ holds, and then $\mathcal{C}$ is satisfied.

Experiment 3: The user can not sleep during the night, so he prefers to switch on and look at the TV: $\mathcal{C} = \texttt{night-time} \sqcap \lambda\texttt{tv-on}_\tau \overleftarrow{\sqcap} \delta\texttt{tv-on}^{f \to t} \overleftarrow{\sqcap} \delta\texttt{bed-pressed}_1^{t \to f} \overleftarrow{\sqcap} \lambda\texttt{bed-pressed}_{1,\tau}$. Here, we do not explicitly specify periods. If $\texttt{history}$ contained,

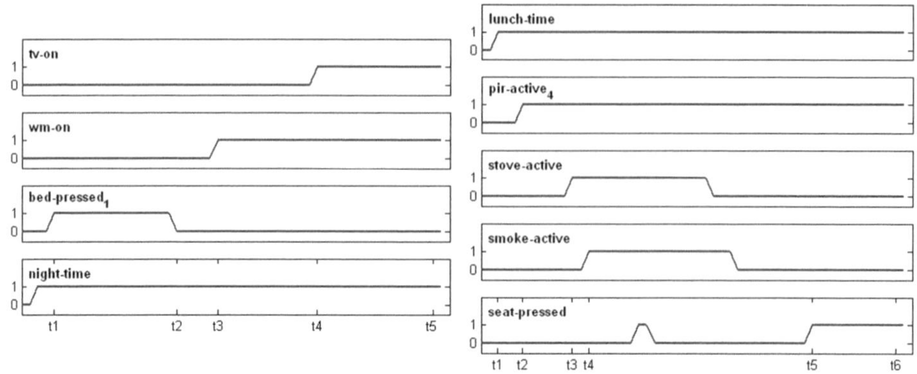

Figure 3. Behavior of relevant predicates in Experiments 3 and 4.

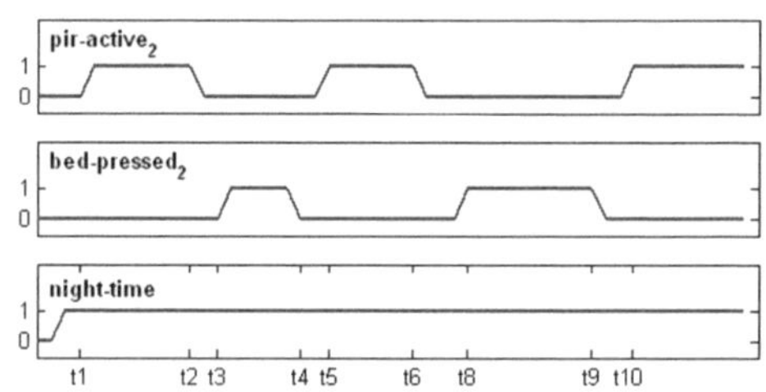

Figure 4. Behavior of relevant predicates in Experiment 5.

among others events, $\lambda\texttt{tv-on}_{\mathcal{T}} \overleftarrow{\sqcap} \delta\texttt{tv-on}^{f\to t} \overleftarrow{\sqcap} \delta\texttt{wm-on}^{f\to t} \overleftarrow{\sqcap} \delta\texttt{bed-pressed}_1^{t\to f} \overleftarrow{\sqcap} \lambda\texttt{bed-pressed}_{1,\mathcal{T}} \sqcap \texttt{night-time}$ (i.e., the washing machine is used during the night) it would be equally subsumed by $\mathcal{C}$. Referring to Figure 3 on the left, it is evident that events related to `wm-on` are irrelevant in subsumption. $\texttt{history} \sqsubseteq \mathcal{C}$ holds because of the events occurring in t_1, t_2, t_4 and t_5.

Experiment 4: The user is taking lunch in the kitchen: $\mathcal{C} = \texttt{lunch-time} \sqcap \lambda\texttt{seat-pressed}_{1,\mathcal{T}} \overleftarrow{\sqcap} \delta\texttt{seat-pressed}_1^{f\to t} \overleftarrow{\sqcap} (\texttt{oven-time} \sqcup (\delta\texttt{smoke-active}^{f\to t} \overleftarrow{\sqcap} \delta\texttt{stove-active}^{f\to t})) \sqcap \delta\texttt{pir-active}_4^{f\to t}$. The user is detected to be *near the table*, then *cooking* and finally *on the seat*. The patterns in Figure 3 on the right are subsumed by $\mathcal{C}$. Other activities in `history`, e.g., *sitting while cooking* (see Figure 3 at the bottom left), possibly interleaved with activities in $\mathcal{C}$, do not invalidate subsumption, since it depends only on events occurring in t_1, t_2, t_3, t_4, t_5 and t_6.

Experiment 5: The user frequently visits the bathroom during the night: $\mathcal{C} = \lambda\texttt{pir-active}_{2,\mathcal{T}} \overleftarrow{\sqcap} \delta\texttt{pir-active}_2^{f\to t} \overleftarrow{\sqcap} \lambda\texttt{bed-pressed}_{2,\mathcal{T}} \overleftarrow{\sqcap} \delta\texttt{bed-pressed}_2^{f\to t} \overleftarrow{\sqcap} \lambda\texttt{pir-active}_{2,\mathcal{T}} \overleftarrow{\sqcap} \delta\texttt{pir-active}_2^{f\to t} \overleftarrow{\sqcap} \lambda\texttt{bed-pressed}_{2,\mathcal{T}} \overleftarrow{\sqcap} \delta\texttt{bed-pressed}_2^{f\to t} \overleftarrow{\sqcap} \lambda\texttt{pir-active}_{2,\mathcal{T}} \overleftarrow{\sqcap} \delta\texttt{pir-active}_2^{f\to t} \sqcap \texttt{night-time}$ (see Figure 4). It is worth noting that here we do not explicitly model how to go from the `bedroom` to the `bathroom`: in principle, the user could go to the `kitchen` first, and do whatsoever activity, but the system would nonetheless infer multiple visits to the bathroom.

3. Conclusion

This paper presented an integrated ontology- and logic-based framework for contexts specification and acquisition. In particular, the system behavior can be formalized using a high-level language, called $\mathcal{CDL}$, which is grounded with respect to an *ontology*. Specifically, *words* belonging to $\mathcal{CDL}$ are mapped into concepts and relationships, which are then used to classify actual patterns of data. Using *subsumption* (i.e., the inference scheme provided by DLs), the system exhibits important desirable features: (i) contexts to be detected do not have to be fully specified, as more complex patterns of events are subsumed by generic context descriptions; (ii) symbolic *false positives* are implicitly filtered away, as patterns with false positives are subsumed by more general descriptions. The experimental results performed in a real set-up confirm these properties.

References

[1] J. C. Augusto and C. D. Nugent. A New Architecture for Smart Homes Based on ADB and Temporal Reasoning. In *Proceedings of the 3rd International Conference on Smart Homes and Health Telematics (ICOST'05)*, Canada, July 2005.

[2] J.C. Augusto, P. McCullagh, V. McClelland and J.A.Walkden. Enhanced Healthcare Provision Through Assisted Decision-Making in a Smart Home Environment. In *Proceedings of the Second Workshop on Artificial Intelligence Techniques for Ambient Intelligence (AITAmI07)*, Hyderabad, India, January 2007.

[3] T. S. Barger, D. E. Brown and M. Alwan. Health-Status Monitoring Through Analysis of Behavioral Patterns. In *IEEE Transactions on Systems, Man, and Cybernetics – Part A*, vol. 35, no. 1, January 2005.

[4] D.J. Cook and S.K. Das. How Smart are our Environments? An Updated Look at the State of the Art. In *Pervasive and Mobile Computing*, vol. 3, no. 2, pp. 53–73, March 2007.

[5] F. Giunchiglia and L. Serafini. Multilanguage Hierarchical Logics. In *Artificial Intelligence*, no. 64, pp. 29–70, 1994.

[6] E. J. Ko, H. J. Lee and J. W. Lee. Ontology-Based Context Modeling and Reasoning for U-HealthCare. In *IEICE Transactions on Information and Systems*, no. 8, pp. 1262–1270, August 2007.

[7] Fan Tian Kong, You-Ping Chen, Jing-Ming Xie and Zu-De Zhou. Distributed Temperature Control System Based on Multi-sensor Data Fusion. In *Proceedings of the 2005 International Conference on Machine Learning and Cybernetics*, China, August 2005.

[8] F. Mastrogiovanni, A. Sgorbissa and R. Zaccaria. An Active Classification System for Context Representation and Acquisition. In *J.C. Augusto and D. Shapiro (Eds.), Advances in Ambient Intelligence*, FAIA Series, November 2007.

[9] J. McCarthy. Notes on Formalizing Context. In *Proceedings of the 13th International Joint Conference on Artificial Intelligence (IJCAI-93)*, pp. 555–560, Chambery, France, 1993.

[10] H.T. Nguyen, J. Qiang and A.W.M. Smeulders. Spatio-Temporal Context for Robust Multitarget Tracking. In *IEEE Trans. on Pattern Analysis and Machine Intelligence*, vol. 29, no. 1, pp. 52–64, January 2007.

[11] I. Satoh. A Location Model for Smart Environments. In *Pervasive and Mobile Computing*, vol. 3, no. 2, pp. 53 – 73, March 2007.

[12] T. Strang and C. Linnhoff-Popien. A Context Modeling Survey. In *Proceedings of the 6th International Conference on Ubiquitous Computing (UbiComp2004)*, Nottingham, England, September 7-10, 2004.

[13] A. Ranganathan and R. H. Campbell. A Middleware for Context-Aware Agents in Ubiquitous Computing Environments. In *Proceedings of the 2003 ACM/IFIP/USENIX International Middleware Conference*, Rio de Janeiro, Brazil, 2003.

[14] I. Horrocks. DAML+OIL: a Reason-able Web Ontology Language. In *Proceedings of the 8th International Conference on Extending Database Technology (EDBT-02)*, Prague, Czech Republic, 2002.

[15] R. Dapoigny and Barlatier. Towards a Context Theory for Context-Aware Systems. In *J.C. Augusto and D. Shapiro (Eds.), Advances in Ambient Intelligence*, FAIA Series, November 2007.

Intelligent Autonomous Systems 10
W. Burgard et al. (Eds.)
IOS Press, 2008

doi:10.3233/978-1-58603-887-8-254

Object Localization using Bearing Only Visual Detection

Kristoffer SJÖ [a,1], Chandana PAUL [a] and Patric JENSFELT [a]

[a] *Royal Institute of Technology (KTH), Centre for Autonomous Systems, SE-100 44 Stockholm, Sweden*

Abstract.

This work demonstrates how an autonomous robotic platform can use intrinsically noisy, coarse-scale visual methods lacking range information to produce good estimates of the location of objects, by using a map-space representation for weighting together multiple observations from different vantage points. As the robot moves through the environment it acquires visual images which are processed by means of a fast but noisy visual detection algorithm that gives bearing only information. The results from the detection are projected from image space into map space, where data from multiple viewpoints can intrinsically combine to yield an increasingly accurate picture of the location of objects. This method has been implemented and shown to work for object localization on a real robot. It has also been tested extensively in simulation, with systematically varied false positive and false negative detection rates. The results demonstrate that this is a viable method for object localization, even under a wide range of sensor uncertainties.

Keywords. Accumulator Grid, Object Detection, Object Localization

Introduction

One of the major thrusts of robotics has been the development of robots which can provide domestic assistance in household tasks. Although this ambitious dream has now come closer to reality, there are still many challenging areas which remain to be addressed. One of the main areas is the development of appropriate representations of the environment which enable the robot to be cognizant of its surroundings and interact in a way that is appropriate to human-centered requirements.

One of the ways in which robots have been made more aware of their surroundings is through the development of spatial maps for navigation. There has been much work in this area, and the field of SLAM, Simultaneous Localization and Mapping has matured significantly. There are now an abundance of approaches to address the problem. For an overview see for example [1].

However, pure navigation maps such as those developed with SLAM, containing no extra structure of the environment, are not sufficient for the robot to deal with complex tasks requiring interaction with objects. For this, more complex representations need to be developed which contain both spatial and semantic information.

[1] Corresponding Author: Kristoffer Sjö, Royal Institute of Technology (KTH), Centre for Autonomous Systems, SE-100 44 Stockholm, Sweden; E-mail: krsj@nada.kth.se

There has been relatively little work which deals with the development of such semantic maps. Kuipers was one of the true pioneers in this area in robotics, with the so called Spatial Semantic Hierarchy [8,9]. Nonetheless, despite the early interest in this topic, the area is just beginning to gain momentum. Galindo *et. al.* have recently developed a hierarchical spatial representation in which simple objects have been detected and added [6]. Vasudevan *et. al.* have developed a representation which combines a topological representation of space with so called object graphs which model the spatial relation between objects within an area [16]. The recognition of objects is based on SIFT features [12]. Ranganathan *et al* have used three different detectors to detect objects: Harris corners, Maximally Stable Extremal Regions (MSER) and Canny edges and using a SIFT descriptor, inserted the features into a constellation graph model to generate a representation of a topological location [14].

Most of this work has dealt with the issue of an ideal structure for the representation of the environment, detached from the requirements of creating such a representation. We bring to light, however, that there is relationship between the accuracy of a representation and the amount of resources used to create that representation. Thus, if the robot performs a detailed inspection of the environment, visually processing every part of the environment from a close distance, it can build an accurate representation. However, if the robot performs a quick perfunctory inspection of the environment, with a fast visual processing algorithm, it can acquire data to build an approximate representation, consisting of hypotheses for potential object locations.

We further bring to light that both these methods can be used in conjunction, leading to a representational structure with two conceptual levels. The first level is an approximate representation of the location of objects in the environment. This representation, which takes very little resources to build, can be used to guide the construction of the second level, which is the accurate spatial and semantic representation of the environment. This multi-tiered approach leads to a much more directed and efficient use of sensory and computational resources in the creation of a semantic representation.

In previous work [13] we developed an accurate semantic representation which consists of four layers. At the first layer, it includes a metric representation created using SLAM methods, at the second a navigation graph of nodes indicating navigable free space and their connectivity, at the third a topological map dividing space into rooms and corridors, and at the fourth a conceptual map of semantic entities corresponding to places and objects and their relationships. In [11] we presented a method for visually recognizing and localizing these objects and inserting them into the map. Such visual procedures are expensive, and require much processing. In this paper, we have focused on the development of a layer of approximate representation regarding the location of objects using fast processing, which can then allow the robot to direct its more expensive visual and computational resources.

Our method of creating an approximate representation from uncertain visual processing techniques has been inspired by the creation of occupancy grids [4]. In an occupancy grid the world is divided into a grid of cells and each cell is assigned a value to reflect the certainty of that cell being occupied. A large number of methods for updating such grids have been proposed, based on Bayesian theory, fuzzy logic, etc. See [15,17] for an overview and comparison of sonar based occupancy grid methods. The sonar sensor provides relatively accurate distance estimates but the angular information is rather vague. Integrating many measurements is therefore necessary to get a good estimate of the structure of the environment. For obstacle avoidance using sonar the Histogram in Motion Mapping approach [2] is sometimes used as it provides a fast way of updating the grid in a time-critical application. In it a simplified update of the occupancy grid is used where only the cells along the acoustic axis of the sensor are updated, by adding and subtracting fixed integer values, thus ignoring the angular uncertainty of the sensor. This is in contrast to the full Bayesian formulation, where every cell will be affected by every observation. Accuracy in the model is instead gained by frequent updates.

However, this work is different from the previous work on occupancy grids, in that fast visual detection is much more uncertain than previous sensing techniques. With a laser or sonar sensor, there is a fairly high likelihood that if a positive reading arises then an obstacle exists within the range of the sensor. However, with a fast visual detection algorithm, many false positives can arise, and thus locating an object with a certain measure of confidence presents a much greater challenge. Furthermore the sensors used for occupancy grids usually provide good range information and moderately precise bearing information. The fast visual detection algorithm used here on the other hand provides no range information, only bearing, which makes it a slightly different problem. Finally, previous techniques have been object non-specific. Thus, every entity in the environment has been considered to be of the same type. In this work, a higher degree of semantic specificity is applied, and only objects recognized by a trained visual algorithm are considered. Thus the entities considered are much sparser in the environment, and the localization problem is different, as detections from the same location are likely to arise from the same object, as opposed to two closely adjacent objects.

Methods akin to the occupancy grid have also been used for global localization of a mobile robot [5], where the grid represents a discretization of the probability density function for the pose of the robot. In this work, however, the robot's position is considered known and objects are what is being localized.

This work is also similar in purpose to the mapping aspect of bearing-only SLAM; however, such methods [3,10] are not designed to cope with the high levels of false positives and false negatives that are present in the visual detection algorithms; rather, they deal with uncertain bearing measurements in addition to missing range data.

In the next section, the accumulator grid algorithm is presented. Following this, in Section 2 a proof-of-concept implementation on a mobile robot is described, and results are demonstrated. Section 3 introduces a simulation environment in which a thorough performance evaluation of the algorithm under varying conditions of false positive and false negative detection rates is performed. The results are presented, followed by discussion and conclusions.

1. Accumulator Grid

In order to create and maintain a distribution across space of the confidence the robot has in the presence of objects, an *accumulator grid* has been developed. This is a grid overlaid on the robot's navigation map, where each grid cell is associated with a confidence value representing the certainty of an object being in the cell. One set of such values is maintained for every distinct object of interest. In the following description, only one object is considered for simplicity.

As the robot moves through the environment and acquires visual data, the accumulator grid cells are updated according to the output of the visual object detection algorithm. As data accumulate from different locations, the confidence values reinforce each other in a way similar to the functioning of the Hough transform [7], embodying an increasingly accurate representation of the true locations of objects in the environment. The robot can then utilize these to direct its motion, or store them for later use. Although the information in the grid is 2-dimensional, it can still be used to guide a directed 3D object search, considering all vertical locations corresponding to the objects 2D location in the map.

1.1. Initialization

The accumulator grid is set up with a specific extent in the X and Y dimensions as well as a cell size. Appropriate values for these parameters are highly dependent on the environment the robot operates in. In general, its extent should equal the size of the operating area and the cell size should be roughly equal to the size of sought objects. If objects of different sizes are to be represented, the smallest size can be used or, alternatively, several grid sizes may be used in parallel.

Initially, the confidence values are all set to 0 if the robot does not possess any prior information about object locations. It would be possible to represent prior knowledge of the likely and unlikely locations of objects during initialization using non-zero values.

1.2. Image processing

The algorithm is designed to work with Receptive Field Cooccurrence Histograms (RFCH). RFCH is an image processing method for object detection based on bulk image properties, as opposed to feature-based methods such as SIFT typically used for recognition. It is used here because it is fast and produces a scalar degree-of-match as output given any part of an image, large or small. RFCH are object-specific and so this algorithm, based on them, will be as well. However, any object detection algorithm with similar properties to RFCH matching could in principle be substituted.

Prior to engaging in the object search, the robot's vision system is trained on each object of interest by performing feature value clustering and histogram extraction on training images of the objects of interest. The clusters and the histogram for the object are stored for the purpose of object detection on the images subsequently acquired.

1.3. Update

As the robot proceeds through its surroundings, it acquires camera images, which are subdivided into small patches for which RFCH are extracted. The resulting histograms are compared to the histogram from the training image of the object being sought, producing a match value for each patch.

It is assumed that the pose of the robot is known at all times. If this is not the case, it will become necessary to take steps similar to those needed for occupancy grids built during mapping under uncertain odometry; such considerations are however beyond the scope of this paper.

The robot's camera is kept horizontal during exploration. As a result, each column of image patches corresponds to a specific interval of bearings, which is projected onto the 2D map from the location of the robot. This produces a "sensor wedge" as shown in Figure 1(a).

The accumulator grid is then updated according to the following principle: Each cell covered by a sensor wedge is incremented by the maximum magnitude of all the RFCH match values associated with it. This value represents the likelihood of the object's presence within that wedge; see Figure 1(b). Note that we do not make any assumptions about range information from the visual detection algorithm, but only bearing information. As the robot continues to acquire images from different vantage points, the wedges intersecting objects will reinforce the grid cells, whereas cells that are only incidentally part of a sensor wedge will not get reinforced; see Figure 1(c).

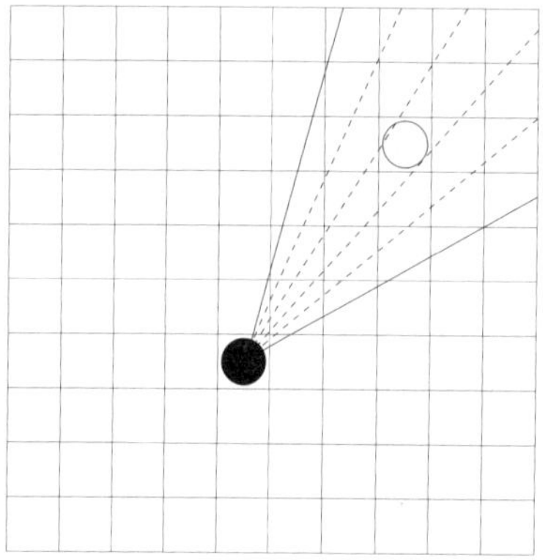

(a) The robot's field of view is subdivided into wedges

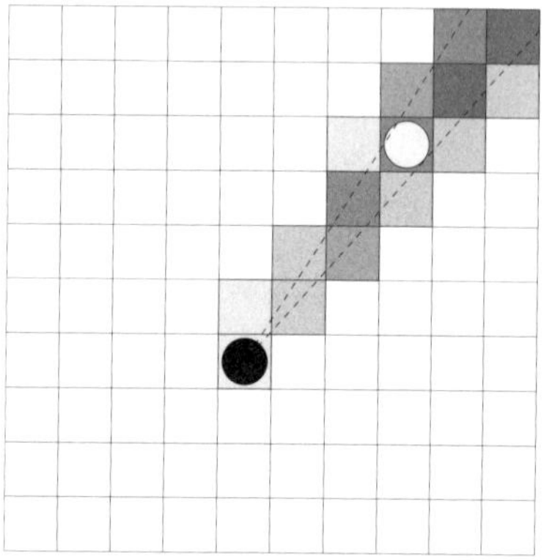

(b) Each wedge is updated by its associated sensor response. Note the smoothing effect of supersampling

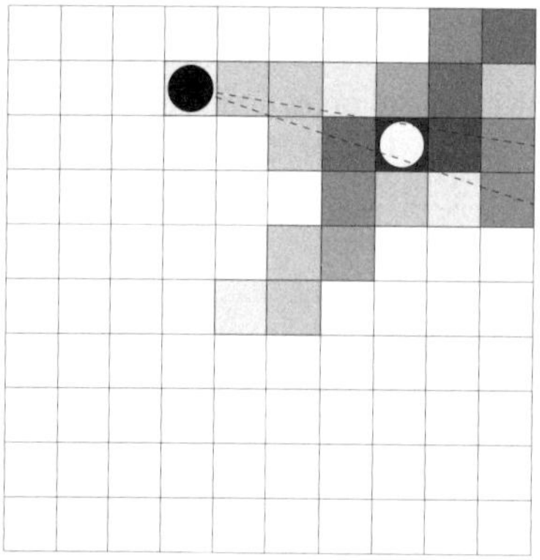

(c) As views are acquired from different vantage points, cells containing objects will be reinforced

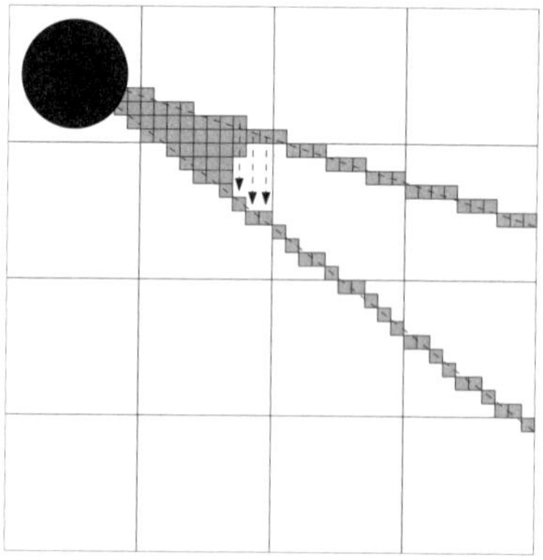

(d) Cells in each wedge are incremented, using supersampling to counter aliasing effects

Figure 1. Principles of the accumulator grid update

In practical terms, the update of the cells in each sensor wedge is carried out by means of line rasterization, performed in parallel for the left- and right-hand rays of the slice, respectively, and by filling in the intermediate cells in rows or columns as appropriate. However, given the potentially very large granularity of the accumulator grid and the aliasing effects this causes, a supersampling scheme is adopted in which the resolution of the grid is augmented by a factor of 10 during the update. In effect, each cell is updated in proportion to how well it is covered; see Figure 1(d). This alleviates aliasing problems without incurring any extra storage costs.

2. Implementation and Experiments

The algorithm described in the previous section was implemented and tested on a Performance Peoplebot mobile robot platform. The robot is approximately 1.2m in height. It is equipped with a Canon VCC4 pan-tilt-zoom camera, a SICK Laser range finder, and ultrasonic sensors. The camera is mounted at a height of 1m above the floor, and has a horizontal field of view of 45 degrees. The camera was used to acquire images at a resolution of 320×240 pixels. In order for the robot to be able to detect objects in a wide field of view as it explores the environment, the camera's zoom was not used. The laser scanner was mounted at 30 cm above the floor, and used to acquire information about the structure of the environment. Features extracted from laser information were then used in a standard EKF based SLAM algorithm, to obtain a metric map of the environment and the position of the robot within it.

The test of object localization was performed in a mockup living room of approximate dimensions 4.5m × 6m, shown schematically in Figure 2(a). The living room consisted of a sofa, a coffee table, a chair, a table, and a bookcase. Other miscellaneous items were also present. The target which was a multi-colored rice box was placed on the coffee table towards one corner of the room. An accumulator grid of 50 × 50 cells was created at a resolution of 0.1m to represent the environment, and all values were initially set to zero. The robot was programmed to visit 5 predefined locations in the room and take pictures in all directions (8 views at a field-of-view of 45°) from each location. RFCH detection was carried out from each viewpoint, and an accumulator grid was updated according to Section 1.3.

After visiting each location and processing the images, the result was the accumulator grid shown in Figure 2(b). The actual location of the sought object, a packet of rice, is plainly visible as the brightest spot in the upper left quarter of the grid. On the center right a chair of somewhat similar color to the object can be made out, and towards the bottom of the grid a bookcase containing many items of varying appearance causes a low-level blur; still, these false positives are clearly less prominent than the true location. A view planning algorithm can use this information to create priorities for regions to investigate, which would lead it to begin with the area that actually contains the object in this case.

3. Simulation

Although the above experiment demonstrates the feasibility of the approach, there are many factors that can affect the reliability of the visual detection on which it is based. These include lighting conditions, the structure of the environment, the saliency of the object and the parameters of the detection algorithm.

In order to perform a broader investigation of how these factors influence the accumulator grid algorithm, the performance of the system was evaluated by means of Monte Carlo simulation in an abstract scenario, in which the false positive and false negative detection rates of the object detection can be varied freely.

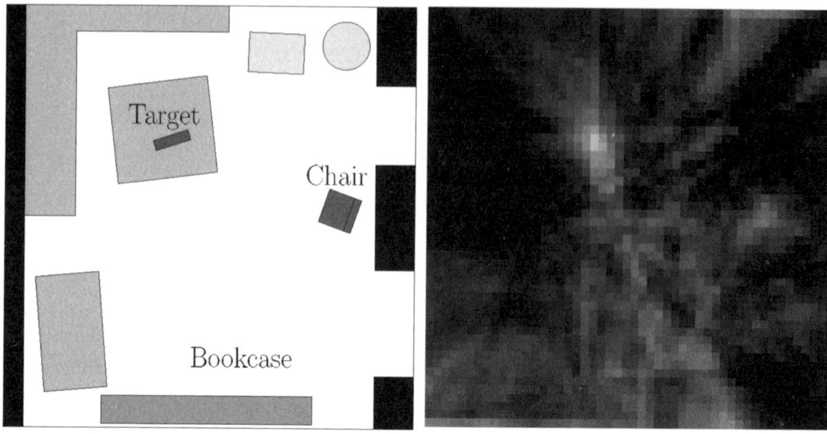

(a) Schematic representation of the room used

(b) Result of grid accumulation. The true object location is at the bright spot on the upper left

Figure 2. Experimental results

The environment is represented as a 20×20 grid. The object is positioned randomly in the environment (excepting the outer edges), and occupies exactly one grid cell. The robot can move to any part of the environment except for the cell containing the object. The robot can also have any orientation in space. In the real world, the robot would follow some continuous trajectory through space, sensing as it went. It would thus see the object from several distinct orientations and views. Here, in order to simulate this fact, while avoiding any bias introduced by a manually selected trajectory, the robot is given a random new orientation and position for every view that is acquired. The robot is assumed to know its pose with perfect accuracy.

For a given robot pose, a field of view is simulated by an angular slice of the map, with its apex at the robot's position, as was shown in Figure 1(a). The field of view is divided into wedges corresponding to the image patch columns described in Section 1.3. For each wedge, the cells it covers are incremented if the object intersects the wedge.

After a set number of random views, the search is terminated and the result is evaluated by locating the maximally-valued cell and comparing its position in the map to the known position of an object. The test is considered successful if the cell with the maximum value in the accumulator grid is the one containing the object or adjacent to it.

The performance of a noisy visual detection algorithm such as RFCH matching was simulated as a detector which gives a binary response on the location of the object, with non-zero false positive and false negative rates P_{fp} and P_{fn} respectively. The performance of object localization with the accumulator grid, under such noisy visual detection conditions, were evaluated in a series of tests. The false positive and false negative rates were varied in intervals of 0.1 between 0 and 1. For each combination of values, tests were performed evaluating the localization with 10, 25, 75 and 150 views. 100 such tests were performed for each parameter setting.

Figure 3 shows the outcomes of the tests. Obviously, a low view count does not provide enough data for a good estimate, but with 10 random views a good guess can be made approximately 40% of the time if the false positive and false negative rates are low. With 25, results are reasonably reliable in the absence of noise. The real-world test

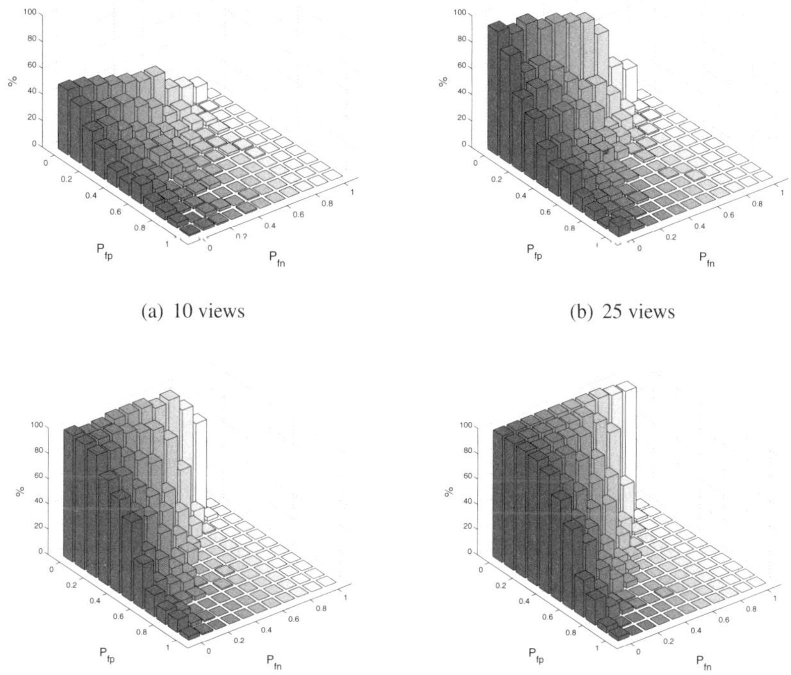

(a) 10 views

(b) 25 views

(c) 75 views

(d) 150 views

Figure 3. Performance of object localization for various probabilities of false positive and false negative rates. The Z axis denotes the percentage correct estimations, and the X and Y axes the rate of false positives and negatives, respectively.

in Section 2 with 40 well-planned views compares to a situation with 75 or 150 random views. At this level the algorithm is able to deal with a large number of false negatives if the false positive rate is low. The converse also holds, though to a somewhat lesser extent. This is because a false negative does not alter the accumulator grid, whereas a false positive introduces flawed data.

Typically most visual processing algorithms are associated with a *Receiver Operating Characteristic* or ROC curve, describing how the false positive rate relates to the false negative rate for that algorithm when varying some discrimination threshold. For RFCH, for instance, a threshold on the degree of match will determine these error rates. Figure 3 shows that good results are achieved whenever either P_{fp} or P_{fn} can be made small, which is the case with many algorithms. Thus, with knowledge of the ROC curve, it is possible to optimize the performance of the accumulator grid for any given algorithm.

4. Discussion

The simulated and real results presented in this paper are promising, and suggest that this algorithm would be a viable method for object localization under various conditions of sensing uncertainty. However, there are several issues which should be considered further.

Robot localization

In the simulation, it was assumed that the robot knew its pose with perfect accuracy. However, in the real world there can be some discrepancy between the robot's real pose and estimated pose. The accumulator grid is dependent upon a good estimate of the position of the robot. If the estimate drifts over time, the grid may be affected through the blurring, offset, or duplication of maxima. Nevertheless, contemporary SLAM methods are typically sufficiently robust for this not to be a problem.

Viewpoint bias correction

In simulation, the robot was given a new position and orientation at every time step. In the real world the robot doesn't move randomly through space, and subsequent locations and views of the object are not entirely independent of previous locations. This is not always a critical problem, as is seen in the successful localization in Figure 2(b), but nonetheless bears consideration.

Extrinsically, the problem can be solved by planning for the robot to move and obtain visual data in a way that provides a good distribution of viewpoints. If this is not possible, intrinsic solutions involving changes to the algorithm itself might be made. Possibilities include weighting of the grid increment based on the amount of sensor data gathered from the current direction, or the creation of multiple grid layers for different viewing directions. Either way would require increasing the amount of data stored.

RFCH

The use of RFCH in this work was due to its ability to produce degrees of match across the entire image, as well as its relatively good object specificity. However, though relatively fast, current implementations still leave something to be desired in terms of speed and especially scalability: at present processing a 320×240 image requires on the order of one second on our platform, and this grows linearly with the number of objects. Faster implementations and alternative methods, especially hierarchical ones, would be a valuable modification.

Furthermore, RFCH doesn't respond equally well to different views of the same object. An easy solution to this is to represent multiple views as multiple objects, but this is costly. Clustering of similar views can help in this regard, as could a hierarchical object detector structure. A hierarchical detection algorithm could also help reduce the memory requirements of the accumulator grid itself, which currently maintains one layer of cells per distinct object.

5. Conclusion

This paper presents a novel method for consolidating noisy bearing-only visual information to achieve object localization. The method uses an accumulator grid to update confidence information through an algorithm that transforms the data from visual into map space. Results are presented of feasibility testing in a realistic scenario as well as an extensive evaluation in simulation. The results indicate that the method performs well in the face of noisy measurements and promises to be useful as a way of obtaining and representing the approximate location of objects in a robot's environment.

Acknowledgements

This work was supported by the EU FP6 IST Cognitive Systems Integrated Project "CoSy" FP6-004250-IP. Kristoffer Sjö was supported in part by the Swedish Research Council, contract 621-2006-4520.

References

[1] T. Bailey and H. Durrant-Whyte. Simultaneous localization and mapping (SLAM): Part II state of the art. *IEEE Robotics & Automation Magazine*, 13(3):108–117, 2006.
[2] J. Borenstein and Y. Koren. Histogram in-motion mapping for mobile robot obstacle avoidance. *IEEE Transactions on Robotics and Automation*, 7(4):535–539, August 1991.
[3] A. J. Davison. Real-time simultaneous localisation and mapping with a single camera. *Computer Vision, 2003. Proceedings. Ninth IEEE International Conference on*, pages 1403–1410 vol.2, 13-16 Oct. 2003.
[4] A. Elfes. Using occupancy grids for mobile robot perception and navigation. *Computer*, 22(6):46–57, 1989.
[5] D. Fox, W. Burgard, and S. Thrun. Markov localization for mobile robots in dynamic environments. *Journal of Artificial Intelligence Research*, 11:391–427, 1999.
[6] C. Galindo, A. Saffiotti, S. Coradeschi, P. Buschka, J.A. Fernández-Madrigal, and J. Gonález. Multi-hierarchical semantic maps for mobile robotics. In *Proc. of the IEEE/RSJ International Conference on Intelligent Robots and Systems (IROS'05)*, pages 3492–3497, 2005.
[7] J. Illingworth. and J. Kittler. A survey of the hough transform. *Computer Vision, Graphics, and Image Processing*, 1988.
[8] B. J. Kuipers. Modeling spatial knowledge. *Cognitive Science*, 2:129–153, 1978.
[9] B. J. Kuipers. The spatial semantic hierarchy. *Artificial Intelligence*, 119:191–233, 2000.
[10] T. Lemaire, S. Lacroix, and J. Sola. A practical 3d bearing-only slam algorithm. *Intelligent Robots and Systems, 2005. (IROS 2005). 2005 IEEE/RSJ International Conference on*, pages 2449–2454, 2-6 Aug. 2005.
[11] D. Galvéz Lopez, K. Sjöö, C. Paul, and P. Jensfelt. Hybrid laser and vision based object search and localization. In *Proc. of the IEEE International Conference on Robotics and Automation (ICRA'08)*, 2008.
[12] D. G. Lowe. Object recognition from local scale-invariant features. In *Proc. of the International Conference on Computer Vision (ICCV 1999)*, pages 1150–57, Corfu, Greece, September 1999.
[13] O. Martínez Mozos, P. Jensfelt, H. Zender, G.-J. Kruijff, and W. Burgard. From labels to semantics: An integrated system for conceptual spatial representations of indoor environments for mobile robots. In *Proc. of the Workshop "Semantic information in robotics" at the IEEE International Conference on Robotics and Automation (ICRA'07)*, Rome,Italy, April 2007.
[14] A. Ranganathan and F. Dellaert. Semantic modeling of places using objects. In *Proc. of Robotics: Science and Systems Conference (RSS06)*, 2003.
[15] M. Ribo and A. Pinz. A comparison of three uncertainty calculi for building sonar-based occupancy grids. In *SIRS*, Coimbra, Portugal, July 1999. A revised version will appear in Journal of Robotics.
[16] S. Vasudevan, S. Gächter, V. Nguyen, and R. Siegwart. Cognitive maps for mobile robots – an object based approach. *Robotics and Autonomous Systems*, 55:359–371, 2007.
[17] O. Wijk. *Triangulation Based Fusion of Sonar Data with Application in Mobile Robot Mapping and Localization*. PhD thesis, Royal Institue of Technology, Stockholm, Sweden, April 2001.

Intelligent Autonomous Systems 10
W. Burgard et al. (Eds.)
IOS Press, 2008

doi:10.3233/978-1-58603-887-8-264

Addressing temporally constrained Delivery Problems with the Swarm Intelligence approach

Silvana BADALONI [a], Marco FALDA [a,1], Francesco SAMBO [a], Leonardo ZANINI [b]
[a] *Dept. of Information Engineering - University of Padova (Italy)*
{silvana.badaloni,marco.falda}@unipd.it, francesco.sambo@dei.unipd.it
[b] *zanini.leonardo@gmail.com*

Abstract. We present an application of Ant Colony Optimization metaheuristic to the Pick-up and Delivery Problem with Time Windows (PDPTW), a variant of the Vehicle Routing Problem with Time Windows (VRPTW) with additional constraints on pairs of source-destination nodes.

We chose to apply the Multiple Ant Colony System (MACS) approach: two ant colonies minimize the number of vehicles to be routed and the travel lengths; cooperation between colonies is performed by exchanging information through pheromone updating. Besides, we studied a novel strategy to bias the attractiveness of a node depending on its nature, called pheromone post-treatment.

The algorithm performances on the number of vehicles needed and the total tour length were comparable to those of the best algorithms in the state of art.

Keywords. Swarm Intelligence, Ant Colony, PDPTW

Introduction

Pick-up and Delivery problems (PDPs) describe common situations in which vehicles must pick-up goods from a variety of places and deliver them [13,15]. They have been widely studied in literature. In 2001 Li and Lim proposed an hybrid meta-heuristic which combines Simulated Annealing and Tabu Search [8]; they adopt a k-starts strategy that reinitializes the Simulated Annealing algorithm after a number of unsuccessful iterations and maintains a tabu list of no-good solutions. Bent and Van Hentenryck [3] apply a two-phases approach: first a Simulated Annealing algorithm optimizes the number of vehicles, then a Large Neighborhood Search algorithm [14] optimizes the total travel time. The work of Ropke and Pisinger [12], based on a variant of the previous one, can be considered the state-of-the-art for PDP.

PDPs can be considered as a standardization of many different problems such as co-ordination of autonomous robots for distributing services, bus routing, airlines schedulings and, in general, logistics problems.

[1]Corresponding Author: Marco Falda. Dept. of Information Engineering - University of Padova (Italy). Email: marco.falda@unipd.it.

In this paper we address a special instance of PDP in which deliveries must satisfy time windows (PDPTW) and we develop an algorithm based on Ant Colony Optimization (ACO), a method inspired by real ants foraging behaviour [5,6].

Our work is an extension of an idea of Gambardella et al. [7]; they apply an ACO algorithm for solving the Vehicle Routing Problem with Time Windows (VRPTW) using two ant colonies, one for minimizing the number of vehicles and the other for the total travel time. Our extension is based on the main fact that PDPTWs can be derived from VRPTWs by adding constraints on source-destination pairs. To this aim, we have included in the ACO algorithm some new techniques to cope with the additional constraints posed by this problem.

PDPTW is a generalization of the Office Delivery service accomplished by autonomous robots, a problem that we addressed previously. In particular, in [1] we used a standard Genetic Algorithm to find an optimal scheduling for documents delivery, while in [2] we enriched the cross-over operators introducing new criteria such as time, capacity, distance, random choice et c. to get the optimal solution. Here we study a similar problem with an Ant Colony algorithm; such an algorithm is more suited for finding optimal paths for problems on graphs, in particular when the underlying structure changes dynamically, since the algorithm runs continuously and adapts to changes as soon as they happen.

The paper is organized as follows. Section 1 gives a precise definition of the Vehicle Routing Problem with Time Windows and presents the application of the Multiple Ant Colony System to the problem. Section 2 describes the Pick-up and Delivery Problem with Time Windows and shows our extension of the Gambardella approach. An experimental setting is discussed in Section 3.

1. Ant Colony Optimization and the Vehicle Routing Problem

In this section we firstly describe the particular variant of the Vehicle Routing Problem with Time Windows and then we present the application to that problem of the Multiple Ant Colony System (MACS) proposed in [7], that we extended for solving the Pick-up and Delivery Problem with Time Windows.

1.1. Problem formulation

The Vehicle Routing Problem can be defined as the problem of transporting a set of goods between a depot and a set of customers by means of a fleet of vehicles. Vehicles travel along a road network, have a limited transport capacity and their tour always begins and ends in the depot. In the Time Windows version of the problem, additional constraints are imposed on periods of the day in which vehicles can serve customers. A solution to this problem is the best tour for every vehicle serving all customers, respecting all constraints and minimizing the number of vehicles employed and the length of each tour.

The problem can be modelled on a weighted graph, in which vertices represent clients and the depot, and edges represent road links, to which is assigned a cost.
To every customer vertex of the graph are associated:

- The quantity of goods which have to be delivered;
- The time window during which the service must take place;

• The service time, needed to deliver goods.

The depot vertex is supposed to have an infinite amount of vehicles and goods, and every vehicle has limited capacity.
The two objectives of the problem are minimizing the number of vehicles employed and the total distance covered by vehicles; the first objective is more important than the second one, since vehicles have additional maintenance costs.

1.2. Multiple Ant Colony System solution

As known, when in Nature ants are searching for food, they start by exploring the surroundings randomly. Once they find the food, they carry it back to the colony and lay a chemical substance, called pheromone, on their paths depending on the quality of the food source. The more pheromone is present, the more ants would be attracted to choose it, and this behaviour implies that pheromone will grow in best paths, since these will be followed more frequently. This paradigm was adopted for developing the Ant Colony Optimization algorithm in which virtual ants explore the search space to find the optimal solution.

The Multiple Ant Colony System (MACS) [7] is applied to this problem with two ant colonies, which simultaneously minimize the two objectives: the first colony, ACS-VEI, tries to minimize the number of vehicles, whereas the second colony, ACS-TIME, optimizes feasible solutions found by ACS-VEI [7]. The main steps are the following: initially, the algorithm identifies a feasible solution ψ^{gb} (*gb* stands for "global best") with a simple heuristic, such as nearest neighbour; the solution is then improved by the two colonies. ACS-VEI tries to find a feasible solution one vehicle shorter than ψ^{gb}, while ACT-TIME tries to optimize the total length of the solutions with the same number of vehicles. Then, ψ^{gb} is updated every time ACS-TIME finds a better solution; if ACS-VEI identifies a solution with less vehicles, the algorithm stops both colonies and reinitializes them with the new constraint on the number of vehicles.

ACS-TIME and ACS-VEI. ACS-TIME is a traditional ACS-based colony, with the objective of finding the trail as shortest as possible, with respect to constraints.

In ACS-TIME, m artificial ants are activated to build the solutions $\psi_1 \ldots \psi_m$; when the solutions are computed, they are compared with ψ^{gb} and, if a better solutions is found, it is sent to the principal procedure MASC-VRPTW and used to update ψ^{gb}. The quality of the best solution is then refined by a local search procedure, based on exchanges of sub-chains of clients in the proposed routes.

In a subsequent step, the pheromone is updated on the edges belonging to the best solution, with the formula:

$$\tau_{ij}(t+1) = (1-\rho)\cdot\tau_{ij}(t) + \rho/J_{\psi}^{gb} \quad \forall(i,j)\in\psi^{gb} \tag{1}$$

where $\tau_{ij}(t)$ is the level of pheromone on the edge (i,j) at time t, ρ is the evaporation factor and J_{ψ}^{gb} is the total length of the best solution.

ACS-VEI, on the other hand, searches for feasible solutions, maximizing the number of visited clients; it starts the computation using $v-1$ vehicles, *i. e.* one vehicle less than the shortest number of vehicles for which a feasible solution was computed.

During the search, the colony builds unfeasible solutions, in which some nodes are not visited, and keeps track of solutions which visit more nodes than $\psi^{ACS-VEI}$, the optimal solution of the preceding step. When a better solution is found, the pheromone is updated on its edges with the formula:

$$\tau_{ij}(t+1) = (1-\rho) \cdot \tau_{ij}(t) + \rho / J_{\psi}^{ACS-VEI} \quad \forall (i,j) \in \psi^{ACS-VEI} \tag{2}$$

Building of solutions. Every artificial ant leaves the depot and iteratively moves to an unvisited node, that does not violate the constraints on time windows and vehicle capacity. When ant k is located at node i, it chooses probabilistically the next node j in the set of the feasible nodes N_k^i that still have to be visited using a mechanism of solutions exploitation and exploration.

With the probability q_0 it chooses the node with the highest $\tau_{ij}(t) \cdot [\eta_{ij}]^{\beta}$, $j \in N_k^i$ (solutions exploitation), where β is a tunable parameter and η_{ij} is the attractiveness of a node, based on the travelling time between nodes i and j, on the time window associated to node j and on the number of times in which node j has not been inserted in a solution.

With the probability $(1 - q_0)$ the ant chooses node j (solutions exploration) with a probability p_{ij}, defined as:

$$p_{ij} = \begin{cases} \frac{\tau_{ij}(t) \cdot [\eta_{ij}]^{\beta}}{\sum_{l \in N_k^i} \tau_{il}(t) \cdot [\eta_{il}]^{\beta}} & \text{if } j \in N_k^i \\ 0 & \text{otherwise} \end{cases} \tag{3}$$

Parameter β weights the relative importance of the heuristic value η_{ij}, while q_0 determines the relative importance of exploitation of previously identified routes versus the exploration of new, promising routes.

When an ant moves from node i to node j, another update of the pheromone is performed on the edge (i, j).

At the end of the building phase the solution could be incomplete, because some nodes may be omitted; the solution is then iteratively completed with the insertion of the remaining nodes, sorted in decreasing order of quantity of goods to deliver and inserted in the routes which allow the shortest travel time.

2. Ant Colony Optimization and the Pick-up and Delivery Problem

In this section we firstly define the Pick-up and Delivery Problem and then we describe our extension of the MACS approach to the problem.

2.1. Problem formulation

Pick-up and Delivery Problem has been formally defined by Salversbergh and Sol [13] in 1995; the main difference between PDP and VRP is that in the former every node is bound to another, which is its source, or its destination, for an exchange of resources that has to be realized by a fleet of vehicles.

To every vertex i of the graph are then alternatively assigned two parameters, S_i and D_i, characterized as follows: goods requested by vertex i must be previously collected from client S_i by the vehicle serving i, while goods collected from i must be delivered to client D_i, which must then be visited after i.

Therefore, the problem consists in determining K minimal cost routes such that:

• every tour visits the depot;
• every client is visited by one and only one tour;
• vehicles load, in every point of the tour, is not negative and must not exceed the total capacity C;
• for every client i, the vertex S_i, if different from the depot, must be visited in the same tour and before i;
• for every client i, the vertex D_i, if different from the depot, must be visited in the same tour and after i.

In the Time Windows version of the problem, clients and the depot have a limited period of time in which they have to be served.

2.2. The proposed solution

We tried to adapt to this problem the approach MACS-VRPTW; we maintained the same architecture of two active procedures: ACS-VEI, which tries to minimize the number of employed vehicles, and ACS-TIME, which tries to minimize the length of the solution.

The formulation of PDP poses some new constraints on the construction of the solution: source/destination couples must belong to the same tour and must be visited in the right order. The problem that emerges is to be able to serve every destination node which remains unserved as the vehicle visits the source nodes during its tour; the temporal constraint of returning to the depot within its time window becomes more and more crucial.

Examining the two situations, it can be seen that: destination nodes have to fulfill only time windows constraints, because during the service the load is lightened. Source nodes instead have to respect also capacity constraints, in addition to the other constraints, and there must be enough time to serve the corresponding destinations.

To deal with the previous problems, we designed a set of enhancements for the algorithm that have been included in the ACO algorithm. They are described in the following.

Pheromone post treatment The pheromone post treatment is a new strategy to heuristically guide the choices of ants towards more promising solutions. The idea is to slightly influence the probabilistic rule for choosing new nodes in the ant trail, biasing the probabilities with the information on the nature of the node.

In practice, in every probability equation the term $\tau_{ij}(t) \cdot [\eta_{ij}]^\beta$ is weighted by a coefficient $\gamma_{source/dest}$ depending on the type of the node j (source or destination); its value will be estimated by experiments, but a reasonable hypothesis is that destination nodes, always feasible, will be more preferable than source nodes.

As will be shown in the experimental results, pheromone post treatment is a good strategy for dealing with the priority problems posed by the two different natures of nodes: the selection of correct routes emerges from the colony behaviour and more complex selection procedures are not necessary.

Initial Heuristic The heuristic for finding the first feasible solution and the unfeasible solution shorter than one vehicle for ACS-VEI generates routes as follows: every destination node is placed in the tour after its source node, after a check on the remaining time of the depot time window. It terminates when every node is visited (in the first case) or when the number of vehicles is equal to that of the best solution minus one (ACS-VEI).

Insertion of remaining nodes The strategy for completing solutions proposed by ant colonies with the insertion of remaining nodes is composed by two methods. The first method tries to insert a destination node in a tour where its corresponding source is present, in a position subsequent to the source and respecting time constraints. The second tries to insert a pair source-destination which is not present in the routes: every tour is checked and, if there is enough time to insert both source and destination, the source is inserted and the previous method is called for the insertion of the destination.

Local search The local search is performed on every single tour, because of the constraint of sources-destinations coupling in tours. Every node swap in the tour is tested, with respect to temporal, capacity and precedence constraints; if a solution with shorter length of the tour is found, the old solution is substituted.

3. Results

The proposed algorithm has been tested on the set of benchmarks problems created by Li and Lin [8], which is composed by 56 instances of 100 nodes. The algorithm is coded in C++ and the runs were performed on a Intel Pentium IV 2.40 GHz.

Parameters of the algorithm were set to

- $m = 10$ (number of ants),
- $q_0 = 0.9$ (greediness coefficient),
- $\rho = 0.1$ (evaporation coefficient),
- $\beta = 0.8$ (exponent of the heuristic estimate for nodes attractiveness),

after a simulation phase, in which we varied each coefficient around the values suggested by Gambardella *et al.* [7] for VRPTW.

The number of ants may seem quite limited, with respect to the large number of individuals found in real ant colonies, but it is perfectly in line with other applications of the ACO approach; a larger set of ants becomes computationally expensive to be managed, thus worsening the performance of the algorithm.

Different techniques and heuristics have been considered for guiding the selection of source nodes by ants, taking into account the service time of corresponding destination nodes; anyway, the simplest strategy, that does not restrict the set of source nodes to be chosen, has proven to be the most effective. This is a further evidence that in swarm intelligence it is often better to leave the collective intelligent behaviour to emerge, rather than guide the search through complex procedures.

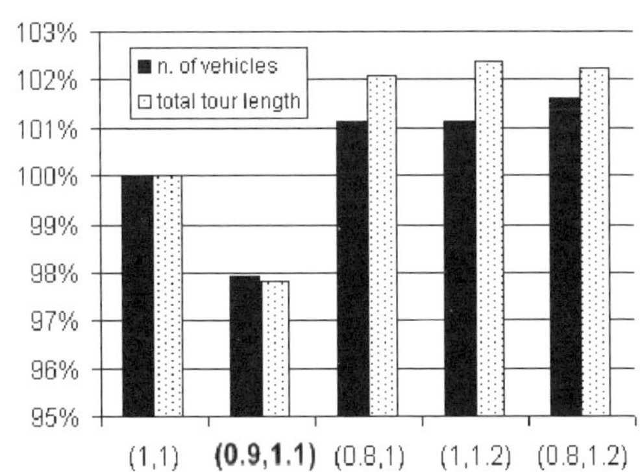

Figure 1.percentage variations of CNV and CTD for different values of $(\gamma_{source}, \gamma_{dest})$ w.r.t. the algorithm performances without post pheromone treatment.

The original algorithms by Li & Lim, Bent & Van Hentenryck and Ropke & Pisinger were executed on a i686 Linux kernel, an AMD 1.2 Ghz and a Pentium 4 1.5 GHz respectively. Despite the differences in computational platforms, performance of our algorithm is comparable with the others: best results in literature differ from our results only by a 2% on the total number of vehicles (CNV) and a 6% on the total length (CTD). We were able to find the optimum path in 24 instances over 56, and this can be considered quite a good result, taking into account the incomplete nature of this type of algorithms.

Furthermore, we tested the algorithm on different combinations of γ_{source} and γ_{dest}, to find the optimal values for Pheromone Post Treatment; we found that the best combination is $\gamma_{source} = 0.9$ and $\gamma_{dest} = 1.1$, which assigns a slightly higher probability to destination nodes w.r.t. source nodes (Figure 1) as had been foreseen in Section 2.2.

To further estimate the effectiveness of Pheromone Post Treatment and of the Local Search Procedure, we run our algorithm with and without these two techniques. Figure 2 shows that the addition of the two methods to the simple ACO algorithm leads to better results for both the number of vehicles and the total tour length.

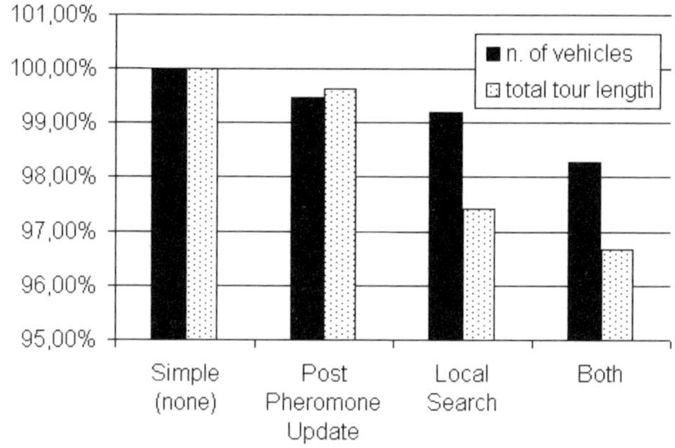

Figure 2. Average number of vehicles (black bars) and total length (gray bars) when adding Pheromone Post Treatment, Local Search and both w.r.t. the simple ACO algorithm.

4. Conclusions

We proposed an application of the Ant Colony Optimization method to the Pick-up and Delivery Problem with Time Windows. This problem has already been dealt with ACO but only in special cases: Doerner et al. applied the ACO methodology to a problem in which only full truckloads have to be considered [4] and in Rizzoli et al. [11] all pick-ups of a tour must take place before deliveries.

In this paper we extended the MACS-VRPTW algorithm by Gambardella et al. [7] to PDPTW and we provided some original enhancements of the algorithm, in particular a strategy for pheromone post-treatment, a novel heuristic for generating initial feasible solutions, a technique for completing partial solutions and, finally, a new method for neighbourhood generation in the local search procedure. These enhancements have been tested on a set of benchmark problems and results are promising, when compared with the state-of-the-art. The proposed algorithm is easy to implement and is able to adapt itself as new constraints arise from the problem.

As far as future directions, a first improvement concerns the refinement of the local search phase, which is fundamental in this kind of problems; for example, we intend to explore a family of heuristics, called λ-opt, which involve shuffling sets of nodes taking into account the structure of the PDPTW problem [8]. Since Ant Colony algorithms can be run continuously and adapt in real time, another promising research direction is their application to dynamic PDPs [10,9], in which the underlying graph structure and the requests change during the execution. This is a common scenario, that arises when a swarm of autonomous robots have to cooperate in a unknown environment and react promptly to external stimuli.

References

[1] S. Badaloni, P. Bison, and G. Costa. A flexible and efficient office-delivery scheduler for mobile robots. In *6th Conf. on Intelligent Autonomous Systems*, pages 825–832, 2000.

[2] S. Badaloni and M. Falda. Mobile robot scheduling using a genetic algorithm enhanced with a credit gain mechanism. In *8th Conf. on Intelligent Autonomous Systems*, pages 485–503, Amsterdam, 2004.

[3] R. Bent and Van Hentenryck. P. A two-stage hybrid algorithm for pickup and delivery vehicle routing problems with time windows. *Comput. Oper. Res.*, 33(4):875–893, 2006.

[4] K. Doerner, R. Hartl, and M. Reimann. Ant colony optimization applied to the pickup and delivery problem. Tech. rep. wp 76, Vienna University of Economics and Business Administration, Vienna, 2000.

[5] M. Dorigo and G. Di Caro. The ant colony optimization meta-heuristic. In D. Corne, M. Dorigo, and F. Glover, editors, *New Ideas in Optimization*, pages 11–32. McGraw-Hill, London, 1999.

[6] M. Dorigo, G. Di Caro, and L. M. Gambardella. Ant algorithms for discrete optimization. *Artificial Life*, 5(2):137–172, 1999.

[7] L. M. Gambardella, E. Taillard, and G. Agazzi. MACS-VRPTW: A multiple ant colony system for vehicle routing problems with time windows. In D. Corne, M. Dorigo, and F. Glover, editors, *New Ideas in Optimization*, pages 63–76. McGraw-Hill, 1999.

[8] H. Li and A. Lim. A metaheuristic for solving the pickup and delivery problem with time windows. In IEEE Computer Society, editor, *13th IEEE Internat. Conf. on Tools with Artificial Intelligence (ICTAI)*, pages 160–167, 2001.

[9] R. Montemanni, L. M. Gambardella, A. E. Rizzoli, and A. V. Donati. Ant colony system for a dynamic vehicle routing problem. *Journal of Combinatorial Optimization*, 10:327–343, 2005.

[10] H. N. Psarftis. *Vehicle Routing: Methods and Studies*. Elsevier Science Publishers B.V., 1998.

[11] A. E. Rizzoli, R. Montemanni, E. Lucibello, and L. M. Gambardella. Ant colony optimization for real-world vehicle routing problems. *Swarm Intelligence*, 1(2):135–151, 2007.

[12] S. Ropke and D. Pisinger. An adaptive large neighborhood search heuristic for the pickup and delivery problem with time windows. *Transportation Science*, 40(4):455–472, 2006.

[13] M. Savelsbergh and M. Sol. The general pickup and delivery problem. In *Transportation Science*, volume 29, pages 17–30, 1995.

[14] P. Shaw. Using constraint programming and local search methods to solve vehicle routing problems. *LNCS*, 1520:417–430, 1998.

[15] P. Toth and D. Vigo. *The Vehicle Routing Problems*. Monographs on Discrete Mathematics and Applications, 2002.

Intelligent Autonomous Systems 10
W. Burgard et al. (Eds.)
IOS Press, 2008

doi:10.3233/978-1-58603-887-8-272

Evaluating Robustness of Coupled Selection Equations Using Sparse Communication

Reinhard LAFRENZ [1], Michael SCHANZ, Uwe-Philipp KÄPPELER, Oliver ZWEIGLE, Hamid RAJAIE, Frank SCHREIBER and Paul LEVI
University of Stuttgart, Germany

Abstract. In this paper we evaluate the robustness of a self-organized mechanism to distribute roles among a team of cooperating robots. In previous work, we showed the principal applicability of this approach for the RoboCup scenario. Now we test the robustness in case of either intentionally sparse communication or the behavior in situations, where the communication is massively disturbed. To overcome the problems caused by this, each robot runs the equations individually, synchronizing from time to time, or when the communication is available.

Keywords. Cooperative Robotics, Self-organization, RoboCup

1. Introduction and Mathematical Foundations

The decision of the role or action for an individual robot is essential for the behavior and the performance of a team of cooperating robots. Besides finite state automata, decision trees [5], Interaction Nets which we use currently for the RoboCup tournaments [11], or combinatorial 2-index assignment strategies like the Hungarian method [4], a fully autonomous method for assignment of robots to tasks using coupled selection equations [10,8] is promising.

A 2-index assignment can be represented by a permutation matrix, where the rows represent e.g. robots and the columns represent possible actions. The coupled selection equations defined by

$$\frac{d}{dt}\xi_{ij} = \kappa\xi_{ij}\left[1 - \xi_{ij}^2 - \beta\sum_{i'\neq i}\xi_{i'j}^2 - \beta\sum_{j'\neq j}\xi_{ij'}^2\right] \qquad (1)$$

are a differential equation based method to solve 2-index assignment problems in the course of time, i.e., the ξ-matrix converges towards a permutation matrix. Hereby the ξ_{ij} describe the current assignment state of robot i to task j, whereas the parameter $\beta > 1/2$ is responsible for the strength of separation, while κ is just a scaling factor

[1]Correspondence to: R. Lafrenz, Univerität Stuttgart, IPVS, Universitätsstr. 38, 70569 Stuttgart, Germany Tel.: +49 711 7816 421; Fax: +49 711 7816 250; E-mail: robocup@informatik.uni-stuttgart.de

controlling the speed of the dynamics. The origin of this approach is a self-organization mechanism adopted from nature [7]. The asymptotic solution of this system of equations is determined by the initial values of the dynamics variables ξ_{ij} in a winner takes it all manner.

In the case of $n \times n$-square matrices their stable asymptotic solutions are permutation matrices. In case of superfluous robots or tasks the $n \times m$ matrices contain additional rows or columns with zero components. Note that this approach can be easily extended to higher-index assignment problems [1].

In realistic applications (like for example the RoboCup-scenario), any assignment strategy has to cope with a highly dynamic environment. In the case of the coupled selection equations, one can either re-solve the problem with different initial values adapted to the current situation or adapt the equations themselves. However re-solving is not an appropriate choice considering the background of self-organization. Therefore the second possibility is chosen here.

In order to use scene information based on sensory input for the control of the assignment and hence to be able to adapt to the dynamic changes, additional parameters are introduced in [9]:

$$\frac{d}{dt}\xi_{ij} = \kappa\,\xi_{ij}\left(\alpha_{ij}\left(\lambda_{ij} - {\xi_{ij}}^2\right) + 2\,\beta\,{\xi_{ij}}^2 - \beta\left(\sum_{i'} {\xi_{i'j}}^2 + \sum_{j'} {\xi_{ij'}}^2\right)\right) \tag{2}$$

Here the situation dependent utility parameters α_{ij} are used to map the scene dynamics to the equations, whereas the activation parameters λ_{ij} are used to control the dynamic by activation or deactivation of targets or actions.

The principle applicability of the extended coupled selection equations is shown in [6].

2. The Software Framework

The software framework is an extension of the successful original RoboCup software of the CoPS (Cooperative soccer playing robots Stuttgart) team. Only few modifications were necessary to use the self-organized role assignment components instead of the usual one [7].

The general software structure of the CoPS framework is shown in Fig. 1

The framework consists of three major components: First, sensor-data acquisition, secondly world modeling with data processing and interpretation on several levels, and thirdly decision making and action execution. With the CoPS framework it is not only possible to control a team of real robots, but also to use a simulator (see [3]), both for sensor data processing and for behavior simulation only. The present paper deals only with simulations, although we will run the system on real robots after some more simulations to figure out parametrization.

The sensor data acquisition provides pre-processed information. The raw sensory data is filtered, pre-processed and in some cases also interpreted before the information at a higher level of abstraction is written into the world model. Such a processing can range from simple collecting distance information with ultrasonic sensors to a complex image processing hierarchy.

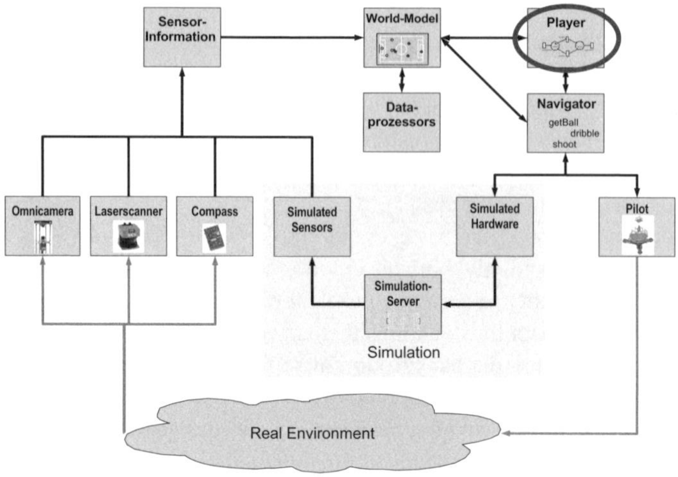

Figure 1. Software architecture of the CoPS framework

The world model stores the data from the sensors and runs data processors, which make the further calculations either periodic with a minimal cycle time or event (data-flow) triggered. These data processors write again into the world model, so that a complete data-driven evaluation cascade can be configured. The mechanisms used and be benefit of the configurability is described in [2].

The top-right component (circled) in Fig. 1, the *player*, is substituted by a decider which uses the above described extended coupled selection equations. The input for calculating the utility parameters α_{ij} of the coupled selection equations is the world-model with its data, here mostly geometrical relations.

3. Experimental Results

In former work, the stability of the system relied on a more or less stable communication. In the present work, we performed first experiments with unreliable communication. To achieve the intended robustness, the system must be able to assign the tasks (temporarily) without communication. In the following, we describe experiments where the communication is only used for the simulation infrastructure data, but the robots do not exchange information about theξ-matrices as in previous work. Also, only local sensory information is used.

In the following, an example scenario is shown, where the communication between two robots is broken since a longer time range (several seconds). For the sake of presentability we use only two robots [2] and five different possible tasks. One task is getting the ball, one is dribbling, where the remaining tasks correspond to fixed strategic positions, in the example the penalty spot and the front corners of the penalty area. The

[2]the matrices own a third row for a spare robot in case of hardware-faults, which is not topic of the present paper

α_{ij} represent an geometry-based utility factor, basically the distance of the i-th robot to the position required to perform task j. The guiding question for the experiments was whether the world model, with a current situation based on local sensor data and the resulting α_{ij} is able to cope with inconsistencies between the assignment decisions of two robots.

Due to the lack of communication and the environmental dynamics, the individual calculation of the assignment matrices ξ_{ij} on each robot diverges after a while. One of the reasons for that is the different sensory information, which stems from different points of view, different noise and different times of the data acquisition.

The figures Fig. 2 to Fig. 5 show three successive situations within one second. The independently calculated matrices of both robots ((a),(b)) and the differences between them ((c) are shown in each case.

In the first situation (Fig. 3), the assignment of robot 2 (shown in the middle row of the matrices) is the same for both calculating robots, whereas robot 1 is differently assigned. This can easily seen in the difference matrix Fig. 3(c).

In the second situation (Fig. 4), the robots have moved towards the target position according to the task assigned in the first situation. After some sensing and processing cycles, the selection equations got too much different input from the local sensors of each robot, so that the relevant tasks are clearly figured out, but the assignment is swapped in the two ξ_{ij} matrices. As a result, both robots try to get the ball, which corresponds to the first task and the leftmost columns of the assignment matrices. Consequently, the difference matrix Fig. 4(c) shows high values in the same columns for both robots. In the meantime, before the third situation, both robots move towards the ball, permanently updating their world model with the local sensor information.

By getting closer to the ball, the used (simulated) camera sensor has less noise and therefore the sensor date become more and more accurate. This results in a better distance estimation for the ball and the other robot, which in turn influences the utility parameters α_{ij}. Finally, in the third situation (Fig. 5), the assignment was corrected by the coupled selection equations and therefor the difference matrix in Fig. 5(c) has only small component values. Summarizing, the distributed system is able to recover from a faulty assignment even in the case of communication breakdown. If the communication is re-established, matrix elements are communicated and the assignment will be consistent immediately.

4. Conclusion and Outlook

We could show that even in case of complete breakdown of the communication channel, the system is able to handle the assignment of robots to tasks. For short times there may be double assignments on the one hand and unassigned robots or tasks on the other hand, but the extended coupled selection equations are able to recover from such a situation quickly. The utility parameters α_{ij}, which are calculated by each robot interpreting local sensory information only, lead to the resolution of a wrong assignment.

In further steps, we want to evaluate systematically the optimal parametrization using sparse communication and avoiding faulty assignments at the same time. One of the next steps will be to run the system on our real hardware, which is relatively easy because of the software architecture framework described in Section 2.

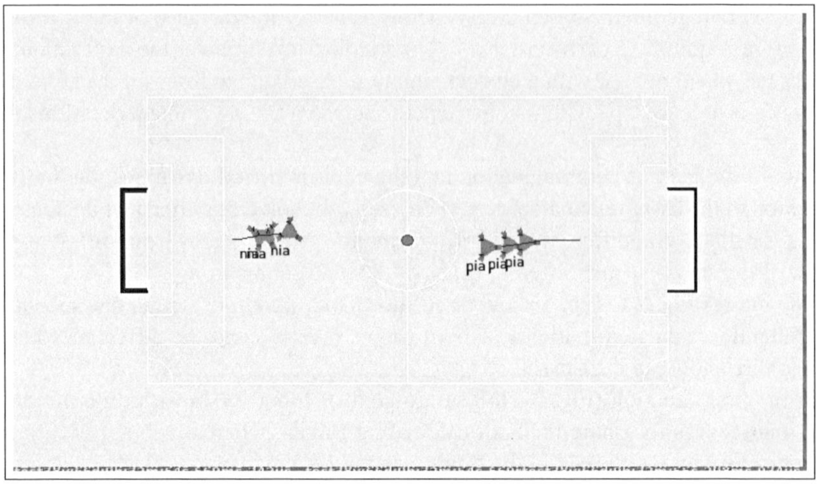

Figure 2. Three consecutive situations in the example. Due to communication breakdown the robots propose different assignments for robot 1. After short time the assignment is re-established. The right robot (pia) is moving from right to left, the other robot is moving forth during the period of wrong assignment and back after the correction.

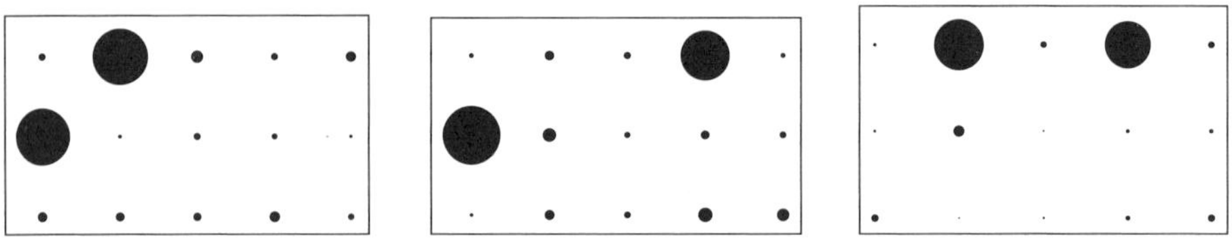

(a) (b) (c)

Figure 3. First situation in the example. Due to communication breakdown the robots propose different assignments for robot 1. (a) shows the ξ-matrix of the first robot, where the rows correspond to the robots (only 2 robots are used in the example) and the columns stand for the different possible actions. (b) shows the ξ-matrix of the second robot, and (c) the differences of the matrix entries of the calculations of robot 1 and 2.

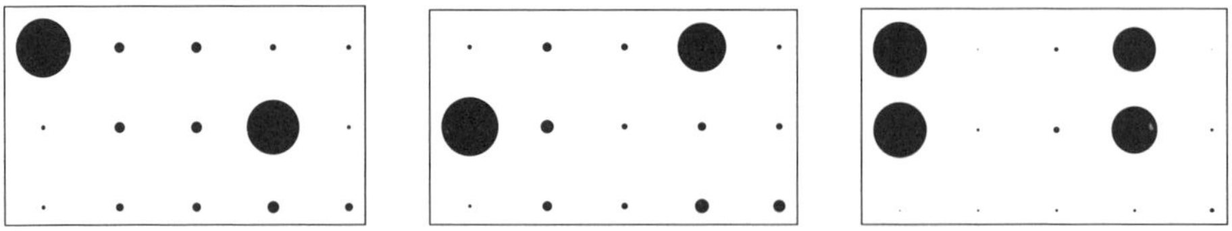

(a) (b) (c)

Figure 4. Second situation: The robots both commit to the same task, which can be easily seen in the difference matrix (c).

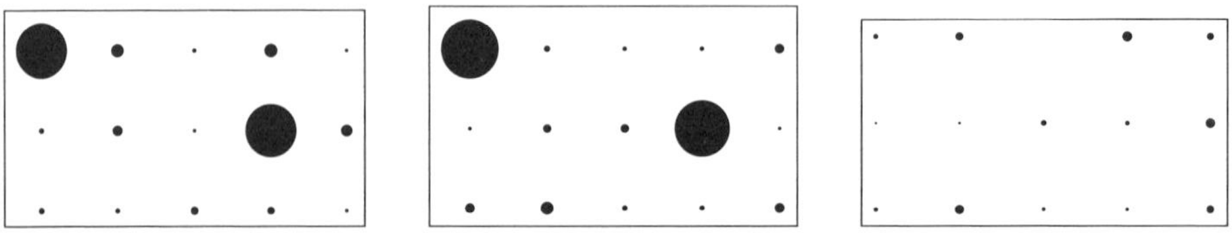

(a) (b) (c)

Figure 5. The system has recovered from the wrong assignment. There are no significant differences between the ξ_{ij} of both robots.

References

[1] M. Becht, T. Buchheim, P. Burger, G. Hetzel, G. Kindermann, R. Lafrenz, N. Oswald, M. Schanz, M. Schulé, P. Molnár, J. Starke, and P. Levi. Three-index assignment of robots to targets: An experimental verification. In E. Pagello, editor, *6th International Conference of Intelligent Autonomous System, IAS-6*.

[2] T. Buchheim, G. Kindermann, and R. Lafrenz. A dynamic environment modelling framework for selective attention. In *IJCAI Workshop: Issues in Designing Physical Agents for Dynamic Real-Time Environments: World modeling, planning, learning, and communicating*. IJCAI, 2003.

[3] A. Kleiner and T. Buchheim. A plugin based architecture for simulation in the f2000 league. 2003. RoboCup Intl. Symposium, Padua.

[4] Harold W. Kuhn. The Hungarian method for the assignment problem. *Naval Research Logistics Quarterly*, 2(1):83–98, 1955.

[5] M. Lötzsch, J. Bach, H.-D. Burkhard, and M. Jüngel. Designing agent behavior with the extensible agent behavior specification language xabsl. In *7th International Workshop on RoboCup 2003 (Robot World Cup Soccer Games and Conferences*, 2003.

[6] M. Schanz, J. Starke, R. Lafrenz, O. Zweigle, M. Oubbati, H. Rajaie, F. Schreiber, T. Buchheim, U.-P. Käppeler, and P. Levi. Dynamic Task Assignment in a Team of Agents. In P. Levi et al., editor, *Autonome Mobile Systeme*, pages 11–18. Springer, 2005.

[7] J. Starke. *Kombinatorische Optimierung auf der Basis gekoppelter Selektionsgleichungen*. PhD thesis, Universität Stuttgart, Verlag Shaker, Aachen, 1997.

[8] J. Starke. Dynamical assignments of distributed autonomous robotic systems to manufacturing targets considering environmental feedbacks. In *Proceedings of the 17th IEEE International Symposium on Intelligent Control (ISIC'02)*, pages 678 – 683, Vancouver, 2002.

[9] J. Starke, C. Ellsässer, and Fukuda. Self-organized control in cooperative robots using pattern formation principles. 2005.

[10] J. Starke and M. Schanz. Dynamical system approaches to combinatorial optimization. In D.-Z. Du and P. Pardalos, editors, *Handbook of Combinatorial Optimization*, volume 2, pages 471 – 524. Kluwer Academic Publisher, Dordrecht, Boston, London, 1998.

[11] Oliver Zweigle, Reinhard Lafrenz, Thorsten Buchheim, Hamid Rajaie, Frank Schreiber, and Paul Levi. Cooperative agent behavior based on special interaction nets. In *Proc. Intelligent Autonomous Systems 9*, 2006.

Intelligent Autonomous Systems 10
W. Burgard et al. (Eds.)
IOS Press, 2008

doi:10.3233/978-1-58603-887-8-278

On the Robustness of Visual Homing under Landmark Uncertainty

Derik Schroeter and Paul Newman
Robotics Research Group, Dept. Engineering Science,
Oxford University, Parks Road, Oxford, OX1 3PJ, United Kingdom
{ds,pnewman}@robots.ox.ac.uk

Abstract. Although a large body of literature exists on methods for landmark-based navigation of mobile robots, little has been said about their robustness in the presence of changing and wrong landmark correspondences as well as erroneous compass measurements. To this end, we investigate the practical implications of different compass-dependent approaches to *2D Bearing-Only Navigation (2D BON)* under such conditions, and we suggest measures to quantify their robustness. The analysis is based on the notion of the *Homing Vector Field (HVF)*, and carried out in simulation, since complete HVFs can not be determined otherwise. However, HVFs provide a generic framework that enables the sound comparison of different 2D BON methods. In addition, we propose a method that is based on the approximation of the gradient of an implicit objective function, which constitutes a gradient descent approach. We present practical results that show the applicability of our approach using natural visual landmarks, omni-directional vision and a compass.

Keywords. visual homing, topological navigation, omnivision

1. Introduction

Enabling mobile platforms to autonomously navigate around their environment is still one of the major challenges in mobile robotics. Although much progress has been made for constrained settings, indoors [4] and outdoors [13,23], the problem remains widely unsolved for more complex scenarios. It is an open question, for example, what kind of perceptual clues are adequate for local navigation, i.e. when moving from one place to another along a given path. For many indoor environments, 2D laser range data seems to be sufficient, although additional sensors are necessary for robust obstacle detection [6]. In outdoors, position measurements from GPS have been successfully utilised together with image and laser range data [13,23]. However, relying primarily on GPS is delicate, since it is not always available, and multi-path GPS signals cause wrong measurements with high confidence. This is particularly true for urban environments, where in addition, the low precision of standard GPS impedes navigation in confined places.

As a consequence, much effort has been put into developing algorithms that solely rely on image data from cameras, e.g. [2,5,8,10,14,15,16]. A common approach, also deployed in the work presented here, is to associate locations in the environment with place descriptors. Based on correspondences between the current and a reference view, the task is to enable the robot to move towards the reference place. For mobile robot navigation the positioning problem is commonly considered to be 2D. That means, the related algorithms determine 2D directions, which when followed, move the robot towards a reference place [2,10,12,15]. A large body of literature covers the interaction of perception and control as well as the stability of the respective control laws [10,18,24]. In [9] Kak gives a comprehensive survey of vision-based navigation and map building. However, a detailed discussion of motion control is beyond the scope of this paper. Here we assume a control algorithm that adjusts the orientation of a mobile robot given a homing direction, and that under perfect circumstances it would converge for all the considered methods[1]. The robot moves with a constant translational velocity.

Our work is motivated by the idea of non-metric topological navigation, that is, enabling mobile robots to perform place traversal in dynamic urban environments using visual clues. Assuming distances between places to be larger than a metre, induces the problem of wide baseline matching, which causes landmarks to be unreliable and time-varying. In addition, dynamic objects frequently cause landmarks to be occluded, which together makes establishing large numbers of reliable landmark correspondences a difficult task. To this end, we investigate approaches to *2D Bearing-Only Navigation*, which potentially work with very few (<20) and changing landmark correspondences. The respective algorithms use only the 2D bearings of landmarks and a compass. They were mainly inspired by studies of insect behaviour, e.g. bees [7,21] and ants [19]. In a seminal paper, Cartwright and Collett [7] proposed the "snapshot" model, where a single-line 360 degree view describes the reference pose. Different schemes have been investigated as to how the two views can be compared to give the homing direction, e.g. [10,12,15,20,22]. In [17] Mallot and Franz give a comprehensive survey.

A large body of established literature deals with the robust estimation of the relative pose between two or more cameras by means of multiple view geometry [11]. Some recent approaches utilise the epipolar geometry between the current and the reference view either by means of the position of the epipoles in the images [16] or by determining the relative pose directly [14]. However, our empirical experience was that to obtain the performance we sought, geometric methods required more reliable view-to-view correspondences than frequently available. We decided to investigate, if we could replace or augment our visual geometry implementation with something that entirely eschewed the metric point of view and demanded fewer stable multi-view feature correspondences.

In this work we present a comparison of different compass dependent approaches to *2D Bearing-Only Navigation* with respect to realistic settings (Sec. 4). This is to investigate the practical implications of such methods, and we suggest measures to quantify their robustness. The analysis is based on the notion of the *Homing Vector Field (HVF)*, which can only be fully evaluated in simulation. However, to the best of our knowledge, such an analysis has not been published before. A brief survey of related methods is given in Sec. 2. In addition, we propose a method that is based on the approximation of the gradient of an implicit objective function (Sec. 3). We present preliminary practical results that show the applicability of our approach in Sec. 5. Finally, Sec. 6 summarises the conclusions and discusses directions of future work.

2. Brief Survey of Methods for 2D-Bearing-Only Navigation

Following the navigation hierarchy as presented by Mallot and Franz [17], only methods of the category *Guidance* are considered in this work. It is important to note at this point, that although their survey is impressively comprehensive, its focus is on the implementation of biologically inspired methods rather than their robustness. In the majority of cases, implementations on mobile robots were used to show the plausibility of the suggested models using artificial landmarks [7,10,15]. Little has been said about the robustness of these approaches in the presence of changing correspondences due to occlusions and dynamic changes, wrong correspondences as well as compass measurement errors. However, if we want to apply these methods in real scenarios, where a robot moves autonomously in an urban environment, robustness in the above sense is a crucial measure for the applicability of any of the given approaches.

[1]Note that stability proofs exist for several such control algorithms and homing behaviours under perfect conditions, see for example [10,18,24]. Our focus, however, is the robustness under landmark uncertainty.

Bekris etal. [2] proposed a purely geometric approach and presented results using a real robot in indoor environments. Although the method is attractive because of not requiring a compass, the paper suggests that it would not work in the presence of changing or wrong correspondences. Other approaches incorporate the size of landmarks, e.g. [3], which assume that objects are approximately cylindrical and their left and right boundaries can be detected. However, these methods are not applicable to the scenario investigated in this paper, which considers 2D point landmarks. Chen et al. [8], propose a qualitative navigation scheme following the teach-and-replay paradigm. Since this approach does not explicitly generate a homing direction, it does not fit into the class of methods investigated here. The following approaches to *2D Bearing-Only Navigation* will be considered in this work:

Online-ALV. The *Average Landmark Vector* (ALV) [15] is computed from the landmark vectors v^i_{lm} that point from the robot pose to landmark i. Given N landmarks, the ALV a is defined as the averaged sum over all landmark vectors: $a = \frac{1}{N}\sum_{i=1}^{N} v^i_{lm}$. The only value stored to describe the reference pose p_{ref} is the respective a_{ref}. For new poses p_{new} in the vicinity of p_{ref} the corresponding a_{new} is computed as above. The homing direction v_{hom} is then given as the difference of the two ALVs: $v_{hom} = a_{new} - a_{ref}$. The ALV model is based on the notion that all landmarks that were visible from p_{ref} are also visible from p_{new}, which implies perfect landmark matching. Not surprisingly, the classic ALV approach performs poorly or not at all in the presence of changing or wrong correspondences. Therefore it is excluded from the following comparison, and a modified online-version of this approach is suggested and used instead. For *Online-ALV* all landmark vectors (or their bearings) are stored together with descriptors to represent p_{ref}. The descriptors are used to establish landmark correspondences between p_{ref} and p_{new}. The homing vector is determined as before, except that both a_{ref} and a_{new} are explicitly computed using only the current correspondences.

Predictive Search Based on the Equal-Distance-Assumption - PredSearchEDA [10]. Assuming that all landmarks have the same distance to p_{new}, new bearings as observed from a pose close to p_{new} can be predicted. This prediction is then compared to the bearings stored for p_{ref}. Practically this means, that the range of possible directions, i.e. 0-360 degree is discretised, and bearings are predicted for small displacement (10-40 cm) in those directions. The sum of bearing differences for landmark correspondences (or the sum of squares thereof) gives a measure of how well the predicted views match with the reference view. The direction that minimises this measure gives the homing vector.

3. Visual Homing using the Direct Gradient Approximation

Considering Fig. 1 (left), the angles α^i_T denote the bearings for a particular landmark i as observed from the reference or target position (subscript T). Accordingly, α^i_R refers to the bearing for the same landmark viewed from the current robot position. Given a particular landmark configuration and target pose, the bearings α^i_T are constant. The goal is to move a robot such that the bearing differences of corresponding landmarks are minimised. Consequently, a natural choice for the respective objective function is the sum of the squared differences between these bearings for all N landmarks:

$$F(\alpha^i_R, i = 1..N) = \sum_{i=1}^{N} (\alpha^i_T - \alpha^i_R)^2 \tag{1}$$

Since in practice this function can only be evaluated for the current position, it can be seen as an implicit objective function that is underlying the problem of *2D-Bearing-Only Navigation* in its above formulation. Fig. 1 (right) depicts an example. Assuming that F is convex within the vicinity of the reference pose p_{ref}, following the gradient of F leads the robot towards p_{ref}. Let (x_R, y_R) and (x_L^i, y_L^i) be the positions of the robot and the i'th landmark, respectively. Then α_R^i can be determined as follows:

$$\alpha_R^i = \arctan(\frac{y_L^i - y_R}{x_L^i - x_R}) \quad (x_L^i, y_L^i) = const, i = 1..N$$

Thus, for a fixed scenario, the implicit objective function F in Cartesian coordinates only depends on the position of the robot:

$$F(x_R, y_R) = \sum_{i=1}^{N} \left[\alpha_T^i - \arctan\left(\frac{y_L^i - y_R}{x_L^i - x_R} \right) \right]^2 \tag{2}$$

The Jacobian of the objective function $F(x_R, y_R)$ from Eq. 2 is given by:

$$\begin{pmatrix} \frac{\partial F(x_R, y_R)}{\partial x_R} \\ \frac{\partial F(x_R, y_R)}{\partial y_R} \end{pmatrix} = f(\alpha_R^i, i = 1..N) = \sum_{i=1}^{N} \left[\frac{2\Delta\alpha^i}{d_{RL}^i} \cdot \begin{pmatrix} -\sin(\alpha_R^i) \\ \cos(\alpha_R^i) \end{pmatrix} \right] \quad \Delta\alpha^i = \alpha_T^i - \alpha_R^i \tag{3}$$

Since the distances d_{RL}^i between the current and the reference pose are unknown, the gradient of F at a certain position (x_R, y_R) can only be approximated. A common approach [10], also applied here, is to assume that all landmarks have the same distance to the current position, i.e. $d_{RL}^i = const \ \forall i$. This is also referred to as the *Equal-Distance-Assumption (EDA)*. We suggest three variants of how to incorporate the angle difference term in Eq. 3:

Angle Weighting (*Gradient-AW*): Uses Eq. 3 directly, but without the $2/d_{RL}^i$ term. This approximation is the closest to the analytical gradient after applying EDA.

No Weighting (*Gradient-NW*): Instead of $\Delta\alpha^i$ only its sign is considered. The rationale is that wrong correspondences are likely to cause large angle differences, thus, biasing the overall homing vector towards a wrong direction. Note that this approximation would also arise, if the sum of squares in F is replaced with the sum of absolute differences.

Kernel Weighting (*Gradient-KW*): Applies a simple non-linear kernel to determine weights w_i from the angle differences $\Delta\alpha^i$, which are then used to weight the vector contributions. The kernel (Eq. 4) maps $\Delta\alpha^i$ to $[0, 1]$, where for $\Delta\alpha^i = k$ the resulting weight is 0.5. In our evaluation k was set to 2 degrees. This variant presents a trade-off between the other two. It still incorporates weights that are proportional to $\Delta\alpha^i$, but it limits the impact of large angle differences from wrong correspondences. The kernel function is given by:

$$w_i = \frac{(\Delta\alpha^i)^2}{k^2 + (\Delta\alpha^i)^2} \qquad \Delta\alpha^i = \alpha_T^i - \alpha_R^i \tag{4}$$

Note that the methods suggested here are very similar to *PredSearchEDA* (Sec. 2). In fact, *PredSearchEDA* presents a numerical approximation of the gradient, whereas the gradient-based approaches give a direct mathematical approximation. Both assume equal distances of landmarks to the current position. However, for *PredSearchEDA* this distance has to be specified explicitly, and should somehow relate to the real settings, e.g. the expected mean. In contrast, for the gradient-based approach the actual distances are irrelevant, if they are assumed to be the same for all landmarks, see Eq. 3. In addition, for *PredSearchEDA* a small displacement has to be specified, which defines the position of the predicted view relative to the current pose. In practice, computing the approximation of the gradient from the true Jacobian is computationally more efficient (more than factor 250) than searching for the best predicted view. In addition, the time needed for *PredSearchEDA* depends linearly on the chosen angle discretisation. However, using a one-degree resolution for *PredSearchEDA*, it can be calculated in a few milliseconds on a state-of-the-art machine.

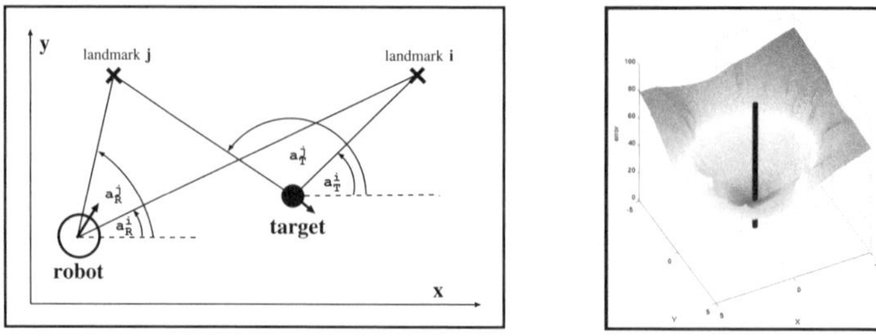

Figure 1. **Left:** The problem of *2D Bearing-Only Navigation*. **Right:** Example of an objective function (Eq. 1)

4. Comparison & Discussion Based on Simulation

All methods considered here produce a homing vector for the current position given landmark correspondences and bearings according to Sec. 2 and 3. In practice, at a specific point in time only the current homing vector can be determined. However, in a simulated environment, the positions of all landmarks are available in a global frame of reference, thus the homing vector can be computed for all possible robot poses. Note that the homing vector for the reference pose is not well defined. The next section describes the simulation setup. The results of the comparison are presented and discussed in Sec. 4.2.

4.1. Simulation Setup

Our analysis is based on the notion of the *Homing Vector Field (HVF)*, which is a discretisation of the space around the reference pose, where each position in the grid gets assigned the respective homing vector. Without loss of generality the reference pose is chosen to be the origin of the coordinate system. In realistic settings, at different, even very close, points in time vastly differing sets of landmarks can be observed, which in general causes the corresponding HVFs to differ. If these variations are small over time, despite wrong correspondences and compass error, the robot would still be able to approach the reference pose to some precision[2]. Thus, the performance of a given method strongly depends on the stability of the underlying HVFs. Fig. 2 shows a scenario where N landmarks are uniformly distributed around the robot, i.e. from 0 to 360 degree, with distances D_i sampled uniformly from a given interval (D_{min}, D_{max}). Although the landmark distributions differ significantly, the HVFs look very similar. To assess the difference between two HVFs, two landmark distributions are randomly generated, and the respective HVFs are computed on a given grid around the reference pose. The difference is measured as the normalised sum of the differences of two homing vectors, i.e. their angles, referring to the same position in the grid:

$$\Delta(hvf_1, hvf_2) = \frac{1}{N_x N_y} \cdot \sum_{i=1}^{N_x} \sum_{j=1}^{N_y} |\Theta(hvf_1(i,j)) - \Theta(hvf_2(i,j)| \tag{5}$$

Where (i, j) are indexes that refer to discrete (x, y) positions in the grid. (N_x, N_y) are the number of grid points in x- and y-direction defined by the size of the area and the resolution. The values of Eq. 5 present the average angular differences of the two HVFs. The angle-function Θ is simply given by:

$$\Theta(hvf(i,j)) = \arctan\left(\frac{v_y(i,j)}{v_x(i,j)}\right) \tag{6}$$

[2]We assume here that a robust control algorithm exists for the case of perfect conditions.

(v_x, v_y) are the components of the respective homing vectors. In addition, each HVF has been compared to an "ideal" HVF using the above measure. Here "ideal" means that all homing vectors point directly towards the reference pose. If in this case the distribution of difference values for a particular setting is well below 90 degree, following any of the homing vectors will move the robot closer to the reference pose. In other words, this measure relates to the convexity of a given HVF. A third measure considers the distribution of target poses as determined from a single HVF for different settings. In particular, we consider the distance of the deduced target pose to the ground truth. The number of landmarks is fixed to 25. While this measure gives additional indications of the convexity of the HVFs, it also shows to which precision the target pose can potentially be reached. For the results presented here, the grid size is 10x10 metres with a resolution of 0.2 metres, thus, N_x=N_y=51. The distance interval for the landmarks is: D_{min}=5 metres, D_{max}=20 metres. 400 Simulations have been carried out for all combinations of: number of landmarks = {5,10,25,50,75,100}, variance of compass noise = {0,2,4,7,10,13,16} degree, percentage of wrong correspondences = {0,10,15,20,25,30,40,50} percent.

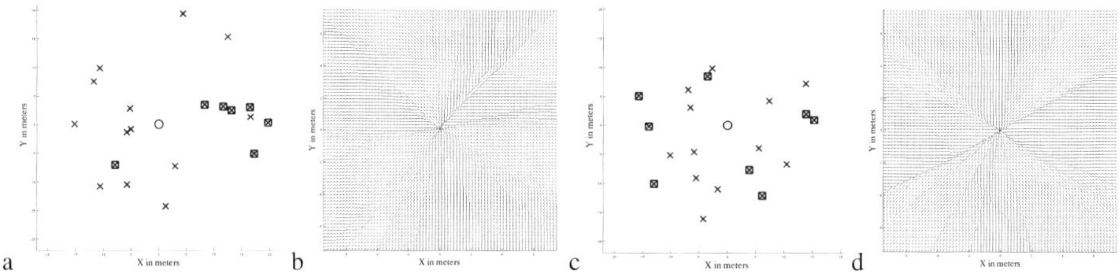

Figure 2. Two example scenarios **(a,b)** and **(c,d)** using the *Gradient-NW* method. 20 landmarks, 30 percent wrong correspondences. **a/c:** Landmark Distribution, **b/d:** *Homing Vector Field (HVF)*. Crosses refer to valid and square markers to wrong landmark correspondences. The circle in the centre marks the reference pose. Note that although the landmark distributions differ significantly, the HVFs look very similar.

4.2. Comparison and Discussion

In case of perfect conditions, i.e. static and correct correspondences and no compass error, all considered methods perform equally well. That means, they potentially steer the robot arbitrarily close to the reference pose. However, in practice, wrong feature correspondences and erroneous compass measurements can not be entirely avoided. The y-axis in Fig. 3 and 4 refers to the mean of the average angular difference in degree of two HVFs computed from two different random landmark distributions and the comparison of one such HVF to an "ideal" HVF, respectively. The vertical lines in all graphs in this section depict the standard deviation.

The first comparison considers only changing correspondence sets over time, but no compass error or wrong correspondences. Fig. 3 shows that in this case the difference between two HVFs becomes smaller with increasing number of landmark correspondences. This tendency holds true despite wrong correspondences and compass error. The comparisons with the "ideal" HVF as described in the previous section give similar results (not shown here). All angle differences are well below 90 degrees, which indicates the convexity of the underlying objective function. Together this implies that the robustness with respect to changing correspondences is comparable for all considered methods.

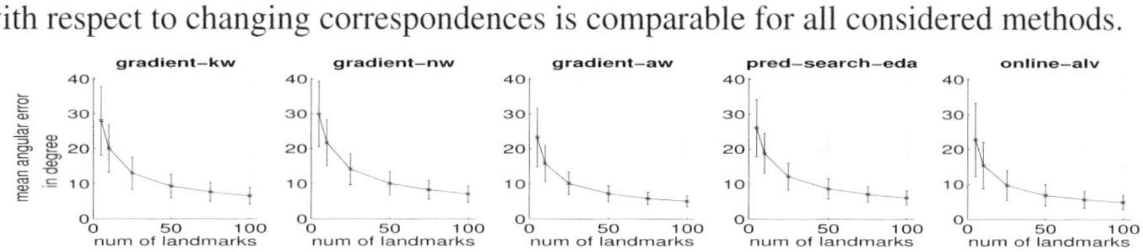

Figure 3. Pairwise comparison of HVFs for randomly generated landmark distributions - no compass error, perfect landmark correspondences.

Fig. 4-A depicts the results when introducing a zero-mean Gaussian compass error. In this case *Online-ALV* outperforms all other methods, with the *Gradient-AW* scheme being almost as good. However, the performance of the other three methods is clearly inferior in the presence of compass errors. Fig. 4-B shows the results when incorporating wrong correspondences, and assuming perfect compass measurements. It is clearly visible that the *Online-ALV* and *Gradient-AW* methods perform poorly in this case. For the latter this is due to the weighting of vector contributions per landmark pair, where some of the weights stem from wrong angle difference. The *PredSearchEDA* and *Gradient-NW* methods show less degradation in this case. For *Gradient-NW* this can be attributed to the vector contributions themselves being less affected by wrong correspondences, see Eq. eq:jac-obj-func. In fact, as long as the sign of the angle differences are correct, the respective vector contributions are perfectly valid. Note that the difference between *Gradient-KW* and *Gradient-NW* is rather small, suggesting that the kernel-weighted scheme does not provide a much better approximation of the gradient. When considering both compass error and wrong landmark correspondences, see Fig. 4-C, it becomes clear that the influence of wrong correspondences dominates the performance. In this case, *Gradient-NW* and *PredSearchEDA* outperform all other methods.

Fig. 5 shows the distributions of distances between the target pose implied by a single HVF to the true target pose. It increases with increasing compass error and percentage of wrong correspondences. Similar to the findings above the robustness is dominated by the percentage of wrong correspondences. Note that the *Gradient-NW* and *PredSearchEDA* methods show only minor degradation in the presence of wrong landmark correspondences.

5. Experimental Results

Following the results of the previous section, we have implemented and tested the *Gradient-NW* method on our research platform "Homer" (Fig. 6, left). A crucial part of such a navigation system is the robust detection and matching of points-of-interest (POIs) preferably with a 360 degree field of view. To this end we utilise an omni-directional sensor[3], the Fast-Hessian POI detector and SURF descriptors [1]. Since in our case the camera does not experience rotation around its optical axis, the rotation invariance of the SURF descriptors has been 'switched off', which leads to more robust feature correlation. In addition, as suggested in [1], the sign of the Laplacian from the POI detection is used to speed up the matching process and improve robustness. The matching is performed using cross-correlation of the descriptor vectors. The Ladybug2 sensor comprises six cameras. Five are arranged in a ring around the vertical axis, and one is looking upwards (currently not used). The navigation system runs at about 6 frames per second on an off-the-shelf PC, which equates to about 30 images (384x512) per second.

As mentioned earlier, the implemented homing behaviour provides a homing direction, which when followed, moves the robot towards the reference pose, as can be seen in Fig. 6 (right). The robot performs autonomous continuous motion using a simple controller that adapts the orientation of the robot according to the homing vector. The translational velocity is constant at about 0.2 m/s. To capture the trajectories, a 2D laser range finder was placed such that it observes the whole scenario. Afterwards, segmentation and tracking techniques have been applied to estimate the robot positions as shown in Fig. 6 (right). The resulting paths are defined by the shape of the implicit objective function F, which in turn depends on the landmark distribution. Given a large number of correct landmark correspondences, e.g. more than fifty equally distributed around the robot, the path would converge towards a straight line. In practice, however, less than twelve corre-

[3]Ladybug2 from Point Grey Research Inc.

A: variable compass error variance in degree (x-axis), perfect correspondences.

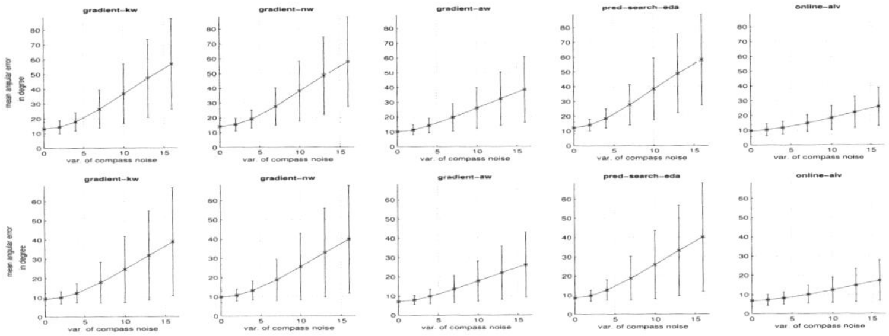

B: no compass error, variable percentage of wrong correspondences (x-axis).

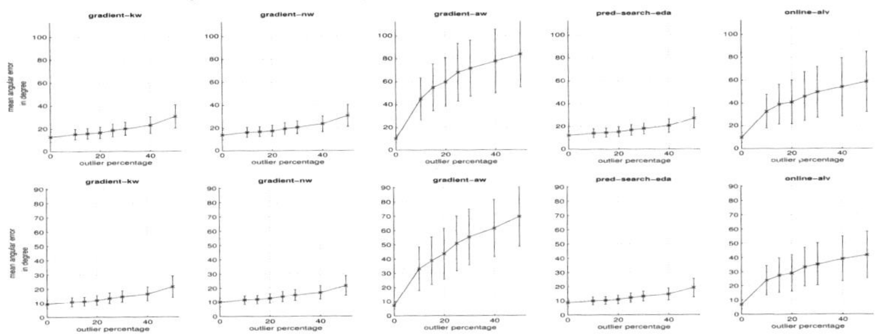

C: 10 degree compass error variance, variable percentage of wrong correspondences (x-axis).

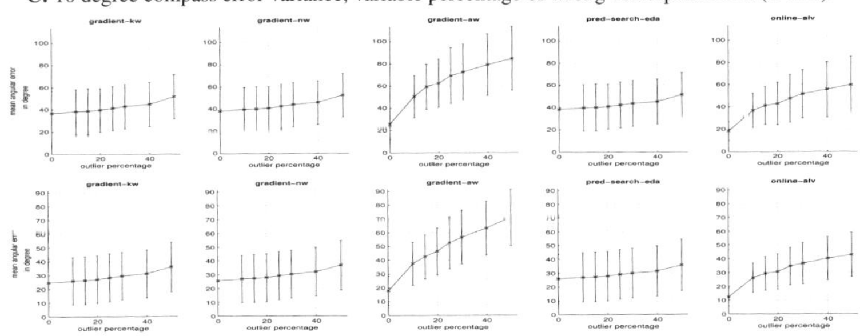

Figure 4. Comparison of HVFs for randomly generated landmark distributions (25 landmarks). The top rows for the three scenarios (A,B,C) refer to the pairwise comparison, the bottom rows to the comparison with the "ideal" HVF. The y-axis in all graphs refers to the average angular difference of the HVFs.

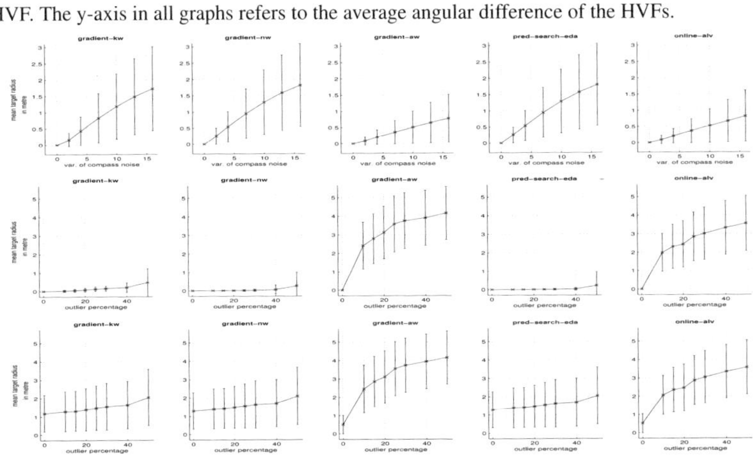

Figure 5. Distance distributions of the target pose (wrt ground truth) as determined from a single HVF (25 landmarks). **Top:** variable compass error, perfect correspondences. **Middle** and **Bottom:** variable percentage of wrong correspondences, no compass error and 10 degree compass error variance, respectively.

spondences were established in average, including varying percentages of wrong correspondences, changing correspondence sets and unsymmetric landmark distributions. As a consequence F changes over time, eventually leading the robot slightly away from the target before returning. It should be noted here that despite these difficult settings the robot successfully approaches the reference pose (centre of the pink circle) from different start poses and with different initial orientation. The maximum final distance to the reference pose is about 0.55 metre (the radius of the circle).

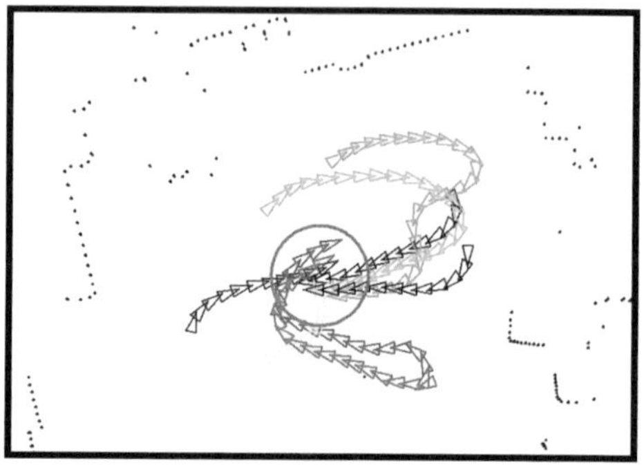

Figure 6. Left: Mobile Platform "Homer" - modified ATRV Junior from iRobot Corporation. The Ladybug2 omni-vision sensor is visible on the top. In addition, an Xsens inertial measurement unit and a 3.2 GHz Pentium 4 (1 GB RAM) form the main components of the presented navigation system. **Right:** Resulting trajectories from running our 2D Bearing-Only Navigation system (*Gradient-NW* method) on our mobile platform. The maximum final distance to the reference pose is about 0.55 metre (radius of the circle).

6. Conclusions & Future Work

In this paper we investigated the robustness of landmark-based, compass-dependent *Visual Homing* methods with respect to changing and wrong landmark correspondences as well as erroneous compass measurements. The considered algorithms derive homing directions from landmark correspondences. Our analysis is based on the notion of *Homing Vector Fields (HVF)*, which provide a generic framework for the comparison of such methods. Three measures have been proposed. (1) The pairwise comparison of HVFs reveals the stability of a given method over time. (2) The comparison with an 'ideal' HVF indicates the degree of convexity that can be expected. (3) The target distribution relates to convexity as well as reachability. Our analysis showed that the robustness of all methods is dominated by the presence and amount of wrong correspondences. While the *Online-ALV* and *Gradient-AW* methods are more robust, if only erroneous compass measurements are present, their performance degrades drastically in the presence of wrong correspondences. *PredSearchEDA* and *Gradient-NW* clearly outperform the other methods in the context of both wrong correspondences and compass error. In comparison to *PredSearchEDA*, the gradient approximations proposed in this paper are based on less strict assumptions. They are simpler to implement and faster to compute. Experiments on our mobile platform show the applicability of our approach using natural visual landmarks, omni-directional vision and a compass.

Given the results of the comparison, it is evident that keeping the compass error within certain bounds, e.g. by means of state estimation techniques, would improve the robustness of all the considered approaches. The same holds for the number of wrong correspondences, which requires elaborate schemes for outlier rejection and robust feature matching. A different route of continuing research is to devise methods that exhibit the same generality and robustness as the approaches considered here, yet are entirely independent of compass measurements. Our main focus, however, is to embed these methods into a non-metric place traversal scheme that effectively constitutes appearance-based topological navigation. Eventually, the homing algorithms are at some point merely used to provide navigational clues that are to be combined with constraints from collision avoidance and dedicated navigation behaviours, e.g. path following.

References

[1] H. Bay, T. Tuytelaars, and L. V. Gool. SURF: Speeded Up Robust Features. In *Proc. of the 9th European Conference on Computer Vision (ECCV)*, 2006.

[2] K. Bekris, A. Argyros, and L. Kavraki. Angle-Based Methods for Mobile Robot Navigation:Reaching the Entire Plane. In *Proc. of IEEE Intl. Conf. on Robotics and Automation (ICRA)*, 2004.

[3] J. Bonora and D. Gallardo. Visual Homing Navigation With Two Landmarks: The Balanced Proportional Triangulation Method. In *Proc. of IEEE/RSJ Intl. Conf. on Intelligent Robots and Systems (IROS)*, 2006.

[4] J. Buhmann, W. Burgard, A. Cremers, D. Fox, T. Hofmann, F. Schneider, J. Strikos, and S. Thrun. The Mobile Robot RHINO. *AI Magazine*, 16(2):31–38, 1995.

[5] D. Burschka and G. Hager. Vision-based control of mobile robots. In *Proc. of IEEE Intl. Conf. on Robotics and Automation (ICRA)*, 2001.

[6] D. Burschka and G. Hager. Stereo-Based Obstacle Avoidance in Indoor Environments with Active Sensor Re-Calibration. In *Proc. of IEEE Intl. Conf. on Robotics and Automation (ICRA)*, 2002.

[7] B. Cartwright and T. Collett. Landmark learning in bees. *Journal of Comparative Physiology*, 151(4):521–543, 1983.

[8] Z. Chen and S. Birchfield. Qualitative Vision-Based Mobile Robot Navigation. In *Proc. of IEEE Intl. Conf. on Robotics and Automation (ICRA)*, 2006.

[9] G. DeSouza and A. Kak. Vision for Mobile Robot Navigation. In *IEEE Transactions on Pattern Analysis and Machine Intelligence*, volume 24, Feb. 2002.

[10] M. Franz, B. Schölkopf, H. Mallot, and H. Bülthoff. Where did I take this snapshot? Scene-based homing by image matching. *Biological Cybernetics*, 79(3):191–202, 1998.

[11] R. Hartley and A. Zisserman. *Multiple View Geometry in Computer Vision*. Cambridge University Press, ISBN: 0521623049, 2000.

[12] J. Hong, X. Tan, B. Pinette, R. Weiss, and E. Riseman. Image-based homing. In *Proc. of the IEEE Intl. Conference on Robotics and Automation (ICRA)*, pages 620–625, 1991.

[13] K. Konolige, M. Agrawal, R. Bolles, C. Cowan, M. Fischler, and B. Gerkey. Outdoor Mapping and Navigation using Stereo Vision. In *Proc. of the Intl. Symposium on Experimental Robotics (ISER)*, 2006.

[14] J. Kosecka and X. Yang. Global Localization and Relative Positioning based on Scale-invariant features. In *Proc. of the 17th Intl. Conference on Pattern Recognition (ICPR)*, pages 319–322, 2004.

[15] D. Lambrinos, R. Möller, R. Pfeifer, and R. Wehner. Landmark Navigation without Snapshots: the Average Landmark Vector Model. In *Proc. Neurobiol. Conf. Göttingen*. Georg Thieme Verlag, 1998.

[16] G. López-Nicolás, C. Sagüés, J. Guerrero, D. Kragic, and P. Jensfelt. Nonholonomic epipolar visual servoing. In *Proc. of the IEEE Intl. Conference on Robotics and Automation (ICRA)*, 2006.

[17] H. Mallot and M. Franz. Biological Approaches to Spatial Representation - A Survey. In *Proc. of the 16th Intl. Joint Conference on Artificial Intelligence (IJCAI)*. Morgan Kaufmann, 1999.

[18] R. Möller. Insect visual homing strategies in a robot with analog processing. *Journal Biological Cybernetics*, 83/3:231–243, 2000.

[19] R. Möller. Modeling the landmark navigation behavior of the desert ant Cataglyphis. Technical Report IFI-AI-00.24, Artificial Intelligence Lab, Dept. Computer Science, University of Zurich, 2000.

[20] H. Neven and G. Schöner. Dynamics parametrically controlled by image correlations organize robot navigation. *Biological Cybernetics*, 75(4):293–307, 1996.

[21] A. Si, M. Srinivasan, and S. Zhang. Honeybee navigation: properties of the visually driven 'odometer'. *Journal of Experimental Biology*, 206(4):1265–1273, 2003.

[22] W. Stürzl and H. Mallot. Efficient visual homing based on Fourier transformed panoramic images. *Robotics and Autonomous Systems*, 54:300–313, 2006.

[23] S. Thrun, M. Montemerlo, H. Dahlkamp, D. Stavens, A. Aron, J. Diebel, P. Fong, J. Gale, M. Halpenny, G. Hoffmann, K. Lau, C. Oakley, M. Palatucci, V. Pratt, P. Stang, S. Strohband, C. Dupont, L.-E. Jendrossek, C. Koelen, C. Markey, C. Rummel, J. van Niekerk, E. Jensen, P. Alessandrini, G. Bradski, B. Davies, S. Ettinger, A. Kaehler, A. Nefian, and P. Mahoney. Winning the DARPA Grand Challenge. *Journal of Field Robotics*, 2006. accepted for publication.

[24] R. Wei, R. Mahony, and D. Austin. A Bearing-Only Control Law for Stable Docking of Unicycles. In *Proc. of IEEE/RSJ Intl. Conf. on Intelligent Robots and Systems (IROS)*, 2003.

Intelligent Autonomous Systems 10
W. Burgard et al. (Eds.)
IOS Press, 2008

doi:10.3233/978-1-58603-887-8-288

Towards Adaptive Multi-Robot Coordination Based on Resource Expenditure Velocity

Dan ERUSALIMCHIK and Gal A. KAMINKA
The MAVERICK Group
Computer Science Department
Bar Ilan University, Israel

Abstract. In the research area of multi-robot systems, several researchers have reported on consistent success in using heuristic measures to improve loose coordination in teams, by minimizing coordination costs using various heuristic techniques. While these heuristic methods has proven successful in several domains, they have never been formalized, nor have they been put in context of existing work on adaptation and learning. As a result, the conditions for their use remain unknown. We posit that in fact all of these different heuristic methods are instances of reinforcement learning in a one-stage MDP game, with the specific heuristic functions used as rewards. We show that a specific reward function—which we call *Effectiveness Index* (EI)—is an appropriate reward function for learning to select between coordination methods. EI estimates the *resource-spending velocity* by a coordination algorithm, and allows minimization of this velocity using familiar reinforcement learning algorithms (in our case, Q-learning in one-stage MDP). The paper analytically and empirically argues for the use of EI by proving that under certain conditions, maximizing this reward leads to greater utility in the task. We report on initial experiments that demonstrate that EI indeed overcomes limitations in previous work, and outperforms it in different cases.

Keywords. Multi-Robot Systems, Reinforcement Learning, Coordination

1. Introduction

This paper begins with a puzzle. In the research area of multi-robot systems, several researchers have reported on consistent success in using heuristic measures—which for the moment we call *coordination cost* measures—to improve loose coordination in teams. Specifically, Goldberg et al. [5], Zuluaga and Vaughan [16], and Rosenfeld et al. [12] all report that minimizing their respective coordination cost measures lead to improved performance.

However, while these heuristic methods has proven successful in several domains, they have never been formalized to a degree that allowed comparison with other methods. Nor have they been put in context of existing work on adaptation and learning. As a result, their optimality and the appropriate conditions for their use remain open questions.

We posit that in fact all of these different heuristic methods are instances of reinforcement learning in a one-stage MDP game [7], with the specific heuristic functions used as rewards. We further argue that the different coordination cost measures are all variations on central theme: Reducing the time and/or resources spent on coordination. These variations can be recast as reward functions within the MDP game.

We show that a specific reward function—which we call *Effectiveness Index* (EI)—is an appropriate reward function for learning to select between coordination methods. EI estimates the *resource-spending velocity* by a coordination algorithm, and allows minimization of this velocity using familiar reinforcement learning algorithms (in our case, Q-learning in one-stage MDP game).

The paper analytically and empirically argues for the use of EI by proving that under certain conditions, maximizing this reward leads to greater utility in the task. We report on initial experiments that demonstrate that EI indeed overcomes limitations in previous work, and outperforms it in different cases.

2. Related Work

Most closely related to our work is earlier work on measures of coordination effort. Rosenfeld et al. [12], presented a method that adapts the selection of coordination methods by multi-robot teams, to the dynamic settings in which team-members find themselves. The method relies on a measuring the resources expended on coordination, using a measure called Combined Coordination Cost (*CCC*). The adaptation is stateless, i.e., has no mapping from world state to actions/methods. Instead, the CCC is estimated at any given point, and once it passes pre-learned (offline learning) thresholds, it causes dynamic re-selection of the coordination methods by each individual robot, attempting to minimize the CCC.

Interference [5] is a closely related measure to CCC, and can be seen as a special case of it: It measures the amount of time spent on coordination. Zuluaga and Vaughan [16] presented an method called *aggression* for reducing interference in distributed robot teams, to improve their efficiency. During movement, multiple robots frequently interfere with each other. When such interference occurs, each of the robots demonstrate its own level of aggression such that the robot with the highest level becomes the winner, while the loser concedes its place. Zuluaga and Vaughan have shown that choosing aggression level proportional to the robot's task investment can produce better overall system performance compared to aggression chosen at random. This result is compatible with Rosenfeld et al.'s conclusions that reducing total resource spending in coordination is highly beneficial.

We formulate and generalize Rosenfeld et al.'s work in terms of reinforcement learning in single-state MDP game (MDG). Based on this generalized formulation, we are able to explain the empirically-observed success of Rosenfeld et al.'s method (as a special case), and suggest novel learning methods that do not require an off-line learning phase.

Indeed, the contribution of our work lies in the introduction of a general reward function for coordination (and only for coordination). This reward function minimizes *the velocity of resource expenditure*. In contrast, most investigations of reinforcement learning in multi-robot settings have focused on other mechanisms (e.g., modifying the basic Q-learning algorithm), and utilized task-specific reward functions. We briefly discuss these below. A recent survey appears in [15].

Balch [1] discusses considerations for task-oriented reward functions for reinforcement learning in multi-robot settings. He shows that the choice of reward function influences the behavioral diversity, and group performance in a variety of tasks, including foraging and soccer. Kok and Vlassis [9] discuss a technique for propagating rewards among cooperative robots, based on the structure of the dependence between the robots. However, they too assume that the reward function is given as part of the task.

Mataric[10] discusses three techniques for using rewards in multi-robot Q-learning: A local performance-based reward (each robot receiving reward for its own performance, and per its own goals), a global performance-based reward (all robots receive reward based on achievement of team goals), and a heuristic strategy referred to as shaped reinforcement. Shaped reinforcement, which was developed by Mataric, provides a heuristic function that combines rewards based on local rewards, global rewards and coordination interference of the robots. in contrast to these investigations, we explore general reward functions, based on minimize resource use, and use them in selecting between coordination behaviors, rather than individual behaviors.

Kapetanakis and Kudenko [7] present the FMQ learning algorithm. This algorithm is intended for coordination learning in one-stage MDP games. FMQ is a modified regular Q-Learning method for one-stage games and this modification is based on the Boltzman's strategy. They then examine how an robot that uses FMQ learning technique may influence other robot's effectiveness of learning, when the latter uses a simple Q-learning algorithm [8]. This method does not use communication or monitoring of other robot's action, but based on the assumption that all of the robots are getting the same rewards.

The Q-learning algorithm used in these works has no states, similarly to the proposed method, but Kapetanakis and Kudenko's works are concentrating on improving effectiveness of the learning algorithm and assume that rewards were pre-defined before and thus, the robot just has to discover them. In opposite, we concentrate on the method of reward determination by the robot. In the real world we do not have predefined rewards and especially when distinguishing between rewards from behaviors with the same goal is needed. Therefore, Kapetanakis and Kudenko's work, like a many other works of Reinforcement Learning is concentrated on Q-learning algorithm modification and assume pre-definition of the rewards, should be considered as a complimentary work instead of an alternative to ours.

3. Maximizing Social Utility by Liming Coordination Costs

We first cast the problem of selecting coordination algorithms as a reinforcement learning problem (Section 3.1). We then introduce the effective index (EI) in Section 3.2. We then discuss the conditions underwhich maximizing it leads to improved task performance, and provide a proof sketch, in Section 3.3.

3.1. Coordination Algorithm Selection as an RL Problem

Multilateral coordination prevents and resolves conflicts among robots in a multi-robot system (MRS). Such conflicts can emerge as results for shared resource (e.g., space), or as a result of violation of joint decisions by team-members. Many coordination algorithms (protocols) have been proposed and explored by MRS researchers [4,5,11,13,14].

Not one method is good for all cases and group sizes [12]. However, deciding on a coordination method for use is not a trivial task, as the effectiveness of coordination methods in a given context is not known in advance.

We focus here on loosely-coupled application scenarios where coordination is triggered by conflict situations, identified through some mechanism (we assume the existence of such mechanism exists, though it may differ between domains; most researchers simply use a pending collision as a trigger). Thus the normal routine of an robot's operation is to carry out its primary task, until it is interrupted by an occurring or potentially-occurring conflict with another robot, which must be resolved by a coordination algorithm. Each such interruption is called *a conflict event*. The event triggers a coordination algorithm to handle the conflict. Once it successfully finishes, the robots involved go back to their primary task. Such scenarios include multi-robot foraging, formation maintenance (coordinated movement), and delivery.

Let $A = \{\ldots, a_i, \ldots\}, 1 \leq i \leq N$ be a group of N robots, cooperating on a group task that started at time 0 (arbitrarily) lasts upto time T (A starts working and stops working on the task together). We denote by $T_i = \{c_{i,j}\}, 0 \leq j \leq K_i$ the set of conflict events for robot i, where $c_{i,j}$ marks the time of the beginning of each conflict. Note that each robot i may have been interrupted a different number of time, i.e., K_i may be different for different robots. For notational uniformity, $c_{i,K_i+1} = T$, and $c_{i,0}$ is defined as time 0.

The time between the beginning of a conflict event j, and up until the next event, the interval $I_{i,j} = [c_{i,j}, c_{i,j+1})$, can be broken into two conceptual periods: The *active* interval $I^a_{i,j} = [c_{i,j}, t_{i,j})$ (for some $c_{i,j} < t_{i,j} < c_{i,j+1}$) in which the robot was actively investing resources in coordination, and the *passive* interval $I^p_{i,j} = [t_{i,j}, c_{i,j+1})$ in which the robot no longer requires investing in coordination; from its perspective the conflict event has been successfully handled, and it is back to carrying out its task. By definition $I_{i,j} = I^a_{i,j} + I^p_{i,j}$. We define the *total active time* as $I^a = \sum_i \sum_j I^a_{i,j}$ and the *total passive time* as $I^p = \sum_i \sum_j I^p_{i,j}$.

Our research focuses on a case where the robot has a nonempty set M of coordination algorithms to select from. The choice of a specific coordination method $\alpha \in M$ for a given conflict event $c_{i,j}$ may effect the active and passive intervals $I^a_{i,j}, I^p_{i,j}$. To denote this dependency we use $I_{i,j}(\alpha), I^a_{i,j}(\alpha), I^p_{i,j}(\alpha)$ as total, active and passive intervals (respectively), due to using coordination method α.

Using this notation, we can phrase the selection of coordination algorithms as determining a policy for selecting between different coordination methods among those in M. We denote a robot i's selection at conflict event $c_{i,j}$ as $\Pi_{i,j}$. A sequence of these selections, for all events $j \leq K_i$, is denoted by Π_i; this defines an individual coordination policy. The set of individual policies of all robots in A is marked Π.

Formally, we define the problem of coordination algorithm selection as a one-stage Markov Decision Process (MDP) game, with a limited set of actions (selectable algorithms), and an individual reward for each robot (player) [7]. Each robot tries to maximize its own reward. Typically, reward functions are given, and indeed most previous work focuses on learning algorithms that use the reward functions as efficiently as possible. Instead, we assume a very basic learning algorithm (a simple Q-Learning variant), and instead focus on defining a reward function. The learning algorithm we use is stateless:

$$Q_t(a) = Q_{t-1}(a) + \rho(R_t(a) - \gamma Q_{t-1}(a))$$

Where ρ is the learning speed factor, and γ is a factor of discounting.

3.2. Effectiveness Index

We call the proposed general reward for coordination *Effectiveness Index* (EI). Its domain independence is based on its using three intrinsic (rather than extrinsic) factors in its computation; these factors depend only internal computation or measurement, rather than environment responses.

The time spent coordinating. The main goal of a coordination algorithm is to reach a (joint) decision that allows all involved robots to continue their primary activity. Therefore, the sooner the robot returns to its main task, the less time is spent on coordination, and likely, the robot can finish its task more quickly. Thus, smaller I_i^a is better.

The frequency of coordinating. If there are frequent interruptions—even if short-lived—to the robot's task, in order to coordinate, this would delay the robot. We assume (and the preliminary results show) that good coordination decisions lead to long durations of non-interrupted work by the robot. Therefore, the frequency of coordination method's use is not less important, than the time spent on conflict resolving. Thus, larger $I_{i,j}^p$ is better.

The cost of coordinating. Finally, in addition to speed of conflict resolution and frequency of calling, careful resource spending is a very important factor for behavior selection. Short-lived, infrequent calls to an expensive coordination method will not be preferable to somewhat more frequent calls to very cheap coordination method. It is thus important to consider the internal resources used by the chosen method. We argue that such internal estimate of resource usage is feasible.

First, some resource usage is directly measurable. For instance, energy consumption during coordinated movement (e.g., when getting out of a possible collision) or communications (when communicating to avoid a collision) is directly measurable in robots, by accessing the battery device before and after using the coordination algorithm.

Second, resource usage may sometimes be analytically computed. For instance, given a the basic resource cost of a unit of transmission, the cost of using a specific protocol may often be analytically computed (as it is tied directly to its communication complexity in bits).

Rosenfeld et al. [12] have defined CCC as the total cost of resources spent on resolving conflicts (re-establishing coordination) before, during, and after a conflict occurs. Their definition of the cost consisted of a weighted sum of the costs of different resources. We denote by U_i^C the utility of coordination, of robot i, of which the cost of coordination, denoted C_i^C is components. By definition, $CCC = C_i^C$. It can be broken into the costs spent on resolving all conflicts T_i, $C_i^C = \sum_j CCC_{c_{i,j}}$.

Let us use a cost function $cost_i(\alpha, t)$ to represent the costs due to using coordination method $\alpha \in M$ at any time t during the lifetime of the robot. The function is not necessarily known to us a-priori (and indeed, in this research, is not).

Using the function $cost_i(\alpha, t)$ we redefine the $C_{i,j}^C$ of a particular event of robot i at time $c_{i,j}$ to be:

$$C^C_{i,j}(\alpha) = \int_{c_{i,j}}^{c_{i,j+1}} cost_i(\alpha, t)\ \mathrm{d}t \tag{1}$$

We remind the reader that $C^C_{i,j}$ is defined as the costs of applying the coordination algorithm during the active interval $[c_{i,j}, t_{i,j})$ and the passive interval $[t_{i,j}, c_{i,j+1})$. However, the coordination costs during the passive interval are zero by definition.

$$\begin{aligned} C^C_{i,j}(\alpha) &= \int_{c_{i,j}}^{t_{i,j}} cost_i(\alpha, t)\ \mathrm{d}t + \int_{t_{i,j}}^{c_{i,j+1}} cost_i(\alpha, t)\ \mathrm{d}t \\ &= \int_{c_{i,j}}^{t_{i,j}} cost_i(\alpha, t)\ \mathrm{d}t \end{aligned} \tag{2}$$

We define the *Active Coordination Cost* (ACC) function for robot i and method α at time $c_{i,j}$, that considers the *active time* in the calculation of coordination resources cost:

$$ACC_{i,j}(\alpha) = \int_{c_{i,j}}^{t_{i,j}} 1 + cost_i(\alpha, t)\ \mathrm{d}t \tag{3}$$

We finally define Effectiveness Index of a particular event of robot i at time $c_{i,j}$ due to using coordination method $\alpha \in M$:

$$EI_{i,j}(\alpha) = \frac{ACC_{i,j}(\alpha)}{I_{i,j}} = \frac{\int_{c_{i,j}}^{t_{i,j}} 1 + cost_i(\alpha, t)\ \mathrm{d}t}{I^a_{i,j} + I^p_{i,j}} \tag{4}$$

That is, the effectiveness index (EI) of an algorithm α during this event is the velocity by which it spends resources during its execution, amortized by how long a period in which no conflict occurs. Since greater EI signifies greater costs, we typically put a negation sign in front of the EI, to signify that greater velocity is worse; we seek to minimize resource spending velocity.

3.3. An Analytical Look at EI

We now turn to briefly sketch the conditions underwhich an EI-minimizing policy Π will lead to greater team performance on its group task. Due to lack of space, we will provide only a sketch of the proof, and refer the reader to [3] for additional details.

Preliminaries. We use the following notations in addition to those already discussed. First, we denote by U_i is the utility of robot i. U_i^T marks its utility due to executing the task, and U_i^C marks its utility due to being coordinated with others at a conflict situation: $U_i = U_i^T + U_i^C$. Each such utility value can be broken into gains G and costs C: $U_i^T = G_i^T - C_i^T$ and $U_i^C = G_i^C - C_i^C$. The social utility U is the sum of all individual utilities of the robots: $U = \sum_{i=1}^N U_i$.

To maximize this sum, the robot can invest effort in maximizing the utility from the task, and/or the utility from coordination. In the same way, to maximize the social utility of the team, each robot can invest effort in maximizing the its own utility and/or the teammates' utility. We are interested in task-independent reward functions, and thus focus our attention on maximizing utility from coordination (social utility).

Let us use a function $cgain_i(\alpha, t)$ to denote the coordination gain at any time t during the lifetime of the robot i that uses method α. When a robot is handling a conflict event, it is not gaining anything from coordination (in fact, it is investing effort in re-establishing coordination). Thus, the $cgain_i(\alpha, t)$ function can be defined as a step function

$$cgain_i(\alpha, t) = \begin{cases} 0 \text{ robot } i \text{ in a conflict situation} \\ 1 \text{ other} \end{cases} \tag{5}$$

Using this function, we redefine the $G^T_{i,j}$ of a particular event of robot i at time $c_{i,j}$ to be:

$$\begin{aligned} G^C_{i,j}(\alpha) &= \int_{c_{i,j}}^{c_{i,j+1}} cgain_i(\alpha, t)\,\mathrm{d}t = \int_{c_{i,j}}^{t_{i,j}} cgain_i(\alpha, t)\,\mathrm{d}t + \int_{t_{i,j}}^{c_{i,j+1}} cgain_i(\alpha, t)\,\mathrm{d}t \\ &= 0 + \int_{t_{i,j}}^{c_{i,j+1}} cgain_i(\alpha, t)\,\mathrm{d}t = \int_{t_{i,j}}^{c_{i,j+1}} cgain_i(\alpha, t)\,\mathrm{d}t = \int_{t_{i,j}}^{c_{i,j+1}} 1\,\mathrm{d}t = I^p_{i,j}(\alpha) \end{aligned} \tag{6}$$

Now, we can define two evaluation functions of coordination policy.

- *Social Utility* of team by using policy Π

$$U(\Pi) = \sum_i^N \sum_j^{K_i} U_{i,j}(\Pi_{i,j}) = \sum_i^N \sum_j^{K_i} U^T_{i,j}(\Pi_{i,j}) + G^C_{i,j}(\Pi_{i,j}) - C^C_{i,j}(\Pi_{i,j}) \tag{7}$$

- *Social ACC* of team by using policy Π

$$ACC(\Pi) = \sum_i^N \sum_j^{K_i} ACCc_{i,j}(\Pi_{i,j}) \tag{8}$$

Based on the above, we would ideally want to show that (1) minimizing EI with each event leads to improved social utility for the team, and that (2) this, in turn, leads to improved overall task performance of the team. The first part is in some sense already given, when we use the FMQ framework. As long as its conditions hold, we can expect individual rewards to be maximized (i.e., the social utility will be greater individually). However, the second part is more challenging.

It is possible to show, that if the social costs for the team are minimized (i.e., the sum of coordination costs for all robots is minimized), then the social utility of the team is greater (Lemma 1).

Lemma 1. *The* Social Utility *for policy* Π' *is better than* Social Utility *for policy* Π'' *if* Social ACC *for* Π' *is lower than* Social ACC *for* Π''.

$$ACC(\Pi') < ACC(\Pi'') \Longrightarrow U^C(\Pi') > U^C(\Pi'')$$

Proof. For space reasons, we provide a proof sketch. See [3] for formal proof. The intuition for the lemma's truth is as follows. If $ACC(\Pi') < ACC(\Pi'')$, then necessarily (by a sequence of rewritings), $I^a(\Pi') + C^C(\Pi') < I^a(\Pi'') + C^C(\Pi'')$. This in turn implies that $I^p(\Pi') - C^C(\Pi') > I^p(\Pi'') - C^C(\Pi'')$. From the definition of $G^C_{i,j}(\alpha)$ above (Eq. 6), we therefore have $G^C(\Pi') - C^C(\Pi') > G^C(\Pi'') - C^C(\Pi'')$, which means that $U^C(\Pi') > U^C(\Pi'')$. □

Overall utility is defined as $U(\Pi) = U^T(\Pi) + U^C(\Pi)$. The question therefore becomes under what conditions does an improved social utility policy leads to improved utility overall; i.e., when does $U^C(\Pi') > U^C(\Pi'') \Rightarrow U(\Pi') > U(\Pi'')$? We consider several cases.

Case 1. $U_i^T(\Pi') \geq U_i^T(\Pi'')$. Here, the conflict solving methods do not affect individual task utility (or make it better), for *all* robots. In this case it is easy to see that the accumulated task utility is greater, and the greater task and social utilities, combined, result in greater overall utility.

Suppose, however, that one robot's task utility under the policy Π' is actually made worse than other the competing policy. Does that automatically mean that the overall utility for the team is worse when using Π'? The answer is no; the robot might in fact be sacrificing its own task utility to maximize the team's (as collaborating robots might be expected to do [6]). The question is whether its sacrifice is compensated for by greater rewards to others.

Case 2. $U_i^T(\Pi') < U_i^T(\Pi'')$, *but* $U^T(\Pi') \geq U^T(\Pi'')$. For all reduction in task utility made by the choice of conflict solving method exists number of compensations in other conflicts of other robots in the team.

Finally, it might still be possible for the team to perform better with policy Π' even when task performance is made worse.

Case 3. $U^T(\Pi') < U^T(\Pi'')$, *but* $U^T(\Pi'') - U^T(\Pi') < U^C(\Pi') - U^C(\Pi'')$. In the case where the loss in team task utility from using policy Π' is smaller than benefit in team coordination utility that policy Π' provides, it is still true that $U^T(\Pi') + U^C(\Pi') > U^T(\Pi'') + U^C(\Pi'')$.

Tying these three cases above together, we now state the concluding theorem:

Theorem 2. *EI is a good individual reward for total social utility,* if *(i) either case 1, 2, or 3 above hold; and (ii) EI minimization policy leads to maximal coordination utility U^C for the team.*

Proof. See discussion above. □

4. Experiments

We now turn to briefly survey a subset of empiric results supporting the use of EI and the stateless Q-learning algorithm in multi-robot team foraging. Here, robots locate target items (pucks) within the work area, and deliver them to a goal region. As was the case in Rosenfeld's work [12], we used the TeamBots simulator [2] to run experiments. Teambots simulated the activity of groups of Nomad N150 robots in a foraging area that measured approximately 5 by 5 meters. We used a total of 40 target pucks, 20 of which where stationary within the search area, and 20 moved randomly. For each group, we measured how many pucks were delivered to the goal region by groups of 3,5,15,25,35,39 robots within 20 simulated minutes. We averaged the results of 16–30 trials in each group-size configuration with the robots being placed at random initial positions for each run.

We compare the EI method with two other coordination algorithm selection methods: Random coordination algorithm selection (RND), and to the method of Rosenfeld

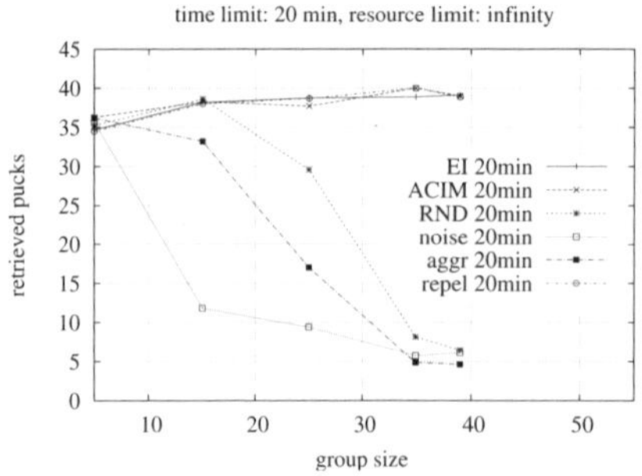

(a) No resource limits.

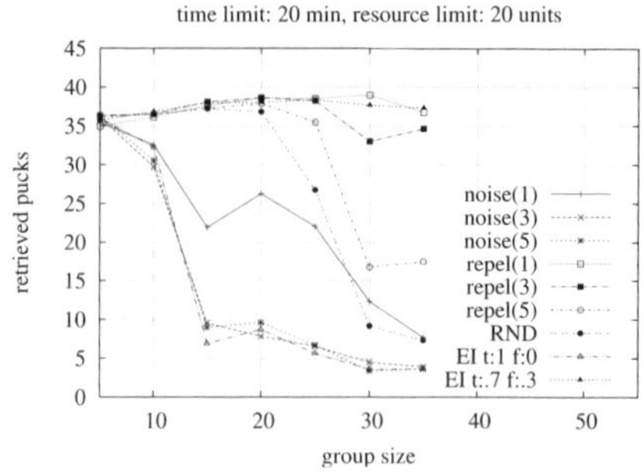

(b) Severe fuel limits.

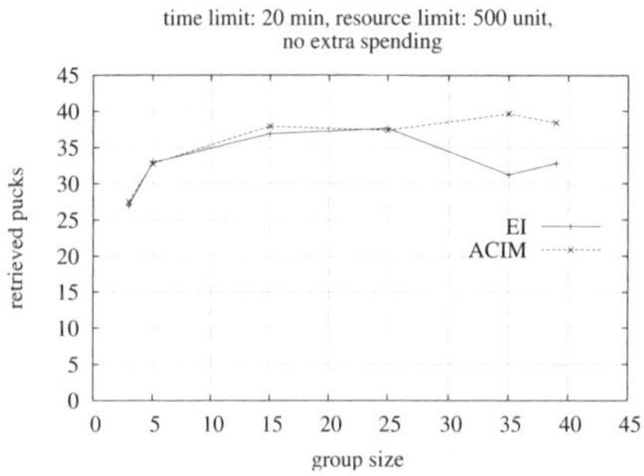

(c) Resource costs known.

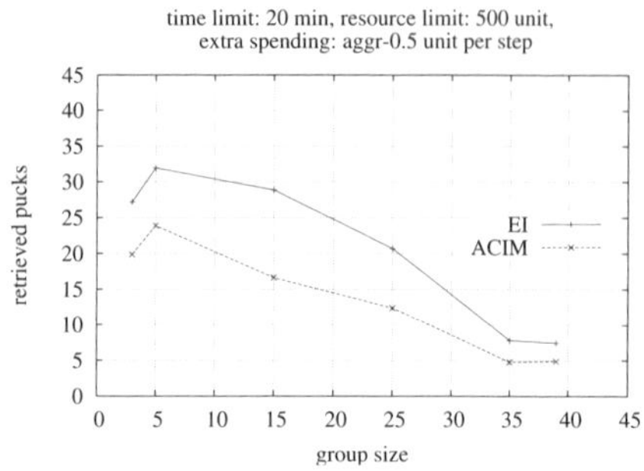

(d) Resource costs unknown.

Figure 1. Experiment results, time limit 20 minutes.

et al. (ACIM). All of these (EI, RND, ACIM) dynamically select between several fixed coordination algorithms, discussed in [12]: Noise (which essentially allows the robots to collide in their motion uncertainty does not prevent collision), Aggression [14], and Repel, in which robots back off a variable distance to avoid an impending collision.

Figures 1(a)–1(d) show a subset of results. In all, the X axis marks the group size, and the Y axis marks the number of pucks collected. Figure 1(a) shows that given no resource limitations, the EI is just as good as ACIM, despite the fact that EI does not use off-line learning. When resource (fuel in this case) limitations are applied (Figure 1(b)), each method spends 1, 3 or 5 units per step (see number inside of brackets after name of method). The EI method can performs as good as any other (given a CCC function which gives time a weight of 70% and fuel 30%), or as worse as any other (with time weight of 100%, i.e., when ignoring fuel costs). When resource limits are known a-priory the ACIM method provides the same result (or slightly superior) as EI (Figure 1(c). But when these resource limits are unknown(Figure 1(d)), and optional methods spend more than advertised (in this case, aggression spend 0.5 extra units of fuel per step), the EI method leads to significantly improved results. In both of the two last figures, the CCC gave time a weight of 70% and fuel 30%. The methods for selection were Noise(5), Repel(5) and Aggression(5.5).

5. Summary

This paper examined in depth the success of previously-report heuristic methods in improving loose coordination in teams, by selecting between different coordination methods. We have shown that these methods can be cast as solving a multi-agent reinforcement learning problem (specifically, a one-stage MDP game), and that existing heuristics can be viewed as rudimentary reward functions.

We have argued for a more principled investigation of appropriate reward functions for this framework, and presented a novel reward function, called Effectiveness Index, which essentially measures the velocity in which resources are spent when reestablishing conflicts. We analytically examine the cases underwhich the use of this reward function leads to improved performance, and then empirically shown that indeed it leads to better performance then existing methods of adaptation. We plan to extend our analysis and empiric investigation to examine additional domains and team tasks.

References

[1] T. Balch. Reward and diversity in multirobot foraging. IJCAI-99 Workshop on Agents Learning, July 1999.

[2] T. Balch. www.teambots.org, 2000.

[3] D. Erusalimchik and G. A. Kaminka. Towards adaptive multi-robot coordination based on resource expenditure velocity: Extended version. Technical Report MAVERICK 2008/02, Bar Ilan University, Computer Science Department, MAVERICK Group, available at http://www.cs.biu.ac.il/~galk/Publications/, 2008.

[4] M. Fontan and M. Matarić. Territorial multi-robot task division. *IEEE Transactions of Robotics and Automation*, 14(5):815–822, 1998.

[5] D. Goldberg and M. J. Mataric. Interference as a tool for designing and evaluating multi-robot controllers. In *Proceedings of the Fourteenth National Conference on Artificial Intelligence (AAAI-97)*, pages 637–642, Providence, RI, 1997. AAAI Press.

[6] B. J. Grosz and S. Kraus. Collaborative plans for complex group actions. *Artificial Intelligence*, 86:269–358, 1996.

[7] S. Kapetanakis and D. Kudenko. Reinforcement learning of coordination in cooperative multi-agent systems. In *Proceedings of the Eighteenth National Conference on Artificial Intelligence (AAAI-02)*, pages 326–331, 2002.

[8] S. Kapetanakis and D. Kudenko. Reinforcement learning of coordination in heterogeneous cooperative multi-agent systems. In *Proceedings of the Third International Joint Conference on Autonomous Agents and Multi-Agent Systems (AAMAS-04)*, pages 1258–1259, 2004.

[9] J. R. Kok and N. Vlassis. Collaborative multiagent reinforcement learning by payoff propagation. *Journal of Machine Learning Research*, 7:1789–1828, 2006.

[10] M. J. Matarić. Reinforcement learning in the multi-robot domain. *Auton. Robots*, 4(1):73–83, 1997.

[11] E. Ostergaard, G. Sukhatme, and M. Matarić. Emergent bucket brigading. In *Proceedings of the Fifth International Conference on Autonomous Agents (Agents-01)*, pages 29–30, 2001.

[12] A. Rosenfeld, G. A. Kaminka, S. Kraus, and O. Shehory. A study of mechanisms for improving robotic group performance. *Artificial Intelligence*, 172(6–7):633–655, 2008.

[13] M. Schneider-Fontan and M. Matarić. A study of territoriality: The role of critical mass in adaptive task division. In P. Maes, M. Matarić, J.-A. Meyer, J. Pollack, and S. Wilson, editors, *From Animals to Animats IV*, pages 553–561. MIT Press, 1996.

[14] R. Vaughan, K. Støy, G. Sukhatme, and M. Matarić. Go ahead, make my day: robot conflict resolution by aggressive competition. In *Proceedings of the 6th int. conf. on the Simulation of Adaptive Behavior*, Paris, France, 2000.

[15] E. Yang and D. Gu. Multiagent reinforcement learning for multi-robot systems: A survey. Technical Report CSM-404, University of Essex, 2004.

[16] M. Zuluaga and R. Vaughan. Reducing spatial interference in robot teams by local-investment aggression. In *Proceedings of the IEEE/RSJ International Conference on Intelligent Robots and Systems (IROS)*, Edmonton, Alberta, August 2005.

Intelligent Autonomous Systems 10
W. Burgard et al. (Eds.)
IOS Press, 2008

doi:10.3233/978-1-58603-887-8-298

Resolving Inconsistencies in Shared Context Models using Multiagent Systems

Uwe-Philipp KÄPPELER, Ruben BENKMANN, Oliver ZWEIGLE, Reinhard LAFRENZ and Paul LEVI
Institute of Parallel and Distributed Systems (IPVS)
University of Stuttgart, Universitätsstraße 38, 70569 Stuttgart

Abstract. Agents acting in physical space use perception in combination with their own world models and shared context models.The shared context models have to be adapted permanently to the conditions of the real world. If a measurement of an agent's sensor does not fit to the corresponding data in the shared context model the system contains an inconsistency. In this case it is necessary to decide whether the reason for the discrepancy is a change in the real world or a measurement error. A solution has to be based on only a limited number of measurements from different agents and a reduction of noise in the sensor data.

The study reported in this paper evaluates procedures that combine a multitude of measurements to a single result that can be integrated in the shared context model. The statistically optimized procedure based on ratings of the participating agents is enhanced using scaled weighted arithmetic means which prevents the system from running into singularities caused by the feedback from the ratings. The method is combined with an additional preprocessing based on fuzzy clustering that detects aberrant measurements which can be excluded from further processing.

Keywords. Sensor Data Fusion, Fuzzy Clustering, Weighted Arithmetic Mean

1. Motivation

The main objective of the Nexus Center of Excellence is the definition and realization of dynamic shared context models for context-aware applications. In this scope, issues concerning communication, information management, methods for model representation and sensor data integration are covered. Based on these digital world models, new innovative applications become possible, which can access information of the real world originating from sensors and additional, aggregated information. We currently witness the rapid proliferation of different kinds of sensor systems. These systems allow the acquisition of context information and make the integration of the sensor data an important research aspect. Open questions are which sensors are suitable for providing context information to the world model with as little redundancy as possible. The problem in updating the world models by sensor measurements is to reduce uncertainty and inconsistency.

If a local world model of an application or agent conflicts with data in the shared context model the system contains an inconsistency. It is hard to decide, whether the data of one single local world model based on sensor data is erroneous or if the shared world model is out of date and needs a correction. If the application needs to be sure about this specific information to work properly, the inconsistency needs to be resolved. One solution to this problem is to repeat the measurement with different sensors and to reduce uncertainty by redundancy. Methods to address other agents with access to sensors and to communicate the sensor data are provided by the Nexus Platform. The study described in this paper examines methods to statistically optimize the reduction of uncertainty where only a few measurements of corresponding physical values are available, which leads to critical and inescapable statistical problems in the rating of the sensors. The reduction of uncertainty is necessary to make relevant contributions to the shared world model, maintaining a high degree of reliability while keeping costs and expenditure of time within an acceptable range by involving only a few number of agents in the process.

2. Scenario - Nexus Platform and Agent Negotiation

2.1. Nexus Platform - Architecture and Applications

The goal of the Nexus Platform is to support all kinds of context-aware applications by providing a shared global world-context model. To achieve this goal, the platform federates local context models from so-called context servers. The local models contain different types of context information: representations of real world objects like streets, rooms or persons and virtual objects that link to digital information spaces. Sensors keep the data of local context models up to date (e.g. the position of a person). There can be many different implementations of a context server. For providing large scale geographical models, we used a spatially enhanced database. It copes with the high update rates of the positions of mobile users [1].Small-scaled sensor platforms like the ContextCube [2] can be used as context server. We also implemented a main memory based sensor context server which is optimized on short response times [3] [4]. More details of the architecture are described in [5]. A context-aware application can use the Nexus Platform in three different ways. First, it can send queries to the federation to get information about its surroundings including infrastructure, points of interest, mobile objects (friend finder), and so on. Secondly, the application can register to the event service [6] to receive a notification when a certain state of the world occurs, e.g. when the user has entered a building or the temperature in a room exceeds a certain threshold. Thirdly, it can use value-added services like a map or navigation service in order to shift common functions to the infrastructure.

2.2. Agent Negotiation

If a measurement of an agent's sensor does not fit to the corresponding data in the shared context model the system contains an inconsistency. In this case a negotiation protocol is initiated to resolve the inconsistency using several sensors. The Nexus platform identifies registered mobile or static agents with corresponding sensors. The more sensors and measurements are available, the easier it is to increase the accuracy of the measurement statistically. But to keep the duration and costs to resolve the inconsistency in an acceptable range we assume that only a few number of measurements are available to the platform. Therefore we try to get the best results from only a few measurements by rating each sensor. After the negotiation has been completed the agent can decide whether the reason for the conflict was its own measurement or erroneous data in the shared context model. If necessary it should update the data on the context server and add meta data describing key data from the negotiation as well as the reliabilities of the participating agents. Calculating the weighted arithmetic mean of all the gathered measurements, the reliabilities of the agents and processing the fuzzy clustering is described in detail in the following sections.

3. Sensor Data Processing

3.1. Simulation of Negotiations

Our first simulations were based on artificially generated sensor data with the advantage of defining standard deviations and known *real* values. For the diagrams in this paper we chose the simulated sensor data of five sensors of which one had a distinct offset and one had a higher standard deviation than the others. The results shown in all the diagrams are given in 10th of degrees.

3.2. Statistically Optimized Approach

After an agent has detected an inconsistency, called other agents and communicated measurements it has to reduce uncertainty of the sensor data combining the measurements to one result. The first approach is based on statistics considering optimized combination of measurements from the book by Dietrich [7]. In use of the open Nexus platform as described above we assume that we can reach only a few agents that are able to execute measurements of one single physical value keeping costs and expenditure of time within an acceptable range. This limitation forbids to rely only on the arithmetic mean of the measurements. One solution to reduce the error in calculating a result from multiple sensor data is to estimate the standard deviation of each sensor, which leads to a maximum likelihood estimator.

A value x is measured by n different sensors giving the results $x_1, x_2, \ldots, x_n$. Assuming the noise in the sensor data is normally distributed and that we know the standard deviations $\sigma_1, \sigma_2, \ldots, \sigma_n$, where q is the most likely value of x the deviations of the results are $q - x_1, q - x_2, \ldots, q - x_n$. So the combined probability of the deviations is the product

$$\prod_{i=1}^{n} \left(\frac{e^{\frac{-(x_i - q)^2}{2\sigma_i^2}}}{\sigma_i \sqrt{2\pi}} \right) = \frac{e^{-\sum_{r=1}^{r=n} \frac{(x_r - q)^2}{2\sigma_r^2}}}{(2\pi)^{\frac{n}{2}} \prod_1^n \sigma_r}$$

The maximum of the combined probability results in the most likely value for q. This leads to

$$q = \frac{\sum_1^n \frac{x_r}{\sigma_r^2}}{\sum_1^n \frac{1}{\sigma_r^2}} \qquad \text{or} \qquad \bar{q} = \frac{\sum_1^n \frac{\bar{x}_r m_r}{\sigma_r^2}}{\sum_1^n \frac{m_r}{\sigma_r^2}}$$

where arithmetic means $\bar{x}_r$ and standard deviations $\bar{\sigma_r}$ are approximated from a limited number m of measurements of the sensors in the past, with

$$\bar{x}_r = \frac{1}{m}\sum_{1}^{m} x_{r_i} \qquad \text{and} \qquad \bar{\sigma_r} = \frac{\sigma_r}{\sqrt{m_r}}$$

To adopt this method to the multiagent system we set

$$\sigma^2 = \sum_{1}^{m} \frac{(x_r - \bar{x})^2}{m - 1}$$

and assume $\bar{x} = 0$ which means that the result of negotiations in the past results in a correct value without noise. In fact each result still contains noise if it is not derived from an infinite number of measurements. But the simulations described later in this section with $m > 200$ show that this assumption is acceptable.

Disadvantage of the Statistically Optimized Approach

The described limitations to only a few participating agents can lead to a situation where every negotiation ends in taking the result from the same agent. This happens if the result of the negotiation is incidentally more than once in a row near the measurement of one single agent. In this case the estimated deviations that improve the result of a single negotiation compared to the simple arithmetic mean can run the system into a singular behavior because of the positive feedback from updating the deviations from the result after each negotiation. A problem occurs when one single agent wins a big part of negotiations that are kept in the buffer for the estimation, which is the more likely the fewer agents are participating in the negotiations.

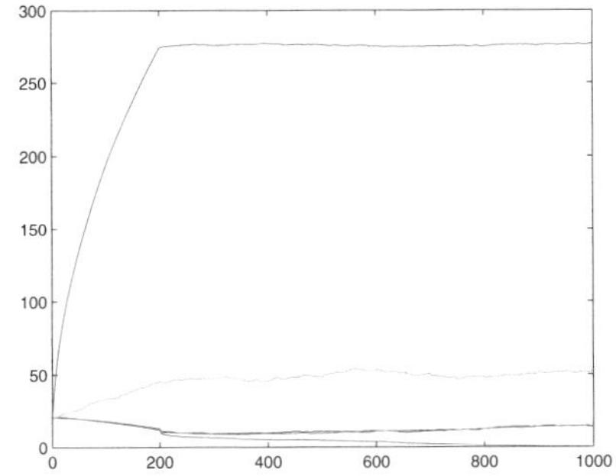

Figure 1. Singular behavior: Absolute deviations in a simulation of 1000 negotiations

Fig. 1 shows the singular behavior of the system in a simulation. The buffer stores 200 negotiation results to estimate the standard deviations which is reflected by the initialization phase visible in the diagram. After several negotiations the standard deviation of a single agent converges to 0.As soon as the standard deviation of one agent is estimated to be 0 the system is not able to leave this state anymore. Each following negotiation of measurements is reduced to an acceptation of the measurement of this agent as a result. Therefore an application of this method in the open nexus platform is not practicable.

Singular Behavior

To analyze the singular behavior of the system we examine its difference equation. To reduce the complexity we assume that the system consists of only three agents x, y and z and we observe their rating and the estimated standard deviation. x_i, y_i, z_i represent the stored deviations beginning at time t where $X_i = x(t-1+i)$ and n is the size of the buffer for the deviations. υ_x, υ_y and υ_z represent the current measurements which are the inputs to the system. One agents estimated deviation in this system is described by

$$x_{n+1} = \upsilon_x - \frac{\frac{n-1}{x_n^2+...+x_2^2+x_1^2}\upsilon_x + \frac{n-1}{y_n^2+...+y_2^2+y_1^2}\upsilon_y + \frac{n-1}{z_n^2+...+z_2^2+z_1^2}\upsilon_z}{\frac{n-1}{x_n^2+...+x_2^2+x_1^2} + \frac{n-1}{y_n^2+...+y_2^2+y_1^2} + \frac{n-1}{z_n^2+...+z_2^2+z_1^2}}$$

where y_{n+1} and z_{n+1} have to be defined correspondingly. Observing agent x the singular behavior can be deduced and is given by $x_{n+1} = 0$ for $x_i = 0$, $y_i \neq 0$ and $z_i \neq 0$ for $i = 1...n$. This means that the weight of one agent leads to taking its measurement for the result of a negotiation and its estimated deviation holds the value of 0 for all the following negotiations. The system cannot leave this state because the ratio of the agents ratings increase to infinity. Extending the buffer of deviations only delays the singular behavior but also lowers the reaction on defective or miscalibrated sensors. Trying to define a minimum value for the deviations only lead to a situation where the deviation of one agent reached the defined minimum and never recovered from this state.

3.3. Normalized Weighted Arithmetic Mean

The singular behavior of the above described negotiation method can not be avoided, but an enhancement enables the system to recover from such a situation. The enhancement is based on a preparatory description of the accuracy of each sensor. A restricted quality description of each sensor prevents an infinite weight of a single measurement. To realize the restriction we normalize the standard deviation of a sensor from the range 0; ∞ to a reliability z_r in the range (0; 1 where it is important to distinguish several good measurements from each other. Therefore we use $f(x) = 1 - \mathrm{e}^{-x}$ for the normalization. In a negotiation the weight for each measurement x_r of a sensor is defined by $w_r = 1 - z_r = \mathrm{e}^{-x}$. And the combination of several measurements to one result is

$$q = \frac{\sum_1^n w_r x_r}{\sum_1^n w_r}$$

Normalizing

It is necessary to use the same normalization for each sensor with the same standard deviation. Consider a negotiation with three participating agents. Using sensors with identical standard deviations but different normalizations the system again runs into a singular behavior and in the end takes the measurements from the agent that benefits from its normalization. But on the other hand, if we have different types of sensors with unequal standard deviations or resolutions it is not practicable to use the same normalization for the sensors reliabilities. Hence individual normalizing is necessary for each sensor and it has to be calibrated once before the sensor data can be included in negotiation processes.

To realize an individual normalization the corresponding function $f(x)$ has to be modified using a coefficient. This leads to

$$f(x) = 1 - e^{-\gamma x} \qquad \text{with} \qquad \gamma = \ln \frac{\eta - 1}{\sigma}$$

where σ is the anticipated standard deviation of the sensor and η the desired normalized value. The anticipated standard deviation of the sensor has to be acquired manually by measuring reference values or can be adopted from the sensor's hardware specification (which has to be reliable). We set the target for the normalized value which represents a deviation of a single measurement in the range of the standard deviation to $\eta = 0.5$ which is in the middle of the range of the weights. The definite value of η can be varied, but it should not be near the boundary of the range of the sensor's reliability to maintain proper differentiation and avoid numerical problems. The target value η has to be the same for every agent participating in a negotiation process.

Results from Simulated Negotiations - Proving Robustness

The results from applying the Normalized Weighted Arithmetic Mean method to the above described simulation based on five agents are shown in Fig. 2. The phase of initialization is visible in both diagrams. It lasts 200 negotiations according to the size of the buffers for the results and measurements.

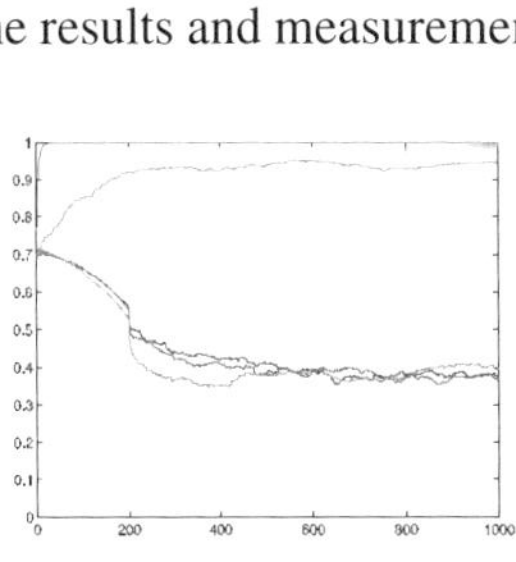

(a) Normalized Reliabilities of the agents

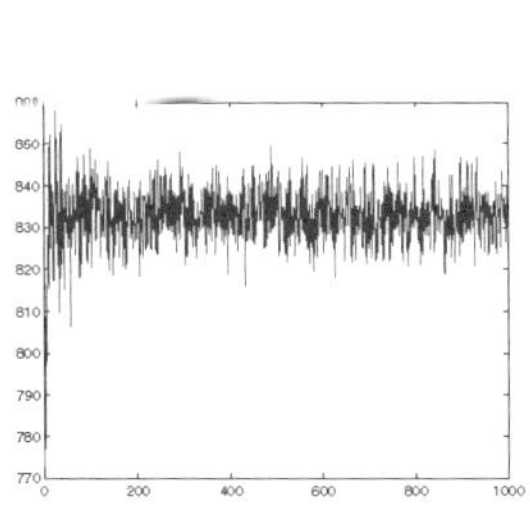

(b) Negotiation results for the measurements in 10th of degree

Figure 2. 1000 consecutive simulated negotiations with 5 agents of similar compass sensors

The system did not run into a singular behavior itself. To prove the robustness and to test the recovery of the system from disturbances we artificially altered the system. First the weighted deviation of one agent with an offset in the measurement results was set to a very low value which results in a big weight of this agent's measurements in the following negotiations. The estimation of the standard deviations of the other agents is also influenced. The disturbance was added in negotiation 400 and the modification is shown in Fig. 3. This test shows that the system can recover from the state that one agent's deviation is estimated to be 0 where the system described before ran into the singular behavior. After the number of negotiations corresponding to the buffer size the system has completely recovered. Similar to the behavior in the initialization process.

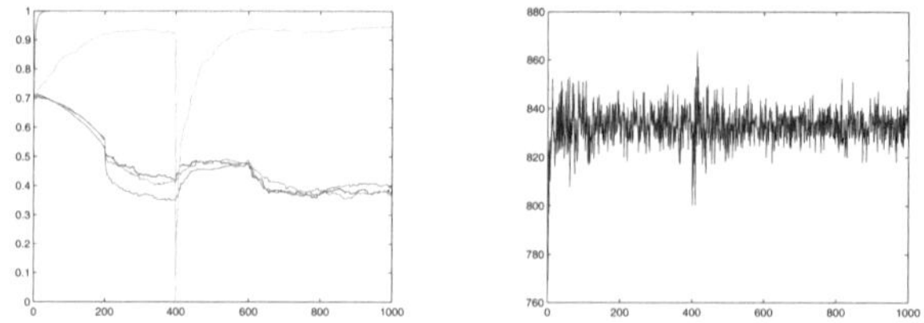

(a) Reliabilities of the agents (b) Corresponding results

Figure 3. 1000 consecutive simulated negotiations with 5 agents. The deviation of one sensor is artificially set to 0 in negotiation 400

The second disturbance to the system is an artificially added noise to the measurements of one sensor. In reality this could be the result of a physical influence to the sensor from which it gets decalibrated. The effect is similar to the one of the first disturbance and shown in Fig. 4. One agent overestimates its own reliability after the disturbance. This influences the results of the negotiations and the self-assessment of the other agents. The system adapts to the new conditions during a number of negotiations according to the described buffer size.

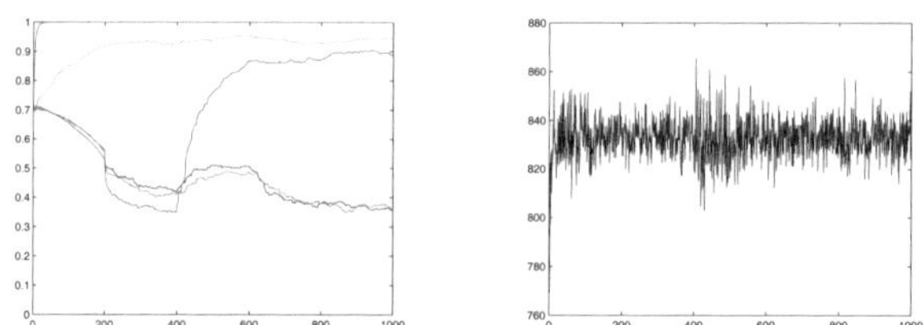

(a) Reliabilities of the agents (b) Corresponding results

Figure 4. 1000 consecutive simulated negotiations with 5 agents. Artificially noise is added to the measurements of one sensor from negotiation 400 onward.

4. Preprocessing to Exclude Erroneous Measurements

Each agent providing sensor data has to execute a preprocessing to improve the quality of the information. The described negotiation process combines the measurements in general to one result without any knowledge about its semantics. But an agent itself can preprocess the sensor data with regards to specific properties of the sensor, e.g. using a low-pass filter. In contrast to the preprocessing by the agent, the preprocessing in the negotiation process can compare the measurements from different sensors to the same phenomenon before combining them to one result.

Clustering of the sensor data is one solution to identify and exclude erroneous or at least abnormal measurements from calculating the result [8]. Clustering works fine if we have lots of measurements from which it has to identify outliers. To prevent the preprocessing from excluding non-outliers from only a few measurements we use Fuzzy Clustering [9]. More details to the algorithm and an evaluation are described in [10].

5. Comparison of the different methods

To compare the results of the different described methods we calculated the average absolute deviations of all negotiation results from the corresponding known real values in the simulations. Table 1 shows the comparison of the described methods, the statistically optimized approach and the normalized weighted arithmetic mean algorithm, to the simple arithmetic mean. The columns normal, disturbed and total describe the mean deviation of the whole simulation without disturbance, the deviation in the moment of the disturbance and the mean deviation at the end of the simulation including the disturbance.

Method	normal	disturbed	total
Arithmetic mean	10,3	13,1	13,1
Statistically Optimized	4,8	5,9	5,7
Normalized	4,7	8,1	5,9
Normalized and Fuzzy	4,7	7,1	6,1

Table 1. Deviations of the negotiation results

Of course the statistically optimized approach leads to better results than the simple arithmetic mean. But since this approach bears the risk of singular behaviors it is not applicable. The enhanced approach of the normalized weighted arithmetic mean algorithm is not as accurate as the first approach but equally better in comparison to the simple arithmetic mean and applicable without restrictions. Using fuzzy clustering for preprocessing the sensor data makes the second approach more robust against erroneous data and changes in the agent's conditions but downgrades slightly the long-term results.

6. Evaluation with real Sensor Data

After the simulations described in section 3 an evaluation of the methods *based on real measurements* was accomplished. For the clarity in the diagrams we repeated the measurement of only one single value although it is possible to apply the methods to non-continuous measurements of varying phenomena. We recorded more than 2000 measurements in a row of eight digital compasses of the same type which were aligned to measure the direction of a building. Fig. 5a shows the sensor data of uncalibrated compasses including different standard deviations and offsets.

The first processing step is the fuzzy clustering. Fig. 5b shows the measurements in red and the main cluster in blue. The results of the negotiations of the clustered measurements is shown in green. Fig. 6a shows the corresponding ratings of all the agents. The diagram shows that the three agents with measurements near the result are rated good where all the others are rated very poor. This indicates the strong influence of the fuzzy clustering on the result of negotiations on measurements of uncalibrated sensors with offsets. In assumption that the real value is near the arithmetic mean of all the measurements this result and the ratings are a good solution.

In a second experiment we wanted to test our method with data of calibrated sensors. Therefore we added a constant correction value to the measurements of each sensor to eliminate the offsets. The ratings of the agents in this experiment are shown in Fig. 6b

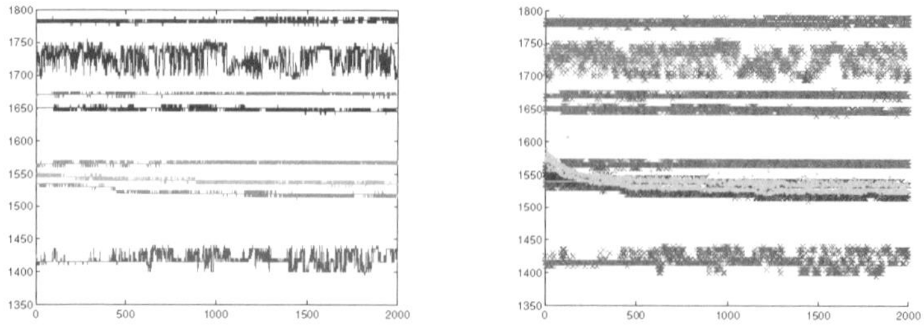

(a) Sensor data: results in 10th of degree, 2000 measurements of eight different digital compasses

(b) Clustering of sensor data and results of negotiations

Figure 5. Evaluation using real sensor data

where two sensors with higher noise in the measurements are rated worse than the others. A third sensor with a drift towards the arithmetic mean can be identified by bad ratings in the beginning.

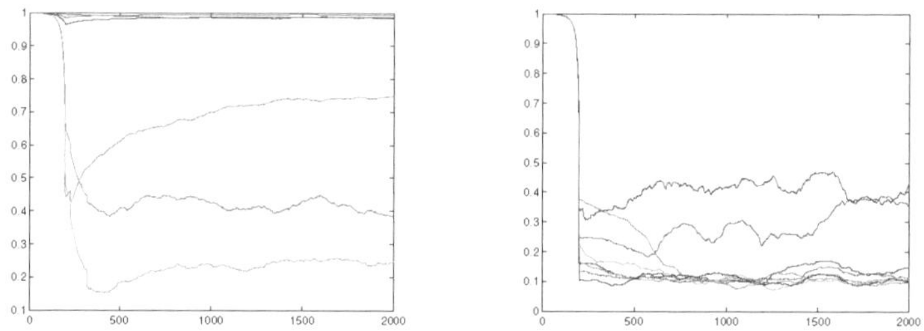

(a) Ratings of uncalibrated sensors

(b) Ratings of calibrated sensors

Figure 6. Ratings of sensors in negotiations based on real sensor data

In an open platform it is necessary to avoid that the system can be tricked by single agents, providers or users. In the simulations we proved that altered ratings or erroneous data can be filtered in the negotiation process. To test the robustness on real sensor data we influenced several measurements in a row of one sensor with a strong magnet. The measurements with a deviation of up to 60 degrees are shown in Fig. 7a, the results of the corresponding negotiations with the other agents in Fig. 7b. The result during the disturbance deviates only about 1.5 degrees in the beginning. Afterwards the affected sensor is rated worse and has less influence to the negotiation result. Additional trust models to single sensors are dispensable when the rating of a sensor is not managed by itself but in the shared context models which prevents it from continuous cheating.

7. Conclusion

An open platform that manages and distributes shared context models needs methods to include sensor data and the possibility of updates based on multiple measurements to eliminate inconsistencies. The statistically optimized approach to combine measure-

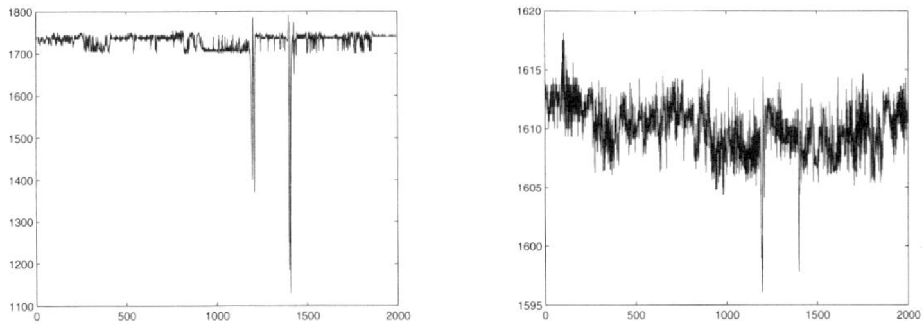

(a) measurements affected by a magnet (b) Corresponding negotiation results

Figure 7. Negotiations based on real sensor data with disturbance

ments of one phenomenon described in [7] and other weighted filters have one big disadvantage - the rating system can run into a singularity of which it never recovers, especially when the data processing is based on only a few measurements. In an open and worldwide system it is not possible to reset the ratings during runtime. Therefore we had to advance the approach and introduced the normalized weighted arithmetic mean algorithm. The enhancement is a restricted quality description of each sensor that prevents an infinite weight of a single measurement, as we proved in experiments on simulated and real sensor data. Combined with fuzzy clustering this approach adopts rapidly to decalibration or erroneous sensor data.

References

[1] A. Leonhardi and K. Rothermel, "Architecture of a large-scale location service," in *22nd International Conference on Distributed Computing Systems (ICDCS '02)*. IEEE, 2002, pp. 465–466.

[2] M. Bauer, C. Becker, J. Hähner, and G. Schiele, "ContextCube – providing context information ubiquitously," in *Proceedings of the 23rd International Conference on Distributed Computing Systems Workshops (ICDCS 2003)*, 2003, pp. 308–313.

[3] U.-P. Käppeler, T. Drosdol, T. Schwarz, and S. Michael, "Sensorcontextserver and sensorclient in der nexus-plattform," University of Stuttgart," Technical Report, 2005.

[4] M. Grossmann, M. Bauer, N. Hönle, U.-P. Käppeler, D. Nicklas, and T. Schwarz, "Efficiently managing context information for large-scale scenarios," in *Proceedings of the 3rd IEEE Conference on Pervasive Computing and Communications*, March 2005.

[5] D. Nicklas, M. Grossmann, T. Schwarz, S. Volz, and B. Mitschang, "A model-based, open architecture for mobile, spatially aware applications," in *Proc. of Symp. on Spatial and Temporal Databases*, 2001.

[6] M. Bauer and K. Rothermel, "An architecture for observing PhysicalWorld events," in *Proceedings of the 11th International Conference on Parallel and Distributed Systems: ICPADS 2005; Fukuoka, Japan, July 20-22, 2005*. IEEE Computer Society, Juli 2005, Artikel in Tagungsband.

[7] C. F. Dietrich, *Uncertainty, Calibration and Probability: The Statistics of Scientific and Industrial Measurement*, 2nd ed. Adam Hilger, 1991.

[8] J. Bacher, *Clusteranalyse, Anwendungsorientierte Einführung*, 2nd ed. Oldenbourg, 1996, in German.

[9] H. Timm, "Fuzzy-clusteranalyse: Methoden zur exploration von daten mit fehlenden werten sowie klassifizierten daten," Ph.D. dissertation, Otto-von-Guericke-Universität Magdeburg, 6 2002.

[10] R. Benkmann, U.-P. Käppeler, O. Zweigle, R. Lafrenz, and P. Levi, "Resolving Inconsistencies using Multi-agent Sensor Systems," in *Proceedings of the 8th Conference on Autonomous Robot Systems and Competition: Robotica 08; Aveiro, Portugal, April 2nd, 2008*, L. S. Lopez, F. Silva, and V. Santos, Eds., Aveiro, April 2008, pp. 93–98.

Intelligent Autonomous Systems 10
W. Burgard et al. (Eds.)
IOS Press, 2008

doi:10.3233/978-1-58603-887-8-308

Robust Vision-Based Semi-Autonomous Underwater Manipulation

Marc HILDEBRANDT [a], Jochen KERDELS [a], Jan ALBIEZ [a] and Frank KIRCHNER [a]
[a] *DFKI Robotics Lab Bremen, 28359 Bremen, Germany*

Abstract. The increasing demand for deep sea resources induces an increased amount of robotic technology to work in this harsh environment. In this paper we present a visual servoing approach to release the strain on operators controlling the currently used remotely operated vehicles, by giving them the possibility to hand some of the tasks over to a semi-autonomous control system. The paper deals with the basic control of a commercially available 7 degrees of freedom hydraulic manipulator, the 3D camera system used and the object recognition and visual servoing control. The presented approach focuses heavily on robustness to augment the acceptance with the offshore industry.

Keywords. Underwater Robotics, Visual Servoing, Dexterous Manipulation, Control, Kinematics

Introduction

An increasing trend in the offshore industry is to mine deep sea resources, e.g. oil or mangan. At the moment this is a very expensive endeavor. Especially, cost-efficient 24h systems for the inspection and maintenance of deep sea production facilities are needed.

The current state of the art for deep-sea system-maintenance done at the well heads and pump stations of offshore oil- and gas-rigs are manipulation tasks like opening or closing of valves, setting up structures or guiding equipment lowered down from the surface. All this work is accomplished by remotely operated vehicles (ROVs) relying on a cable connection to a surface vessel for control and energy. The demands on an operator team of such a vehicle are high. They have to control the vehicle in all six dimensions, while trying to interpret two dimensional video feeds as three dimensional data and accomplishing the aforementioned manipulation tasks. The control interface for these systems are classical tele-operation interfaces relying on a master-slave control with limited feedback ability.

Due to the high costs for operating a ROV and its parent ship the offshore industry is extremly conservative in terms of new, untried technology. The environmental conditions for the technical systems are extremly harsh. Up to 600 bar of pressure, low temperatures and a highly corrosive environment while deployed is added to rough handling onboard the support vessels due to sometimes harsh weather conditions resulting on a high strain on every part of the equipment. In this paper we present a visual servoing approach which therefore relies mainly on components well known in the offshore industry like the Orion 7P hydraulic manipulator and extends its capabilites by adding semi-autonomous components. So far there has been some work done in underwater vision for controlling AUV or ROV (e.g. [13]), but in the case of autonomous underwater manipulation closed-loop vision-based control has not been applied to this degree.

1. Manipulator Control

Underwater manipulator control is a very current topic. A number of significant differences to traditional manipulator control complicates the transfer of established techniques and algorithms to this specific new environment [3]. Some of the major differences and problems are the following:

- **Instable Base**
  Underwater manipulator systems are usually not ground-based but operate from platforms like heavy-workclass-ROV systems. These systems inherently have 6 DOF and are subject to effects of currents, swaying motions induced by active vehicle movement, and the 'tail wags the dog'-problem [4], placing the manipulator system in a very unstable configuration.
- **Low Control Frequency**
  Since Underwater manipulators are built for manual operator control, their control frequency is rather low (e.g. 60Hz forward/12Hz response for the Orion7P). This poses a problem on e.g. smooth visual servoing.
- **Limited Feedback**
  Due to the difficulties of pressure-resistant sensor systems, most manipulators provide very little sensory feedback. The only two established types of feedback are position and force. Multi-dimensional sensor-arrays for slip-detection in the jaws or similar sensors used in traditional robotics are not available in any form. Additionally, the visual feedback is only available through the on-board camera systems, which usually have mediocre image quality and, due to their high costs, are only used in small numbers.
- **Highly Dynamic Environment**
  The underwater environment is very dynamic. The workspace cannot be cleared of marine life (which is usually surprisingly interested in new light-sources and movement), and movements of the base systems causes changes in size and shape of the workspace.

The consequence of these problems is that the modus operandi of underwater manipulators still is manual operator control. In the CManipulator project, computer-based manipulator control is evaluated. The basic setup of the used system is shown in figure 1.

At this point, various benefits of computer-based manipulator control could already be tested and will be addressed in depth in the following paragraphs.

1.1. Improved sensory feedback

In the original deployment of the used Orion7P manipulator system the only sensory feedback was the joint angles as alphanumeric data in the control panel. In a situation where only parts of the manipulator arm are visible, this makes it very hard to estimate the current manipulator configuration. To improve this shortcoming, a 3D-visualization

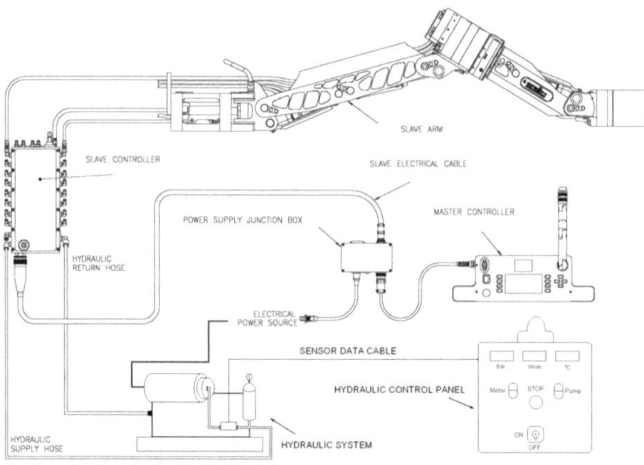

Figure 1. Communication connections in a typical manipulator system.

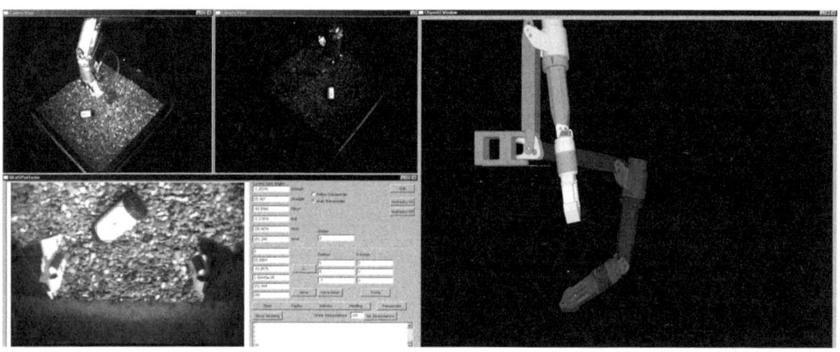

Figure 2. View of the Orion7P control center. On the left side the three camera images can be seen, on the right side the 3D representation of both the current (solid) and future (transparent) positions of the manipulator.

of the system's current state was generated. It allows real-time movement and rotation in order to give the operator the best observation position. Additionally, a second 3D visualization of the manipulator arm is projected as semi-transparent overlay. Now it is possible to first visualize a desired manipulator configuration, and allows a much safer movement after validation. This can be seen in figure 2 on the right side.

To improve the ability to precisely manipulate even small objects, a wrist-camera was mounted onto the last joint of the Orion7P. While this restricts its otherwise freely rotable wrist to $\pm 180°$, it provides valuable close-object information and eye-in-hand visual servoing [5]. Currently, different approaches to mount the camera in a way not restricting continuous rotary motion are evaluated.

Another area of interest is jaw-force-feedback. The original Orion7P only provides different rates of jaw movement but does not provide any feedback information about position or force acting on the brackets. This results in always applying its peak force of 4400 N [6]. To prevent damage to manipulated objects and to detect object movement while in the manipulator's jaws, a 2D-sensor matrix is being applied on the jaws. First experiments with a piezo-resistive sensor matrix proved very promising and further research into this area is planned.

1.2. Cartesian manipulator control

A number of tasks requires manipulator control in the Cartesian coordinate system rather than the axial coordinate system intrinsic to 6DOF manipulators. The design of a fast algorithm for computing this transformation was given a high priority at the beginning of the CManipulator project. The inherent flaws of numerical algorithms (namely speed and computational stability) made them unsuitable for this application, and a closed-form geometric/analytic solution for the inverse kinematics problem of the Orion 7P was developed by solving (1) for θ_1 to θ_6 for a given $\mathbf{R}_6^0$ (the transformation from base coordinate system to tip coordinate system) containing the rotational (r_{1n}, r_{2n} and r_{3n}) and translatory (p_x, p_y and p_z) components of the desired end-effector position. $\mathbf{T}_{i+1}^i$ are the transformations along the kinematic chain of the manipulator according to Denavit-Hartenberg (see [15]).

$$\mathbf{R}_6^0 = \begin{bmatrix} r_{11} & r_{12} & r_{13} & p_x \\ r_{21} & r_{22} & r_{23} & p_y \\ r_{31} & r_{32} & r_{33} & p_z \\ 0 & 0 & 0 & 1 \end{bmatrix} = \mathbf{T}_1^0(\theta_1) \cdot \mathbf{T}_2^1(\theta_2) \cdot \mathbf{T}_3^2(\theta_3) \cdot \mathbf{T}_4^3(\theta_4) \cdot \mathbf{T}_5^4(\theta_5) \cdot \mathbf{T}_6^5(\theta_6) \quad (1)$$

With highly optimized implementation, the time for computation of one random inverse configuration could be reduced to 64μs on a desktop computer, enabling it to be calculated in real-time without consuming too much computational power. This is vital since resulting from the dynamic nature of the system a precalculation of position is not possible, so it needs to be done on-line. At the same time it has to be computationally stable to prevent undesirable movement, which is covered by using a closed-form solution.

One of the benefits of Cartesian manipulator control is the possibility to compensate much more intuitively for vehicle movement, which can be seen as rotation and translation of the manipulator's base. If this movement is known it could directly be compensated by a simple inverse movement of the end-effector position prior to the inverse-kinematics calculation. So in world coordinates this would lead to equation (2) with $\mathbf{P}_{TIP}$ (the absolute transformation to the tip coordinate system from world origin, including the ROV-position $\mathbf{P}_{ROV}$) staying constant at both times t_0 and t_1 as long as the ROV's movement between these two time-frames $\Delta\mathbf{P}_{ROV}$ is known.

$$\begin{aligned} &\mathbf{P}_{ROV}(t_0) \cdot \mathbf{R}_6^0 = \mathbf{P}_{TIP} \\ &\mathbf{P}_{ROV}(t_1) \cdot (\Delta\mathbf{P}_{ROV}(t_1 - t_0))^{-1} \cdot \mathbf{R}_6^0 = \mathbf{P}_{TIP} \end{aligned} \quad (2)$$

This idealized equation will not be directly applicable in practice. The exact measurement of vehicle movement is the subject of intensive ongoing research, but to date the results are of rather poor long-term stability. The consequence is that $\Delta\mathbf{P}_{ROV}$ is not exact enough to allow such direct compensation of vehicle movement. The approach used in the CManipulator project to meet this problem is visual servoing.

1.3. Visual Servoing

The basic idea of visual servoing is to use the visual information from a camera system to control a manipulator's movement. A good overview of basic visual servoing methods can be found in [7]. Usually, the goal is to change the end-effector's position relative to the object of interest (e.g. moving closer in order to manipulate it). Visual servoing assumes that the object of interest has already been detected in the camera image and will stay recognizable during the whole visual servoing process. This already poses an aggravated problem in the underwater environment, as described in section 2. Having surpassed this challenge the second problem is the dynamics of the system. Usually the object will not only move because of manipulator movement, but also because of vehicle or environment movement. Fortunately for a eye-in-hand-system this poses only minor problems, as the necessary transformation usually is not calculated for execution in one single step, but iteratively optimized in a closed-loop control fashion. This way small vehicle/object movements are compensated directly by the visual servoing algorithm.

A close analysis of a common underwater manipulation task, i.e. gripping of an object in the field of view, yields another problem: the object might be out of manipulator reach, or become out of manipulator reach due to vehicle movement during the servoing process. To prevent this from happening, not only the manipulator has to be controlled, but the vehicle as well. This would involve using a second visual servoing control loop which moves/keeps the vehicle in a desired position relative to the object, while the object is gripped by the manipulator. First steps into this direction were already made in [2]. Up to this point the two systems were operated separately, but for the future a coupled system is planned. The necessity for this coupling arises from the problem, that a system with two actively controlling components, which are connected in a kinematic chain as in the vehicle-manipulator system, tend to over-compensate: the manipulator-visual servoing loop may issue a movement command along e.g. the x-axis. At the same time the vehicle loop may issue a similar command along the same axis, resulting in a situation where the wrist-camera of the manipulator looses the object due to much larger movement than estimated. This can either be addressed by very small control steps, simply minimizing the probability of the occurrence of such a situation, or by the aforementioned coupling of the two systems. Since smaller control steps would need higher control rates as available in the Orion7P system, the second approach will be investigated.

Finally another underwater-specific problem has to be addressed for visual servoing. The used system was characterized with the term eye-in-hand system, which computationally is absolutely correct. Unfortunately the design of the Orion7P does not allow the mounting of the wrist-camera in a place where the object of interest will be visible through the whole visual servoing process. The best way to compensate for this would be the use of a second wrist-camera, which is not applicable due to a number of reasons. The only currently available solution is a visual servoing up to a certain point, and a subsequent teached-in final movement. This last movement should be as small as possible to minimize changes in the system's state. In our hitherto experiments this last correctional movement has not posed major problems.

2. Underwater Computer Vision

The underwater environment poses some challenges with respect to computer vision. For example, the optical properties of water are dependent on various parameters like salinity, temperature, and pressure (depth). Changes in these parameters cause a change in the density of the water which modifies the projection of the image in the optical system, e.g., at a depth of several thousand meters in cold saltwater, objects appear to be closer to the camera.

Another optical effect is caused by the different absorption rates of the particular wavelengths of light in water. In the visible spectrum the absorption rate rises with the wavelength, i.e. the red spectrum is absorbed first, then the green spectrum and after that the blue spectrum [11]. This effect causes the perceived color of an object to change depending on the distance to that object.

A third challenging property of the underwater environment is water turbidity. The turbidity reduces the visibility range and the contrast of the images. In addition, the lights used to illuminate the field of vision of an underwater vehicle are usually located near the cameras due to mechanical restrictions. Due to the small angle between the light and camera axis, most of the light is directly reflected back by small particles in the water (i.e. marine snow, etc.), which further degrades the visibility conditions.

Further details on these challenges in underwater computer vision are given in e.g. [13] and [14].

2.1. Stereo Camera System

In addition to the wrist-mounted camera a stereo camera system is used in the CManipulator project. The two high-quality digital cameras are capable of 200 FPS noise-free image transmission at their full resolution. This enables very good image qualities even in low lighting conditions often encountered in underwater environments. This stereo camera system allows 3D-reconstruction of object positions in observed scenes. In order to make these calculations the system has to be calibrated both intrinsically and extrinsically. Intrinsic calibration includes the determination of the camera-lens parameters, e.g. distortion, for each individual camera. Extrinsic parameters describe the camera's position and orientation. Since the optical properties of water change with its chemical composition and pressure as described above, the intrinsic calibration has to be conducted just before the usage of the system, on site. The same applies to the extrinsic calibration, on the one hand since it relies on the intrinsic calibration, on the other hand since its accuracy depends on an unchanged system after calibration, which would not be certain if conducted prior to submersion.

The conclusion is, that a method had to be developed which calibrates the camera system at any specified time without external operator actions necessary. Our solution involves a tool-change, where the Orion7P grabs a 2D calibration rig and moves it into the FOV of the cameras in a pre-defined fashion. The images acquired in this fashion are then calibrated using the algorithm described in [12]. The result is a fully calibrated camera system with all the benefits described above.

2.2. Object Recognition

In one of the application scenarios of the CManipulator project the operator of the system selects the object which should then be picked up. The operator selects the object by clicking on it in one of the images coming from the stereo vision system. The object then is searched in the second image by scanning along the corresponding epipolar line based on the previously described camera calibration. With these object positions in the images, a position in 3D space is calculated and the robotic arm is moved to an intermediate position above the object, so that the object is visible in the image coming from the wrist-camera. The object is then tracked during the visual servoing process (s. section 1.3).

Figure 3. The Orion 7P in the Underwater Testbed of the DFKI-Lab Bremen.

In the current stage of development, the object which is picked up is a transponder mock-up. The experimental setup is shown in fig. 3. The transponder is of cylindrical shape, 75mm in diameter, and 200mm in length. It has a yellow-colored body with black caps at both ends. Although this object seems to be easily identifiable in an image, the different perspectives, varying lighting conditions, and the usage of different cameras make it quite hard to find a stable detection method. Our current approach scans the area with a scan grid to detect the borders of the transponder. To detect the borders, an unsharp mask filter is applied to the scan line prior to the use of a standard edge detection filter. The collected border points are then checked against the shape properties of the transponder, e.g., parallel edges and aspect ratio. From the remaining set of points the orientation and size of the transponder is extracted. This straightforward approach turns out to be very stable under various lighting conditions and on different camera systems.

However, the final object recognition approach will use a more sophisticated method which allows the detection and tracking of arbitrary objects. The particular algorithms were developed in a parallel project which implements a vision-based station-keeping on a ROV [2]. The approach will search for interesting features in the region surrounding the point where the operator clicked on the target object. The interesting features are detected by an improved version of the Harris corner detector [8]. The features are then described with the SIFT descriptor [10], and subsequently the corresponding set of features is searched along the corresponding epipolar line in the second image. After the robotic arm has reached the intermediate position above the object, the set union of features from both stereo camera images is used to identify the object in the wrist-camera image. Once the object is detected, the features of the object are tracked using a pyramidal implementation of the Lucas Kanade feature tracker [9].

3. Conclusion and Outlook

In this paper we presented an underwater manipulation system which is, at the current stage of development, capable of autonomously grasping a transponder-like object in a static underwater environment where neither the base of the robotic arm nor the object are moving. In the next development stage we will enhance the system to be capable of grasping various objects in a dynamic underwater environment. In this *dynamic* case, the base of the robotic arm as well as the target object will move. Furthermore we will introduce the concept of skills to represent other common underwater manipulation tasks, e.g., opening/closing of valves or coupling of underwater connectors.

Acknowledgements

This work is funded in the CManipulator project by the German Ministry of Economics (BMWI), grant number 03SX231

References

[1] D. Spenneberg and J. Albiez and F. Kirchner and J. Kerdels and S. Fechner, *CManipulator: An Autonomous Dual Manipulator Project for Underwater Inspection and Maintenance*, Proceedings of OMAE 2007 ASME 2007 International Conference on Offshore&Mechanics and Arctic Engineering June 10-15, 2007, San Diego, USA.

[2] J. Kerdels and J. Albiez and F. Kirchner, *A Robust Vision-Based Hover Control for ROV*, Proceedings of Oceans 2008 April 8-11, 2008, Kobe, Japan.

[3] M. Hildebrandt and J. Albiez and F. Kirchner, *Computer-Based Control of Deep-Sea Manipulators*, Proceedings of Oceans 2008 April 8-11, 2008, Kobe, Japan.

[4] A.J. Healey and A.M. Pascoal and F. Lobo Pereira, *Autonomous underwater vehicles for survey operations: theory and practice*, Proceedings of the American Control Conference 1995 June 21-23, 1995, Seattle, USA.

[5] G. Chesi and K. Hashimoto, *Static-eye against hand-eye visual servoing*, Nippon Robotto Gakkai Gakujutsu Koenkai Yokoshu Vol. 19, 2001, Japan.

[6] Schilling Robotics, *Orion7P Technical Manual*, 1st ed. Davis, CA: Schilling Robotics 2007.

[7] F. Chaumette and S. Hutchinson, *Visual Servo Control*, IEEE Robotics and Automation Magazine, vol. 13, no. 4, 2007.

[8] C. Harris and M. Stephens, *A combined corner and edge detector*, Proceedings of The Fourth Alvey Vision Conference, 1988, S. 147-151

[9] J.Y. Bouguet, *Pyramidal Implementation of the Lucas Kanade Feature Tracker*, Intel Corporation Microprocessor Research Labs, 2000

[10] D.G. Lowe, *Distinctive image features from scale-invariant keypoints*, International Journal of Computer Vision, 60, 2 (2004), pp. 91-110.

[11] R. M. Pope and E. S. Fry, *Absorption spectrum (380-700 nm) of pure water. II. Integrating cavity measurements* Applied Optics, vol. 36, pp. 8710-8723, 1997.

[12] Z. Zhang, *A flexible new technique for camera calibration*, IEEE Transactions on Pattern Analysis and Machine Intelligence, 22(11):1330-1334, 2000.

[13] G.L. Foresti, *Visual inspection of sea bottom structures by an autonomous underwater vehicle*, IEEE Trans. Syst. Man and Cyber, Part B 31, 691-705, 2001.

[14] Y.Y. Schechner, N. Karpel, *Clear Underater Vision*, Proceedings of the 2004 IEEE Computer Society Conference on Computer Vision and Pattern Recognition, 2004.

[15] J. Denavit and R. S. Hartenberg,*A Kinematic Notation for Lower- Pair Mechanisms Based on Matrices*,Journal of Applied Mechanics, Trans. Of the ASME, June 1955.

Intelligent Autonomous Systems 10
W. Burgard et al. (Eds.)
IOS Press, 2008

doi:10.3233/978-1-58603-887-8-316

Classifying Efficiently the Behavior of a Soccer Team

José Antonio IGLESIAS [a] Agapito LEDEZMA [a] Araceli SANCHIS [a] and Gal KAMINKA [b]
[a] *Carlos III University of Madrid. {jiglesia, ledezma, masm}@inf.uc3m.es*
[b] *Bar-Ilan University, Israel. galk@cs.biu.ac.il*

Abstract. In order to make a good decision, humans usually try to predict the behavior of others. By this prediction, many different tasks can be performed, such as to coordinate with them, to assist them or to predict their future behavior. In competitive domains, to recognize the behavior of the opponent can be very advantageous.

In this paper, an approach for creating automatically the model of the behavior of a soccer team is presented. This approach is an effective and notable improvement of a previous work. As the actions performed by a soccer team are sequential, this sequentiality should be considered in the modeling process. Therefore, the observations of a soccer team in a dynamic, complex and continuous multi-variate world state are transformed into a sequence of atomic behaviors. Then, this sequence is analyzed in order to find out a model that defines the team behavior. Finally, the classification of an observed team is done by using a statistical test.

Introduction

Knowledge about others is very beneficial for coordination, collaboration and adversarial planning. In order to make a good decision, humans usually try to predict the behavior of others. There are new theories which claim that a high percentage of the human brain capacity is used for predicting the future, including the behavior of other humans [1]. In adversarial domains, Riley [2] supposes that human players observe the behavior of the opponent during a game and try to match the observed playing style to previously encountered ones; then, the player selects the best performed strategy against the previous opponent. Therefore, in order to act efficiently, agents (software agents, robots or human beings) should try to recognize the behavior of other agents. Different techniques have been used in agent modeling in very different areas; for instance, opponent-modeling in soccer domain simulation, intelligent user interface, and virtual environment for training. In particular, this paper focuses on the soccer domain simulation.

In this paper we propose an efficient approach for recognizing and classifying an observed team behavior. The goal of this research can be divided in 2 parts: Firstly, the different *team behavior models* (classes) are created by observing the behavior of a team during a game. Then, a team behavior is observed and classified into the predefined models (classes).

In a soccer team (multi-agent environment), a group of agents cooperates to achieve a common goal. Therefore, the changes made to the environment are not the result of the behavior of a single agent, but the interaction of the agent with each other and the world in which they act. As a consequence, the emergent behavior is usually hard to understand because the global behavior is not the sum of the local behaviors of the agents. We propose an approach for modeling the behavior of a team as a global behavior defined by the sum of the more relevant actions of its player members.

Because of any behavior has a sequential aspect, we consider that a team actions are usually influenced by its past actions. The presented approach is an notable and successful improvement of our previous work [4]. However, in this work, a soccer *agent* team behavior is represented as a distribution of its relevant atomic behaviors. Also, an effective statistical method is used in the classification method. This approach has been fully implemented and empirically evaluated in a complex and noisy multi-agent environment: the *Soccer Server System* [5] as used in the *RoboCup Soccer Coach Simulation* [6].

The rest of this paper is organized as follows. First, Section 1 provides a brief overview of the background and related work relevant to this research. The approach is detailed in section 2. The experimental results obtained in the proposed environment are shown in Section 3. Finally, Section 4 contains future work and concluding remarks.

1. Background and Related Work

In many areas is very useful to model and recognize the behavior of other agents. Agent modeling has been applied in real-time domains with continuous state and action spaces. In competitive domains, the knowledge about the opponent can be very advantageous. For example, in the *RoboCup Simulation League* [7] the modeling and classification of the opponent team is necessary to adapt the own team behavior. In this area, Carmel and Markovitch [8] propose a method to infer a model of the opponent's strategy which is represented as a Deterministic Finite Automaton. Tambe et al. [9] present an approach for tracking agent models based on a plan recognition task. Ledezma et al. [3] present an approach to modeling low-level behavior of individual opponent agents. Time series and decision tree learning are used by Visser and Weland [10] to induce rules that describe a team behavior. Riley et al. [2] propose a classification of the current adversary into predefined adversary classes. Han and Veloso [11] recognize behaviors of robots using *Hidden Markov Models*.

Similar to our approach, Kaminka et al. [12] recognize basic actions based on descriptive predicates, and detect the relevant actions of a team using a statistical approach and a *trie* data structure. This structure is also used in [13] to create frequent patterns in dynamic scenes. However, these previous works focused on unsupervised learning, with no ability to classify behaviors into classes.

2. Soccer Team Classification. Our Approach

In this section, our approach for modeling and classifying the behavior of a soccer agent team is described. This approach is divided in two different parts: **Off-Line Phase**: For each game observed, the behavior of a team is analyzed, modeled and stored in a behavior-library (*LibTBM*) (Section 2.1). **On-Line Phase:** The goal is to classify *on-line* the behavior of an observed team into the team behavior models previously stored in the *LibTBM* (Section 2.2).

2.1. ***Off-line phase:*** *Construction of the Team Behavior Models*

The actions performed by an agent soccer team are inherently sequential, and this sequentiality is considered in the modeling process proposed. This section describes the two stages for constructing the model of a soccer agent team from its observation: First, the observation stream is transformed into an ordered sequence of recognized atomic behaviors (Section 2.1.1), then the model is created and stored in *LibTBM* (Section 2.1.2).

2.1.1. *Obtaining Atomic Behavior Sequences*

In this phase, each observation is a snapshot of the soccer field that do not offer any information about its actions. Therefore, the actions taken by the agent team should be estimated by contrasting consecutive snapshots. The procedure to identify atomic behaviors (events) in a soccer game is based on a work of Kuhlmann et al. [14] and eight atomic behaviors are inferred: *Pass, Dribble, Intercept pass, Steal, Goal, Missed shot, Foul* and *Hold*. Each event is characterized by the players that have executed the action. The result of this phase is a sequence of atomic behaviors ordered by time.

As a sample sequence of events, lets consider we are observing an agent team and its behavior (during a small period of time) is represented by the following sequence: {*Pass1To2* → *Pass2To1* → *Pass1To2* → *Pass2To1* → *Dribble1*}.

2.1.2. *Creating the behavior model*

In this approach, the temporal dependencies are very significant and it is considered that a current event depends on the events that have happened before and it is related to the events that will happen after. Taking this aspect into account, to get the most representative set of sequential events (subsequences) from the acquired sequence, the use of a *trie* data structure [15] is proposed. For constructing a *trie* from a single sequence of events, the sequence is processed in three steps explained below:

a. Segmentation of the sequence:

In this step, the sequence is segmented into several subsequences. This segmentation is divided by defining an appropriate length and obtaining every possible ordered subsequence of that specific length. In the proposed sample sequence, let 3 be the subsequence length, then it is obtained: {*Pass1To2* → *Pass2To1* → *Pass1To2*} and {*Pass2To1* → *Pass1To2* → *Pass2To1*} and {*Pass1To2* → *Pass2To1* → *Dribble1*}

b. Storage of the subsequences:

The subsequences of events are stored in a *trie*, such that all possible subsequences are accessible and explicitly represented. In a *trie*, every node represents an event, and the node's children represent the events that have appeared following this event. Also, each node keeps track of the number of times an event has been inserted on to it. As the dependencies of the events are relevant in an agent behavior, the subsequence suffixes (subsequences that extend to the end of the given sequence) are also inserted.

Considering the previous example, the first subsequence ({*Pass1To2* → *Pass2To1* → *Pass1To2*}) is added as the first branch of the empty *trie* (Figure 1 a). Each event is labeled with the number 1 that indicates that it has been inserted in the node once (in Figure 1, this number is enclosed in square brackets). Then, the suffixes of the subsequence ({*Pass2To1* → *Pass1To2*}) and {*Pass1To2*}) are also inserted (Figure 1 b). Finally, after inserting the three subsequences and its remaining suffixes, the completed *trie* is obtained (Figure 1 c).

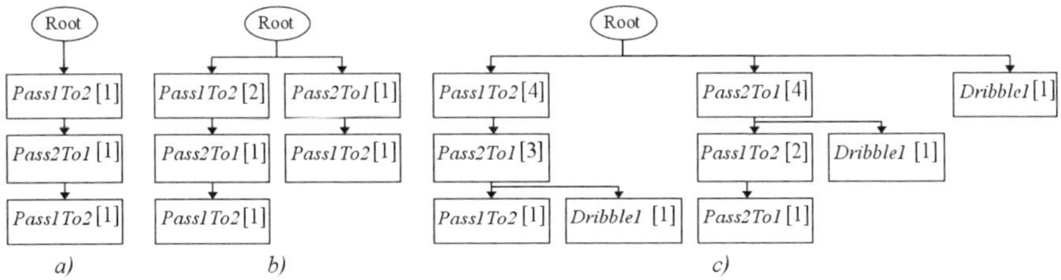

Figure 1. Steps of creating a *trie*

c. Creation of the behavior model: distribution of atomic behaviors

Once the *trie* is created, it is traversed to calculate the relevance of each subsequence (path from the *root* node to any other node of the *trie*). For this purpose, frequency-based methods are used. In particular, in this approach, to evaluate the relevance of a subsequence, its relative frequency or support [16] is calculated. Therefore, in this step the *trie* is transformed into a set of subsequences labeled with a value (support).

After labeling all the subsequences of the *trie*, the model of an agent behavior is represented by the distribution of its subsequences. In the previous example, the *trie* consists of 9 nodes; therefore, the model consists of 9 different subsequences which are labeled with its support. Figure 2 represents the distribution of these subsequences.

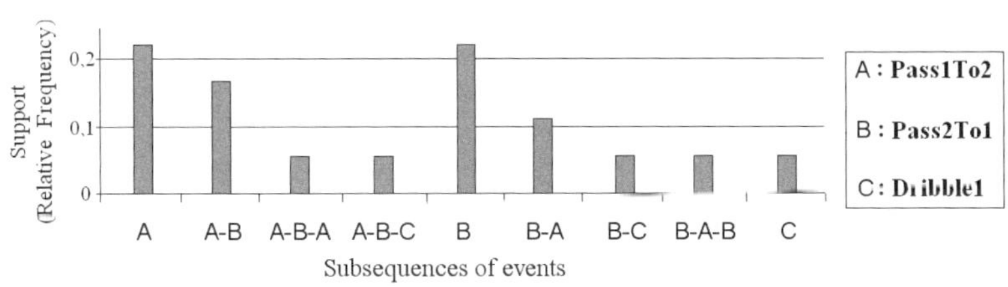

Figure 2. Distribution of subsequences.

However, in the environment in which we will apply our approach, the behavior of a team must be inferred by using two different games: *Base Strategy:* Game in which a general strategy of the team is played. And *Play Pattern:* Game in which a specific behavior of several team players is activated using the previous *Base Strategy*.

Therefore, in this case, the behavior of the team is modeled by creating the two distributions (*tries*) of atomic behaviors of the two games (*base strategy* and *play pattern*) and then, comparing them. This comparison is needed because the relevant subsequences appears only in the *play pattern* or with a low value in the base strategy. Therefore, the result of this comparison is a distribution with the subsequences of the *Play Pattern* distribution labeled by subtracting (*Play Pattern* subsequence support) from (*Base Strategy* subsequence support).

Finally, once a behavior model has been created, it is stored in *LibTBM* as a *trie* for a good and effective handling. The different models created are stored and labeled in the library with a *name* that identifies them.

2.2. ***On-line phase:*** *Team Behavior Classification*

In this second phase, a game is observed and the set of behavior models that the team follows must be reported. Firstly, the observations of the agent team to classify are collected and its behavior model is created. Then, this model is matched with all the behavior models stored in *LibTBM*. As both models are represented by a distribution of events, a statistical test is applied for comparing these distributions. The proposed test applied for matching two behaviors is a modification of the *Chi-Square Test* for two samples. A non-parametric test (or distribution-free) is used because this kind of test does not assume a particular population distribution. To apply this test, the behavior model to classify is considered as an observed sample and all the behavior models stored in *LibTBM* are considered as the expected samples.

The *Chi-Square Test* is the comparison of two sets of support values in which *Chi-Square* is the sum of the terms $\frac{(Obs-Exp)^2}{Exp}$ calculated from the observed (*Obs*) and expected (*Exp*) distributions (models). This test has not been used in our classification process because only the expected values are compared and if an observed value is not represented in the expected distribution, it is not considered. Therefore, in order to solve these problems, the way to compare the two distributions is modified to the sum of the terms $\frac{(Exp-Obs)^2}{Obs}$, what we call *Chi-Square-Obs Test*. By applying this test, a value that indicates the deviation of the two distributions is obtained. The lower the value, the closer the similarity between the two behaviors. Figure 3 represents graphically the idea of the proposed novel comparing method.

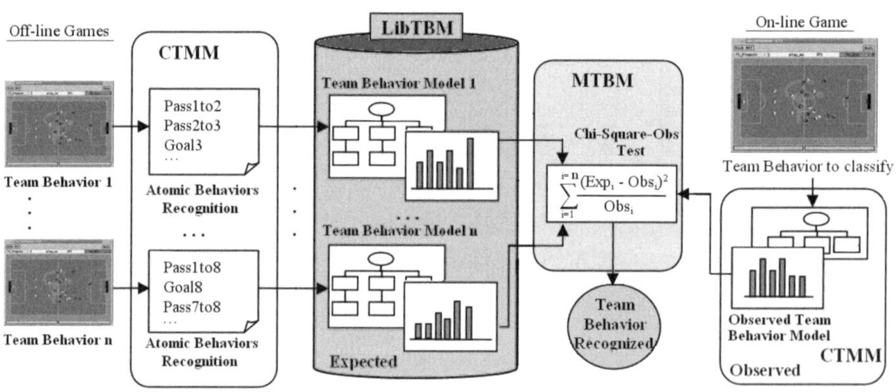

Figure 3. Team Behavior Classification Process

An important advantage of the proposed test is its rapidity because only the observed subsequences are evaluated. However, there is no penalty for the expected relevant subsequences which do not appear in the observed distribution. The *Chi-Square-Obs Test* is applied once for each behavior model stored in *LibTBM*. The number of terms to sum in each comparison is always the same: number of subsequences in the observed behavior model. It means that the degrees of freedom (*dof*) are the same in all the comparisons and a normalization of the results according to the *dof* is not necessary. The model which comparison obtains the lowest value is considered as the most similar one.

As example, lets consider the sequence that represents the observed behavior is: $\{Pass1To2 \rightarrow Pass2To1 \rightarrow Goal1\}$. Figure 4 shows the comparison between the previous expected distribution (*Expected Team Behavior Model 1*) and the observed distribution (*Observed Team Behavior Model*). The comparison value in this example is: $\frac{(0{,}22-0{,}16)^2}{0{,}16} + \frac{(0{,}16-0{,}16)^2}{0{,}16} + \frac{(0{,}22-0{,}16)^2}{0{,}16} + \frac{(0-0{,}16)^2}{0{,}16} + \frac{(0-0{,}16)^2}{0{,}16} + \frac{(0-0{,}16)^2}{0{,}16} = 0{,}525$.

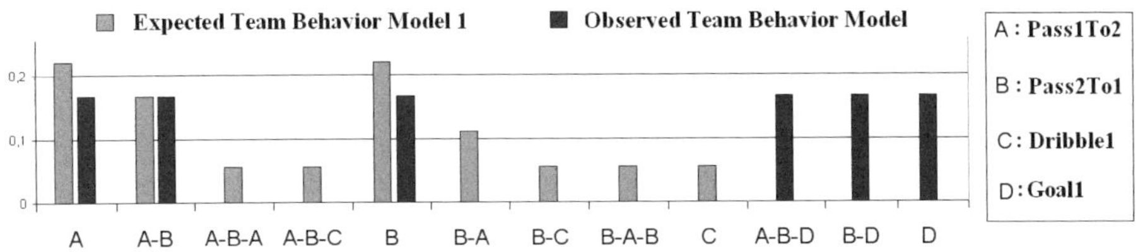

Figure 4. Observed and Expected Comparison Example

3. Experiments. RoboCup Soccer Team Behavior Classification

The *Soccer Server System* [5] is a server-client system which simulates a soccer game in a discrete fashion. Eleven players (agents) can only perceive objects that are in their field of vision and interact in a complex and noisy multi-agent environment. In particular, this research uses the environment of the *RoboCup Soccer Coach Competition* [6].

3.1. RoboCup Soccer Coach Simulation

The main goal in this environment is to classify a soccer simulation opponent team behavior by observing its actions (the behavior of the players does not vary significantly over the course of the game). The construction of models is done by analyzing several game *log files*. Each game represent a *team behavior*. Then, it is observed a new game in which several *team behaviors* are activated at the same time. The classification is done by reporting the *team behaviors* activated in the observed game.

For the experiments in this domain, we have used the rules from the *RoboCup 2006 Coach Competition* [17] and the experiments have been performed in the same way that this competition. However, although the competition consists of 3 different rounds with different *team behaviors* to analyze and classify, we only show the results of the first two rounds obtained using our approach. In the first round, 17 different *team behaviors* are analyzed (download from [6]) and stored in *LibTBM*. The games analyzed in this case are around 1000 to 3000 cycles games (in our case, the atomic behaviors identified for each game are around from 100 to 200). Then, in each iteration of the round, a different game where 4 or 5 different *team behaviors* have been activated is observed (*team behavior*).

3.2. Results

Table 1 shows the ranking list obtained for the 3 iterations of the first and second rounds (because of lack of space, only the 8 first *team behaviors* have been represented). The number of *team behaviors* activated in each iteration is indicated in brackets. In the first round, the *team behaviors* are identified with a number (from pattern00 to pattern16) and in table 1 the *team behaviors* that have been activated are marked with an asterisk. As we can see, the result are very promising since in the six first places of the 6 iterations, there are at least 3 *team behaviors* that have been correctly identified as activated.

Analyzing these results, the *player behaviors* named *pattern04* and *pattern16* have been classified correctly in first position. Although the way to define these *player behaviors* is using a special language called *CLang* [18] (with which the behavior of the simulated soccer player can be modified), we describe these *player behaviors* as follows:

Table 1. Results for the *RoboCup Coach Competition*. Round1

Round1 (17 different team behaviors)			Round2 (16 different team behaviors)		
Iter1 (4)	*Iter2* (5)	*Iter3* (5)	*Iter1* (5)	*Iter2* (4)	*Iter3* (4)
pattern04 *	***pattern16*** *	***pattern04*** *	***patternJ*** *	patternB	patternF
pattern16	***pattern01*** *	***pattern02*** *	patternL	patternF	***patternE*** *
pattern00 *	pattern00	pattern13	***patternF*** *	patternA	patternJ
pattern12	***pattern13*** *	pattern05	patternE	***patternE*** *	patternD
pattern15 *	pattern05	***pattern00*** *	patternD	patternJ	***patternB*** *
pattern05	***pattern07****	***pattern12*** *	***patternA*** *	***patternP*** *	patternH
pattern09	pattern03	pattern01	patternG	patternD	patternA
pattern03	pattern09	***pattern06*** *	patternP	***patternK*** *	patternP
...	...	...	...	...	...

- ***Pattern04:*** *Players 6, 7 and 8 pass the ball to an specific point in the field.*
- ***Pattern16:*** *Player 1 pass to player 3 or 4. Player 3 and 4 dribble to a fix space.*

However, the following *player behaviors* are not recognized correctly:

- ***Pattern14:*** *The player 3 dribbles to the space where the player 5 is situated. The player 5 dribbles to the space where the player 9 is situated. The player 9 dribbles to the space where the player 3 is situated.*
- ***Pattern08:*** *If the ball is situated in a defined area, player 8 dribbles to a fix space. Otherwise, player 8 passes to player 0.*

As we can see analyzing the results, our approach works successfully when the behavior of the team to recognize is related to the actions of the players. Any other kind of *team behavior* (related to the different field regions in which the action occurs or the cycle when it occurs) could not be detected. However, there are some *team behaviors* that are not related to actions but the way the players play makes that it can be recognized.

These results are very interesting for the official goal of the *RoboCup* (*to create a team of fully autonomous humanoid robot soccer players*) because, as humans do, the robot players should recognize the opponent behavior in order to act optimally.

4. Conclusions and Future Works

This paper presents an approach for modeling and classifying a soccer team behavior from observation based on a previous work [4]. The underlying assumption in this approach is that the observed team behavior can be transformed into a sequence of ordered atomic behaviors. This sequence is segmented and stored in a *trie*, then the relevant subsequences are evaluated by using a frequency-based method. The main aspect in this approach is that the model of an agent behavior is represented by a distribution of relevant subsequences. Also, for classifying a given team behavior, a modification of the *Chi-square Test* for two samples is proposed.

This approach has been evaluated in the real-time and multi-agent domain *RoboCup Soccer Coach Simulation* by conducting a large set of experiments. Although the results depend of the behavior to recognize, the obtained results by using the proposed approach are very satisfactory. The approach is evaluated only in one domain; however the only step of the approach which is domain-dependent is the transformation from observation to a sequence of atomic behaviors. Therefore, the approach can be generalizable to modeling and classifying other behaviors represented by a sequence of events (such as GUI events, network packet traffic and so on).

In this research, we have considered that the behavior of a team does not change over a game. However, in order to construct a soccer team model more realistic, it should be frequently revised to keep it up to date. This aspect could be solved by using *Evolving Systems* and it is proposed for future work. The use of the classification result for carrying out useful actions in the proposed environment is also considered as future work.

Acknowledgments. This work has been supported by the Spanish Ministry of Education and Science under project TRA-2007-67374-C02-02.

References

[1] N. J. Mulcahy and J. Call. Apes save tools for future use. *Science*, 312(5776):1038–1040, May 2006.

[2] P. Riley and M. M. Veloso. On behavior classification in adversarial environments. In *DARS*, pages 371–380.

[3] A. Ledezma, R. Aler, A. Sanchís, and D. Borrajo. Predicting opponent actions by observation. In *RobuCup*, volume 3276 of *Lecture Notes in Computer Science*, pages 286–296. Springer, 2004.

[4] J. A. Iglesias, A. Ledezma, A. Sanchis. A comparing method of two team behaviours in the simulation coach competition. InV. Torra, Y. Narukawa, A. Valls, J. Domingo-Ferrer (Eds.), MDAI, Vol. 3885 of LNCS, Springer, 2006, pp. 117–128.

[5] I. Noda, H. Matsubara, K. Hiraki, and I. Frank. Soccer server: a tool for research on multiagent systems. In *Applied Artiificial Intelligence*, volume 12, pages 233–258. Taylor and Francis Ltd, 1998.

[6] The robocup 2006 coach competition web page (http://ssil.uni-koblenz.de/rc06/), December 2006.

[7] H. Kitano, M. Tambe, P. Stone, M. Veloso, S. Coradeschi, E. Osawa, H. Matsubara, I. Noda, and M. Asada. The robocup synthetic agent challenge 97. In *IJCAI97*, 1997.

[8] D. Carmel and S. Markovitch. Opponent modeling in multi-agent systems. In *Adaptation and Learning in Multi-Agent Systems*, pages 40–52. Springer-Verlag: Heidelberg, Germany, 1996.

[9] M. Tambe and P. S. Rosenbloom. Architectures for agents that track other agents in multi-agent worlds. In *ATAL*, pages 156–170, 1995.

[10] U. Visser and H.-G. Weland. Using online learning to analyze the opponent's behavior. In *RoboCup*, pages 78–93, 2002.

[11] K. Han and M. Veloso. Automated robot behavior recognition applied to robotic soccer. In *Proceedings of IJCAI-99 Workshop on Team Behaviors and Plan Recognition*, 1999.

[12] G. A. Kaminka, M. Fidanboylu, A. Chang, and M. M. Veloso. Learning the sequential coordinated behavior of teams from observations. In *RoboCup*, volume 2752 of *Lecture Notes in Computer Science*, pages 111–125. Springer, 2002.

[13] Z. Huang, Y. Yang, and X. Chen. An approach to plan recognition and retrieval for multi-agent systems. In *Proceedings of AORC*, 2003.

[14] G. Kuhlmann, P. Stone, and J. Lallinger. The champion UT Austin Villa 2003 simulator online coach team. In *RoboCup-2003: Robot Soccer World Cup VII*. Springer Verlag, Berlin, 2004.

[15] E. Fredkin. Trie memory. *Comm. A.C.M.*, 3(9):490–499, Sept. 1960.

[16] R. Agrawal and R. Srikant. Mining sequential patterns. In *Eleventh International Conference on Data Engineering*, pages 3–14, Taipei, Taiwan, 1995.

[17] R. Committee. Robocup 2006 official rules for the competition (http://ce.sharif.edu/ m_sedaghat/ robocup/orga/rc06/rules0.0.pdf, December 2006.

[18] M. Chen, E. Foroughi, F. Heintz, Z. Huang, S. Kapetanakis, K. Kostiadis, J. Kummeneje, I. Noda, O. Obst, P. Riley, T. Steffens, Y. Wang, and X. Yin. Soccerserver manual ver. 7, 2 2002.

Intelligent Autonomous Systems 10
W. Burgard et al. (Eds.)
IOS Press, 2008

doi:10.3233/978-1-58603-887-8-324

Audio-Visual Detection of Multiple Chirping Robots

Alexey GRIBOVSKIY and Francesco MONDADA
Laboratoire de Systèmes Robotiques (LSRO), Ecole Polytechnique Fédérale de Lausanne, Switzerland

Abstract. Design, study, and control of mixed animals-robots societies are the fields of scientific exploration that can bring new opportunities for study and control of groups of social insects and animals and, in particular, for improvement of welfare and breeding conditions of domestic animals. Our long-term objective is to develop a mobile robot, socially acceptable by chickens and able to interact with them using appropriate communication channels. For interaction purposes the robot has to know positions of all birds in an experimental area and detect those uttering calls. In this paper, we present an audio-visual approach to locate the robots and animals on a scene and detect their calling activity. The visual tracking is provided by a marker-based tracker with help of an overhead camera. Sound localization is achieved by the beamforming approach using an array of sixteen microphones. Visual and sound information are probabilistically mixed to detect the calling activity. The experimental results demonstrate that our system is capable to detect the sound emission activity of multiple moving robots with 90% probability.

Keywords. Microphone arrays, sound localization, audio-visual multi-source information fusion.

Introduction

Study of animal-robot interaction can bring new opportunities to analyze and control groups of social insects and animals and, in particular, to improve welfare and breeding conditions of domestic animals. After the promising results of the European project Leurre, where a mixed society of cockroaches and robots was created and successfully controlled [1], we focused our interest on more complex social animals – chickens. Our goal is to develop a mobile robot socially acceptable by chickens and able to interact with them. Design of this type of robots has to take into account first of all the communication channels, relevant for the animals' social interactions. For birds in general and for chickens in particular there are two major communication channels: visual and auditory. Hence in order to interact with chickens, a robot has to be able to use efficiently both these modalities. One of the tasks that is addressed in this paper is the use of visual and audio information to obtain positions of all birds in an experimental area and to detect among them the birds uttering calls.

Recently, the field of robotic audition has received increasing attention [2–5]. The majority of studies focus on sound sources localization, separation and recognition. In a number of studies these problems are solved by using robotic auditory system of only two microphones [2]. However use of microphone arrays, containing a large number of sensors, provide considerably better results, as they allow to improve the resolution of the localization procedure and its robustness to ambient noise [3]. In [4] a robot equipped with an array of eight microphones localizes and tracks multiple moving speakers using beamforming and particle filtering techniques. In [5] the beamforming and Kalman filters are used for the same task by the SIG2 robot with an array of eight microphones. The method of audio-visual data fusion for human tracking and speaking activity detection was proposed in [6] and tested on the HRP-2 robot equipped with two arrays of eight microphones each. In the biological studies sensor networks, where each sensor includes a microphone array, are applied for detection, recognition and localization of bird songs [7].

In this paper, we present a system for analysis of audio-visual information that serves to determine positions of animals and robots in the experimental area and to detect sound emission activity. The positions of robots and animals are obtained by the marker-based visual tracker, while sound localization is provided by the interpolation beamforming approach using an array of sixteen microphones. These data are then probabilistically mixed to detect the calling activity. The system was tested to track the positions and sound emission activity of e-puck robots, playing a prerecorded bird's call, and in future will be tested on the real chickens.

Although the visual tracking and sound localization methods, used in our system are not new and have already been employed in robotics and biological study, there are no reported results known to authors where such methods were combined and applied for tracking animals and detecting their calling activity. Thus we believe that this approach provides an useful tool for the field of behavioral studies and interactions with vertebrates.

The paper is organized as follows. Section 1 presents a brief overview of the system. Sections 2.1 explains how the sound localization system works and Section 2.2 presents the visual tracker that we are using. Section 3 describes the probabilistic calls detector, followed by the experimental results in Section 4.

1. System Overview

The system is shown in Figure 1. It is composed of the following parts:

- a microphone array;
- a sound localization subsystem;
- a camera;
- a markers based visual tracking subsystem;
- a probabilistic calls detection subsystem.

The circular microphone array is formed by sixteen omnidirectional microphones. The input audio data are processed by the beamforming based localizer that provides directions to potential sound sources around the microphone array. The visual tracker processes the input video data and detects positions of all the object of interest in the experimental area. Then the calls detector uses results from these two subsystems to associate the sound sources with the tracked objects.

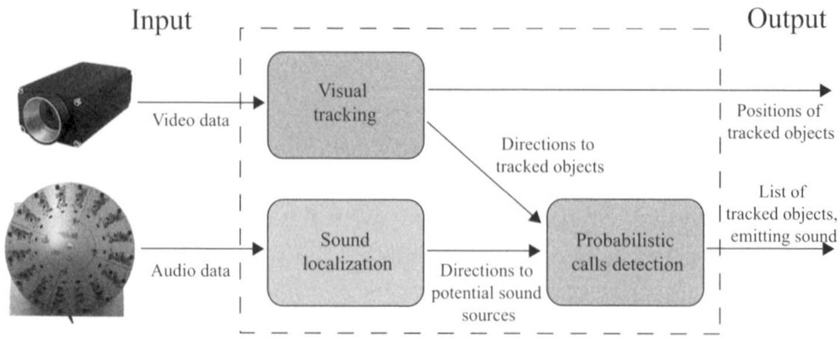

Figure 1. The overview of the system.

2. Features Extraction From Audio and Video Data

2.1. Sound Localization

In our experiments we use an array of sixteen microphones with a diameter of 15 cm (Figure 3b). We developed it taking into account that to minimize the spatial aliasing effect the number of microphones M has to satisfy the inequality

$$M \geq \frac{4\pi r f_{max}}{v_s}, \tag{1}$$

where v_s denotes the speed of sound (344 m/s), r is the radius of the array (in our case 7.5 cm) and f_{max} denotes the maximal frequency of a signal. As our microphone array is designed to be used in experiments with chickens, which calls consist mainly of frequencies between 400 and 6000 Hz [8], hence $M = 16$ satisfies (1) for the majority of calls.

2.1.1. Sound Localization by Beamforming

In sound localization systems based on the beamforming approach microphone arrays are steered to multiple directions and maximums of output signals' energy are then searched. From the mobile robotics applications point of view, an advantage of this approach over other sound localization methods, such as those based on time-difference of arrival information and techniques adopting spectral analysis concepts, is that beamforming can be used for wideband signals and a multi-speaker model [4, 9, 10].

In our system we use the interpolation beamforming method that is a modification of the classical delay-and-sum beamforming method [11]. Let $x_m(t)$ be the input signal for the m-th microphone, $m = 1, 2, ..., M$, and let θ be an arbitrary direction in the plane of the microphone array, $\theta \in [0, 2\pi)$. The output of the delay-and-sum beamforming steered to the direction θ has the following form:

$$y^{\theta}(n) = \sum_{m=1}^{M} x_m(n - \tau_m^{\theta}),$$

where τ_m^{θ} is the time of arrival difference between the m-th and the first microphones:

$$\tau_m^\theta = \frac{(\mathbf{r}_m - \mathbf{r}_1, \mathbf{e}^\theta)}{v_s}.$$

Here $\mathbf{r}_m$ is the position vector of the m-th microphone, $\mathbf{e}^\theta$ denotes the unit vector pointing in direction θ, and $(\cdot,\cdot)$ denotes the inner product of two vectors. The output energy of the beamformer steered to direction θ over a frame of the length N is thus given by:

$$E(\theta) = \sum_{i=1}^{N} y^\theta(i)^2. \tag{2}$$

If a sound source is located in a direction θ', then the beamformer's output energy $E(\theta)$ will have the local maximum at the point θ'.

The main drawback of the delay-and-sum beamforming lies in the sampling procedure: to approximate time delays required for beam steering with reasonable precision, signals have to be sampled at a rate much greater than the Nyquist frequency. The interpolation beamforming method overcomes this constraint by temporally interpolating the sensor data before the beamforming operation; this allows to sample input data at the Nyquist rate. The interpolation is performed by zero padding the sensor data and passing the resulting sequence through a finite impulse response (FIR) low-pass filter.

In order to match channels and consequently increase the noise suppression and localization precision of the beamforming we use the microphones' gain self-calibration procedure proposed in [12]. The procedure is based on projection of microphones' coordinates on direction of sound arrival line and approximation of received energy levels.

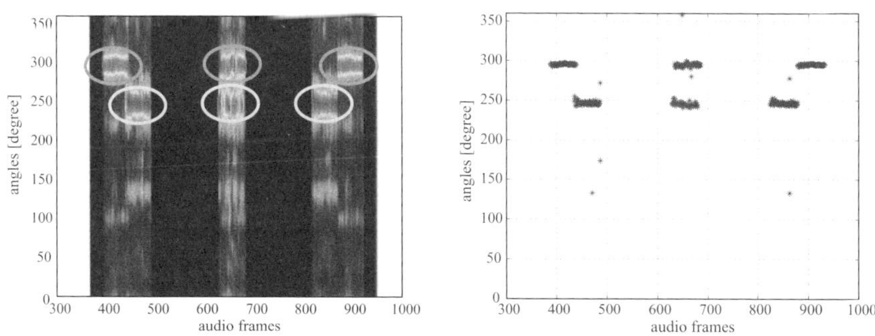

(a) The result of applying the sound localization system. Ovals mark robots producing sound (better see in colors).

(b) The detected sound sources' directions. Stars correspond to the maximal peaks of the beamformer output energy.

Figure 2. The localization of multiple sound sources using the interpolation beamforming.

Figure 2a shows the result of applying the described sound localization system to the sample experimental audio data. Potential directions to sound sources can be estimated from maximal peaks of the output energy (2) (Figure 2b). After potential sound sources are localized, the set of corresponding directions is transferred to the calls detection subsystem.

2.2. Visual Tracking

For visual tracking we use the SwisTrack software [13]. SwisTrack is an open source tracking software that was initially developed for the Leurre project to track cockroaches' and robots' displacements. The advantage of this software is its flexibility: by combining different image processing algorithms it allows to track both marked and marker-less objects. The average absolute calibration error of SwisTrack, reported in [13] is 2.3mm (standard deviation 1.4mm).

(a) Two e-puck robots with red markers.

(b) The microphone array with three blue markers.

Figure 3. The marked robots and the microphone array.

For tracking purposes we equipped two e-puck robots and the microphone array with color markers (Figure 3): each robot has one red marker and the microphone array has three blue markers. This allows to obtain positions of the robots on the arena and both the position and orientation of the microphone array (Figure 4), that are further used by the calls detection subsystem.

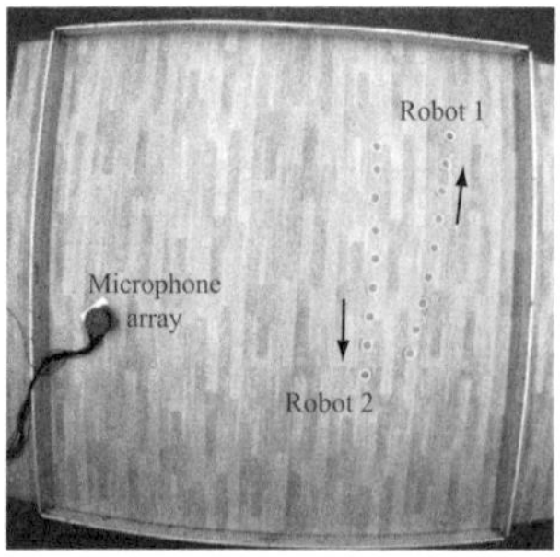

(a) The experimental video data recorded by the overhead camera.

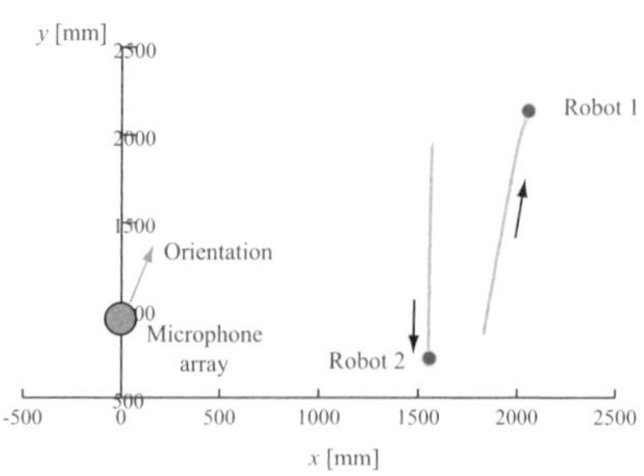

(b) The trajectories of the robots and the position and orientation of the microphone array, obtained by the visual tracker.

Figure 4. The visual tracking of the microphone array and the robots.

3. Probabilistic Calls Detection

The role of the calls detection subsystem is to superpose the acoustical events with visually found robots' positions and determine for each time frame the robots, emitting sound. Taking into account the fact, that the visual tracking algorithms have better spatial resolution than acoustic localization techniques, it is natural to choose the robots' positions, obtained by visual tracking, as an initial guess for the analysis.

We model each robot in the angle domain as a single Gaussian distribution, with an experimentally determined constant variance and mean, relative to the visually determined robot's position (in angles). We denote as $P(r_i|\Theta_{detected})$ the probability that the i-th robot is emitting sound given the known directions to the sound sources $\Theta_{detected}$, received from the sound localization subsystem. The robots that emit sounds are obtained by evaluating $P(r_i|\Theta_{detected})$ for all the robots in the experimental area. A similar approach has been used in [14], where it was applied to the problem of localization and identification of the participants at a meeting using audio-visual sensors.

4. Experimental Results

The experimental set-up is composed of a square arena with sides of three meters in length, an overhead camera and the microphone array placed on the arena. The camera used in our experiments is the color Basler scout Gigabit Ethernet camera scA1000-30gc.

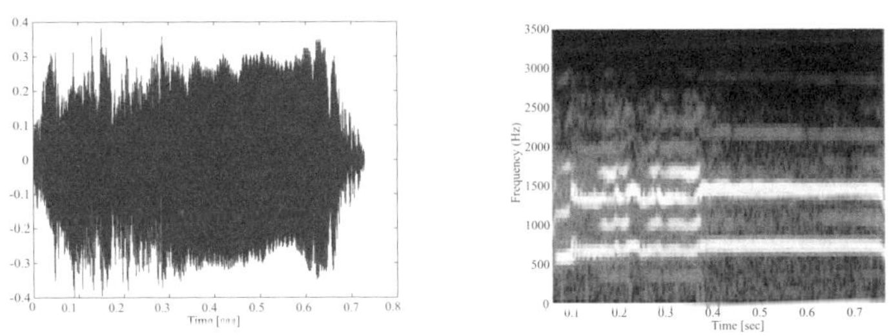

(a) The call's waveform. (b) The call's spectrum.

Figure 5. The prerecorded chicken's call used in the experiments.

For experiments two e-puck robots were programmed to emit the prerecorded chicken call (Figure 5); one robot every six seconds and the other every five seconds. The emission frequency difference allows us to test robustness of the sound detection for non-overlapped, slightly overlapped and strongly overlapped calls.

All results presented in this paper were obtained through off-line computations using audio and video records of the experiments. We computed the beamformer output energy over a frame of length $N = 1024$ samples at 44.1 kHz, scanning all directions around the microphone array with step size 1°. Video data were recorded at 10 fps. Estimated parameters of Gaussian distribution are the following: mean is 1.0544° and variance is 2.4252°. They were evaluated in the preliminary experiments by placing one e-puck robot producing sound in the different parts of experimental arena at distances from the microphone array between 1.2 and 2.5 meters and estimating differences between the target directions, obtained by sound localization and visual tracking. The non-zero mean can be explained by minor errors in the marker placement on the microphone array, that entails systematic bias in the estimation of the microphone array's orientation.

In the main series of experiments we evaluated the ability of the system to detect the robots' calling activity. We used two criteria: detection probability, which we esti-

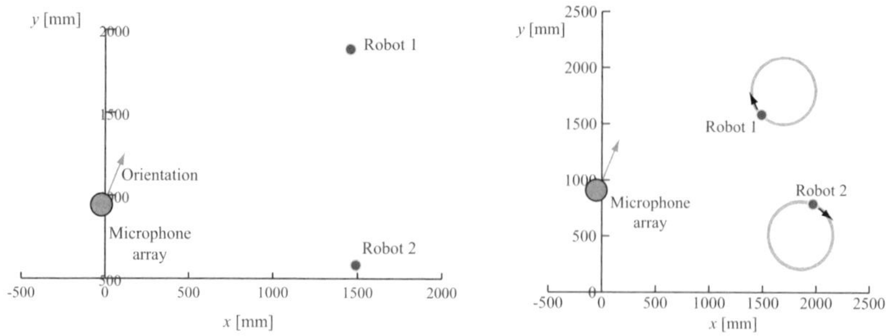

(a) The positions of the robots and the microphone array in the experiment with the stationary robots.

(b) The trajectory of the robots and the position of the microphone array in the experiment with the moving robots.

Figure 6. The location of the robots and the microphone array in the calling activity detection experiment.

Table 1. The results of calling activity detection for stationary and moving robots

Type of experiment	**Detection probability (%)**	**False alarm rate (%)**
Stationary robots	91.45	1.09
Moving robots	86.76	1.24

mate as a fraction of time frames, where the sound emission is correctly detected for the robot, from the whole number of frames of its calling activity; and a false alarm rate that is estimated as a fraction of time frames, where the sound emission is detected but the robot doesn't produce sound, from the whole number of frames of absence of its calling activity. We performed eight experiments: four with two stationary robots and four with two robots, moving on the arena (Figure 6). Each of eight experiments was one minute long that gave us more than 20000 audio frames to analyze. As it can be seen from the Table 1, the calls detection is fairly robust achieving around 90% of detection probability and a false alarm rate of around 1%. The difference in detection probabilities between stationary and moving robots can be explained by the limited resolution of the beamformer. If two calls are emitted at the same time close to each other, the beamformer produces a single energy peak centered in-between the directions of sound sources and thus the calling activity can not be detected. Another evident issue of the detection method is that when two robots are placed close enough to each other and one emits a sound, both robots will be marked as producing sounds thus increasing false alarm rate.

5. Conclusion

In this paper, we have described a system that is able to localize, using audio-visual information, the positions and calling activities of multiple moving robots with good accuracy.

Several issues remain open for improvement. First, in the current system we work with prerecorded data, however, our goal is to make our system work in real time. We are currently developing a 24 channels audio acquisition board to be able to process the audio data in real time. Second, as our robot has to understand the meaning of chickens calls, the sound separation and birds' calls recognition system have to be integrated into our system. Finally, the evaluation of our approach on real animals will be a part of future work.

Acknowledgements

This work is supported by the Swiss National Science Foundation grant no. 112150 "Mixed society of robots and vertebrates". We thank Michael Bonani (LSRO), Daniel Burnier (LSRO) and Dr. Roderich Groß (LSRO) for discussions and technical support.

References

[1] J. Halloy, G. Sempo, G. Caprari, C. Rivault, M. Asadpour, F. Tache, I. Said, V. Durier, S. Canonge, J.M. Ame, C. Detrain, N. Correll, A. Martinoli, F. Mondada, R. Siegwart, and J.-L. Deneubourg. Social integration of robots into groups of cockroaches to control self-organized choices. *Science*, 318(5853):1155–1158, 2007.

[2] H.-D. Kim, K. Komatani, T. Ogata, and H. G. Okuno. Auditory and visual integration based localization and tracking of humans in daily-life environments. In Kazunori Komatani, editor, *International Conference on Intelligent Robots and Systems (IROS 2007)*, pages 2021–2027, 2007.

[3] J.-M. Valin, F. Michaud, B. Hadjou, and J. Rouat. Localization of simultaneous moving sound sources for mobile robot using a frequency – domain steered beamformer approach. In *IEEE International Conference on Robotics and Automation*, volume 1, pages 1033–1038, 2004.

[4] J.-M. Valin, F. Michaud, and J. Rouat. Robust localization and tracking of simultaneous moving sound sources using beamforming and particle filtering. *Robotics and Autonomous Systems*, 55(3):216–228, 2007.

[5] M. Murase, S. Yamamoto, J.-M. Valin, K. Nakadai, K. Yamada, K. Komatani, T. Ogata, and H. G. Okuno. Multiple moving speaker tracking by microphone array on mobile robot. In *European Conference on Speech Communication and Technology (Interspeech)*, pages 249–252, 2005.

[6] I. Hara, F. Asano, H. Asoh, J. Ogata, N. Ichimura, Y. Kawai, F. Kanehiro, H. Hirukawa, and K. Yamamoto. Robust speech interface based on audio and video information fusion for humanoid HRP-2. In F. Asano, editor, *International Conference on Intelligent Robots and Systems (IROS 2004)*, volume 3, pages 2404–2410, 2004.

[7] V. Trifa, L. Girod, T. Collier, D.T. Blumstein, and Taylor C.E. Automated wildlife monitoring using self-configuring sensor networks deployed in natural habitats. In *International Symposium on Artificial Life and Robotics (AROB07)*, 2007.

[8] Lesley J. Rogers. *The Development of Brain and Behaviour in the Chicken*. Oxford University Press, 1996.

[9] M. S. Brandstein and D. B. Ward, editors. *Microphone Arrays: Signal Processing Techniques and Applications*. Springer-Verlag, 2001.

[10] Y. Tamai, Y. Sasaki, S. Kagami, and H. Mizoguchi. Three ring microphone array for 3d sound localization and separation for mobile robot audition. In *IEEE/RSJ International Conference on Intelligent Robots and Systems*, pages 4172– 4177, 2005.

[11] Roger G. Pridham and Ronald A. Mucci. A novel approach to digital beamforming. *The Journal of the Acoustical Society of America*, 63(2):425–434, 1978.

[12] N. Tashev. Gain self-calibration procedure for microphone arrays. *IEEE International Conference on Multimedia and Expo (ICME 2004)*, 2:983–986, 2004.

[13] N. Correll, G. Sempo, Y. Lopez de Meneses, J. Halloy, J-L. Denebourg, and A. Martinoli. Swistrack: A tracking tool for multi-unit robotic and biological systems. In *IEEE/RSJ International Conference on Intelligent Robots and Systems*, pages 2185–2191, 2006.

[14] C. Busso, P.G. Georgiou, and S.S. Narayanan. Real-time monitoring of participants' interaction in a meeting using audio-visual sensors. *IEEE International Conference on Acoustics, Speech and Signal Processing (ICASSP 2007)*, 2:II–685–II–688, 2007.

Intelligent Autonomous Systems 10
W. Burgard et al. (Eds.)
IOS Press, 2008

doi:10.3233/978-1-58603-887-8-332

Sliding Autonomy for Peer-To-Peer Human-Robot Teams[1]

M. Bernardine DIAS, Balajee KANNAN, Brett BROWNING, E. Gil JONES,
Brenna ARGALL, M. Freddie DIAS, Marc ZINCK,
Manuela M. VELOSO, and Anthony J. STENTZ
School of Computer Science, Carnegie Mellon University

Abstract. The vision of humans and robots working together as peers to accomplish complex tasks has motivated many recent research endeavors with a variety of applications ranging from lunar construction to soccer. However, much of this research is still at an early stage, and many challenges still remain in realizing this vision. A key requirement for enabling robustness and efficiency in human-robot teams is the ability to dynamically adjust the level of autonomy to optimize the use of resources and capabilities as conditions evolve. While sliding autonomy is well defined and understood in applications where a single human is working with a single robot, it is largely unexplored when applied to teams of humans working with multiple robots. This paper highlights the challenges of enabling sliding autonomy in peer-to-peer human-robot teams and extends the current literature to identify and extend six key capabilities that are essential for overcoming these challenges. These capabilities are *requesting help*, *maintaining coordination*, establishing *situational awareness*, enabling interactions at different levels of *granularity*, *prioritizing* team members, and *learning* from interactions. We demonstrate the importance of several of these characteristics with results from a peer-to-peer human-robot team engaged in a treasure hunt task.

Keywords. Sliding autonomy, human-robot teams, peer-to-peer teams, pickup teams, autonomous teamwork, multi-agent coordination, adjustable autonomy.

Introduction

The vision of humans and robots working together to accomplish complex team tasks is driving much of the current research in the area of autonomous teamwork. As robots become more capable they can handle increasingly complex tasks and highly uncertain environments, but the robotic capabilities in many domains are still insufficient to execute these tasks robustly and efficiently. In these scenarios, robots can still accomplish the tasks with human assistance as human capabilities are often better-suited for some tasks and complement robot capabilities in many situations. Thus, for robots to become an integral part of society, human-robot teams must be effective in a variety of settings.

[1] This work is primarily funded by the Boeing Company Grant CMU-BA-GTA-1. The content of this publication does not necessarily reflect the position or policy of the Boeing Company and no official endorsement should be inferred. This work was also enabled in part by funding from the Qatar Foundation for Education, Science and Community Development and from the U.S. Army Research Laboratory, under contract Robotics Collaborative Technology Alliance (contract number DAAD19-01-2-0012).

In most of the published work on human-robot teams, the human's role is limited to either a supervisor [6] or end-user [9] when interacting with a single or multiple robots or agents [10]. While this hierarchical relationship between human and robot is appropriate in some domains, there are many applications where a peer-to-peer relationship enables more effective use of the complimentary capabilities of humans and robots. In this work, we focus on *peer-to-peer human-robot teams* where humans and robots can assign tasks to each other through direct requests/commands, or through automated task allocation systems. Several research efforts are emerging in this area [7] with applications ranging from lunar construction [4] to soccer [2]. Within the topic of human-robot teams, we are especially interested in *pickup teams* [5] where the composition of the team is not previously known and where members joining the team can vary in their capabilities, expertise, and knowledge of the task. Pickup teams should absorb this wide variety of members to quickly form effective teams, and improve over time as the strengths and weaknesses of different members are discovered and accounted for in the team strategy.

An important aspect of enabling effective human-robot pickup teams is allowing the team to adjust its level of autonomy as necessary. *Sliding autonomy*[2] was introduced to optimize performance by allowing a system to adapt its level of autonomy during execution to accommodate dynamic conditions. The agent and robotics literature are populated with many studies on sliding autonomy applied in different scenarios. However, this work has not been extended to peer-to-peer human-robot teams to date. This paper explores the challenges in applying sliding autonomy in peer-to-peer human-robot team settings and proposes a set of guidelines for accomplishing this task. The proposed guidelines are used to implement a system of humans and robots engaged in a treasure hunt task.

1. Sliding Autonomy in Peer-To-Peer Human-Robot Teams

"Autonomy" is defined in terms of a system's ability to function effectively without human intervention. For example, a fully autonomous system (or "pure autonomy") is said to require no human intervention to complete a task [6]. "Sliding autonomy" is similarly defined in terms of the system's ability to incorporate human intervention when needed (and to otherwise operate independently) [10]. Both of these definitions must change when humans are a part of the "system" or team and where the humans and robots interact as peers. We extend the definition of autonomy presented by Maheswaran et al. [9] where the ability to decide transfer (or sharing) of control governs the level of autonomy. Thus, s*liding autonomy in peer-to-peer teams means that members of the team (humans, robots, and software agents) can actively decide if and when to transfer control to (or share control with) another member of the team or, in some cases, to some entity outside the team.* Because the team members are heterogeneous, some team members may not be capable of making their own decisions. Hence, the decision-making control can shift between different members of the team as

[2] Sliding autonomy and adjustable autonomy are used interchangeably in parts of the literature. We prefer the term "sliding" autonomy because it implies that the level of autonomy can be dynamically adjusted during execution.

needed, but all team members may not possess this capability equally. We also allow prioritization of different team members such that higher priority members can seize control from lower priority members if deemed necessary.

Our work builds on the methodology for sliding autonomy in multi-agent teams proposed by Sellner et al. [6] and work on mixed-initiative teams reported by Bruemmer and Walton [11]. Specifically, we identify six necessary capabilities for enabling sliding autonomy in peer-to-peer human robot teams[3]. The first three of these capabilities are extended forms of characteristics presented by Sellner et al. [6] as major issues that affect human awareness in multi-agent teams. They are the ability to request help, the ability to maintain team coordination during interventions, and the ability to provide situational awareness. The second set of characteristics are augmented versions of capabilities identified by Bruemmer and Walton [11] in the context of robots in mixed-initiative teams. They are the ability to interact at different granularities, the ability to prioritize team members, and the ability to learn from interactions. We next discuss how these six characteristics enable sliding autonomy in a peer-to-peer team setting of humans and robots.

In peer-to-peer teams no single member is necessarily aware of the entire team state. Hence, these teams are more effective when individual agents and sub-teams can identify situations where they need to ***request help*** from other members of the team. These requests for help primarily occur in situations where an agent discovers a failure it cannot rectify without the help of another agent and hence must adjust its level of autonomy [8]. In some situations a team member may not be capable of asking for help or assisting in a recovery process from a failure [8]. In this event, other team members will need to recognize this failure and adapt the team strategy as needed. Monitoring teammates becomes more difficult in pickup teams since team composition can change over time and unfamiliarity with identifiable characteristics that indicate faults in new team members can impede the process of fault recognition and identification. Another type of help request can occur when an agent is assigned a task that requires resources or capabilities that the agent doesn't possess. In this case, the agent needs to recruit others to assist with the task and, in some cases, may need to adjust its level of autonomy to accomplish the task.

Maintaining coordination during interventions is important for effective team performance. For example, if one robot in the team suffers a failure during the execution of a team task, the rest of the team should maintain their coordination while re-strategizing to assist in the recovery from the failure. The team may continue to execute the task despite the failure, or discover that they are unable to complete the task due to the failure and hence request further assistance from an entity outside the team. Therefore, autonomy may be adjusted for some tasks that require external intervention while other tasks are carried out autonomously. Furthermore, if the team is able to overcome the failure without external intervention, it is important that coordination is maintained while the failure is addressed by the relevant team members and that other members of the team continue with their tasks.

Gaining and maintaining situational awareness is perhaps the biggest relevant challenge in a team setting. Situational awareness is a key factor in executing early and successful interventions, and in decisions for adjusting autonomy. In teams with

[3] We assume the teams can also be pickup teams [5] and that the team composition and capabilities can change during execution.

multiple mobile humans, it is not sufficient to capture information in a single graphical interface (GUI), and customization of the state information for the different members of the team may be required. Furthermore, the state of the humans and the dialog and gestures that are a natural part of human-to-human communication must be captured and made transparent to the robots on the team since interventions might be carried out by robots. Situation awareness in pickup teams is also difficult because we must be able to accommodate new capabilities and resources as members join the team, and we must be able to expose the state of the current team to the new member quickly and effectively. Several research efforts are focused on a variety of communication strategies for human-robot teams that include tools such as GUIs, 3-D interactive environments, dialog, and gestures ([1], [3], [4], [5], [6], [11]). However, there is still much to be accomplished in this area of research.

The ***granularity of interaction*** must often be flexible in peer-to-peer teams and this also translates to sliding autonomy. This primarily impacts exposure to information and interactions between team members. The level of granularity of the information presented to any team member will often need to be adaptable for individual components of the system for effective comprehension by different team members. This directly correlates to the previous requirement for effective situational awareness. In terms of interactions among team members, the most effective teams allow for some members to interact in a tightly-coordinated manner to accomplish some tasks, while others act independently. In the case of sliding autonomy, this is also important because autonomy might be adjusted by a single agent's intervention or by a sub-team intervention. For example, two or more agents may coordinate to assist a robot that is stuck in the mud by tightly coordinating in a sub-team, but once the robot is out of the mud, they may coordinate more loosely to accomplish other tasks.

Explicit ***prioritization of team members*** is important in peer-to-peer teams because we cannot assume an inherent hierarchy. The prioritization of the team members is ideally adjustable and specifiable along different dimensions. For example, human members might be prioritized in safety considerations, but a robot with powerful computing capability might be prioritized for planning tasks. These priorities can change when the team composition changes, and also due to other dynamic conditions and should also translate to sliding autonomy. Higher priority members of the team might be eligible for earlier interventions when there is contention for limited resources. Another possibility is that higher priority members can override the autonomy of lower priority members and temporarily seize control for certain tasks.

Finally, ***learning from interactions*** is important for effective team performance. For example, in pickup teams, the prioritization of new members for different tasks and concerns may not be initially known and instead have to be learned based on interactions over time. Also, in terms of sliding autonomy, team members may need to learn to detect indicators of failure so that autonomy can be adjusted when necessary.

In summary, we have discussed six necessary capabilities for sliding autonomy in human-robot peer-to-peer teams: R*equesting help*, *maintaining coordination*, establishing *situational awareness*, enabling interactions at different levels of *granularity*, *prioritizing* team members, and *learning* from interactions. Next, we describe an initial implementation of some of these capabilities for a peer-to-peer human-robot team engaged in a treasure hunt task.

2. Approach and Implementation Details

Our implementation (described in further detail in previous publications [3] and [5]) supports two different *granularities* for tasking human-robot teams. First, a high-level task objective is issued to the system. The system responds by autonomously selecting a plan that is potentially multi-agent and tightly-coordinated to accomplish the objective. A pickup sub-team is created from available team members with the necessary capabilities to efficiently execute this plan. Once the pickup team has been selected, the agents *coordinate* via simple communication protocols during execution, handle errors, and report status to each other. Occasionally errors occur or new information is discovered that cannot be addressed by the robotic agents alone. In these cases, the robots *request help* from a human peer who may or may not have been part of the original sub-team team. The human can join the team and seize control of the system, physically intervene, or issue low-granularity commands to the robot participants. Currently, we use a fixed *prioritization* technique on the robots where low-level commands from humans override robot commands. We do not yet address *learning* (an area of future work). Finally, the flow of information between robots and humans is directed through a GUI application. *Each human* can use a tablet PC that runs the somewhat customizable GUI so that they are *aware* of the current system state. Human actions are explicitly communicated to robots in the current implementation.

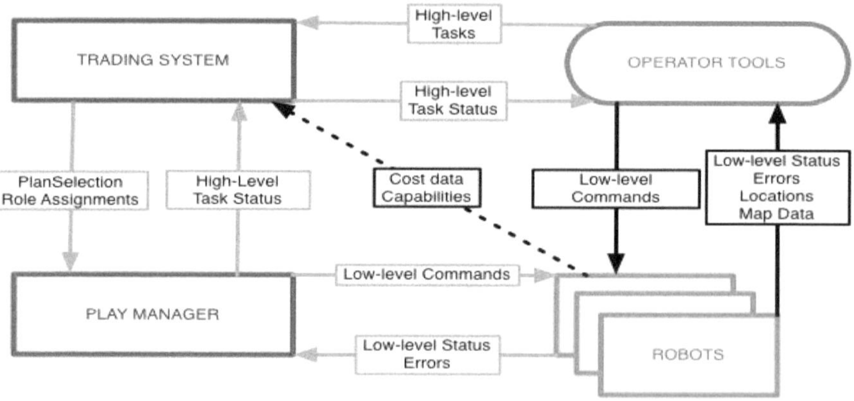

Figure 1: The four components of our system, with arrows indicating the flow of information. The pathway for high-level tasks and resulting high-level status is shown in grey arrows. The pathway for low-level commands directly to robots for error recovery is shown by black arrows. The dotted arrow represents information supplied by the robots to be used in high-level planning and allocation.

Our system consists of four main components, as shown in Figure 1: a human interface tool (OPERATOR TOOLS), a distributed market-based planning and allocation system (TRADING SYSTEM), a component for synchronized tightly-coordinated multi-agent plan execution (PLAY MANAGER), and robot software that supports sensing and acting in the environment (ROBOTS). The operator tools, allow an operator or human peer to issue both high- and low-level tasks and to process status messages in addition to displaying state information of the robots. These are our primary method for supporting visual situational awareness for the humans. Tasks can be autonomously allocated via the Trading System, or can be issued directly to specific agents. The Trading System uses an instantaneous allocation approach, where agents will only participate in the formation of a new sub-team if they are not actively involved in another high-level task. Multiple sub-teams can simultaneously address different tasks and errors are reported to both humans and robots. We use a tiered auction approach, where individual agents attempt to generate plans and recruit other agents' participation in those plans; the trading system selects the most efficient plan and allocation from the submitted bids. Agents provide information about their capabilities and other cost data that helps the trading system determine plan efficiency. Capability information is used to determine which agents can best fill particular roles in a possible plan. Cost data is used to differentiate between different capable agents when determining which agents can most efficiently fill a particular role. The trading system provides the ability to dynamically form sub-teams that will be maximally efficient in addressing high-level objectives and maintaining coordination even if team composition changes during operation. Currently, humans do not participate in the auctions for tasks but instead are explicitly enrolled in relevant tasks.

Once a plan is selected and roles assigned the information is passed to the Play Manager to coordinate the execution of actions by the sub-team of multiple agents. The Play Manager sends a series of low-level commands to the agents assigned to participate in the coordinated plan. If execution concludes successfully, status is reported to all participating agents. In some cases, due to the highly unstructured and dynamic nature of the environment and the realities of robot hardware, agents may fail and may have no contingency plan. In this case they report errors and directly request help. Help requests primarily take the form of visual cues on the operator GUI (as shown in Figure 2), but we have also experimented with dialog-based error notification [3]. Intervention to recover from failures can take two primary forms – physical interaction and direct low-granularity commands. When resolving failures physically, a human directly interacts with robot hardware. For instance, a robot may experience a problem with its laser range finder that can only be resolved by power cycling the laser unit; if a human performs this action in a timely manner, execution of the original plan resumes. In direct robot command, the human can use the operator tools to issue low-granularity commands to a particular robot. For instance, a robot may become trapped or lost. In these situations a human can issue a series of relative waypoints to free the robot or to move it back to a known area, after which plan execution can continue. Thus, sliding autonomy increases robustness and adaptability.

The final component of our system is the robot/agent software. The pickup team formulation depends on abstracting away many elements of robot software implementation in order to support the seamless integration of new team members. We represent robots in terms of their capabilities, the actions they can perform, and the sensor data and errors they can produce. If agents can represent themselves in this manner our system can easily accommodate pickup teams, with members fluidly participating in sub-team formation and execution of tasks.

3. Experiments, Results, and Discussion in the Treasure Hunt Domain

We demonstrate the effectiveness of our approach in the "Treasure Hunt" domain [5], which is motivated by applications such as de-mining where human exposure to danger must be minimized but humans are needed to deal with safe maneuvering of the discovered items. The task requires a human-robot team to locate and retrieve items of interest or "treasure" (visual fiducials) in an unknown environment. The key tasks include exploration, mapping, search and localization of treasure, and retrieval of treasure to a "home" location. We use heterogeneous platforms with complimentary capabilities: Pioneer IIDX robots equipped with SICK LiDar and fiber optic gyros, and Segway RMPs and ER1s equipped with cameras. Pioneers build maps and maintain an accurate pose while Segways can locate Pioneers and treasure, and localize based on the observed position and report location of the Pioneers. ER1s are similar in capabilities to the Segways but are teleoperated. Humans cannot directly observe the operational area from the home location, interact with robots via GUIs, and perform retrievals by following Pioneers to the treasure location and back home. The team *requires* coordination to achieve the task since no team member can perform all operations. Figure 2 shows a screen shot from the GUI, which provides *situational awareness* to the humans. Three types of errors are identified and reported. Laser errors relate to a problem with the Pioneer laser, pose errors occur when a robot's localization becomes corrupted, and arc errors occur when a robot cannot independently discover a safe path. The experiments were performed in a large, complex, cluttered, and dynamic indoor environment (see Figure 3).

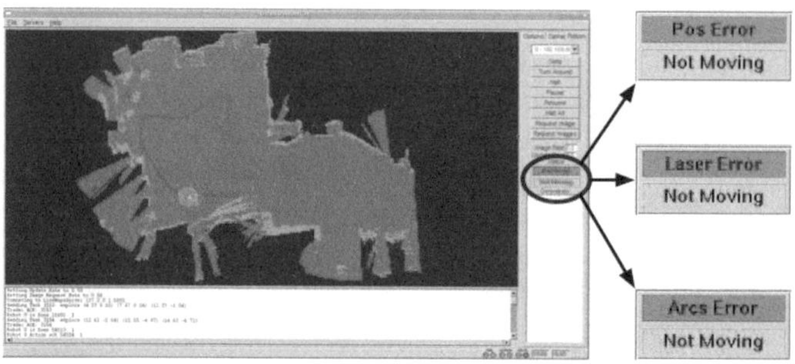

Figure 2: A screenshot from a GUI showing the fused map built from the Pioneer robots. Shown is the recent trajectory of a pioneer (red trail), with a pose/laser error. Other errors show up in a similar manner. A human can issue commands and monitor the state of the team.

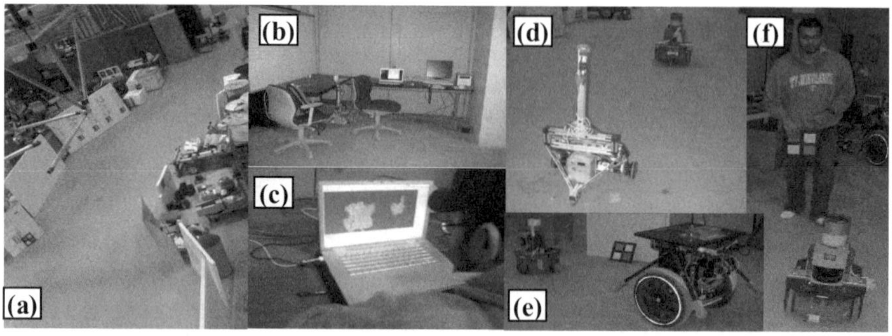

Figure 3: (a) Overhead view of the operating environment where 7 "treasures" are randomly placed, (b) "Home" location, (c) A human team member observing the map being built by the Pioneer robots, (d) An ER1 robot teleoperated to follow a Pioneer robot to search for treasure, (e) A Segway robot autonomously following a Pioneer robot to search for treasure (an item of "treasure" is seen between the two robots), (f) A human being lead back to the "home" location after successfully retrieving an item of treasure.

We measure performance based on the number of successfully identified and retrieved treasure items in a limited time-frame. The first two experiments compare team performance with sliding autonomy enabled versus disabled, while keeping the task parameters constant. In the third and fourth configuration of the system, we replace the autonomous Segway with a teleoperated ER1 robot, and increase the number of humans to evaluate the adaptability of the approach.

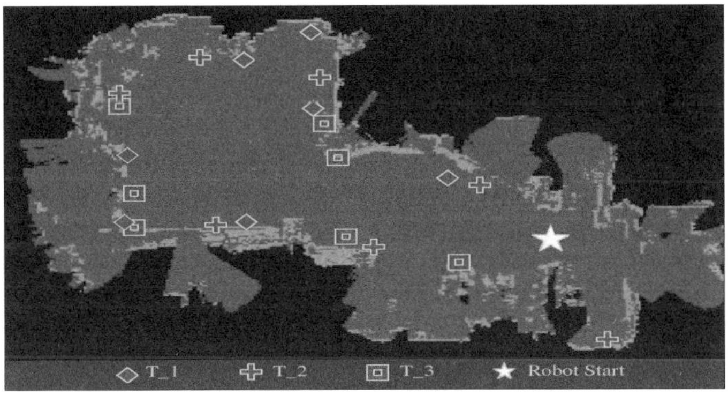

Figure 4: Laser map of Highbay area with different treasure configuarations

We perform initial experiments with 3 different treasure configurations (see Figure 4). Each run is conducted over a fixed time period of 15 minutes with a total of 7 "treasure" items scattered throughout the environment. The first set of experiments was performed for a team consisting of 2 humans, 1 pioneer and 1 Segway robot. Table 1 shows the experimental results with sliding autonomy enabled. During these runs, *requests for help* were generated and handled by the system, while maintaining *coordination during intervention*. In contrast, Table 2 shows the experimental results with sliding autonomy disabled. For this experiment, no *requests for help* were generated by the system. The time at which each error occurred is also shown. For all experiments, a combination of errors was either randomly artificially generated (G) or occurred naturally over the course of operation (N). A comparison of the results in Table 1 and Table 2 show that the productivity of the team, measured by the number of treasure items identified and retrieved, decreases in the absence of sliding autonomy.

The third set of experiments demonstrated an alternative human-robot sub-team capable of performing the treasure hunt task. The Pioneer robot in this experiment was autonomous while the ER1 robot was teleoperated by a human. Table 3 shows the experimental results with a sub-team consisting of 3 humans, one Pioneer and one ER1. Table 4 shows the experimental results with error handling turned off for the ER1-Pioneer team. In order to avoid human-biasing as a result of familiarity with the environment and the system, the experiments were performed by two-different humans with no prior experience and one human with prior experience dealing with the robots. During these runs, *requests for help* were generated and handled by the system, while maintaining *coordination during intervention*. In the case of experiments where the sliding autonomy was turned off, no human intervention was provided when the robots fail to autonomously handle errors.

Table 1: Results of 3 runs with sliding autonomy enabled. Type of Errors – Arc (A), Laser (L), and Pose (P). # Error Generated Type – Artificial/manually induced (G) or naturally occurring as part of the system/environment (N) # Robots – R1: leader/explorer Pioneer, R2: retriever Pioneer.

Run	Treasure seen (recovered)	Error Types	Error Source	Error per Robot
T_1	4 (2)	**Total: 5** [L(1), A(2), P(2)]	N(5)	R1(2), R2(3)
T_2	3 (2)	**Total: 6** [L(4), A(1), P(1)]	G(2), N(4)	R1(2), R2(4)
T_3	2 (0)	**Total: 2** [P(1), L(1)]	N(2)	R1(1), R2 (1)

Table 2: Results of 3 runs with sliding autonomy disabled

Run	Treasure seen (recovered)	Error Types	Error Source	Error per Robot
T_1	2 (2)	**Total: 1** [L (6.5 min)]	G(1)	R1(1)
T_2	1 (1)	**Total: 2** [P (2 min), L (5 min)]	G(2)	R1(1), R2(1)
T_3	0 (0)	**Total: 1** [P (7.5 min)]	G(1)	R1(1)

Table 3: Results of 3 runs with a sub-team of humans, a Pioneer and an ER1. Labeling as in Table 1.

Run	Treasure seen (recovered)	Error Types	Error Source	Error per Robot
T_1	4 (4) (Skill level - Novice)	**Total: 2** [L(1), P(1)]	G(2)	R1(1), R2(1)
T_2	6 (3) (Skill level - Expert	**Total: 5** [L(1), A(3), P(1)]	G(2), N(3)	R1(3), R2(2)
T_3	4 (2) (Skill level - Novice)	**Total: 3** [L(1), A(1), P(1)]	G(2), N(1)	R1(2), R2(1)

Table 4: Results of 3 runs with for Pioneer, ER1, human team with sliding autonomy disabled.

Run	Treasure seen (recovered)	Error Types	Error Source	Error per Robot
T_1	3 (2) (Skill level - Novice)	**Total: 1**[L (1)]	G(1)	R1(1)
T_2	0 (0) (Skill level – Novice)	**Total: 1** [A (1)]	N(1)	R1(1)
T_3	4 (2) (Skill level - Novice)	**Total: 1** [L(1)]	G(1)	R1(1)

Overall, these experiments primarily demonstrate that sliding autonomy can improve team performance and that the implemented system can be flexible in accommodating different team configurations for accomplishing the same task.

4. Conclusions and Future Work

The ability to dynamically adjust the level of autonomy during execution can enhance the performance of human-robot teams. This paper extends the framework for sliding autonomy to address peer-to-peer human-robot teams. We highlight six important characteristics in this context: *requesting help*, *maintaining coordination*, establishing *situational awareness*, enabling interactions at different levels of *granularity*, *prioritizing* team members, and *learning* from interactions. We implement several of these characteristics and demonstrate them in a peer-to-peer human-robot team engaged in a treasure hunt task. While initial experiments show promising results, the current implementation can be improved in several ways. Situational awareness should be enhanced for the robots and humans by capturing human state and communication among human team members. The prioritization of team members should be dynamically adapted to allow for changes in team composition and task priorities. Finally, the system does not currently incorporate any learning. If team members can learn to perform better based on their interactions and other performance metrics, the overall team performance should improve. Ongoing work addresses these extensions.

Acknowledgements

We acknowledge the assistance of Victor Marmol, Sarah Belousov, Ferdinand Justus, and Murali Krishnamoorthy in carrying out the reported experiments. We thank Thomas Harris, Alexander Rudnicky, and Matthew Marge of Carnegie Mellon University, Scott Smith, Ronald Provine, Richard Wojcik, and Michael Kerstetter from the Boeing Company, and the members of the rCommerce and the CORAL groups.

References

[1] F. Driewer, M. Sauer, and K. Schilling, Discussion of Challenges for User Interfaces in Human-Robot Teams, Proceedings of the Third European Conference on Mobile Robots, 2007.

[2] B. Argall, Y. Gu, B. Browning, and M. Veloso, "The First Segway Soccer Experience: Towards Peer-to-Peer Human-Robot Teams," Proceedings of the First Annual Conference on Human-Robot Interactions, 2006.

[3] M. B. Dias, T. K. Harris, B. Browning, E. G. Jones, B. Argall, M. Veloso, A. Stentz, and A. A. Rudnicky, "Dynamically Formed Human-Robot Teams Performing Coordinated Tasks," AAAI Spring Symposium "To Boldly Go Where No Human-Robot Team Has Gone Before," 2006.

[4] T.W. Fong, J. Scholtz, J. Shah, L. Flueckiger, C. Kunz, D. Lees, J. Schreiner, M. Siegel, L. Hiatt, I. Nourbakhsh, R. Simmons, R. Ambrose, R. Burridge, B. Antonishek, M. Bugajska, A. Schultz, and J.G. Trafton, "A Preliminary Study of Peer-to-Peer Human-Robot Interaction," International Conference on Systems, Man, and Cybernetics, IEEE, 2006.

[5] E. Jones, B. Browning, M.B. Dias, B. Argall, M. Veloso, and A. Stentz, "Dynamically Formed Heterogeneous Robot Teams Performing Tightly-Coordinated Tasks," Proceedings of the International Conference on Robotics and Automation, 2006.

[6] B. Sellner, F. W. Heger, L. M. Hiatt, R. Simmons, and S. Singh, "Coordinated Multiagent Teams and Sliding Autonomy for Large-Scale Assembly," Proceedings of the IEEE, Vol. 94, No. 7, 2006.

[7] F. Tang and L. E. Parker, "Peer-to-Peer Human-Robot Teaming through Reconfigurable Schemas," AAAI Spring Symposium on ``To Boldly Go Where No Human-Robot Team Has Gone Before", 2006.

[8] M. B. Dias, R. Zlot, M. Zinck, and A. Stentz, "Robust Multirobot Coordination in Dynamic Environments", Proceedings of the IEEE International Conference on Robotics and Automation, 2004.

[9] R. T. Maheswaran, M. Tambe, P. Varakantham and K. Myers, "Adjustable Autonomy Challenges in Personal Assistant Agents: A Position Paper," in *Agents and Computational Autonomy: Potential, Risks, Solutions,* M. Nickles, M. Rovatsos and G.Weiss (Eds.), Lecture Notes in Artificial Intelligence, vol. 2969, pp. 187-194, Springer-Verlag, 2004.

[10] P. Scerri, K. Sycara, and M. Tambe, "Adjustable Autonomy in the Context of Coordination," AIAA 3rd "Unmanned Unlimited" Technical Conference, Workshop and Exhibit, 2004.

[11] D. J. Bruemmer and M. Walton, "Collaborative tools for mixed teams of humans and robots," Proceedings of the Workshop on Multi-Robot Systems, 2003.

Intelligent Autonomous Systems 10
W. Burgard et al. (Eds.)
IOS Press, 2008

doi:10.3233/978-1-58603-887-8-342

An Adaptive Multi-Controller Architecture for Mobile Robot Navigation

Lounis ADOUANE [a]
[a] *LASMEA, UBP-UMR CNRS 6602, France*

Abstract. This paper proposes a hybrid control architecture dedicated to the navigation of autonomous mobile robots in presence of obstacles. The proposed control structure is flexible, adaptive and takes into account the different possible interactions between the multitude of controllers composing the architecture of control. This is enabled by the use of stable elementary controllers and to specific mechanisms of coordination and adaptation which guaranties the stability of the overall architecture.

Keywords. Mobile robot navigation, Hybrid architecture of control, Adaptive systems, Lyapunov controller synthesis.

Introduction

The navigation of autonomous mobile robots in a cluttered and open environment is a fundamental problem that has received a large amount of attention. A part of literature in this domain considers that the robot is fully actuated with no control bounds and focuses their contribution on path planning. Voronoï diagrams and visibility graphs [1] or navigation functions [2] are among these roadmap based methods. However, the other part of the literature considers that to control a robot with safety, flexibility and reliability, it is essential to accurately take into account: robot's structural constraints (e.g., nonholonomy); avoiding command discontinuities and jerking set points, etc. Nevertheless, even in this method, there are two schools of thoughts, the one which uses the notion of planning and re-planning to reach the target e.g., [3] and [4] and those which are more reactive (without planning) like in [5] or [6]. Our proposed architecture of control/management is linked to this last approach.

The architecture that will be able to guaranties multi-objective criteria can be elaborated with modular and bottom-up manner as introduced in [7] and so called behavioral architectures [8]. They are based on the concept that a robot can achieve a global complex task while using only the coordination of several elementary behaviors. It is considered in a lot of studies (the proposed paper is among them), the investigation of the potentialities of the hybrid systems controllers [9] which provide a formal framework to demonstrate the robustness and the stability of such architecture. Thus, this formalism permits a rigorous automatic's control analysis on the performances of the architecture of control [10].

The rest of the paper is organized as follows. Section 1 gives the specificities of the proposed control architecture. In section 2, the architecture of control is applied to the task of navigation in presence of obstacles. Details of the proposed and implemented elementary controllers are given also in this last section. Section 3 is devoted to the description and analysis of the simulation results. This paper ends with some conclusions and further works.

1. Control architecture

The proposed structure of control (cf. figure. 1) has an objective to manage the interactions between different elementary controllers while guaranteing the stability of the overall control. It will also obtain a suppleness of the control command when the switch between controllers occurs. It will permit for example to an autonomous applications of travelers transportation [11] to have more comfortable displacements of the passengers. The specific blocks composing this control are detailed below.

1.1. Hierarchical action selection

The activation of one controller in favor of another is achieved here completely with a hierarchical manner like the principle of the subsumption proposed by Brooks in [7].

1.2. Controllers

Every controller is characterized by a stable nominal law which is represented by the function: $F_i(P_i, t) = \eta_i(P_i, t)$ with: P_i perceptions useful to the controller "i". Otherwise, in order to avoid among others, the important jump of commands at the time of the switch between controllers (for example here from the controller "j" toward the controller "i" at the instant t_0), an adaptation of the nominal law is proposed:

$$F_i(P_i, t) = \eta_i(P_i, t) + G_i(t) \tag{1}$$

with: $G_i(t)$ a monotonous function that tends to zero at the end of a certain time"T" and

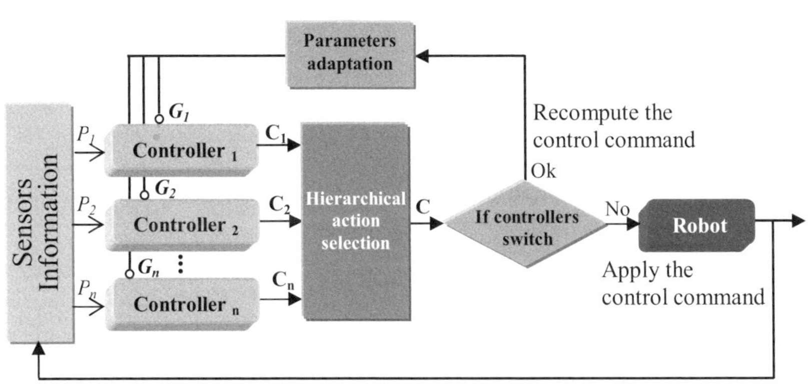

Figure 1. Hybrid control architecture for mobiles robots navigation

$$G_i(t_0) = F_j(P_j, t_0 - \Delta t) - \eta_i(P_i, t_0) \quad (2)$$

where: Δt represents the sampling time between two control commands.

The definition of $G_i(t)$ allows us to guarantee that the law of control (cf. equation. 1) tends toward the nominal control law after a certain time "T", i.e., $F_i(P_i,\ t_0 + T) = \eta_i(P_i,\ t_0 + T)$.

The function of adaptation $G_i(t)$ is updated by "the parameters adaptation block (see section below)" every time the switch of control toward the "i" controller occurs. The main constraint emanating from this structure of control is to guarantee the stability of the updated control law (cf. equation. 1) during the period where $G_i(t)$ is not zero.

1.3. Parameters adaptation

This block has as input the "conditional block" (cf. figure. 1) that verifies if the controller which should be active in the current instant "t" is different than the one active at the instant "t-Δt". If it is the case then it must update its function of adaptation (cf. equation. 2).

2. Navigation in presence of obstacles

The investigated task that is used to illustrate the application of the proposed control structure corresponds to the navigation of an unicycle robot in presence of obstacles (cf. figure. 2). The controlled position of the unicycle can be off-center as it is given in figure 2 by (l_x, l_y).

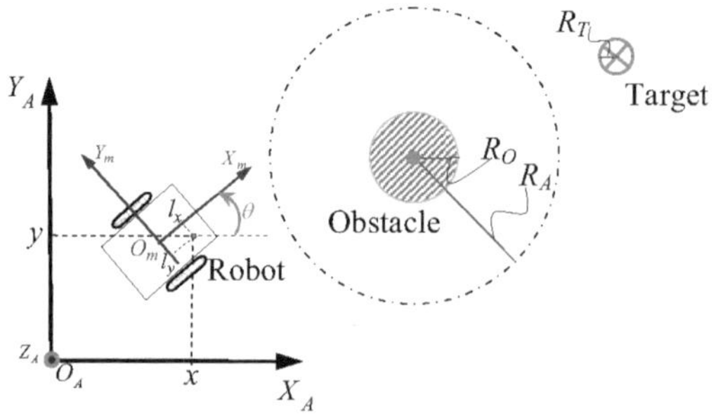

Figure 2. Mobile robot and the task parameters for the navigation in presence of obstacles

One presents below the proposed elementary controllers to achieve the desired navigation task.

2.1. Attraction to the objective controller

This controller guides the robot toward the target to reach. Classical techniques of linear systems stabilization can be used here to asymptotically stabilize the error $e_x = x - x_T$ and $e_y = y - y_T$ to zero, where: $(x,\ y)$ corresponds to the position of the robot and (x_T, y_T) corresponds to the position of the target (cf. figure. 2). We use a simple proportional command which is given by:

$$\begin{pmatrix} v \\ w \end{pmatrix} = -K \begin{pmatrix} \cos\theta & -l_x \sin\theta \\ \sin\theta & l_x \cos\theta \end{pmatrix}^{-1} e = -K \begin{pmatrix} \cos\theta & \sin\theta \\ -\sin\theta/_{l_x} & \cos\theta/_{l_x} \end{pmatrix} \begin{pmatrix} e_x \\ e_y \end{pmatrix} \quad (3)$$

With: v and w are respectively linear and angular speed controls; $K > 0$; $l_x \neq 0$ (cf. figure. 2). To guaranty the right transition between controllers as described in sections (1.2) and (1.3), the modification to the above command becomes:

$$\begin{pmatrix} v \\ w \end{pmatrix} = -K \begin{pmatrix} \cos\theta & \sin\theta \\ -\sin\theta/_{l_x} & \cos\theta/_{l_x} \end{pmatrix} \begin{pmatrix} e_x \\ e_y \end{pmatrix} + \begin{pmatrix} G_{A\_v}(t) \\ G_{A\_w}(t) \end{pmatrix} \quad (4)$$

While using the following Lyapunov function $V_1 = \frac{1}{2}d^2$ (with $d = \sqrt{e_x^2 + e_y^2}$), we therefore obtain $\dot{V}_1 = d\dot{d}$.

To guaranty that the proposed controller is asymptotically stable we must have $\dot{V}_1 < 0$. Therefore, after some simplification we can deduce that:

$$K > \frac{-(G_{A\_v}(t)e_x + G_{A\_w}(t)e_y)}{e_x^2 + e_y^2} \quad (5)$$

Where $G_{A\_v}(t)$ and $G_{A\_w}(t)$ functions must be chosen with respect to the constraints given in sections (1.2) and (1.3) and to the fact that they decrease more quickly to zero than d^2 (to have bounded K).

2.2. *Obstacle avoiding controller*

To implement this controller, the method of the limit-cycle was used [12]. The control law proposed to follow these trajectories (which tends toward a circle of R_A radius (cf. figure. 2)) is a control of orientation. The robot is controlled in relation to the center of its axle. The desired robot orientation θ_d is given by $\theta_d = \arctan(\frac{\dot{y}}{\dot{x}})$ where ($\dot{x}$ and $\dot{y}$) are given by a differential equation describing the limit-cycle [12], and the error by $\theta_e = \theta_d - \theta$.

One can control the robot to move to the desired orientation by using the following control law: $w = \dot{\theta}_d + K_p\theta_e + G_O(t)$, where $K_p > 0$ and $G_O(t)$ the function which guaranties the right transition between controllers (cf. sections (1.2) and (1.3)). It is noted that the nominal linear speed of the robot v for this controller is considered as a constant.

The derivative of θ_e is given by $\dot{\theta}_e = -K_p\theta_e - G_O(t)$. One takes the following Lyapunov function $V_2 = \frac{1}{2}\theta_e^2$. Therefore, $\dot{V}_2 = \theta_e\dot{\theta}_e = -K_p\theta_e^2 - G_O(t)\theta_e$. To guaranty that the proposed controller is asymptotically stable we must have $\dot{V}_2 < 0$, so:

$$K_p > -\frac{G_O(t)}{\theta_e} \quad (6)$$

Where $G_O(t)$ function is chosen with respect to the constraints given in section 1.3 and the fact that it decreases more quickly to zero than θ_e.

3. Simulation results

Before giving comments on the achieved simulation, it must be noted that $G_{A\_v}(t)$, $G_{A\_w}(t)$ and $G_O(t)$ were implemented using decreasing linear functions. Figures 3(a) and 3(b) show respectively the clockwise and counter-clockwise obstacle avoiding when the proposed control architecture is used.

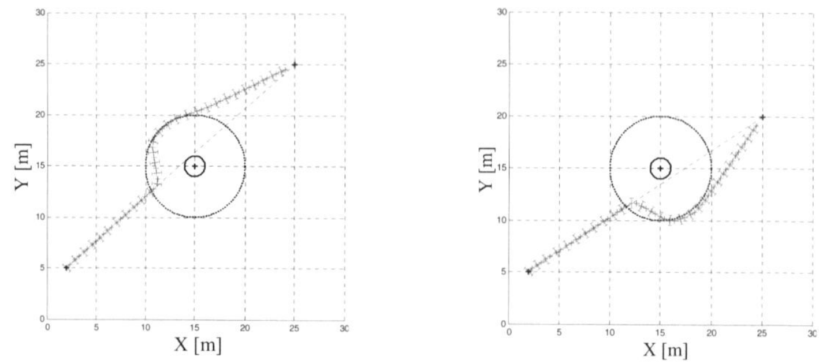

(a) Clockwise skirting (b) Counter-clockwise skirting

Figure 3. Application of the proposed control architecture for different target position

Figure 4 shows the effect of the use of adaptive parameters mechanism on v and w control commands (here the obstacle is clockwise skirting). v and w control becomes thus less abrupt and smoother when the switch between controllers occurs.

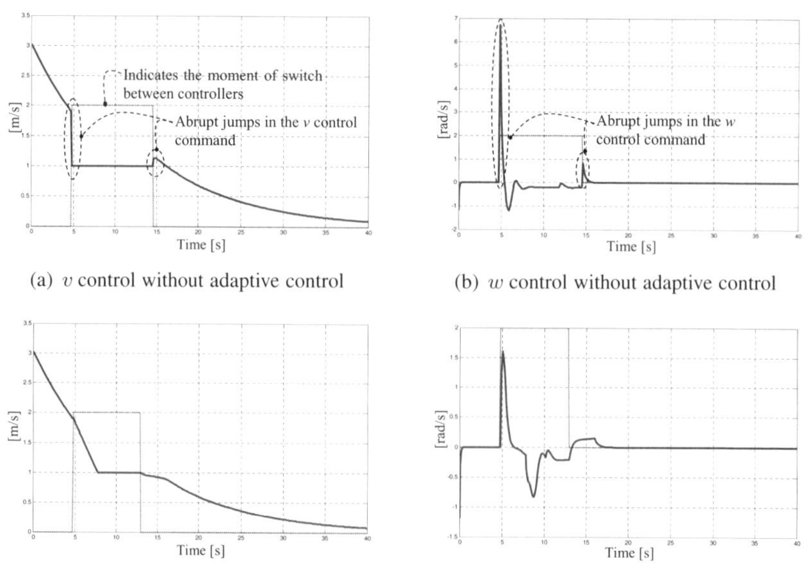

(a) v control without adaptive control (b) w control without adaptive control

(c) v control with adaptive control (d) w control with adaptive control

Figure 4. Effect of the adaptive parameters mechanism on v and w control commands

Figure 5 shows that the overall proposed structure of control is stable, and here the Lyapunov function attributed to each controller $V_i|_{i=1..2}$ decreases always asymptotically to the equilibrium point.

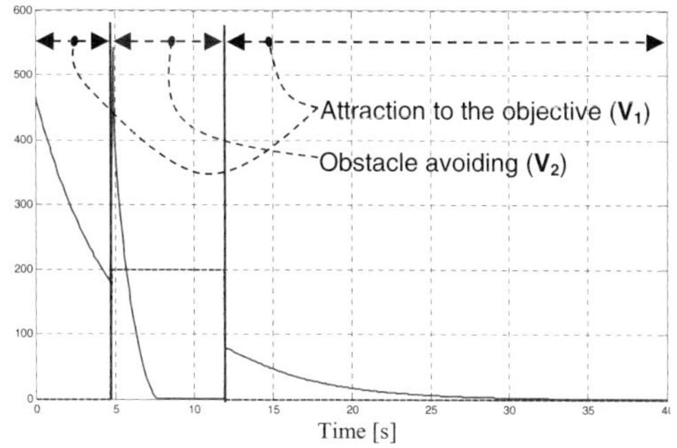

Figure 5. Evolution of V_1 and V_2 Lyapunov functions

4. Conclusion and future works

In this paper, an adaptive multi-controller architecture of control is proposed and applied on the task of mobile robot navigation in presence of obstacles. The proposed adaptive mechanism permits to achieve the switch between controllers while guaranteeing the smoothness of the control commands and the stability of the overall system. Future works will test first the proposed architecture of control on the CyCab vehicle [11]. The second step is to apply the proposed structure of control on more complex tasks like navigation in highly dynamical environment.

References

[1] J.-C. Latombe, Robot Motion Planning, Kluwer Academic Publishers, Boston, MA, 1991.

[2] E. Rimon, D. E. Koditschek, Exact robot navigation using artficial potential fields, IEEE Transactions on Robotics and Automation 8(5) (Oct. 1992) 501–518.

[3] C. Belta, V. Isler, G. J. Pappas, Discrete abstractions for robot motion planning and control in polygonal environments, IEEE Transactions on Robotics 21(5) (Oct. 2005) 864–874.

[4] D. C. Conner, H. Choset, A. Rizzi, Integrated planning and control for convex-bodied nonholonomic systems using local feedback, in: Proceedings of Robotics: Science and Systems II, MIT Press, Philadelphia, PA, 2006, pp. 57–64.

[5] M. Egerstedt, X. Hu, A hybrid control approach to action coordination for mobile robots, Automatica 38(1) (2002) 125–130.

[6] J. Toibero, R. Carelli, B. Kuchen, Switching control of mobile robots for autonomous navigation in unknown environments, in: IEEE International Conference on Robotics and Automation, 2007, pp. 1974–1979.

[7] R. A. Brooks, A robust layered control system for a mobile robot, IEEE Journal of Robotics and Automation RA-2 (1986) pp.14–23.

[8] L. Adouane, N. Le Fort-Piat, Hybrid behavioral control architecture for the cooperation of minimalist mobile robots, in: International Conference On Robotics And Automation, New Orleans-USA, 2004, pp. 3735–3740.

[9] M. Zefran, J. W. Burdick, Design of switching controllers for systems with changing dynamics, in: IEEE Conference on Decision and Control CDC'98, Tampa, FL, 1998, pp. 2113–2118.

[10] M. S. Branicky, Multiple lyapunov functions and other analysis tools for switched and hybrid systems, IEEE Transaction on Automatic Control 43(4) (1998) 475–482.

[11] C. Pradalier, J. Hermosillo, C. Koike, C. Braillon, P. Bessière, C. Laugier, The cycab: a car-like robot navigating autonomously and safely among pedestrians, Robotics and Autonomous Systems 50 (1) (2005) 51–68.

[12] D.-H. Kim, J.-H. Kim, A real-time limit-cycle navigation method for fast mobile robots and its application to robot soccer, Robotics and Autonomous Systems 42(1) (2003) 17–30.

Intelligent Autonomous Systems 10
W. Burgard et al. (Eds.)
IOS Press, 2008

doi:10.3233/978-1-58603-887-8-348

Decentralized cooperative exploration: Implementation and experiments

Antonio FRANCHI [1], Luigi FREDA, Luca MARCHIONNI, Giuseppe ORIOLO and Marilena VENDITTELLI
Dipartimento di Informatica e Sistemistica
Università di Roma "La Sapienza", Italy

Abstract. We present an implementation of the SRG method, a decentralized cooperative exploration strategy. In this method, a roadmap of the explored area with the associate safe region is stored in a compact data structure, called Sensor-based Random Graph (SRG). No task decomposition and/or allocation is performed. The roadmap is incrementally built through a simple decentralized mechanism: each robot moves towards its local frontier to access areas that, on the basis of the available information, appear to be unexplored by the rest of the team. A detailed description of the software architecture used to implement the strategy is given. Preliminary experiments with a team of Khepera III robots are presented to show the performance of the proposed technique.

Keywords. Cooperative exploration, multi-robot systems, decentralized algorithms

Introduction

Exploration of unknown environments is one of the most challenging problems in robotics. This task typically requires a mobile robot to cover an unknown area while learning, at the same time, a model of the environment or locating a given object. A wide range of applications are conceivable, including automated surveillance, search-and-rescue operations, map building and planetary missions.

In general, a number of advantages come from the use of a multi-robot system [1,2]. In exploration, a team of robots typically reduce the time required to complete the task. If a map is to be acquired, the redundant information provided by multiple robots can be also used to increase the final map accuracy and the quality of the localization [3]. In order to achieve these objectives, some sort of task decomposition and allocation are required. In practice, strategies to conveniently distribute robots over the environment should be accurately devised in order to reduce the occurrence of spatial conflicts [4] and actually reap the benefits of a multi-robot architecture. Communication plays a crucial role in achieving a truly cooperative behavior [5].

In most exploration strategies, the boundary between known and unknown territory (the *frontier*) is approached in order to maximize the information gain. For the multi-

[1]Corresponding Author: Dipartimento di Informatica e Sistemistica, Università di Roma "La Sapienza", Via Ariosto 25, 00185 Roma, Italy; E-mail: franchi@dis.uniroma1.it

robot case, a pioneering work in this direction is [6]: the robots merge the acquired information in a global gridmap of the environment, from which the frontier is extracted and used to plan the individual robot motions. While this basic scheme lacks an arbitration mechanism preventing robots from approaching the same frontier region, in [7] it is proposed to negotiate robot targets by optimizing a utility function which takes into account the information gain of a particular region, the cost of reaching it and the number of robots currently heading there. In [8], the utility of a particular frontier region from the viewpoint of relative robot localization (and hence, of the accuracy of map merging) is also considered. In the incremental deployment algorithm of [9], the robots approach the frontier while retaining visual contact with each other. An interesting multi-robot architecture in which the robots are guided through the exploration by a market economy is presented in [10], whereas [11] proposes a centralized approach which uses a frontier-based search and a bidding protocol assign frontier targets to the robots.

In this paper, we present an implementation and preliminary experiments of the SRG method, a decentralized strategy for cooperative robot exploration presented in [12]. In particular, we detail the structure of the communication threads which were not discussed in [12]. The reported experiments are conducted with a team of real robots and a detailed description of the software architecture is given.

1. Problem assumptions

For simplicity, the SRG multi-robot exploration method is described under the following assumptions, which are verified for our experimental setup (see Sect. 4):

1. The robots move in a planar workspace $\mathcal{W} \subseteq I\!R^2$.
2. Each robot is a disk, whose configuration q is described by the cartesian position of its center. The disk is *path controllable*, i.e., it may follow any path in its configuration space with arbitrary accuracy. This assumption is verified for free-flying as well as (most) nonholonomic mobile robots.
3. Each robot is equipped with a sensory system which provides the *Local Safe Region* LSR(q), a (possibly conservative) estimate of the free space surrounding the robot at q within a perception range R_p (Fig. 1, left).
4. Each robot can broadcast/receive data to/from other robots within a communication range R_c.

However, 3D workspaces, higher-dimensional configuration spaces and heterogeneous robots can be accommodated within the same framework. Also, path controllability can be replaced with simple controllability.

2. The Sensor-based Random Graph

The Sensor-based Random Graph (SRG) is a data structure that represents the workspace area explored by the team of robots. Nodes (arcs) of the SRG represent collision-free configurations (paths) that have been visited (traversed) by at least one robot. The Local Safe Region LSR(q) (Fig. 1, left) is also included in the description of node q.

Exploration actions are actually planned using the *Local Reachable Region* LRR(q), defined as the set of configurations that can be reached from q with the robot staying

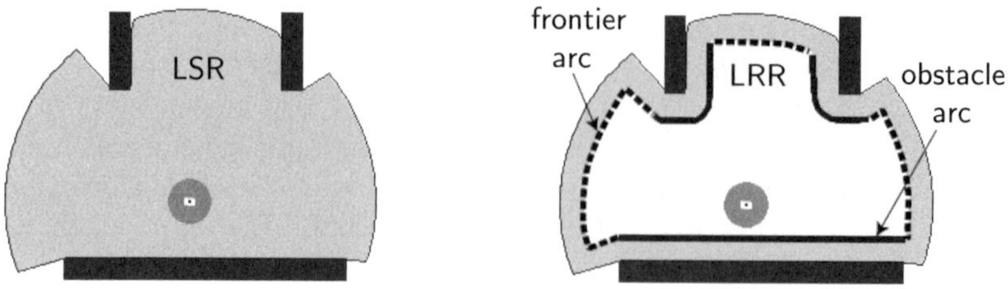

Figure 1. The Local Safe Region (left) and the Local Reachable Region (right) at a certain configuration. Also shown are the obstacle (solid) and frontier (dashed) arcs of the LRR boundary.

within LSR(q) (see Fig. 1, right). Under Assumption 2, the LRR(q) is obtained by *eroding* the LSR(q) with the robot disk as structuring element, and then *extracting* the connected component containing q. The boundary of an LRR can be partitioned in *obstacle*, *free* and *frontier* arcs (see Fig. 1, right). The first correspond to configurations at which the robot would graze a locally detected obstacle, and are easily identified from the range scan. Free arcs (not present in Fig. 1 where a single LRR is shown) do not correspond to obstacles but fall within the interior of other LRRs. Arcs that are neither obstacle nor free are classified as frontier; their union is the *frontier* of the LRR and, by extension, of the associated node. By construction, the frontier is the portion of the LRR boundary leading to (so far) unexplored areas. Depending on the available memory, the LRR and the frontier associated to a node may be stored with the node or computed when necessary from the LSR.

Additional structures, called *bridges*, are continuously attached to the SRG to improve its connectivity by incorporating shortcuts. A bridge is an arc-node-arc sequence (which may degenerate to a single arc) that is added to the SRG to connect any pair of nodes v and w such that *(i)* the graph distance between v and w is greater than a certain threshold *(ii)* the robot can move from v to w while remaining in $\mathrm{LSR}(v) \cup \mathrm{LSR}(w)$. At the time of its creation, a bridge has not been crossed by any robot of the team.

It should be emphasized that the SRG is a 'virtual' data structure, which is actually represented in a distributed fashion in the team. In fact, the generic i-th robot has its own version of the graph, which will be called SRG_i in the following. SRG_i is built by the robot on the basis of data acquired either by the robot itself or via communication with other robots. The SRG, which is the union of all the SRG_i's, is not explicitly represented at any level.

3. Functional architecture

The architecture of the software running on each robot of the team is shown in Fig. 2. Blocks with thick edges represent processes, those with thin edges represent threads, and dashed rectangles represent data. Arrows indicate an information flow: thick for inter-process communication, thin for communication between threads, dashed for read/write operation on data structures.

The *robot explorer* implements the SRG exploration algorithm, while the *robot driver* provides low-level primitives for motion, localization and perception. The two processes communicate through the TCP protocol, allowing a distributed instantiation of the architecture and providing a flexible integrated environment for simulation and

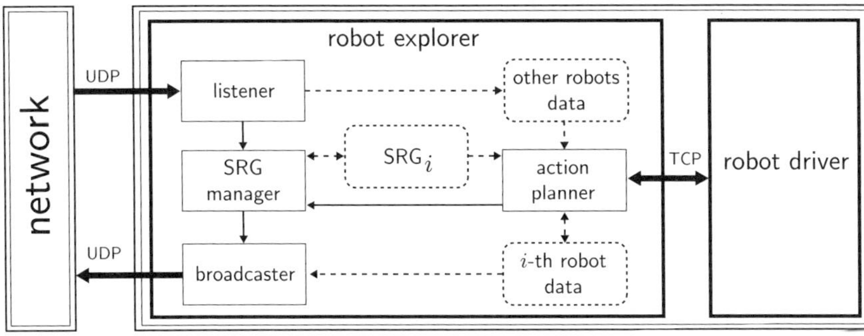

Figure 2. Functional architecture of the software running on the i-th robot.

experimental validation. With this architecture, in fact, the explorer and the driver do not need to run on the same machine, and the latter can be a real or a simulated robot.

The robot explorer is realized by four threads: the *action planner*, the *SRG manager*, the *broadcaster* and the *listener*. The action planner represents the core of the robot explorer: it is in charge of choosing the next exploration action in a cooperative and coordinated way. The task of SRG manager is to elaborate and continuously update the data stored in the SRG_i on the basis of the information received from the action planner or, through the listener, from the rest of the team. In particular, the action planner makes available the LSR acquired by the robot, while the listener provides the LSRs acquired by other robots of the team with which communication has taken place. The SRG manager merges these data so as to maintain the consistency of local representations. To this end, self-localization and mutual localization information coming from the robot driver (and, through the listener, from other robots) is used. Finally, the broadcaster transmits all the information currently available to the robot.

3.1. Action planner

The algorithm implemented by the action planner on the i-th robot is an instantiation of the SRG multi-robot exploration strategy (see [12] for details).

At the start of each iteration of the algorithm, the robot is stationary in a position corresponding to a node of the SRG_i. The action planner sends a 'perceive' command to the robot driver, receives the current LSR and makes it available to the SRG manager. If the frontier of the current LRR, computed by the SRG manager, is not empty, the action planner selects a new 'view' configuration on this frontier according to a randomized mechanism[2]. A path reaching the configuration is then planned inside the current LRR. If the frontier of the current LRR is empty, the action planner plans a path on SRG_i towards the closest node with a non-empty local frontier. A new iteration of the algorithm is started as soon as the robot reaches the first node (after the start node) on this path, so that planning is always performed on the basis of the most recent available information.

Interlaced with perception, target selection and path planning there are synchronization steps necessary to achieve coordination. Before choosing the next view configura-

[2]A deterministic version of this planner can be envisaged in which a specific configuration on the closer frontier is selected according to some criterion. However, given the large branching factor of the exploration problem, a biased random procedure can be considered a valid (and easy to implement) alternative to more computationally expensive greedy strategies.

tion, the robot has to identify the *Group of Engaged Agents* (GEA), i.e, the other agents of the team with which cooperation and coordination are necessary, based on the information stored in its SRG_i and on the other robots data. Formally, two robots are *GEA-coupled* when their LSRs overlap. The GEA of the robot is composed by all the robots to which it can be connected through a chain of GEA couplings. Since the other robots of the team may be stationary as well or moving, a synchronization phase is needed. To this end, the robot first builds the *Group of Pre-engaged Agents* (GPA), i.e., the robots which are candidate to belong to the GEA. Two robots are *GPA-coupled* if the distance between their targets is at most $2R_p$. The GPA of the robot is formed by all the robots to which it can be connected through a chain of GPA couplings. To achieve synchronization, the GPA is computed and updated with the data made available by the listener until all its members are stationary.

The actual GEA is built when the LSRs of all the robots in its GPA have been received by the listener and integrated in SRG_i by the SRG manager. If the GEA is composed only by the robot itself, the action planner sends a motion command to the robot driver and the robot moves to its target. Otherwise, it waits for receiving all the prospective paths of the robots in the GEA and checks them for mutual collisions. Collision paths are then classified in feasible and unfeasible according to a deterministic rule implemented by each robot. If the planned path is not feasible, the robot's move is forbidden by resetting the target to its current configuration. Last, a motion command leading the robot towards its target is sent to the robot driver.

Exploration iterations are repeated until there are no reachable nodes in the SRG_i with non-empty frontier. This means that the boundary of the explored region reachable by the robot is only composed by obstacle arcs. Hence, the robot is unable to further expand the graph and it has to go back to its starting position (*homing*).

3.2. SRG manager

The SRG manager receives data from the action planner and the listener and updates SRG_i accordingly. Two kind of data can be received: a node or an arc. The node comes with its associated LSR(q), while the arc consists in a couple of nodes to be joined by a feasible path. On reception of a node (either from the action planner or from the listener), the SRG manager 1) adds the node to SRG_i or updates the info associated to the preexisting node 2) computes the LRR(q) and the associated frontier 3) updates the frontiers of the nodes in SRG_i whose LRRs have a non-empty intersection with LRR(q). On reception of an arc, new nodes[3] are added to the SRG_i if they are not already stored in the graph. Then, the new arc is attached to the two nodes. Each time a new node v is added to SRG_i, the SRG manager creates a bridge between v and any node w in SRG_i satisfying the two conditions given in Sect. 2.

3.3. Broadcaster and listener

On each robot, the broadcaster and the listener respectively transmit and receive information. This is organized in three possible messages containing 1) the current robot configuration, the current target, the step of the exploration algorithm currently executed by the action planner, the robot's GPA/GEA[4]; 2) the nodes between which an arc is to be

[3] To reduce bandwidth consumption, the nodes do not come with an associated LSR.

[4] This information is necessary only if the communication range R_c is limited.

created; 3) the node, either new or already present in the SRG_i, with the associated LSR. Messages of the first type are transmitted synchronously by the broadcaster, while the other two kind of messages are transmitted as new data are made available from the SRG manager. The listener receives messages asynchronously from the network and makes them available to the action planner (other robots data) and the SRG manager.

4. Experiments

A preliminary simulation study in [12], performed with teams of various cardinality, has shown that both the exploration time and the mean traveled distance quickly decrease in the beginning with the number of robots, then asymptotically tend to a constant value. In the case of scattered start this asymptotic value is zero while in the case of clustered start it depends on the size of the environment. In the present work, experiments have been conducted with a team of 4 Khepera III robots (http://www.k-team.com). In addition to the standard equipment, each robot has been provided with a wireless card, for communication between robots of the team and/or with a remote computer, and a Hokuyo URG-04LX laser rangefinder, used for the construction of the LSR, with the following characteristics: angular resolution of 0.36°, radial resolution of 1 mm, scanning angle equal to 240°, maximum perception range of 4 m. In the adopted experimental setup, a robot driver runs on each Khepera III, whereas each explorer process can flexibly run on any separate remote computer. During the exploration, an additional process, called *visualizer*, is in charge of 'sniffing' and storing all the packets exchanged among explorer processes in order to visualize and monitor the exploration progress. Although all the robot explorers and robot drivers communicate with each other through a single LAN, hence sharing the same bandwidth, the limited set of exchanged messges prevented any relevant loss of communication packets in all the conducted experiments.

Each robot is provided with a basic self-localization module in which incremental scan matching [13] is used to correct odometry localization. A preliminary calibration of robot odometry and sensor parameters was performed beforehand as described in [14]. In these preliminary experiments, the robots know their relative configurations at the start of exploration. Hence, mutual localization is maintained on the basis of this initial knowledge and the above mentioned local registrations.

Figure 3 show two exploration environments built inside in a rectangular arena whose area is approximately 8 m^2. In these experiments, for each robot, the laser perception range has been limited to 1 m and the maximum cartesian velocity was 0.15 m/sec. The robot communication range was not artificially limited[5]. At the end of each experiment, the robots return back to their start positions (homing). Figures 4 and 5 show the progress of the exploration as reconstructed by the visualizer. The relevant numerical data associated to the experiments are reported in Fig. 6.

Note that the environment considered in the first experiment is simply connected and its topology is correctly captured by the resulting SRG in the form of a tree (see Fig. 4). Analogously, the multiply connected environment of the second experiment is efficiently represented as a graph by the resulting SRG (see Fig. 5). In this case, the number of bridges is higher than in the first experiment to accommo-

[5] In itself, the SRG method does not require an unlimited communication range. See [12] for further details.

Figure 3. *Left*: the first environment. *Right*: the second environment.

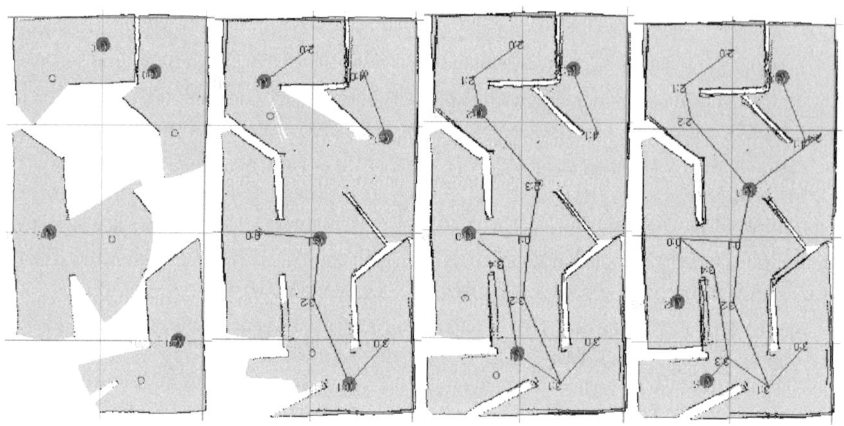

Figure 4. Exploration of the first environment as reconstructed by the visualizer. Bridges are shown in blue.

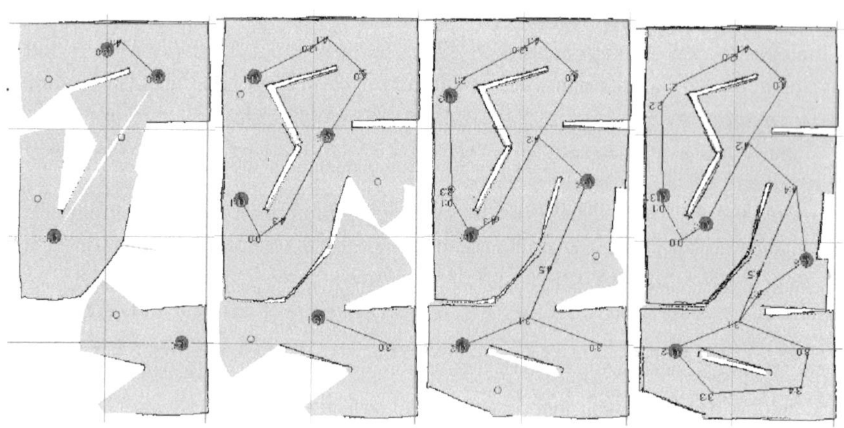

Figure 5. Exploration of the second environment as reconstructed by the visualizer. Bridges are shown in blue.

date the presence of loops. Additional experiments, including movies, are available at `http://www.dis.uniroma1.it/~labrob/research/multiSRG.html`.

	first experiment							second experiment						
	robots				aggregate data			robots				aggregate data		
	1	*2*	*3*	*4*	*mean*	*st dev*	*total*	*1*	*2*	*3*	*4*	*mean*	*st dev*	*total*
# nodes	5	7	2	3	4	2	17	4	5	7	3	5	2	19
# total arcs	5	7	1	5	5	3	18	4	5	9	3	5	3	21
# bridge arcs	3	4	0	0	2	2	7	2	2	6	2	3	2	12
traveled distance (m)	3,34	2,65	1,28	2,35	2,41	0,86	9,62	3,42	4,05	3,98	1,27	3,18	1,30	12,72
exploration time (sec)	240	228	92	174	184	67	240	265	259	264	179	242	42	265
homing error (m)	0,053	0,055	0,036	0,022	0,042	0,016	0,166	0,020	0,074	0,018	0,156	0,067	0,065	0,268

Figure 6. Table of numerical results.

5. Conclusions

In this paper we have presented the implementation and preliminary experiments of the SRG method, a decentralized cooperative exploration strategy for multiple mobile robots. Future work include: design of a procedure for global alignment of scans, based on the work in [15], to be performed each time a loop closure is detected; quantitative study of the robustness and scalability properties of the method; the development of an efficient mutual localization method.

References

[1] Y. Cao, A. Fukunaga, and A. Kahng, "Cooperative mobile robotics: Antecedents and directions," *Autonomous Robots*, vol. 4, no. 1, pp. 7–27, 1997.

[2] G. Dudek, M. Jenkin, E. Milios, and D. Wilkes, "A taxonomy for multi-agent robotics," *Autonomous Robots*, vol. 3, pp. 375–397, 1996.

[3] I. Rekletis, G. Dudek, and E. Milios, "Multi-robot collaboration for robust exploration," *Annals of Mathematics and Artificial Intelligence*, vol. 31, no. 1, pp. 7–40, 2001.

[4] D. Goldberg and M. Mataric, "Interference as a tool for designing and evaluating multi-robot controllers," in *14th AAAI/9th IAAI*, 1997, pp. 637–642.

[5] T. Balch and R. Arkin, "Communication in reactive multiagent robotic systems," *Autonomous Robots*, vol. 1, no. 1, pp. 27–52, 1994.

[6] B. Yamauchi, "Frontier-based exploration using multiple robots," in *2nd Int. Conf. on Autonomous Agents*, 1998, pp. 47–53.

[7] W. Burgard, M. Moors, C. Stachniss, and F. Schneider, "Coordinated multi-robot exploration," *IEEE Trans. on Robotics and Automation*, vol. 1, no. 3, pp. 376–386, 2005.

[8] J. Ko, B. Stewart, D. Fox, K. Konolige, and B. Limketkai, "A practical, decision-theoretic approach to multi-robot mapping and exploration," in *2003 IEEE/RSJ Int. Conf. on Intelligent Robots and Systems*, 2003, pp. 3232–3238.

[9] A. Howard, M. Mataric, and S. Sukhatme, "An incremental deployment algorithm for mobile robot teams," in *2003 IEEE/RSJ Int. Conf. on Intelligent Robots and Systems*, 2003, pp. 2849–2854.

[10] R. Zlot, A. Stenz, M. Dias, and S. Thayer, "Multi-robot exploration controlled by a market economy," in *2002 IEEE Int. Conf. on Robotics and Automation*, 2002, pp. 3016–3023.

[11] R. Simmons, D. Apfelbaum, W. Burgard, D. Fox, M. Moors, S. Thrun, and H. Younes, "Coordination for multi-robot exploration and mapping," in *17th AAAI/12th IAAI*, 2000, pp. 852–858.

[12] A. Franchi, L. Freda, G. Oriolo, and M. Vendittelli, "A decentralized strategy for cooperative robot exploration," in *First Int. Conf. on Robot Communication and Coordination*, 2007.

[13] A. Censi, "An icp variant using a point-to-line metric," in *2008 IEEE Int. Conf. on Robotics and Automation*, 2008.

[14] A. Censi, L. Marchionni, and G. Oriolo, "Simultaneous maximum-likelihood calibration of robot and sensor parameters," in *2008 IEEE Int. Conf. on Robotics and Automation*, 2008.

[15] F. Lu and E. Milios, "Globally consistent range scan alignment for environment mapping," *Autonomous Robots*, vol. 4, pp. 333–349, 1997.

Intelligent Autonomous Systems 10
W. Burgard et al. (Eds.)
IOS Press, 2008

doi:10.3233/978-1-58603-887-8-356

Structure Verification toward Object Classification using a Range Camera

Stefan GÄCHTER, Ahad HARATI and Roland SIEGWART [1]
Autonomous Systems Lab (ASL)
Swiss Federal Institute of Technology, Zurich (ETHZ)
8092 Zurich, Switzerland

Abstract. This paper proposes an incremental object classification based on parts detected in a sequence of noisy range images. Primitive parts are jointly tracked and detected as probabilistic bounding-boxes using a particle filter which accumulates the information of the local structure over time. A voting scheme is presented as a procedure to verify structure of the object, i.e. the desired geometrical relations between the parts. This verification is necessary to disambiguate object parts from potential irrelevant parts which are structurally similar. The experimental results obtained using a mobile robot in a real indoor environment show that the presented approach is able to successfully detect chairs in the range images.

Keywords. Object part detection, object structure verification, range camera

Introduction

Object recognition and classification are long standing issues in computer vision. They date even before mobile robotics became popular. In the current literature, most of the approaches rely on appearance of the objects. Although appearance gives useful hints about the nature of the objects, yet alone, it is not enough to achieve object classification under view point and illumination changes as demanded in mobile robotic applications. More importantly is the ability of abstraction needed in object classification, where structural variations within a certain class of object should be handled. As one step from object recognition toward classification, 3D perception seems necessary for dealing with objects of the real world.

Structure variability within a class of objects may be well explained using a geometric grammar, preferably probabilistic to take care of uncertain and incomplete measurements. In such an approach, object classification is reduced to detection of some object parts and verification of the required geometric relations among them. However, robust detection of complex object parts has more or less the same nature as object classification itself. Therefore, primitive parts are used to represent the overall structure of object parts as bounding-boxes. Such simplified geometric models are easier to detect in point cloud observations and deliver independence with regard to appearance, making the object classification more practical. Considering a chair for example, stick-like shapes are

[1]This work was partially supported by the EC under the FP6-IST-045350 robots@home project.

primitive parts which may correspond to a chair leg or something structurally similar. Having a proper probabilistic geometric grammar as discussed in [1], it would be possible to obtain the most probable hypotheses of the object and its parts from the set of detected primitive parts.

In recent years, a novel type of range camera emerged on the market to capture 3D scenes for mobile robotic applications [7]. Although very compact, light and capable of measuring distances up to several meters at high frame rate, lower measurement quality in general [5] poses great challenges in using such devices in an object classification framework. Therefore, the main goal and contribution of this paper is to bring well grounded approaches from different domains together, extend and adapt them to a novel object classification framework which can work with poorly observed scenes using a mobile robot equipped with a range camera.

The proposed approach can account for different views of the same object and for variations in structure, material, or texture of the objects of the same kind as far as the decomposition of the objects into its parts is known. The decomposition itself, that is the grammar, is given here and the focus remains on dealing with part detection, object localization and structure verification in realistic cases.

When dealing with noisy measurements, the segmentation for each primitive part can be challenging. Therefore, a track-before-detect scheme is implemented using a particle filter, which accumulates information of the local object structure, a local disparity measure – represented in form of shape factors – and estimates pose and extension of the potential primitive parts over time. This is a common approach in radar applications, where a target has to be jointly tracked and classified in highly noisy data [8]. To realize the observation function of the particle filter, a classifier is trained using support vector machine for each part category. Therefore, different types of primitive parts are detected in parallel while a structure verification procedure based on a voting scheme periodically applies the considered grammar to come up with the location of potential objects in the scene. Thus, although single observations are too poor to infer the presence of any object directly, the presented approach incrementally collects evidences from the sequence of range images and tracks the hypothetical primitive parts leading to object hypotheses.

The approach presented here is quite general in handling different object parts with simple geometry. However, throughout this paper, chairs consisting of legs, a seat and a back support are chosen as example objects to demonstrate the method.

1. Particle Filter based Primitive Part Detection

In this approach, part detection is formulated as tracking hypothetical bounding-boxes in a sequence of voxelized point clouds using a particle filter. The details of the algorithm can be found in [3]. Here, a brief summary is given.

1.1. Incremental State Estimation

For a sequence of registered and voxelized range images $\mathbf{Z}_k$ up to step k, the part detection process is given by

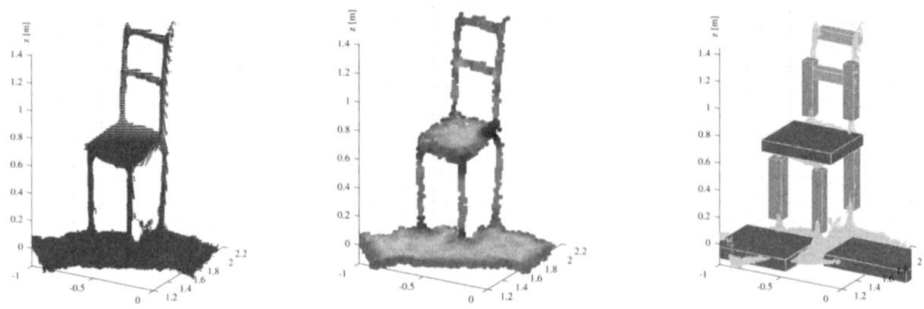

(a) Point Cloud (b) Voxel Set (c) Estimated Parts

Figure 1. (a) Single point cloud and (b) a quantized version of a sequence of five registered range images at step $k = 25$. The colors in (b) indicate the shape factors: red for *linear* like, green for *planar* like, and blue for *spherical* like local structures. (c) Estimated primitive parts at step k = 25. The color indicates the number of hypothetical parts encoded by a particle: green for 2, blue for 3, and crimson for 5 states.

$$\begin{aligned} p(\mathbf{y}_k|\mathbf{Z}_{k-1}) &= \int p(\mathbf{y}_k|\mathbf{y}_{k-1})p(\mathbf{y}_{k-1}|\mathbf{Z}_{k-1})d\mathbf{y}_{k-1} \\ p(\mathbf{y}_k|\mathbf{Z}_k) &\propto p(\mathbf{z}_k|\mathbf{y}_k)p(\mathbf{y}_k|\mathbf{Z}_{k-1}), \end{aligned} \tag{1}$$

where $\mathbf{y}_k = [R_k, \mathbf{x}_{1,k}^{\mathsf{T}} \dots \mathbf{x}_{r_k,k}^{\mathsf{T}}]^{\mathsf{T}}$ is the augmented state, which contains the current estimate of number of object parts present in the view R_k and their bounding-box parameters $\mathbf{x}_{i,k}$ at step k. Similar to [2], the number of primitive parts R_k is modeled by a Markov chain with a predefined transition matrix, where the state value at step k is a discrete number $r_k = \{0, \dots, M\}$ with M being the maximum number of parts expected in each view. Here, M varies between 4 and 8 depending on the part to detect.

1.2. Feature Vector

The shape factors characterize the local part structure by its linear, planar, or spherical likeliness. They are calculated for each voxel using its surrounding spatial voxel distribution by the decomposition of the distribution into the principal components:

$$r_l = \frac{\lambda_1 - \lambda_2}{\lambda_3 + \lambda_2 + \lambda_1}, \quad r_p = \frac{2(\lambda_2 - \lambda_3)}{\lambda_3 + \lambda_2 + \lambda_1}, \quad r_s = \frac{3\lambda_3}{\lambda_3 + \lambda_2 + \lambda_1}. \tag{2}$$

where λ_i are the ordered eigenvalues $\lambda_1 \geq \lambda_2 \geq \lambda_3$. Shape factors r_l, r_p, and r_s express local similarity to linear, planar, and spherical shapes respectively. Figure 1(a) depicts the original point cloud acquired with a range camera and figure 1(b) depicts the corresponding shape factor colored voxel set of a chair, where for each voxel the shape factor was computed according to (2).

The shape factor distribution in the region of interest defined by the bounding-box is approximated by a histogram to obtain a unique feature vector that models the object part. This approach is inspired by the work done in [2], while color is replaced by shape factor. In addition, dimensionality reduction is applied on the histogram to obtain a compact representation. The feature vector is composed of the compact histogram and six simple geometric features to account for the occupancy and eccentricity of the voxel distribution in the bounding-box.

1.3. Integration of the Observations

The observation likelihood function generates the importance weights used to incorporate the measurement information $\mathbf{z}_k$ in the particle set. A support vector classifier [4] is trained to detect the primitive parts in the sparse and noisy data. In figure 2(a), twelve different chairs are depicted which are used to obtain voxel images from different viewpoints and to build a manually labeled training set.

In the detection framework, the observation likelihood is usually defined as a ratio of the probability that an object part is present to the probability of its absence. This is equivalent to the ratio of the classification probabilities computed with the learned classifier. Assuming that the classification can be done independently for each hypothetical part and considering the probability as a distance $a_{i,k}$ in the range of $[0, 1]$, the importance weight $\tilde{\pi}_k$ for each particle is computed as:

$$\text{if } R_k = 0, \tilde{\pi}_k = 1 \text{ and if } R_k > 0,\ \tilde{\pi}_k = \exp\left(-\frac{1}{b}\sum_{i=1}^{r_k}(1 - 2a_{i,k}) + r \cdot c\right), \quad (3)$$

where b is a parameter to adjust the observation sensitivity and c accounts for the a-priori knowledge. Figure 3(a) depicts the outcome of the leg classifier for each voxel considering a fixed bounding-box. This is treated as the observation likelihood in the particle filter framework. Different primitive parts are detected with particle filters working in parallel using classifiers trained on different training sets. For the demonstrated chair example, leg, seat, and back classifiers are considered which are mainly detecting vertical stick, horizontal plate, and stick like structures.

2. Object Structure Verification

The detection algorithm provides different primitive parts as can be seen in figure 1(c). The encoded knowledge of the structure in the grammar is used to disambiguate primitive parts belonging to the object from the rest. In the case presented in figure 1(c), this means to reject planar patches detected on the ground that are structurally similar to the seat, or leg like structures supporting the back. This process is called object localization and structure verification, which applies the grammar constraints. It is implemented as a voting scheme similar to the *implicit shape model* as presented in [6], but extended to the 3D case.

2.1. Structure Model

The implicit shape model as used here is a probabilistic, non-parametric model which encodes 3D structure of the object in terms of relative location of every part with respect to a pre-defined reference point, here the center of the seat. The implicit shape model consists of a set of object specific parts $\mathcal{S} = \{s_j\}$ and a set of corresponding votes $\mathcal{V} = \{v_j\}$. This model is learned by memorizing all relative locations of the parts as depicted in figure 2(b) for the chair collection.

The votes can be seen as a sample-based representation of a spatial probability distribution $p(\mathbf{x}_o|o, s_j, \mathbf{x}_j)$ for each s_j at a given position $\mathbf{x}_j$, where $\mathbf{x}_o$ denotes the refer-

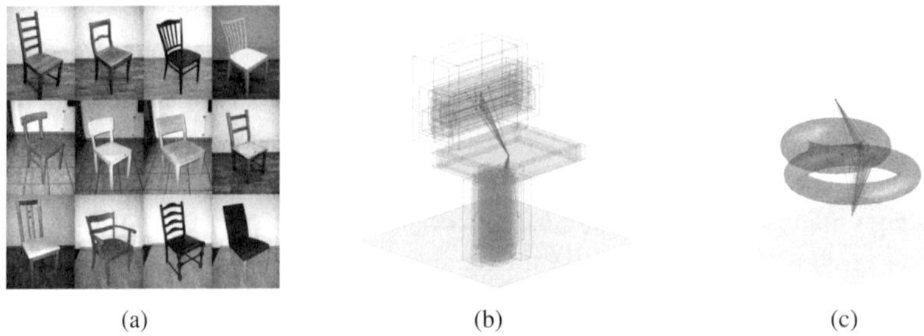

Figure 2. (a) Training set of twelve chairs. (b) Extracted object parts and their relative position with respect to the chair center. (c) Implicit shape model of the chair taking into account the part symmetry. Voting vectors for the leg are depicted in red and the ones for the back in blue. The voting vectors for the seats are zero.

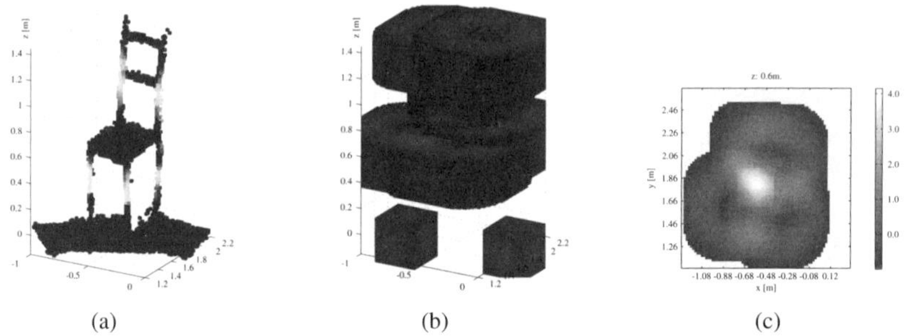

Figure 3. (a) Observation likelihood for the leg classifier using a fixed bounding-box size at step k = 25. Brighter colors indicate higher classification probabilities. (b) Snapshot at step k = 25 of the 3D voting space and (c) a slice where the maximum detection likelihood occurs. The detection likelihood is normalized such that it reflects best the contribution of each primitive part.

ence point of the object o. In the present case, the object is a chair and the primitive parts s_j are of leg-, seat-, or back-like structure. Thus, the implicit shape model represents a set of a-priori known part types, where the object reference point with respect to each part position is represented by a probability distribution.

Here, the spatial probability distribution for each part is assumed to be Gaussian which is represented by the collected votes during the training. Since the orientation of the object relative to the observer is arbitrary, some symmetry is added to the probabilistic model. In other words, each learned Gaussian is rotated along the z-axis and for each part a donut shape is obtained in the voting space. The resulting implicit shape model for the chair collection of figure 2(a) is depicted in figure 2(c).

2.2. Structure Verification

In the original implicit shape model approach, each interest point that match with an entry in the codebook casts its votes. Similarly, each detected primitive part casts its votes from the estimated position of the bounding-box center. The estimated position $\hat{\mathbf{x}}_k^\tau$ for each part type is obtained as

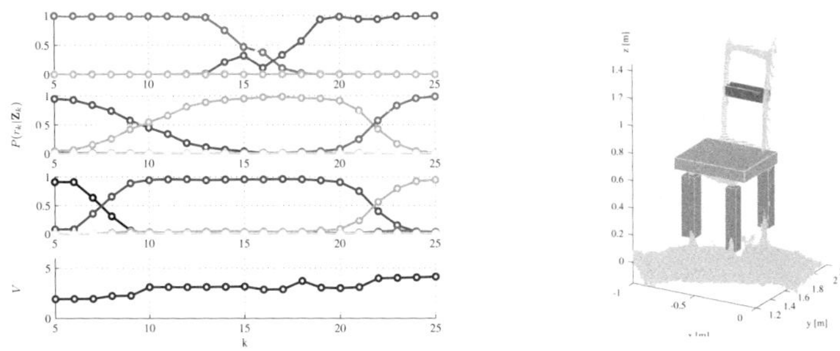

(a) (b)

Figure 4. Results of the first experiment. (a) Evolution of the part presence probabilities over time for the leg, seat, and back like parts in the first three rows. The last row indicates the maximum detection likelihood. The color indicates the number of hypothetical parts encoded by a particle: black for 0, red for 1, green for 2, blue for 3, yellow for 4, magenta for 5, and cyan for 6 primitive parts. Other colors indicate a higher number of parts. (b) Object parts after the localization and structure verification at step $k = 25$. Legs are depicted in red, seat in green, and back in blue.

$$\hat{r}_k = \arg\max_i \frac{\sum_{n=1}^{N} \delta(R_k^{(n)}, i)}{N}, \qquad \hat{\mathbf{x}}_k^r = \frac{\sum_{n=1}^{N} \mathbf{x}_k^{(n)} \delta(R_k^{(n)}, \hat{r}_k)}{\sum_{n=1}^{N} \delta(R_k^{(n)}, \hat{r}_k)}, \tag{4}$$

where $\hat{r}_k$ is the maximum a-posteriori estimate of the number of present primitive parts at step k. For each estimated part, votes are cast according to the learned distribution, see figure 2(c). In the experiments presented here, all votes are considered with equal weights. A sample snapshot of the voting space for the estimated parts of figure 1(c) is depicted in figure 3(b). The presented slice in figure 3(c) clearly shows the aggregation of the votes in the center of the seat as desired.

Once the voting space is populated, the local likelihood maxima indicate potential object locations. The object localization is completed by searching for these maxima. The current implementation uses a non-maxima suppression technique. Given a potential object position, the object structure is verified by checking the part constellation in the neighborhood of the detection likelihood maxima according to the chair category definition: stick-like parts below a seat are seen as chair legs, plate-like parts with some sticks underneath as chair seats, and horizontal stick-like parts above a seat as chair backs.

3. Experiments

The above discussed structure verification method is exemplified with a chair. Chairs in reality are designed with various shapes and structures. Here, they are modeled with three different kind of bounding-boxes to cover the stick- and plate-like structures of the chair legs, seats, and backs. For each object part class, an independent particle filter is used for the detection. The parameter setting follows the recommendations of [3]. The outcome of the part detection undergoes the localization and structure verification.

Two experiments are performed with the *range camera* mounted on a *robot* at height of about 1.1 m facing downward with a tilt angle of about 15°. In the first experiment, only one chair is in the scene while in the second experiment the robot is observing a round dining table, two chairs and a coffee table in the cafeteria of our lab. In all exper-

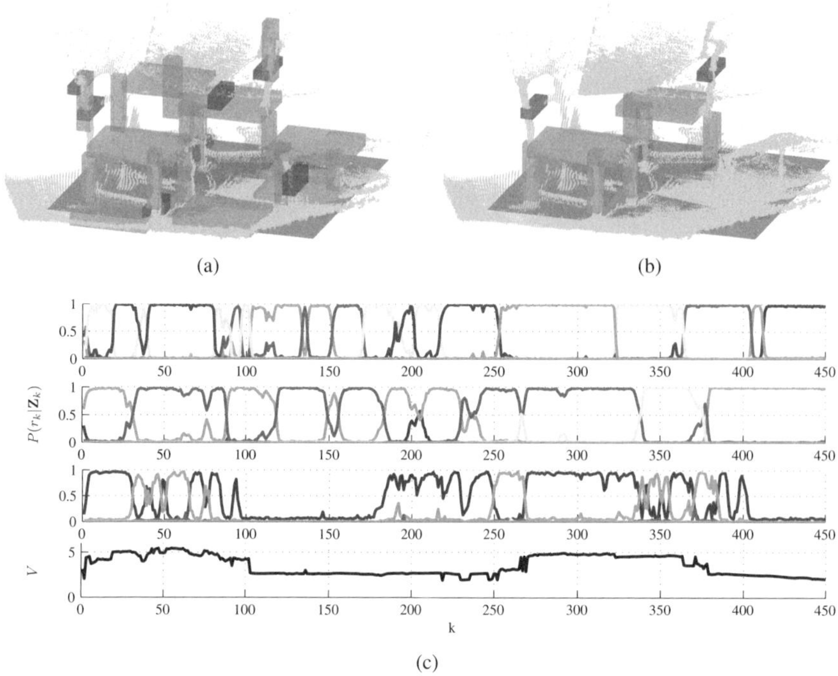

Figure 5. Results of the second experiment. (a) Detected primitive parts before the structure verification step for scene with a round dining table, two chairs and a coffee table. (b) Object parts after localization and structure verification step. Legs are depicted in red, seat in green, and back in blue. (c) Evolution of the part presence probabilities over time for the leg, seat, and back like parts in the first three rows. The last row indicates the maximum detection likelihood. Colors are explained in figure 5.

iments, the robot approaches the objects in the scene at speed of 2 cm/s while recording range images and odometry at about 10 Hz. Totally 200 and 450 range images are captured in the first and second experiment respectively. Because of occlusions and the narrow field of view of the camera, the number of hypothetical chair parts in the view varies considerably when the robot moves through the scene. Hence, the algorithm should dynamically adapt to what momentarily is present in the view.

The result of the first experiment is depicted in figure 4. Figure 4(a) depicts the evolution of the part presence probabilities for the three primitive part types over time. The number of all potential parts – leg, seat, and back – increase over time. Accordingly, the maximum vote strength depicted in the last row increases too. Thus, the evidence of having a chair present increases when having more parts present that vote for the same object center. All detected part like structures at step 25 are depicted in figure 1(c). The corresponding votes cast for each part are depicted in figure 3(b). The chair structure is verified with respect to the position of the maximum detection likelihood depicted in figure 3(c). The resulting chair parts are depicted in figure 4(b). Chair legs, seat, and back are extracted correctly. The back consist of one stick like part as memorized by the implicit shape model depicted in 2(b).

In the second experiment with a more realistic scenario, the robot is faced with the challenge of object detection in the cafeteria. In figure 5(a), the detected primitive parts before the structure verification are depicted overlaid with two original point clouds. In figure 5(b), the object parts after the structure verification are depicted. The verification was done for hypothetical objects with detection likelihood maxima higher than 2.9. The two chairs are successfully detected despite missing parts and considerable number of detected additional primitive parts. The algorithm has to deal with many appearing and disappearing parts as can be seen in figure 5(c). The upper three graphs depict the probabilities of the number of primitive parts present in the view for leg-, seat-, and back-like parts. The probabilities oscillate where the scene changes considerably. As in the first case, the evidence of having a chair present, depicted in the last row, is correlated with the number of primitive parts present in the view.

4. Conclusion

This paper presented an incremental object classification based on parts. Primitive parts are detected using a particle filter with a support vector classifier based observation function. A voting scheme is applied to localize the objects and a bounding-box check to verify their structure. The provided experimental results show that the approach detects successfully the chairs even though the hypothetical parts vary considerably in the view.

However, the method needs further testing and improvements for its robust application in robotics. The structure verification step has to be refined to cope with higher intra and inter class variabilities. Therefore, the probabilistic grammar has to be extended and a sophisticated parser should be designed. The implicit shape model presented here, then can be used as an initialization for such a parser.

Acknowledgements

Thanks to Luciano Spinello for the fruitful discussions on the implicit shape model approach.

References

[1] M. A. Aycinena. Probabilistic geometric grammars for object recognition. Master's thesis, Massachusetts Institute of Technology - Department of Electrical Engineering and Computer Science, 2005.

[2] J. Czyz, B. Ristic, and B. Macq. A color-based particle filter for joint detection and tracking of multiple objects. In *Proceedings of the IEEE International Conference on Acoustics, Speech, and Signal Processing (ICASSP '05)*, volume 2, pages 217–220, 2005.

[3] S. Gachter, A. Harati, and R. Siegwart. Incremental object part detection toward object classification in a sequence of noisy range images. In *Proceedings of the IEEE International Conference on Robotics and Automation (ICRA)*, 2008.

[4] C.-W. Hsu, C.-C. Chang, and C.-J. Lin. A practical guide to support vector classification. Technical Report July 18, 2007, Department of Computer Science - National Taiwan University, 2007.

[5] T. Kahlmann. *Range Imaging Metrology: Investigation, Calibration and Development*. PhD thesis, Eidgenössische Technische Hochschule Zürich, ETHZ, Diss ETH No 17392, 2007.

[6] B. Leibe, A. Leonardis, and B. Schiele. An implicit shape model for combined object categorization and segmentation. In *Towards Category-Level Object Recognition*, pages 496–510. Springer, 2006.

[7] MESA Imaging AG. Switzerland, `http://www.swissranger.ch/` (13.9.2007).

[8] B. Ristic, S. Arulampalam, and N. Gordon. *Beyond the Kalman Filter - Particle Filters for Tracking Applications*. Artech House, 2004.

Intelligent Autonomous Systems 10
W. Burgard et al. (Eds.)
IOS Press, 2008

doi:10.3233/978-1-58603-887-8-364

Coevolutionary modelling of a miniature rotorcraft

Renzo DE NARDI and Owen E. HOLLAND
University of Essex,
Department of Computing and Electronic Systems,
Colchester CO4 3SQ, United Kingdom;
rdenar@essex.ac.uk,owen@essex.ac.uk

Abstract. The paper considers the problem of synthesising accurate dynamic models of a miniature rotorcraft based on minimal physical assumptions, and using the models to develop a controller. The approach is based on the idea of building models that predict accelerations, and is implemented using evolutionary programming in a particularly efficient co-evolutionary framework. Both the structure and the parameters of the nonlinear models are jointly identified from real data. The modelling method is demonstrated and validated on a miniature quadrotor rotorcraft, and a controller based on the model is then developed and tested.

1. Introduction

For decades, the wide range of potential applications of unmanned aerial vehicles (UAVs) has made them the subject of academic and commercial research. Quadrotor rotorcraft like the one used in this study (Figure 1) are typical examples of very small and capable flying machines. Mechanically very simple to build and maintain, robust to crashes, lightweight, powerful and manoeuvrable, and readily available in the marketplace, these flying machines are becoming the platforms of choice for many research teams ([7],[9],[12],[13]).

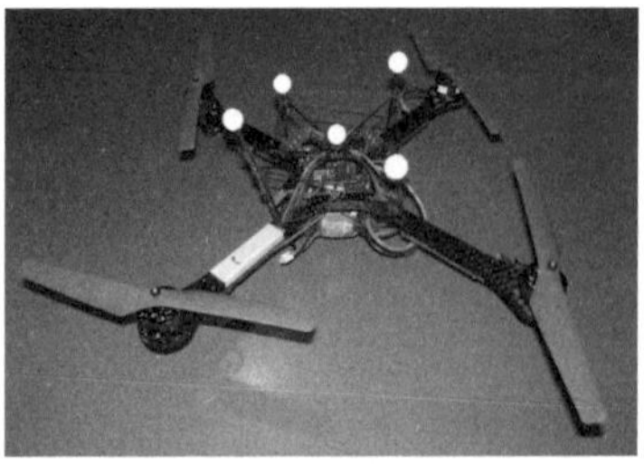

Figure 1. Quadrotor rotorcraft. Note the reflective markers used for tracking.

The challenge now is in fully exploiting the potential of these machines; in this paper we are seeking a methodology able to identify a system using only data collected during normal piloted flight, and without requiring any prior formal knowledge of the structure of the system model. The advantages of this approach are clear: as well as eliminating the need for in-depth knowledge of a specialist engineering domain, such as aeronautics, it offers the possibility of automatically identifying unknown and novel systems. Such systems are still being developed by talented craftsmen who build novel flying machines (or variations of existing concepts) relying almost entirely on their insight [14], [18],[15].

A second motivation is the possibility of using automatic modelling for automatic damage recovery (e.g. [6]), where after automatically relearning or adapting the system model, new control strategies that cope with the damage can be obtained.

In any exercise in modelling, it is first necessary to specify the domain of use for the model in question, as this allows us to define the types and extents of discrepancies from the real system that can be tolerated. In our case we plan to use the model to produce a dynamic simulator that, within the normal envelope of use of the machine, is sufficiently accurate for the purpose of developing and testing a controller. In this simulator-based approach, the form of the model is not important as long as the model replicates the input-output behaviour of the actual process. However, for comparability with other methods, we will develop a model in the form of a set of nonlinear differential equations (i.e. in state space form), a common way of representing dynamic systems:

$$^{1}\dot{\mathbf{x}}(t) = f\left(\mathbf{x}(t), \mathbf{u}(t - \boldsymbol{\tau})\right). \quad (1)$$

The next two sub-sections will give overviews of evolutionary modelling, and of modelling quadrotors. Our own approach is explained in detail in Section 2, and Section 3 reports the results of the experiments conducted to validate the approach.

1.1. Quadrotor Modelling and Control

Many publications deal with quadrotors, and most of them are dedicated to modelling and control. The control techniques analysed range from simple PID methods [8] to more complex techniques like LQR, sliding mode or integral backstepping ([7],[9]). A notable exception is [19] where the vertical dynamic of the quadrotor is learned using Locally Weighted Linear Reagression and a controller is trained using reinforcement learning. Most authors begin by proposing models based on first principles. A model for the thrust of each rotor is generally used, and the balance of forces and moments at the quadrotor's centre of mass allows the computation of the dynamic behaviour of the machine. More sophisticated models ([13],[16]) can include the aerodynamic effects of blade flapping, or simple models for the ground effect ([7]). Model parameters are mainly derived from static tests; if a specific component is known to be critical for control, specific tests are devised to model it correctly. Ultimately the engineer uses his insight and the results of control experiments to decide which effects need to be modelled, and which estimated, so that the final controller will yield the desired performance.

1.2. Canonical and Evolutionary System Identification

System identification in aeronautics is of course a well developed field of research. Most established methods tackle the problem of estimating the parameters of a model with a known structure, one that is generally based on first principles. Methodologies that can

[1]The vector $\mathbf{x} = [\phi, \theta, \psi, u, v, w, p, q, r]$ represents the state (ϕ, θ, ψ are the quadrotor's attitude angles, u, v, w and p, q, r are its linear and angular velocity in the body frame of reference), $\mathbf{u} = [u_P, u_R, u_Y, u_T]$ is the input vector made up by the pitch, roll, yaw, and throttle commands and $\boldsymbol{\tau}$ is the vector of input time delays.

also determine the model structure, for example by using such techniques as artificial neural networks, have also been developed (e.g. [11]), but the general need for a model to be transparent and understandable makes those techniques of limited practical use.

The use of evolutionary computation for system identification is not new, but it is only recently that it has been applied to the identification of models of the type and size useful in a domain like ours. For example, Bongard and Lipson have successfully demonstrated the use of co-evolution to regress the dynamics of complex nonlinear systems [4]. The idea behind the techniques is simple: a set of fitness maximizers is evolved, but the tests against which the performance of the maximizers is measured are also under evolutionary control. Each period of evolution of the models is interleaved with a period of evolution of the tests; the fitness of the models is obtained by testing them with the evolved tests, while the fitness of the tests is the variance that each produces in the fitness of the models. The tests therefore evolve to be as discriminative as possible when measuring the performance of the models. In [5] Bongard and Lipson also suggest partitioning a complex problem by regressing each modelled dimension independently of the others; this allows the methodology to scale up and approach problems out of the reach of standard techniques.

Common to any system identification approach there is the fact that any effect not present in the dataset cannot be learned. In our case, the experimenter has to make sure that the flight envelope of interest is adequately covered by the collected data.

2. Our approach

The only domain knowledge we use is the assumption that the system can be sufficiently well approximated as a 6DoF rigid body. This might seem to be rather limiting, but in practice a wide range of vehicles, from wheeled robots to aircraft, helicopters and even ships can be treated in this way. Physics tells us that any motion that our system exhibits is due to the effects of forces and moments (i.e. linear and angular accelerations) applied to the rigid body. If we can relate these instantaneous accelerations to the state and inputs of the system, we will be able to predict the motion of the object under study. This concept paves the way for the two main ideas of our approach:

- using a general and computationally efficient co-evolutionary method to infer the structure and the parameters of the nonlinear relationships between the inputs, the state, and the accelerations;
- directly modelling the accelerations in the body coordinates using the laws of physics to propagate the state forward in time so that the effects of the translation and rotation of the body's frame of reference can be explicitly taken into account.

The sizes of our state and input vectors (12 and 4 respectively) mean that the number of possible nonlinear functions that could relate them to the accelerations is too large for any method of extensive search, and so an evolutionary approach is indicated. We have chosen genetic programming (GP) for its ability to handle both the structure and the parameters of the model, and to deal with both linear and nonlinear functions. The efficiency of the coevolutionary setup is also important, since every function evaluation involves integrating the full 6 dimensional state of the model, which is computationally expensive.

Given that all the state variables and inputs are available from the data, and that we can precompute the relevant accelerations offline, an obvious approach is to search for the model that minimizes the error between the predicted and the computed accelerations. However, when the acceleration prediction is integrated forward in time, any errors will be accumulated, creating an obvious risk of divergence. A better approach is to integrate the predicted acceleration for a specified number of time steps, and select the model that minimizes the error between the predicted and the computed value of the state variable itself. Any divergence will produce a higher error, ensuring a search for models with good long term prediction ability. As the predicted acceleration is successively integrated through time, it will affect not just its own, but also the other state variables. Prediction errors will propagate through the system just as they would when the model is used as a simulation tool. In effect, our evolutionary algorithm will produce a model that can cope with the effects of its own errors.

A vehicle travelling in a given direction, and at the same time rotating, will experience a sideslip force as a result of inertia. These effects are nonlinear [3], which makes the model learning problem even harder. The discrete time update of the state variables can be written in body coordinates as:

$$\begin{bmatrix} u \\ v \\ w \end{bmatrix}_{t+1} = R_{b_t}^{b_{t+1}} \left(\begin{bmatrix} u \\ v \\ w \end{bmatrix}_t + \begin{bmatrix} a_x \\ a_y \\ a_z \end{bmatrix}_t \Delta t \right) \quad \begin{bmatrix} p \\ q \\ r \end{bmatrix}_{t+1} = \left(\begin{bmatrix} p \\ q \\ r \end{bmatrix}_t + \begin{bmatrix} \dot{p} \\ \dot{q} \\ \dot{r} \end{bmatrix}_t \Delta t \right) \quad (2)$$

where u, v, w are the velocities in the quadrotor body frame of reference, p, q, r are the rotational velocities about the axes, $R_{b_t}^{b_{t+1}}$ is the matrix transformation that rotates the body frame from its orientation at time t to that at $t+1$; $\mathbf{a} = [a_x, a_y, a_z]$ and $\boldsymbol{\alpha} = [\dot{p}, \dot{q}, \dot{r}]$ are respectively the linear and angular accelerations in body coordinates. Equation 2 shows that the linear accelerations are not simply the derivatives of the linear velocities, and so an additional nonlinear transformation is needed to compute them. By performing the transformation $R_{b_t}^{b_{t+1}}$ in our integration routine, we simplify the learning task.

2.1. Data filtering, computation of derivatives and integration

The first step in our method is the data collection. The 6DoF pose of the vehicle, and the input from the pilot, are recorded at $100Hz$. However, we require not only the absolute pose of our vehicle, but also its first and second derivatives (i.e. the angular and linear velocities, and the angular and linear accelerations). We first apply a low pass filter to reduce the effects of noise. A transfer function frequency plot from our data showed the dynamics of our system to be quite slow, with most of the frequency content below $4Hz$; this allowed us to set the cutoff frequency to $5Hz$. To avoid phase distortion, and to allow for delay compensation, we use a finite impulse response (FIR) filter with 200 taps.

The need to integrate the state equations forward in time is a significant computational drawback of any time-domain method. However, given the limited bandwidth of our system, we can mitigate this problem by downsampling our data. We then compute the first and second order derivatives as first order differences. During the computation of the derivatives the necessary changes of coordinates are performed in order to obtain the mathematical counterpart of the integration procedure described next.

As a final check, we integrate the computed accelerations forward in time to verify that the resultant time series matches the original time history. With a requirement for a relative squared error (RSE) lower than 5% within a time span of 750 steps[2] we can

[2]The length normally used in the regression algorithm, equivalent to $30s$ of clock time.

safely achieve a downsampling factor of 4, which brings the sampling frequency to $25Hz$. The trim values (i.e. the control values corresponding to the hover condition) for each of the control data series are then subtracted; this is standard practice in time domain methodologies.

To produce the development over time of the state variables, the accelerations predicted by the model need to be integrated from some initial state. As well as executing the discrete time update equation for linear and angular velocities (i.e. equation (2)), the integration routine also needs to compute the position and attitude. In order to avoid the gimbal lock problem arising from the use of Euler angles, and also to improve accuracy, the attitude is represented in quaternion form [3].

2.2. Co-evolutionary Setup

The coevolutionary setup of our algorithm consists of two main steps that are interleaved in time: in the first, the models are evolved, and in the second, the tests are evolved.

In our case a model is simply a GP expression tree with a maximum depth of 5. The inner nodes of the trees are the usual functionals $\{+, -, *, /\}$ while the leaves can be any of the inputs or a constant value. To reduce the computational complexity we have opted for using 14 parallel hill climbers in place of a population based strategy; at each generation the parent is replaced if the offspring generated by mutation performs better. Mutation is the only evolutionary operator used, and either a macro or a micro mutation is used with equal probability. A macro mutation randomly selects a node in the tree and replaces it with a newly generated one, removing or generating child nodes as needed. A micro mutation selects a node and replaces it with a newly generated subtree. To generate a new subtree, nodes are drawn at random with a probability of 0.7 of picking a leaf. Following a macro or micro mutation, one of the constants in the tree is perturbed with a probability of 0.5 by adding a random value from a Gaussian distribution with mean 0 and variance 0.1. The initial value of a constant node is chosen from a uniform distribution ranging from -5 to 5. Finally, mutation is applied to the delay of each input: with probability 0.1, a delay can be replaced with a new random value between 0 and 5 (the maximum allowed delay). All delays are originally initialised to 0.

In our algorithm, tests are simply short subsets of consecutive data chosen from the training dataset. The first sample of the set represents the initial conditions, and the subsequent samples are used by the integration routine. The average squared error of the model computed from those samples is summed over all the tests in the suite to yield the model fitness. The position at which a subset of points is taken from the dataset is controlled by evolution; mutation is applied by adding a randomly generated value to the current starting index of the data in the test. In the case of the tests, the population of 5 individuals is evolved using five parallel hill climbers. The fitness function used for a test is the variance that each test produces in the current population of models, thereby choosing for identification the parts of the dataset not well described by the population of models. This stops the optimization from focusing too much on dynamic states that are overrepresented in the data (e.g. hovering). To avoid cyclic oscillations of fitness, we do not use the evolved tests directly for the evaluation of the models, but instead maintain a suite of 6 tests to which we add the last best test generated. If the test overlaps one of the tests already in the suite, it replaces it, otherwise it replaces the oldest test in the suite.

[3] Although the core integration routine uses quaternions, the attitude is expressed in Euler angles when passed to the modelling equations.

Bloat tends to increase the size of the models significantly, so after each cycle of the algorithm, we try to simplify each model by replacing a randomly selected subtree with a constant. We then try to reconverge the tree by evolving it using only the Gaussian mutation operator. If the resulting tree performs no better than the original, it is discarded.

An iteration of the algorithm typically consists of 600 generations of model evolution, 80 generations of test evolution, and one instance of tree simplification. A typical run of the full algorithm contains 30 iterations.

3. Experiments

3.1. The quadrotor

The quadrotor used in the study (see figure 1) is a commercially available model [1] powered by four brushless motors fitted with 8" propellers. The only modification made for these experiments was the addition of the 5 infrared tracking markers. For data collection the quadrotor was manually flown in our flying arena which is equipped with a Vicon MX [2] infrared motion capture system; this system tracks and resolves the quadrotor's 3D position and orientation in real time with high accuracy and precision (of the order of millimetres) with a sampling frequency of $100Hz$. A standard RC transmitter was used to fly the quadrotor manually; the commands were recorded at $100Hz$ and synchronised to the flight data. The transmission delay of the RC transmitter is $18ms$, and the data capture delay is typically under $10ms$; both delays were ignored since at $25Hz$ (the post filtering data sampling rate) they are smaller than one time step.

A quadrotor is by design a mechanically unstable system since even if the four rotors are driven at the same speed, the rotational moments produced by each rotor will not cancel out exactly due to minor differences in drag and lift. MEMS gyros are commonly used on this type of rotorcraft to provide active stabilisation based on rotational speed feedback. In [12] Gurdan et al. describe the low level stabilisation algorithm used in our quadrotor; it is basically a set of three independent PD loops, one for each rotational axis. No dynamic models were used to design the machine or the controller, and the authors relied solely on their insight and experience to tune the controller empirically.

During the flight we did not have direct access to the low level motor inputs; only the control inputs to the stabilisation loops were recorded. In contrast to the 'first principles' models discussed in 1.1 our system identification technique will provide a lumped model of the quadrotor plus the stabilisation controller. We do not consider this as a limitation, but instead we see it as illustrating the flexibility of our approach. The user can decide at what level to model the system, without the need to understand either the requirements or the implications for the structure of the model.

3.2. Identification

After filtering and downsampling, a series of 13800 data points was extracted; 11800 of them were used for training with the remainder reserved for validation. Each sample point consists of 3 arrays, the state $\mathbf{x}$, the computed accelerations $\boldsymbol{a}$,$\boldsymbol{\alpha}$ and the control input $\mathbf{u}$. During evolution, the individual tests were conducted using short chunks of 750 consecutive sample points[4]. The state information in the dataset is used only to define the initial

[4]Longer time windows would be able to to enforce a search for even better long term prediction abilities but at the expenses of increased computational needs; the length of 750 was empirically found to be a good trade off.

condition of the system; from that moment onward, the state is computed using the integration routine. At each timestep, the model under test is fed the control inputs, and the current state. Since the delay of each input is controlled by the evolutionary algorithm, it is important to remember that some of the control data might not be the "current" ones. The predicted accelerations and the remaining "true" accelerations from the dataset form the inputs to the integration routine.

The fitness of a given model is the average absolute error between the true state variable and the one predicted by the integration routine; for example, for a model that predicts the variable $\dot{e}$, the fitness ($f_{\dot{e}}$) would be:

$$f_{\dot{e}} = -\frac{\sum_{n=0}^{N}\left(\sum_{t=0}^{T}|e_t - \hat{e}_t|\right)_n}{N * T} \qquad e \in \{u, v, w, p, q, r\}$$

where T is the number of samples in each of the tests, N is the number of tests in the test suite, e_t is the true and $\hat{e_t}$ is the predicted state variable. The algorithm also delivered a very similar performance using the classic squared error fitness metric.

We can now move towards the production of a full model. Each of the six equations was first identified independently, and then, for each one, the best model found after 30 repetitions of the coevolutionary algorithm was selected. Then we simply provided the predictions of the set of six models as the input to the integration routine. After setting the state to some initial value from the dataset we propagated the system forward in time with the recorded control input as the only input. To have a better understanding of the algorithm's precision and reliability we repeated the procedure of producing a full model 30 times. A qualitative example of the evolution in time of the state variables for a window of $7s$ (randomly selected from the validation data) is shown in Figure 2. The pre-

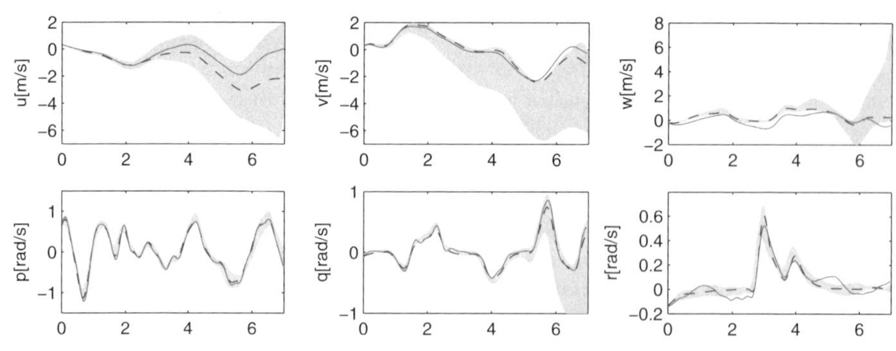

Figure 2. Prediction over a $7s$ time window; true data (continuous line), best model (dashed line), envelope of the min and max values predicted within the whole population of 30 models (shaded area).

dictions appear to be in good agreement with the recorded data; this is especially so for the angular velocities since they depend directly on the control input. The linear velocities suffer a larger error as consequence of the accumulation of the angular velocity error. As expected the error increases as a function of time, but even the worst models still appear well behaved.

In table 1 we can see the equations[5] of the model with the median error performance produced by the 30 runs of the GP process (we will refer to the median model as *modelM*

[5]We have simplified the expressions and rounded the constant terms to one decimal place for better readability.

from now on); we chose this as a good representative of the full set of models, but a similar structure can also be seen in the best model.

Median model
$a_x = 10.0\boldsymbol{\theta} - 0.6\boldsymbol{u} + 0.18 - 0.5q - 0.56pr + 0.1v - 0.6u_Y$
$a_y = \mathbf{0.3} - 9.2\boldsymbol{\phi} - 0.6\boldsymbol{v} + v/(4.3 + \phi - r)$
$a_z = 28.2\boldsymbol{u_{T-5}} - \boldsymbol{w} + 2.1u_P^2 - 0.02$
$\dot{p} = 10.3\boldsymbol{u_{R-1}} - 4.15\boldsymbol{p}$
$\dot{q} = -9.8\boldsymbol{u_P} - 3.3\boldsymbol{q} - 1.2\theta + 0.2u + 9.8qu_Ru_Y + (\psi - qu_P)(-u_Y + qu_Ru_Y)$
$\dot{r} = -7.8\boldsymbol{u_{Y-2}} - 1.9\boldsymbol{r} + 2qu_Y - r^2u_{Y-2}$

Table 1. Model with median error performance from 30 repetitions of the algorithm. Terms in bold appear (directly or after an appropriate series expansion) in every single one of the 30 models. The symbol u_{X-k} indicates the input X delayed by k timesteps.

Although understanding the model produced by our evolutionary algorithm is far from simple, it is interesting and instructive to try to analyse the structures present in all 30 models. (They are highlighted in bold in table 1). First, in the expression of the linear accelerations a_x and a_y, we can recognize something like the small angle approximation to the projection of gravity (i.e. 10.0θ and 9.2ψ). At hover the thrust is exactly equal to the weight, so any pitch or roll movements will project the accelerations $\sin(\theta)$g and $\sin(\phi)\cos(\theta)$g in the forward and lateral direction; again, our terms are approximations of these. The reader familiar with aerodynamics will also spot the Stokes' drag (the expression of the viscous drag appropriate for relatively slow speeds) appearing in the terms $0.6u$ for a_x and $0.6v$ for a_y; this also appears in the expression of a_z as $-w$. The vertical acceleration is seen to be proportional to the throttle input u_T; this makes sense because in a quadrotor the throttle directly controls the mean speed of the rotors, which is proportional to the thrust (and our brushless motor controller can exactly maintain the required speed thanks to rotational speed feedback). Of course, the inertia of the motors and rotors reduces the bandwidth of the propulsion group; our GP expression models this as a control delay u_{T-5}. The angular accelerations show the presence of the PID controller loop. We know from [12] that the controller produces a change in rotor speeds proportional to the control input; this explains the terms $10.3u_{R-1}$,$9.8u_P$ and $7.8u_{Y-2}$ (in $\dot{p}$,$\dot{q}$ and $\dot{r}$ respectively). The controller output is also inversely proportional to the rotational speed in the controlled axis, giving rise to the terms $-4.15p$, $-3.3q$ and $-1.9r$

Although they were derived from real and noisy data, the models produced here are completely deterministic; a way of simulating the disturbances naturally produced in rotors suggested in [9] is to add some form of noise to the control inputs. In all our simulations we have used Gaussian noise with a standard deviation equal to 10% of the maximum value of the control input in the data used for identification.

3.3. Control

We then moved on to test the model in conjunction with a controller. As a first step we manually tuned a PID controller on the test quadrotor to perform waypoint navigation. The position and attitude estimation provided by the Vicon system were used as the feedback signals for a series of nested PID loops, with the inner layer consisting of three PD loops controlling pitch, roll, and yaw angle using the attitude and angular speed infomation from the motion capture system. The outer layer consisted of three PD loops, two of which, given the lateral and longitudinal distance from the next waypoint (i.e. the current error in body coordinates), would output a pitch and a roll angle to be tracked by the loops of the inner layer. The third PID loop controlled the throttle input as a function of the altitude error.

The next step was to investigate the possibility of producing a controller directly based on the model.We were not attempting at this stage to produce an optimal controller (indeed, better tracking results than those presented here can be found in the literature e.g. [17]); a simple but effective controller was all that was required. Our previous experience in evolving controllers had been successful ([10]) so we decided to evolve from scratch the parameters of the PID controller we had already developed. We used a simple approach based on evolutionary strategies: each candidate controller would attempt to fly *modelM* on a randomly generated course. The controller's fitness was defined as proportional to the distance covered in the allotted time, and inversely proportional to its deviation from the course (for a detailed explanation we refer to [10]).

To compare the performance of the controller in the simulated and in the real system, we used one of the trajectories generated during the evolution of the PID controllers. We first recorded the path followed in simulation, and then recorded the path of the real quadrotor flying the same trajectory while controlled by the evolved PID. The recorded trajectories are plotted in Figure 3. The high similarity between the paths followed by the

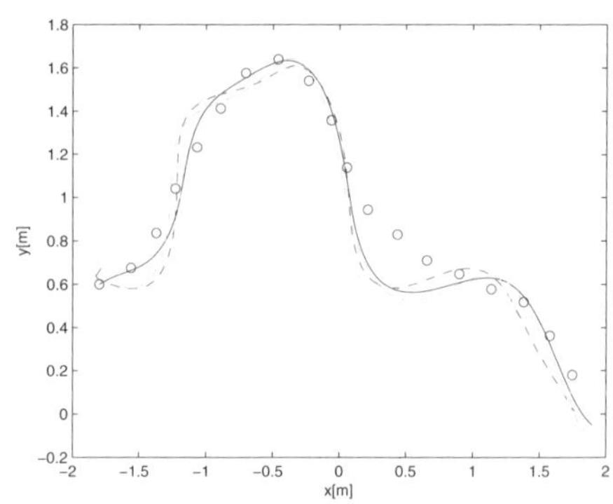

Figure 3. Evolved PID controlling the quadrotor and its model; simulated using the model (blue continuous line) and three repetition of the task on the real quadrotor (red,green and cyan dashed lines).

real quadrotor and the simulation experimentally validates our modelling technique, and shows that a PID controller evolved in simulation can be transferred to the real system. Both the model and the quadrotor behave poorly (but more important similarly) right after the sharpest bend; we believe this to be a result of our suboptimal choice of controller structure.

4. Conclusion

The methodology presented here has been shown to be capable of identifying the dynamics of a non-trivial platform without the need for any specialised domain knowledge. Stages traditionally left to the abilities of the skilled engineer, such as the choice of the model structure or of the relevant inputs, have been automated. The use of genetic programming permits the nonlinearities in the model to be handled naturally, along with the identification of all its parameters, enabling the results achieved in simulation to be transferred to the real platform.

Although we have successfully demonstrated the principle, this work will need to be further validated and extended to be of any practical use. At present the technique does not provide any adequate characterisation of the parameters in the model, nor does it include a sensitivity analysis of the model output to parameter changes.

Future research will address the complementary problem of automatically building a controller; we are hopeful that this will reach the same levels of performance as more traditional approaches, with the advantages of being a fully automatic methodology.

5. Acknowledgment

Our thanks go to Richard Newcombe and Julian Togelius for many insightful discussions, and to Swarm Systems Ltd. for financial support.

References

[1] Ascending technologies GmbH. http://www.asctec.de/main/index.php?id=4&pid=2&lang=en&cat=pro.

[2] Vicon MX homepage. http://www.vicon.com/products/viconmx.html.

[3] P. Abbeel, V. Ganapathi, and A. Y. Ng. Modeling vehicular dynamics, with application to modeling helicopters. In *Neural Information Processing Systems*, December 2006.

[4] J. Bongard and H. Lipson. Nonlinear system identification using coevolution of models and tests. *IEEE Transaction on evolutionary computation*, 9(4):361–384, August 2005.

[5] J. Bongard and H. Lipson. Automated reverse engineering of nonlinear dynamical systems. *PNAS*, 104(24):9943–9948, June 2007. p.

[6] J. Bongard, V. Zykov, and H. Lipson. Resilient machines through continuous self-modeling. *Science*, 314:1118–1121, 2006.

[7] S. Bouabdallah. *Design and control of quadrotors with application to autonomous flying*. PhD thesis, EPFL, 2007.

[8] S. Bouabdallah, A. Noth, and R. Siegwart. PID vs LQ control techniques applied to an indoor micro quadrotor. In *Proceeding of IROS 2004*, 2004.

[9] I. D. Cowling, O. A. Yakimenko, J. F. Whidborne, and A. K. Cooke. A prototype of an autonomous controller for a quadrotor UAV. In *Proceedings of ECC 07*, 2007.

[10] R. De Nardi, J. Togelius, O. Holland, and S. Lucas. Neural networks for helicopter control: Why modularity matters. In *IEEE Congress on Evolutionary Computation*, July 2006.

[11] W. E. Faller and S. J. Schreck. Neural network: Applications and opportunities in aueronautics. *Progress in Aerospace Sciences*, 32:433–456, 1996.

[12] D. Gurdan, J. Stumpf, M. Achtelik, K. Doth, G. Hirzinger, and D. Rus. Energy-efficient autonomous four-rotor flying robot controlled at 1khz. In *The 2006 International Conference on Robotics and Automation.*, September 2006.

[13] G. M. Hoffmann, H. Huang, S. L. Waslander, and C. J. Tomlin. Quadrotor helicopter flight dynamics and control: Theory and experiment. In *Proceedings of the AIAA Guidance, Navigation, and Control Conference*, 2007.

[14] P. Muren. The proxflyer. http://www.proxflyer.com.

[15] K. Nakamura. Mr. Kimio NAKAMURA's Coaxis Micro Helicopter. http://liaison.ms.u-tokyo.ac.jp/agusta/coaxis/nakamura.html.

[16] P. Pounds, R. Mahony, and P. Corke. Modelling and control of a quad-rotor robot. In *ACRA 2006*.

[17] M. Valenti, B. Bethke, G. Fiore, J. How, and E. Feron. Indoor multi-vehicle flight testbed for fault detection, isolation, and recovery. In *AIAA Conference on Guidance, Navigation and Control*, 2006.

[18] A. Van de Rostyne. The pixelito. http://pixelito.reference.be.

[19] S. L. Waslander, G. M. Hoffmann, J. S. Jang, and C. J. Tomlin. Multi-agent quadrotor testbed control design: Integral sliding mode vs. reinforcement learning. In *IEEE/RSJ International Conference on Intelligent Robots and Systems*, 2005.

Intelligent Autonomous Systems 10
W. Burgard et al. (Eds.)
IOS Press, 2008

doi:10.3233/978-1-58603-887-8-374

Design and Implementation of Humanoid Programming System Powered by Deformable Objects Simulation

Ryohei Ueda, Takashi Ogura, Kei Okada and Masayuki Inaba [b]
ueda@jsk.t.u-tokyo.ac.jp
[b] Graduate School of Information Science and Technology, The University of Tokyo, 7-3-1 Hongo, Bunkyo-ku, Tokyo 113-8656, Japan

Abstract. This paper describes the design and implementation of a humanoid robot system that is capable of simulating deformable objects such as a cloth or a towel. Two key issues of the program system design are discussed: 1) adding dynamics to selected objects at the selected time. 2) the simple API for mesh creation of deformable objects. This system helps users to develop deformable object manipulation and recognition. Finally, a cloth recognition and motion generation carrying a mat using the developed system is presented.

Keywords. Humanoids, Deformable Objects Simulation, Robot Simulator, Recognition of Cloth

1. Introduction

Humanoid robots are expected to assist human activities in daily life. Therefore humanoid robots need to deal with various kinds of objects. Especially, it is important for humanoid robots to be able to deal with deformable objects like cloth.

In order to simulate flexible objects in a humanoid robot simulator, the function of deformable objects simulation is needed. However the existing humanoid robot simulators, such as OpenHRP[1], FAST[2], Robotics Studio[3], do not incorporate soft body simulation. The existing humanoid robot simulation systems need to consider how to embed deformable objects simulation into them.

When humanoid robots recognize deformable objects such as a cloth, it is difficult to recognize deformable objects with visual processing using the only real robot's view images because of its deformation and occlusion. It may be efficient to apply flexible models for recognition of deformable objects. Kita et al. shows the effectiveness of a deformable model for estimating the states of clothing[4]. Felzenszwalb adopted deformable template models to detect deformable shapes[5].

In grasping field, Howard et al. used 3D physically-based model for grasping soft body objects by two cooperative manipulators[6].

In this paper, the design and implementation of a deformable objects simulation embedded into the existing humanoid robots programming environment. The simulator is applied to a humanoid robot for recognition and handling of deformable objects.

2. Humanoid Robot Simulator Embedded into a Robot Brain

The robot simulator presented in this paper is embedded into a robot brain and applied for recognition and motion generation.

A robot simulation should enable the user to realize an efficient software development. A virtual robot model is used to test new algorithms before using them for experiments on the real robot. The following requirements have to be fulfilled:

I The robot models in the simulator can be controlled with the same interface as the real robot.
II The sensors of the robot can be emulated in the simulator.

Robot simulators are also expected to be embedded into a robot brain. They are used to predict a situation or to plan a specific task. In order to meet these requirements, a robot simulator additionally needs to have the following functions:

1. Addition of an object to the virtual world at any time.
2. Adding or removing dynamics to an object at any time.

1. is required when a robot finds a new object. The object is added to the virtual world of the robot brain.

2. is needed to reduce the computational load. When a robot execute a task in a complex environment, such as a room, the robot needs to deal with many objects. However, if the dynamics of every object are simulated the computational load will be enormous. It is therefore important to compute only the dynamics of the task objects the robot deals with.

Ogura et al. developed EusDyna[7], which meets the requirements described above. In this paper the incorporation of a deformable objects simulation into EusDyna is discussed.

Fig.1 illustrates the system configuration of EusDyna. EusDyna is implemented in EusLisp[8], which is a robot system description language. EusLisp is a Lisp implementation capable of dealing with three dimensional shapes and multithreading.

The humanoid robots in EusDyna have the same interface to the real robots and the software resources developed in EusLisp can be used in EusDyna. Thus it is easy to embed a simulation engine into a robot brain.

EusLisp is capable to deal with three dimensional shapes. The robot and environment models are represented by non-dynamic three dimensional objects. EusDyna adds dynamics to these objects using an external simulation engine. EusDyna is also capable of switching the external simulation engine between ODE[9],PhysX[10],and Math Engine of Vortex.

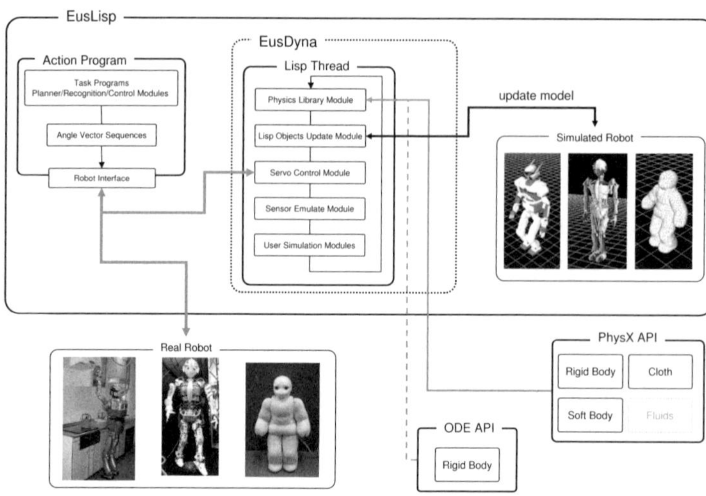

Figure 1. Configuration of Humanoid Robot Programming System and EusDyna

3. Implementation of a Simulation Environment Incorporating Deformable Objects

3.1. Mass-Spring System Model of Deformable Objects in PhysX

It is effective to use the simulator for action generation. An application example is when the robot recognizes objects of the real world and this information is sent to the simulation where a virtual robot performs actions on the objects and learns how to handle them. This virtual learning process allows the robot brain to acquire a knowledge of its environment and can be used by the real robot encounting these same objects in the real world. If this learning process shall be performed online the simulation speed is important.

We use the game dynamics engine "PhysX", because PhysX is capable to simulate the behavior of rigid bodies in a fast time by PPU (Physics Processing Unit) and is equipped with a deformable objects simulation that can treat with 2-D flexible objects (Cloth model), 3-D flexible objects (Soft body model) and liquids (Fluids model). In this paper, soft body model and cloth model are focused on.

In PhysX, the modeling method of Position Based Dynamics[11] is used for the deformable objects model.

Fig.2 shows the components of the Soft Body and Cloth model of PhysX. A tetrahedra element is used for the Soft body model. A tetrahedron mesh is used to represent the different shapes of 3-D deformable object. The Cloth model component is a triangle.

3.2. Adding Dynamics at a Selected Time

EusDyna is capable of adding dynamics to a selected object at any time as described above. This means that the dynamic calculation of every object has to be started first.

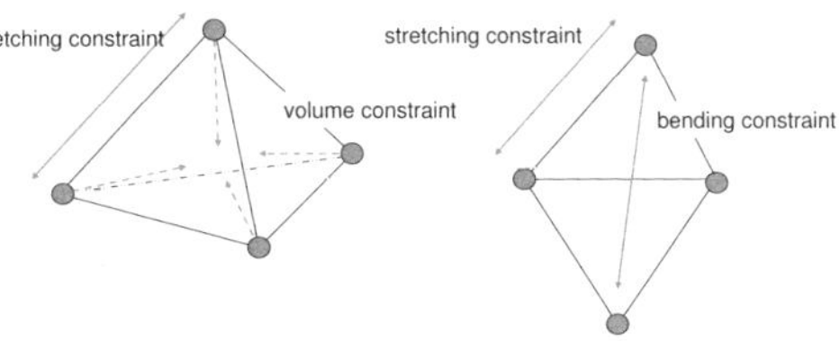

Figure 2. Tetrahedra : Component of Soft Body and Triangle : Component of Cloth

In the case of a deformable object its deformation is only computed when it is made to be simulation target beforehand. Otherwise, it corresponds to a rigid body and it conserves its initial shape. Fig.3 shows a cloth model before and after applying the dynamic calculation. By treating deformable objects in this way the software development efficiency increases because the position and orientation of the deformable objects can be defined before the dynamics calculation is started. There is another advantage in embedding into a robot brain. When a robot recognizes deformable objects, it is difficult to determine whether rigid or soft a motionless deformable object is. It is not until a robot can find deformation that a deformable object is distorted. Therefore it may be efficient to recognize a motionless flexible object as a rigid body first and recognize it as a deformable object when it is deformed. In this recognition way, treatment of deformable objects in EusDyna described above is functional.

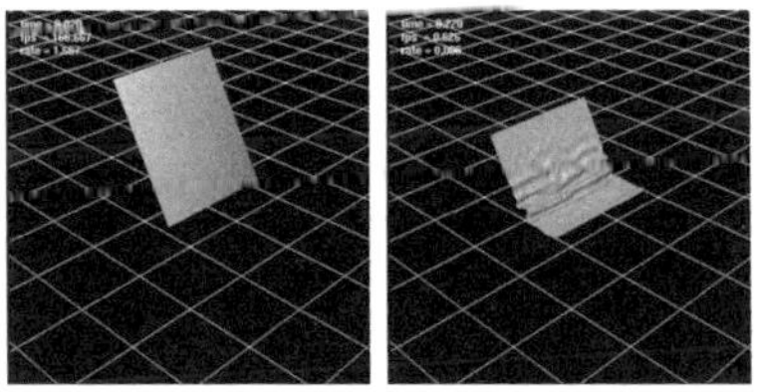

Figure 3. Left : Before Adding DynamicsRight : After Adding Dynamics

3.2.1. API for Mesh Creation of Deformable Objects

When using a deformable object model the mesh creation is a difficult problem. Furthermore, the mesh of a soft body is one of the major elements which determines its behavior. Taking into account the facility of programming and being embedded into a robot brain a simple API for mesh creation is required. Fig.4 shows the constitution of a deformable cube which is a primitive shape of deformable objects of the API of EusDyna. The shape of a soft cube is defined by the edge length in x, y and z axis and the mesh resolution in x, y, and z axis. The cube is a three dimensional deformable object and consists of a tetrahedron mesh. It is assembled of vowels and each of the vowels is composed of five tetrahedron. Fig.5 shows the images of a deformable cube simulation using this system. Fig.6 is the source code of Fig.5 simulation.

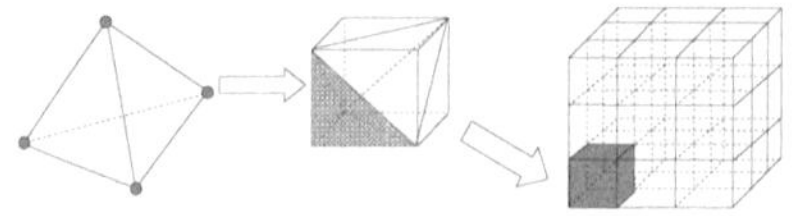

Figure 4. Constitution of Soft Body Cube

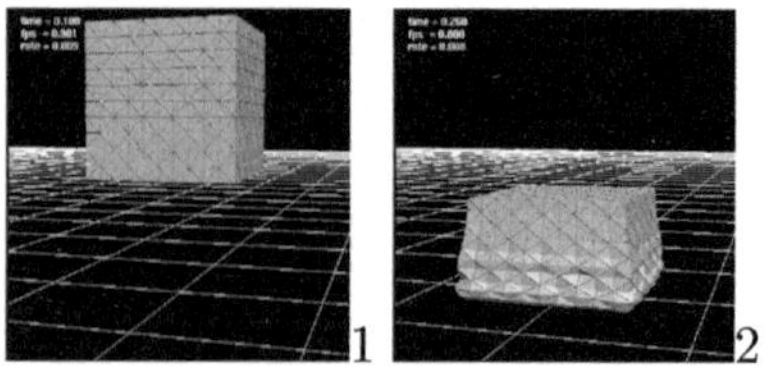

Figure 5. A Deformable Cube in EusDyna

```
(d-init) ;;initialize dynamics
(setq *sc* (instance soft-body
    :init :cube (list 200 200 200 10 10 10)))
        ;;create soft-body cube
(setf (get *sc* :face-color) :gray)
(send *sc* :locate #f(0 0 300))
(d-make *sc*) ;;add dynamics to *sc*
(send *sc* :draw-tetra)
    ;;select the mode of drawing
(objects (list *sc*)) ;;draw *sc*
(d-start)    ;;start simulation
```

Figure 6. The Source Code of Fig.5

4. Experiments

4.1. Recognition of Deformable Objects

4.1.1. Estimating the state of a cloth

In this section a cloth recognition method using this system is presented. The problem situation is the following:

1. Determining where a cloth is picked up.
2. The robot is given only two images(Fig.7).
3. The environment information such as the height of the desk is known.

In order to solve this problem the EusDyna system is used. The method is described below.

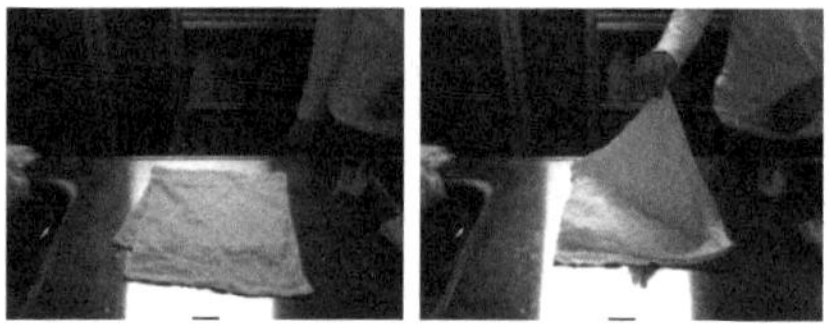

Figure 7. Picking Up a Cloth

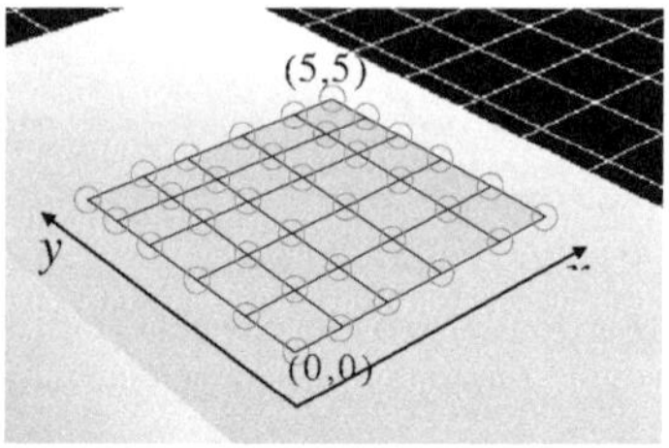

Figure 8. the Grid of a Cloth model

First, the cloth model is created. The parameters of the cloth shape are obtained by the left of Fig.7, and the height of the desk.

Second, the simulated cloth model is picked up. The candidates of the picking-up point are 6×6 grid points of the cloth model(Fig.8). Each of the cloth grid

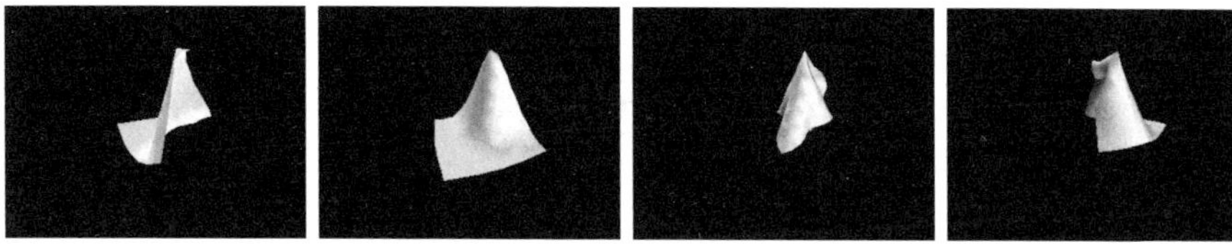

Figure 9. The Simulated Cloth Images

points are moved in the simulator towards the position of the hand, which is obtained by stereo vision.

Third, the simulated cloth model is compared with the cloth in the real world. The image of the cloth in the real world, the right of Fig.7, is obtained from the view of the real robot. The images of the cloth model in the virtual world are created from the robot model view placed in the virtual environment.The similarities between the simulated view images and the real view image are calculated.

The similarity is calculated using the cloth contour. The definition of the similarity is eq.(1).

$$Similarity = \frac{And(SimulatedImge, RealImage)}{PixelNum} + \frac{And(Not(SimulatedImage), Not(RealImage))}{PixelNum} \quad (1)$$

As a result, the most similar image of the simulated cloth is the second one of Fig.9. The image is picked up at $(5,0)$ in the grid coordinates of Fig.8.

4.1.2. Replaying the Cloth Folding Motion

In the next experiment, we control a cloth model in the simulator according to the behavior of the cloth in the real environment (Fig.10).

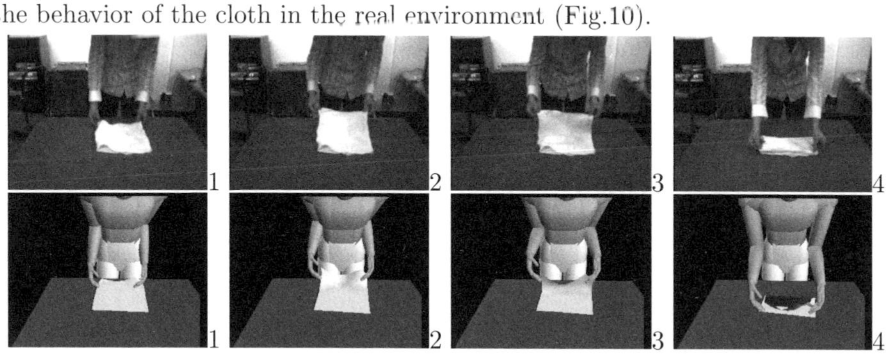

Figure 10. The Real and Simulated Images of Cloth Folding

In this experiment, the cloth model is created from the robot's view of a cloth before being folded. When the cloth is folded, the cloth model in the simulator is moved according to the position of the hands(Fig.11), which is obtained using stereo view of the real robot. When a robot recognizes the flexible object which deforms largely, it is difficult to recognize it based on the visual processing using the only real robot's view image. This experiment shows the effectiveness of recognition using a deformable objects simulation.

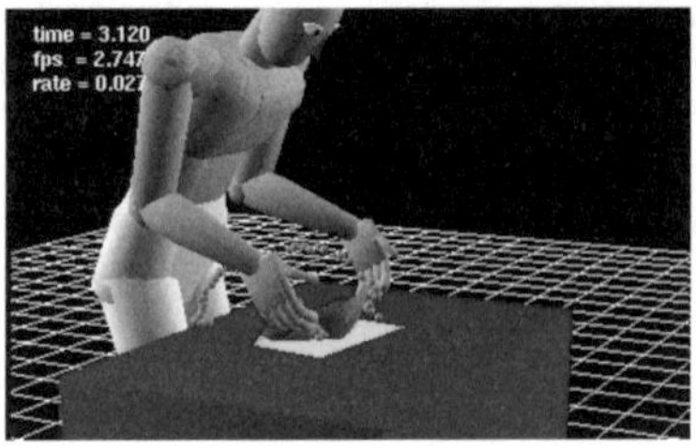

Figure 11. The Path of the Man's Hands

In addition, the action for a playback of the motion may be planned using this system.

4.2. Motion Generation of Carrying a Mat

In the next experiment, we present the method to acquire a stable motion for carrying a mat using the system developed. In full search, the trial programs are executed after the initial condition is defined based on each of the parameters.

The motion parameters are the distance of the robot's hands in x and y direction from the center of gravity of a mat(Fig.12). *ArmDistance* is changed in 10 steps and each of $x_{DeformabelObjects}$ and $y_{DeformabelObjects}$ is varied in 3 steps.

The value function is eq.(3). The less the value of F is, the more stable the motion is. In addition, the threshold in order to detect whether the mat is fallen down(Fig.13).

$$dF = \left(\frac{\sum_{v \in V} v}{\sum_{v \in V}} - P_{\text{humanoid}} - D_0 \right) \quad (2)$$

$$F = \sum_{\text{for all simulation step}} dF dt_{sim} \quad (3)$$

Where V are the vertices of the deformable objects, P_{humanoid} is the position of the robot and D_0 is the initial distance between the robot and the mat.

As a result, the most stable motion is shown in the left images of Fig.14. The right images of Fig.14 are one of the unsuccessful motions which is valued less.

5. Conclusion

In this paper, a system configuration consisting of a deformable objects simulation was presented. The simulation is embedded in a humanoid robot programming environment. The applicability of the method was shown at the examples of cloth recognition and mat carrying motion generation.

The key feature of the system is:

1. Adding dynamics to the selected objects at any time

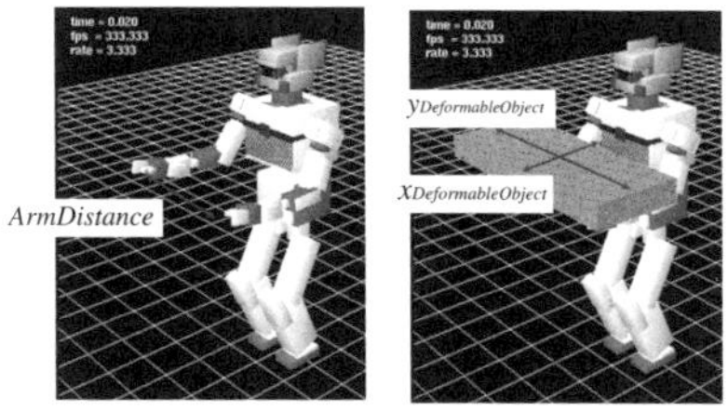

Figure 12. Simulation Parameters

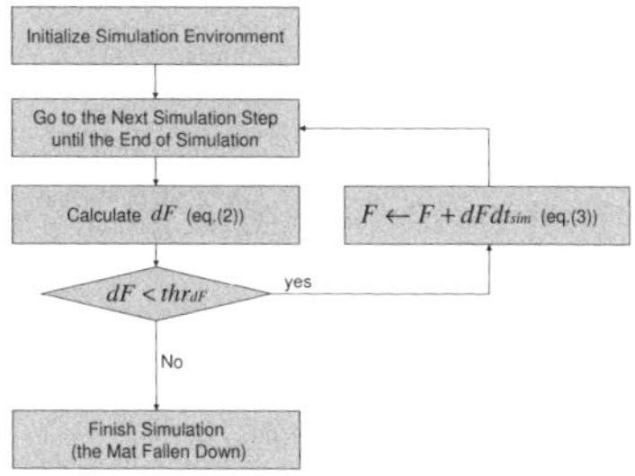

Figure 13. The Flow Chart of a Trial in Simulation

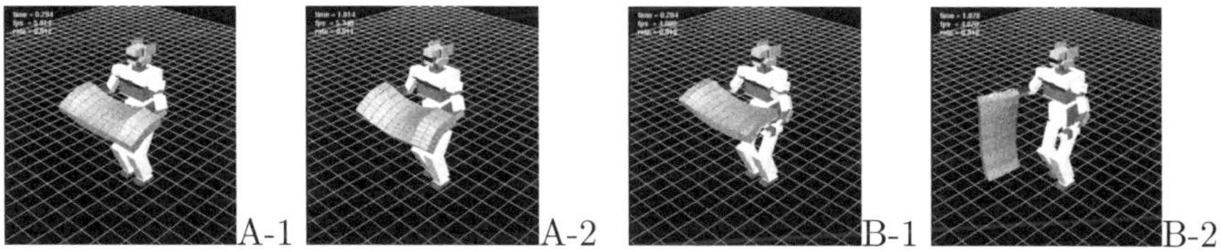

Figure 14. The Temporal Images of Carrying a Mat. A-1 and A-2 are the Most Stable Motion. B-1 and B-2 are One of the Unsuccessful Motion.

2. Simple API for mesh creation of deformable objects

For future research, it is planned to use this system for teaching motions such as cloth folding.

References

[1] Fumio Kanehiro, Kiyoshi Fujiwara, Shuuji Kajita, Kazuhito Yokoi, Kenji Kaneko, Hirohisa Hirukawa, Yoshihiko Nakamura, and Katsu Yamane. Open architecture humanoid robotics platform. In ICRA, pages 24–30. IEEE, 2002.

[2] Kei Okada, Yasuyuki Kino, Fumio Kanehiro, Yasuo Kuniyoshi, Masayuki Inaba, and Hirochika Inoue. Rapid development system for humanoid vision-based behaviors with real-virtual common interface. In Proceedings of the 2002 IEEE/RSJ Intl. Conference on Intelligent Robots and Systems (IROS'02), pages 2515–2520, 9 2002.

[3] Microsoft. Robotics Studio. http://www.microsoft.com/.

[4] Y. Kita and N. Kita. A model-driven method of estimating the state of clothes for manipulating it. pages 63–69, 2002.

[5] P.F. Felzenszwalb. Representation and detection of deformable shapes. Pattern Analysis and Machine Intelligence, IEEE Transactions on, 27(2):208–220, Feb. 2005.

[6] A.M. Howard and G.A. Bekey. Recursive learning for deformable object manipulation. Advanced Robotics, 1997. ICAR '97. Proceedings., 8th International Conference on, pages 939–944, 7-9 Jul 1997.

[7] T. Ogura, K. Okada, and M. Inaba. Realization of Dynamics Simulator Embedded Robot Brai for Humanoid Robots. In IEEE International Conference on Robotics and Automation (ICRA2007), pages 2175–2180, 2007.

[8] Toshihiro Matsui and Masayuki Inaba. Euslisp: An object-based implementation of lisp. Journal of Information Processing, 13(3):327–338, 1990.

[9] Open Dynamics Engine ODE. http://ode.org/.

[10] AGEIA. PhysX. http://www.ageia.com/.

[11] M. Müller, B. Heidelberger, M. Hennix, and J. Ratcliff. Position based dynamics. J. Vis. Comun. Image Represent., 18(2):109–118, 2007.

Intelligent Autonomous Systems 10
W. Burgard et al. (Eds.)
IOS Press, 2008

doi:10.3233/978-1-58603-887-8-382

Sensor-Behavior Integration in Wheelchair Support by a Humanoid

Shunichi NOZAWA, Toshiaki MAKI, Mitsuharu KOJIMA,
Shigeru KANZAKI, Kei OKADA, Masayuki INABA

nozawa@jsk.t.u-tokyo.ac.jp
the Graduate School of Information Science and Technology, The University of Tokyo,
7-3-1 Hongo, Bunkyo-ku, Tokyo 113-8656, Japan

Abstract. In this paper, we deal with wheelchair support by a life-sized humanoid robot. In wheelchair support, integration between sensory information and behavior is important to adapt the humanoid robot to the physics of the real world, environment and the user. Contributions of this paper is (1)whole-body motion including pushing motion using offset of ZMP and observation of attitude outlier, (2)recognition of a wheelchair using particle filter and (3)human-interface behind the person using face detection and recognition of gesture. This paper aims at the support that the humanoid robot pushes the wheelchair and the user can transfer to anywhere only by instructing where to go.

Keywords. Wheelchair Support by a Humanoid Robot, Whole-Body Motion, Environment Recognition and Human-Interface

1. Introduction

It is important for humanoid robots expected to assist human daily life and added to various functions [1, 2] to be capable of accomplishing nursing care tasks. Currently nursing care tasks require comprehensive support. Humanoid robots usually behave according to sensor information in various levels e.g. to adapt itself to environment. Especially in nursing care tasks, humanoid robots must conform with the humans according to the sensor information.

This paper deals with wheelchair support, including support for heavy works and interaction with people that are the common elements in nursing care tasks. Here, we select such principal elements as follows from among many tasks in a wheelchair support scenario: (1) whole-body motion for controlling a wheelchair (2) recognition of environment (3) human-interface with the user from behind the wheelchair. In this paper, we describe the integration of sensor into behavior in (1), (2) and (3).

2. Sensor-Behavior Integration in Wheelchair Support

Almost all behavior of humanoid robots use sensor information in various levels in a broad sense and it is difficult for the robots to behave without sensor information.

In this paper, the humanoid robot integrates sensor information into behavior as follows(Fig.1): First, the robot counters the real world physics, by off-setting the ZMP(Zero Moment Point) based on somatic sensors and by preventing from tumbling based on its attitude. Next, the robot detects obstacles by using visual sensors and recognizes its surrounding, like the wheelchair, to adapt itself to the environment. Finally, the robot recognizes the human's face and gestures to accommodate to the person.

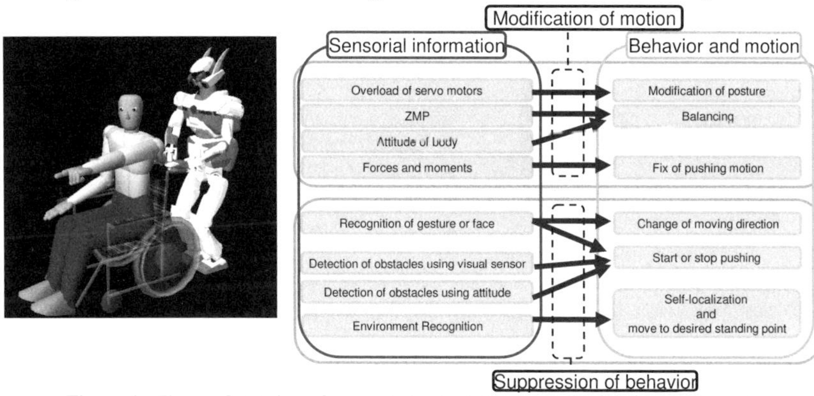

Figure 1. The configuration of sensor-behavior integrated wheelchair support system

3. Whole-Body Motion in Wheelchair Support

In this paper, the humanoid robot obtains the pushing force by offsetting the ZMP. Furthermore, suppose the robot keeps on pushing when the wheelchair would not move, motors of the robot may suffer from overload or the humanoid robot could possibly tumble down. Thus, the robot has to interpret whether the wheelchair moved or not, according to the sensory information.

3.1. Determination of ZMP-offset Appropriate to Hand External Forces

We introduce the Cart-Table-Model [3], the dynamic model which the humanoid robot follows while pushing the wheelchair.

The ZMP is obtained by solving the balance between gravity, inertia force and external forces on the robot hands [4].

$$x_{ZMP} = x_{COG} - \ddot{x}\frac{z_{COG}}{g} + \frac{z_H F}{Mg} \tag{1}$$

Let x_{COG} be the direction of the wheelchair motion. z_{COG} is the height of the COG(Center Of Gravity) and $\ddot{x}$ is the acceleration of the COG. F is the x-component of the reflected forces working on the robot hands. Equation.1 implies the offset required to be added to the ZMP, to enable the robot to move the wheelchair.

3.2. Obstacle Detection According to Estimation of Attitude

In wheelchair support, the robot can detect obstacles by the following two ways; the robot can recognize a visible obstacle using visual sensors, or by detecting the collision between the wheelchair and hindrances at a dead angle by utilizing somatic sensors.

In this section, we describe the latter method.

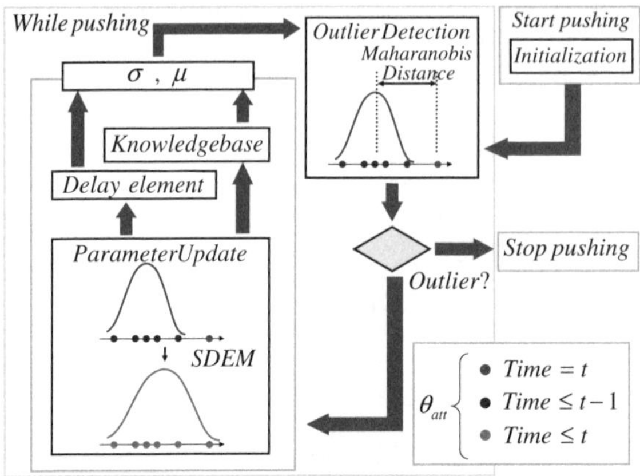

Figure 2. Concept of threshold-decider using the SDEM algorithm and the Maharanobis distance

In the following we manage the pitch-component of the robot's attitude(θ_{att}), for the pitch-component is the principal cause of tumble while transferring the wheelchair. When θ_{att} is over the threshold, the robot detects collision with an obstacle.

This section introduces the method for learning the threshold. In case that the threshold is determined off-line, it is necessary to configure the threshold every time when the robot's behavior change. Therefore, we modified the method to decide the threshold.

The outline of the method is as follows:

1. Learn parameters of attitude distribution for a period of time after the robot starts walking.
2. After the outlier of attitude is detected, stop walking.

1 : In order to estimate parameters of the attitude distribution, we employ the SDEM algorithm(Sequentially Discounting Expectation and Maximizing) [5]. The SDEM algorithm is adopted for calculation of distributive parameters in Gauss mixture model and has the feature of being capable of incremental measurement.

2 : The robot assumes the Maharanobis distance for the criterion of whether or not the current attitude is outlier, after parameters of attitude distribution have been calculated.

$$Maharanobis\ distance = \frac{|\theta_{att}^{(t)} - \mu^{(t-s)}|}{\sigma^{(t-s)}} \tag{2}$$

Let σ be the variance of the attitude. Let μ be the mean of the attitude. We use delay element in calculating the Maharanobis distance; for example, when the robot judges attitude at the frame t, it adopts parameters at the frame $t - s$(a few seconds before).

4. Recognition of Environment in Wheelchair Support

4.1. Self-Localization using Recognition of Wheelchair

In pushing the wheelchair, it is necessary for the robot to perform self-localization. This paper deals with the recognition of the wheelchair as an example of self-localization. There are many cases that the robot is separated from the wheelchair; when the robot returns to the wheelchair, it is necessary to recognize the wheelchair before holding the grip-bar of it.

Our work employs the model-fitting method using multiple visual cues and matches the three dimensional model of the wheelchair with the real wheelchair. We introduce the particle filter technique for model-fitting [6]. The particle filter provides a robust recognition of time-varying state vectors. Okada et al. researched the integration between recognition using multiple visual cue and behavior of the humanoid robot [7]. The detail of the recognition according to [7] is as follows.

To recognize the wheelchair we employ two visual cues: 3D feature points and 2D straight edges. The two visual cues above give the following probability:

$$p(\boldsymbol{z}_t|\boldsymbol{x}_t) = p_{point}(\boldsymbol{z}_t|\boldsymbol{x}_t)\, p_{edge}(\boldsymbol{z}_t|\boldsymbol{x}_t) \tag{3}$$

Let z_t be the measurement vector using some visual cue at the frame t. x donates the state vector representing position and rotation of an object. $p(x|z_t)$ stands for the posterior distribution.

In this case the state vector is defined as $\boldsymbol{x}_t = [x_t\ y_t\ \theta_t]^T$. These variables represent the estimated horizontal position($x_t,\ y_t$) and yaw rotation(θ_t) at the frame t.

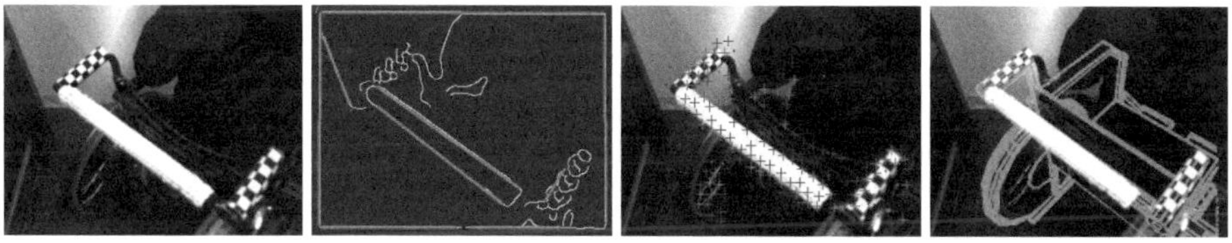

Figure 3. The wheelchair model fitting. First : the robot's view :: Second : 2D straight edge :: Third : 3D feature points :: Fourth : the result image in which the wheelchair is located in the estimated position and rotation, superimposed with red lines.

Fig.3 illustrates the model-fitting experiment. It is seen that the robot can recognize the wheelchair.

5. Human-Interface behind the User in Wheelchair Support

In many cases of nursing care the humanoid robot is located near a person. In wheelchair support the robot stands behind a person.

In this section, we treat human-interface behind the user(back-interaction) as the following communication:

- From the person to the robot
  Face detection and gesture recognition behind the person
- From the robot to the person
  Informing the person about the behavior of the robot

5.1. Face Detection behind the Person

While pushing the wheelchair the face of the person is not usually visible to the robot. However, it is important that the robot detects when the person is looking back and recognize him/her. This looking-back motion will be the cue when the person wants to communicate with the robot. In this subsection we describe the face-detection method using particle filter.

In this section, we apply particle filter to face detection as follows. The state vector is the 3D position of the face($\boldsymbol{x}_t = [x_t\ y_t\ z_t]^T$). We employ 3D position and color histogram [8] of the face at the frame t, to calculate the likelihood of the face detection.

Fig.4 is the face detection example.

Figure 4. Face detection using particle filter. Left : the view image of the robot :: Center : KLT feature points :: Right : the color-extracted view image. At the image to the right, the white pixel has higher similarity with the face color histogram. Also, small green crosses express particles and the large green cross displays the estimated center of the face. Suppose the sum of the particles' weights exceed the threshold, "FIND FACE" is displayed.

5.2. Gesture Recognition behind the Person

In this section, we treat following instructions: emergency stop indication and direction indication(go forward, turn left, turn right etc).

This paper assumes the behavior of a humanoid robot as instructed by the operator: while transferring the wheelchair, the operator tells the robot the destination. The robot must recognize human gesture when the person instructs the robot not only to start and to suspend, but also to move to the desired directions.

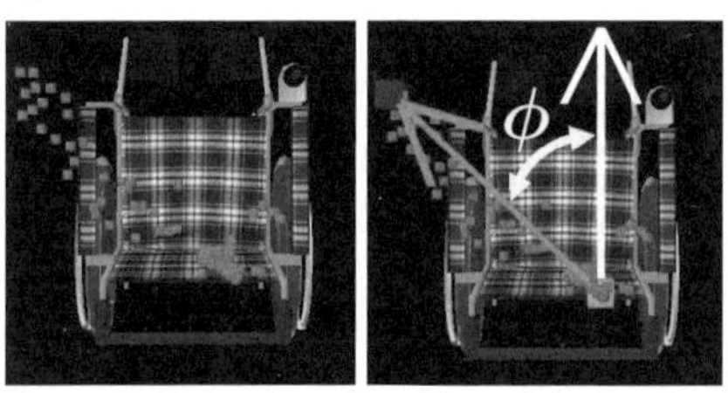

Figure 5. Estimation of position of the hand and the shoulder(left : before estimation :: right : after estimation). At the right image the blue rectangle displays hand position, and the green rectangle shows shoulder position.

The robot knows gestures by estimating hand position and shoulder position according to the arrangement of 3D feature points as follows: first of all, 3D feature points over the wheelchair are extracted. The hand position is calculated as the center of the points far from the robot, and the shoulder position as the center of the points near the robot.

Fig.6 is the gesture-recognition example.

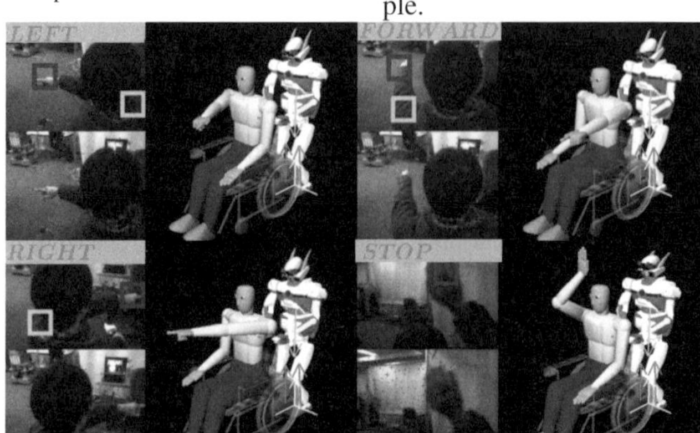

Figure 6. Examples of gesture recognition. Left top : "Left" gesture :: Right top : "Forward" gesture :: Left bottom : "Right" gesture :: Right bottom : "Stop" gesture.

6. Experiment and Result

This section shows the wheelchair support experiments using the above-mentioned technique.

6.1. Obstacle-Detecting Experiment

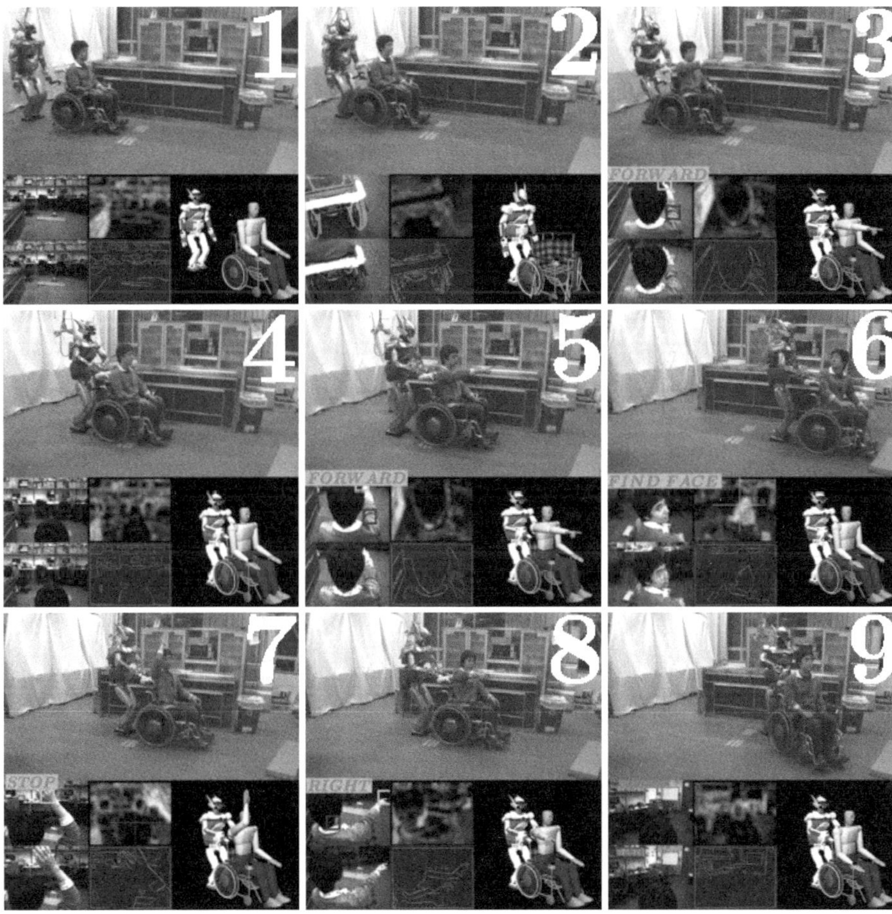

Figure 7. Obstacle detection experiment

In this experiment(Fig.7), we employ five sensor-behavior integrations: pushing motion using compensation of the ZMP(Section 3.1), suspension of walking by detecting outlier of attitude(Section 3.2), face detection(Section 5.1), gesture recognition(Section 5.2) and wheelchair recognition(Section 4.1).

In (1), the robot approaches the wheelchair and the robot recognizes the wheelchair using the particle filter in (2). After recognition, the robot holds (3) the grip-bar of the wheelchair and recognizes gestural instruction to go forward by the operator and starts pushing the wheelchair. The pushing force in this case is about $50[N]$. Therefore, the robot slides its waist forward $76[mm]$ (according to the third term of 1). Fig.8 is pitch of

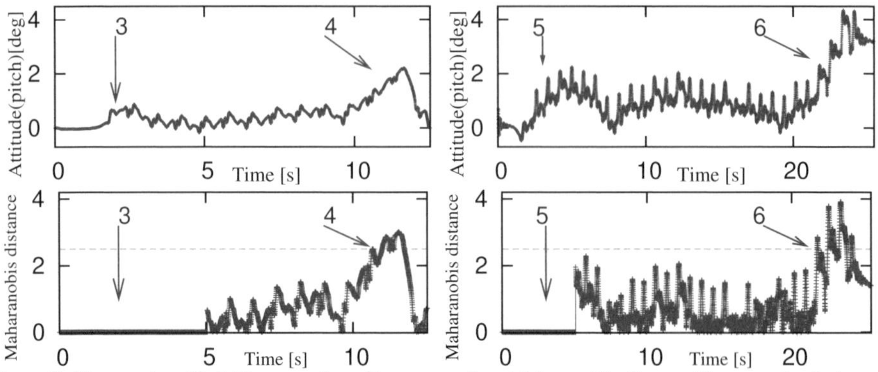

Figure 8. Top graphs : Pitch-Time graphs :: Bottom graphs : Maharanobis distance-Time graphs. In bottom graphs, the green line represents the threshold(2.5) with the period from 0[s] to 5[s] as the learning period.

attitude and Maharanobis distance graphs. Since the robot learns parameters of attitude distribution on an even ground, on the rough terrain the Maharanobis distance will exceed the threshold(in this experiment the threshold is 2.5) and the robot will suspend walking in (4). The learning of parameters is performed during the first few steps when the robot starts walking. In (5), the robot recognizes gestural instruction to go forward and the robot resumes walking and learns parameters on the rough terrain. After the wheelchair collides with an obstacle in (6), the Maharanobis distance exceed the threshold again and the robot suspends walking. After the collision, the robot finds the face of the human and recognition of the face represents the instruction to go backward. Therefore, the robot pulls the wheelchair backwards, until it recognizes gestural instruction to suspend walking in (7). The human points at the right in (8) and the robot can avoid the obstacle in (9). In this experiment, during switching of behavior the robot informs the content of the behavior; for example, in (3) the robot speaks, "I will go forward".

This experiment shows that recognition of the wheelchair enables the robot to behave apart from the wheelchair and to resume control of the wheelchair. This experiment shows that the integration of recognition of human instructions enables the robot to avoid the obstacle. Without indications by the user the robot must investigate the obstacle closely after collision, and has to judge whether to avoid it or to go directly forward. Human instructions enable the robot not only to transfer the wheelchair to the desired direction, but also to avoid obstacles.

7. Conclusion

This paper takes up several tasks in wheelchair support and presents the following integration of sensor information into behavior to achieve each task.

1. Whole-body motion
   Compensation of the ZMP in proportion to reflected forces enables the robot to push the wheelchair. While conveying the wheelchair, the robot discovers abnormality such as collisions with obstacles according to the outlier-detection of the attitude.

2. Environment recognition
   Recognition of the wheelchair using particle filter allows the humanoid robot to continues its task after being separated from the wheelchair.
3. Human-interface behind the user
   The robot follows the user's instruction by face-detection using particle filter and recognition of gesture estimated by arrangement of 3D feature points. During the switching of behavior, the robot tells the content of its behavior to the user.

In our work, the humanoid robot integrates sensory information into behavior to adapt itself to the physics of the real world(1), environment(2) and the user(3). Without adjustment to the real world physics, the humanoid will not be able to control the wheelchair. It is also necessary for the robot to recognize its environment, to allow it to resume its task after separated from the wheelchair . Furthermore, human-interface behind the user allows the robot to perform human-led behavior; for example, to move to the desired direction and to avoid obstacles.

Future challenges to be taken is (1) expansion of mobility and (2) evaluation of the instruction system by many people.

References

[1] Y.Sakagami, R.Watanabe, C.Aoyama, S.Matsunaga, N.Higaki, , and K.Fujimura. The intelligent ASIMO: System overview and integration. In *Proceedings of the 2002 IEEE/RSJ International Conference on Intelligent Robots and Systems (IROS'02)*, pages 2478–2483, 2002.

[2] M.Inaba, S.Kagami, F.Kanehiro, Y.Hoshino, , and H.Inoue. A Platform for Robotics Research Based on the Remote-Brained Robot Approach. In *The International Journal of Robotics Research*, pages 933–954, 2000.

[3] S.Kajita, F. Kanehiro, K. Kaneko, K. Fujiwara, K. Harada, K. Yokoi, and H. Hirukawa. Biped Walking Pattern Generation by using Preview Control of Zero-Moment Point. In *Proceedings of the 2003 IEEE International Conference on Robotics and Automation*, pages 1620–1626, 2003.

[4] Mike Stilman, Koichi Nishiwaki, , and Satoshi Kagami. Learning object models for whole body manipulation. In *Proceedings of IEEE-RAS International Conference on Humanoid Robots(Humanoids2007)*, 2007.

[5] K.Yamanishi, J.Takeuchi, G.Williams, and P.Milne. Online Unsupervised Outlier Detection Using Finite Mixtures with Discounting Learning Algorithms. In *Proceeding of ACMSIGKDD Int'l. Conf. on Knowledge Discovery and Data Mining*, pages 320–324, 2000.

[6] Michael Isard and Andrew Blake. Condensation – conditional density propagation for visual tracking. In *International Journal of Computer Vision*, pages 5–28, 1998.

[7] Kei Okada, Mitsuharu Kojima, Satoru Tokutsu, Toshiaki Maki, Yuto Mori, and Masayuki Inaba. Multicue 3D Object Recognition in Knowledge-based Vision-guided Humanoid Robot System. In *Proceedings of the 2007 IEEE/RSJ International Conference on Intelligent Robots and Systems (IROS2007)*, pages 3217–3222, 2007.

[8] Patrick Perez, Carine Hue, Jaco Vermaak, , and Michel Gangnet. Color-based probabilistic tracking. In *ECCV '02: Proceedings of the 7th European Conference on Computer Vision-Part I*, pages 661–675, 2002.

Intelligent Autonomous Systems 10
W. Burgard et al. (Eds.)
IOS Press, 2008

doi:10.3233/978-1-58603-887-8-390

Learning by Observation of Furniture Manipulation in Daily Assistive Tasks for Humanoid Robots

Mitsuharu KOJIMA [a], Kei OKADA [a] and Masayuki INABA [a]
[a] *the Graduate School of Information Science and Technology, The University of Tokyo, 7-3-1 Hongo, Bunkyo-ku, Tokyo 113-8656, Japan*

Abstract. Discussion about the elements of learning of manipulation in humanoid robots, a method of learning of grasping points and some experiments are presented. In order to realize humanoid robots which assist human activities in daily life, it is an important ability to learn the manipulation of objects from humans or by themselves. It is thought that the manipulation of objects and the recognition of humans, objects and relation between them are necessary for the learning of manipulation. The manipulation and recognition of the objects must be based on the knowledge which could be learned by the robots. We have developed the object recognition and manipulation system whose knowledge could be learned by the robots and now have the interest on recognition of humans and the relation between humans and objects. We discuss the elements which are necessary to realize the learning ability and describe the recognition and manipulation system of objects whose knowledge could be learned, then as a first step present a method of learning of grasping points from humans and some experiments.

Keywords. Humanoid Robot, Learning of Manipulation, Learning by Observation of Manipulation

Introduction

Humanoid robots which assist human activities in daily life are expected. Many researches have been done to realize humanoid robots in a daily environment [1,2]. In order to achieve these tasks in the real world, the ability for these robots to use some objects, tools and furniture which are designed for human is important. In order to realize such humanoid robots, we have developed knowledge based humanoid system that integrates both motion planning system and visual object recognition system [3]. And now one of our main interests is that the robots have the ability to learn by themselves or from humans the knowledge which is necessary to use these objects. Some researches are done to realize such learning ability [4–7].

In this paper, we discuss the ability for humanoid robots to learn manipulation and the elements which are necessary to realize the ability. We describe the

Figure 1. Learning of Grasping Points towards Learning of Manipulation (In the top left image, the robot observes the person's hand, the refrigerator and the relation between them. In the top right image, the eye's view of the robot is shown and the hand extraction and object recognition result is shown. In the bottom left image, the recognized 3d position of the hand, which is the learned grasping point, is shown in the red cube. In the bottom right image, the robot opens the drawer, based on the learned grasping point.)

implemented ones of these elements and the learning of grasping points, which is a first and important step to realize learning of manipulation.

In section 1, we discuss the ability to learn manipulation and the elements of the ability and describe the implemented ones. Section 2 describes the process of learning of grasping points. Section 3 shows the experiments of the learning of grasping points. Section 4 sums up this paper.

1. Learning of Manipulation in a Humanoid Robot System

1.1. Learning of Manipulation

1.1.1. Elements of Learning of Grasping Points

Figure 1 shows learning of grasping points. In the top left image, the robot observes the person's hand, the refrigerator and the relation between them. In the top right image, the eye's view of the robot is shown. The hand extraction result is shown as a blue ellipse. The object recognition result is shown as the green super imposed lines. In the bottom left image, the recognized 3d position of the hand is shown in the red cube. The robot takes it as the learned grasping point. In the bottom right image, the robot opens the drawer, based on the learned grasping point.

There are the elements of learning of manipulation in Figure 1 as follows. (1) Object recognition about which the robot learns manipulation (2) Recognition of a person's hand (3) Relation between the object and the hand.

Moreover, it is thought that those process are necessary for more general learning of grasping points as follows.

1. The robot searches the more proper grasping points around the taught points. It is necessary to check whether the robot gets the hand stuck in the handle of the object or not with force sensor or something.
2. The robot verifies the result of its motion which is generated on the learned grasping point, with vision sensor or something. If the result is bad, the robot must try the other more proper motion.

Learning of grasping direction is also necessary.

1.1.2. Elements of Learning of Manipulation

A principle learning methodology of the systems following the programming by demonstration paradigm is discussed in [5]. We summarize the main basic elements which are necessary for learning of manipulation as follows.

1. Manipulation of Objects : The manipulation must be done based on the knowledge which could be learned by the robots.
2. Recognition of Objects : Recognition of objects' positions and postures, based on the knowledge which could be learned.
3. Recognition of Person : Recognition of person's whole body movements and gaze is necessary.
4. Recognition of Relation between Person and Objects : The key point in recognition of person and objects is the recognition of relation between them as follows (a) Where a person gazes on an object? (b) Where a person's hand or body touches on an object?

In order to realize these elements, we have developed the system for the recognition of objects [3]. In the next section, we describe them briefly.

1.2. A Knowledge Based Humanoid Robot System for Learning of Manipulation

We have developed knowledge based humanoid system that integrates both motion planning system and visual object recognition system [3]. In this section, recognition and motion planning sub systems of the integrated system are briefly described. The knowledge in the system could be learned by the robots.

1.2.1. Multi-cue Integrated 3D Object Recognition System

In the recognition sub system, the objects are recognized by the visual 3D object recognition method with multi-cue integration using particle filter technique [3, 8]. In the recognition system, "*3D Edges*", "*Color histogram*" and "*3D shape*" are used as the visual feature knowledge and the likelihoods between them and the extracted features from input images are integrated by the particle filter technique. The likelihoods are calculated as follows.

- **Likelihood between 3D edges and 2D edge segments:** "*3D edges*" are visual feature knowledge and "*2D edge segments*" are extracted from input image. The smaller the distance and the angle between them are, the higher the likelihood is.

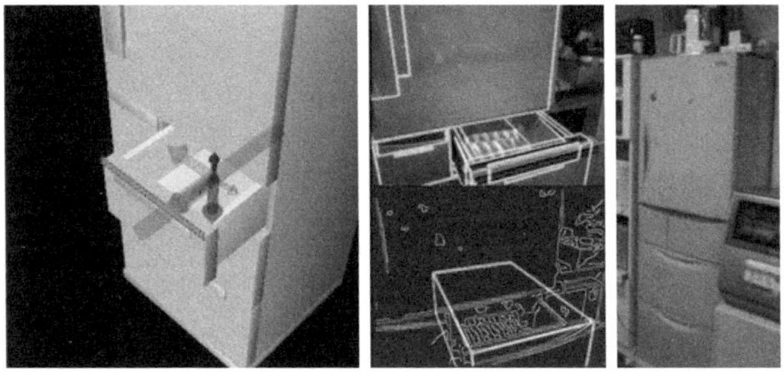

Figure 2. The Object Recognition of a Refrigerator's Drawer by the Visual 3D Object Recognition Method with Multi-cue Integration Using Particle Filter Technique (In the left image, the red lines are "*3D edges*" of visual feature knowledge and the red cylinder shows the "*search areas*" of the "*linear joint*" of a drawer. In the middle image, the recognition results are shown. In the right image, the scenes of the experiment is shown.)

- **Likelihood between 3D shape and 3D feature points:** "*3D shape*" is visual feature knowledge and "*3D feature points* " are extracted from input image. "*3D feature points*" are extracted as followings.
  (1) The 2D feature points are generated by using the KLT feature extraction method [9]. (2) The correlation based stereo matching is applied to calculate the disparities of the feature points. (3) The 3D distances of the points from the camera origin are calculated by assuming that the internal and external camera parameters are calibrated.
  The smaller the distance between the faces of "*3D shape*" and "*3D feature points*" is, the higher the likelihood is.
- **Likelihood between color histograms:** the likelihood is calculated using the bhattacharyya distance [10].

In Figure 2, an example of object recognition by the recognition system is shown. A refrigerator's drawer shown in right image of Figure 2 is recognized. As visual feature knowledge, the "*3D edges*" shown as the red lines in left image of Figure 2 are used. In the upper image of the middle images of Figure 2, the recognized model is super imposed in the green lines and the red lines show the candidates of joint recognition results. The redder the lines are, the higher the likelihood of the candidate is. The lower image of the middle images is the edge segment extraction result and the green lines are the estimated result. Appropriate likelihoods of the candidates and correct recognition result are shown.

These visual feature knowledge could be learned by 3d modeling technique using cameras on the robots.

1.2.2. Motion Planning System

In the recognition system, "*Handle*" (target posture and position of the robot's end effector to manipulate the object, which is the Grasping Points), "*Attention*", "*Spot*" (standard coordinates for tasks) and "*Joint*" (movable mechanism of objects) are used as the manipulation knowledge.

Using manipulation knowledge, the motion planner generates motion in the following steps. (1) Move the object to the intended posture and position or the intended joint angle. (2) Update "*handle*" to follow the object motion. (3) Solve inverse kinematics to move the end effector to the "*handle*".

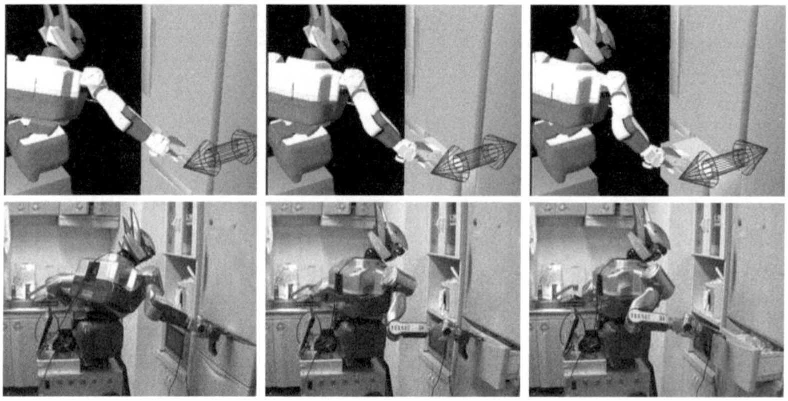

Figure 3. Motion planning to manipulate the refrigerator's drawer and the experiment on a real robot (Top row images are planning results and bottom row images are the real motions.)

In Figure 3, an example of motion planning is shown. The robot opens the refrigerator's drawer. The top images are planning results. As described above, in the planning, (1) Update drawer position by updating the position of joint, which is shown in the blue cylinder with bidirectional arrow. (2) Update the handle, which is shown green triangles, to follow the drawer. (3) Solve inverse kinematics.

Based on the planned motion, the robot opens the drawer in the bottom images. These manipulation knowledge could be learned mainly by human observation.

2. Learning of Grasping Points

2.1. Learning Process

The learning process of grasping points of some objects from human is described as follows. (1) Recognize the target object about which the robot learns manipulation. (2) Recognize where the human hand touches on the object.

We discussed the elements which are necessary for the learning of manipulation in the Section 1.1, and the recognition of object and human hand's is necessary for learning of grasping points as above.

2.2. Recognition of Objects and Person's Hand in Learning of Grasping Points

The object is recognized, by the method described in Section 1.2.1 The image processing to recognize the human hand position is shown in Figure 4. The image processing is as follows.

1. Extract skin color region from a robot's eye input image. Figure 4-2 shows the input image at Figure 4-1 setting. Figure 4-3 shows the extracted skin color region. HSI color space model is used in the extraction. The blue ellipse in Figure 4-4 shows the skin color region.
2. Calculate the hand's 3d position. Based on the skin color extraction, the grasping point is calculated as follows. (a) The point which is on the ellipse

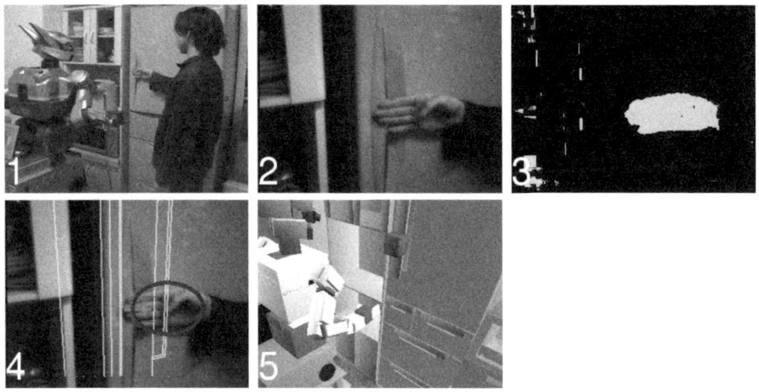

Figure 4. Image Processing of Learning of Grasping Points (1: External View of Learning, 2: Robot's Eye View at 1, 3: Skin Color Extraction Result of 2, 4: Super Imposed of 3, 5: Calculated 3D Position of Hand (Learned Grasping Point))

and its major axis and far from human body, is extracted (The red cross point in Figure 4-4). (b) The template region is defined, based on the point (The red rectangle in Figure 4-4). (c) The 3d position of the hand is calculated by stereo block matching of the template region (The red cube in Figure 4-5).

3. If the calculated hand's 3D position is near to the object and unchanged for a while, the robot takes its position as a grasping point of the object.

3. Experiments of Learning of Grasping Points

Some experiments of learning of grasping points are shown in Figure 5, Figure 6.

3.1. An Experiment of Learning of Grasping Points of A Refrigerator's Door

Figure 5 shows an experiment of learning of grasping point of a refrigerator's door. Figure 5-2 shows the object recognition result. The red lines are super impose of the refrigerator at the candidates' positions and postures. In Figure 5-3,4, the robot learns the grasping point. Figure 5-4 shows the result of the image processing to recognize the human hand position at Figure 5-3 setting. Figure 5-6 shows the motion planning result, based on the learned grasping point. Figure 5-7 shows the real motion, based on the planned motion. In Figure 5-9,10, the robot tries to open the door and succeeds in Figure 5-11.

3.2. An Experiment of Learning of Grasping Points of A Refrigerator's Drawer

Figure 6 shows an experiment of learning of grasping point of a refrigerator's drawer. Figure 6-2 shows the object recognition result. The red lines are super impose of the refrigerator at the candidates' positions and postures. In Figure 6-3∼6, the robot tracks the human hand's position. Figure 6-8 shows the result of the image processing to recognize the human hand position at Figure 6-7 setting. Figure 6-10 shows the motion planning result, based on the learned grasping point. Figure 6-11 shows the real motion, based on the planned motion. In Figure 6-13,14, the robot tries to open the drawer and succeeds in Figure 6-15.

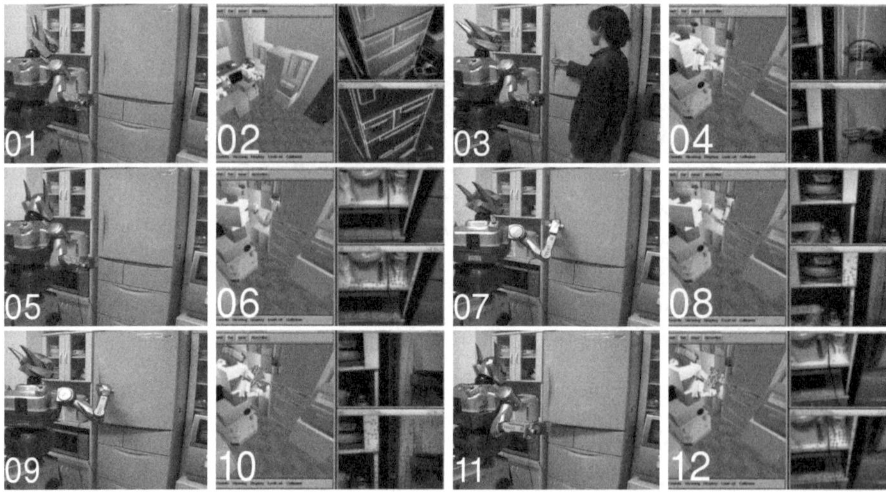

Figure 5. Learning of Grasping Point of Refrigerator's Door and Manipulation of the Door, based on the learned Grasping Point (An odd number image is external view for the next number. 1,2 show the object recognition. 3,4 show the learning of grasping point. 6 shows the planned motion, based on the learned grasping point. 7,8 show the real motion, based on the planned motion. 9,10 show the robot tries to open the door. 11,12 show the result.)

4. Conclusions

The important elements of learning of manipulation could be summarized as follows (1) Recognition of person (2) Recognition and manipulation of objects, based on the knowledge which could be learned by the robots (3) Recognition of relation between person and objects

We described the recognition and manipulation system whose knowledge could be learned and present the method and experiments of learning of grasping points as first and important step towards the learning of manipulation.

To realize learning of manipulation in humanoid robots, more ability of recognition of human is necessary. In the recognition and manipulation of objects, the ability to manipulate and recognize more various objects such as water, cloth and string is required.

References

[1] Y. Sakagami, R. Watanabe, C. Aoyama, S. Matsunaga, N. Higaki, and K. Fujimura. The intelligent ASIMO: system overview and integration. In *Proceedings of the 2002 IEEE/RSJ International Conference on Intelligent Robots and System (IROS 2002)*, Vol. 3, pp. 2478–2483, 2002.

[2] T. Asfour, K. Regenstein, P. Azad, J. Schroder, A. Bierbaum, N. Vahrenkamp, and R. Dillmann. ARMAR-III: An Integrated Humanoid Platform for Sensory-Motor Control. In *Proceedings of the 6th IEEE-RAS International Conference on Humanoid Robots (Humanoids 2006)*, pp. 169–175, 2006.

[3] Kei Okada, Mitsuharu Kojima, Satoru Tokutsu, Toshiaki Maki, Yuto Mori, and Masayuki Inaba. Multi-cue 3D Object Recognition in Knowledge-based Vision-guided Humanoid Robot System. In *Proceedings of the 2007 IEEE/RSJ International Conference on Intelligent Robots and Systems (IROS 2007)*, pp. 3217–3222, 2007.

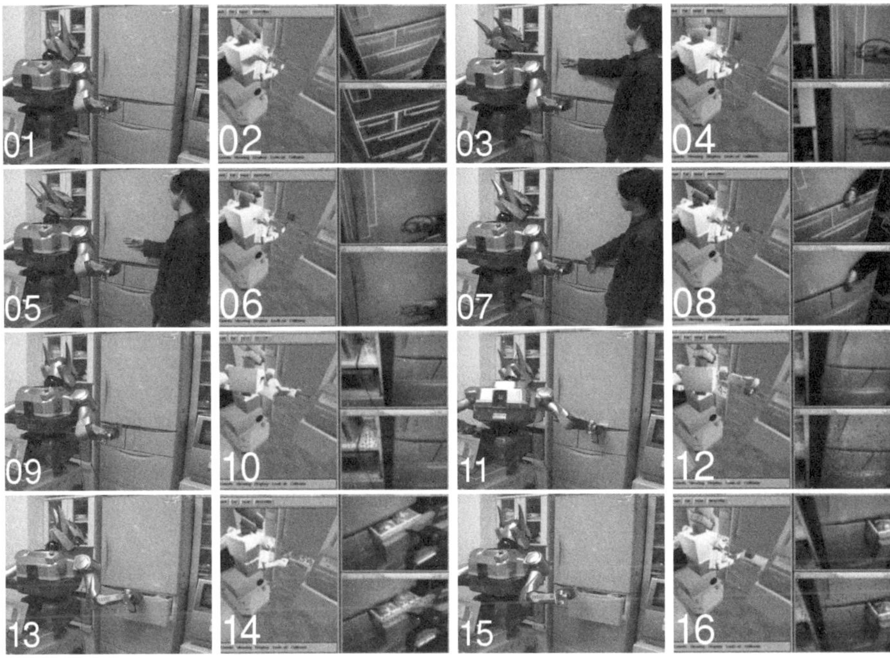

Figure 6. Learning of Grasping Point of Refrigerator's Drawer and Manipulation of the Drawer, based on the learned Grasping Point (An odd number image is external view for the next number. 1,2 show the object recognition. 3∼6 show the tracking of human hand. 7,8 show the learning of grasping point. 10 shows the planned motion, based on the learned grasping point. 11,12 show the real motion, based on the planned motion. 13,14 show the robot tries to open the drawer. 15,16 show the result.)

[4] Yasuo Kuniyoshi, Masayuki Inaba, and Hirochika Inoue. Learning by Watching: Extracting Reusable Task Knowledge from Visual Observation of Human Rerformance. *IEEE Transactions on Robotics and Automation*, Vol. 10, No. 6, pp. 799–822, 1994.

[5] R. Dillmann, O. Rogalla, M. Ehrenmann, R. Zollner, and M. Bordegoni. Learning Robot Behaviour and Skills Based on Human Demonstration and Advice: the Machine Learning Paradigm. In *Proceedings of the 9th International Symposium of Robotics Research (ISRR 1999)*, 1999.

[6] D. Omrcen, A. Ude, K. Welke, T. Asfour, and R. Dillmann. Sensorimotor Processes for Learning Object Representations. In *Proceedings of the 7th IEEE-RAS International Conference on Humanoid Robots (Humanoids 2007)*, 2007.

[7] O. Stasse, D. Larlus, B. Lagarde, A. Escande, F. Saidi, A. Kheddar, K. Yokoi, and F. Jurie. Towards Autonomous Object Reconstruction for Visual Search by the Humanoid Robot HRP-2. In *Proceedings of the 7th IEEE-RAS International Conference on Humanoid Robots (Humanoids 2007)*, 2007.

[8] M. Isard and A. Blake. CONDENSATION – Conditional Density Propagation for Visual Tracking. *International Journal of Computer Vision*, Vol. 29, No. 1, pp. 5–28, 1998.

[9] J. Shi and C. Tomasi. Good features to track. In *Proceedings of the IEEE Conference on Computer Vision and Pattern Recognition (CVPR 1994)*, pp. 593–600, 1994.

[10] Thomas Kailath. The Divergence and Bhattacharyya Distance Measures in Signal Selection. *IEEE Transactions on Communications*, Vol. 15, No. 1, pp. 52–60, 1967.

Intelligent Autonomous Systems 10
W. Burgard et al. (Eds.)
IOS Press, 2008

doi:10.3233/978-1-58603-887-8-398

Scenario controller for daily assistive humanoid using visual verification, task planning and situation reasoning

Kei OKADA [a,1], Satoru TOKUTSU [a], Takashi OGURA [a], Mitsuharu KOJIMA [a], Yuto MORI [a], Toshiaki MAKI [a], and Masayuki INABA [a]

[a] *7-3-1 Hongo, Bunkyo-ku, Tokyo, Japan*

Abstract.
This paper describes scenario description and control method for a daily assistive humanoid robot. High level representation of a task scenario help user to develop various daily assistive tasks of a humanoid robot. Four key techniques are presented. 1) Three layered architecture integrating a vision based behavior control layer, a task planning layer, a situation based scenario generation layer. 2) Visual verification system for adapting symbol level description into real-world situation. 3) Task level planner providing high-level action primitives for scenario description, 4) Vision and auditory based situation interpretation for controlling the scenario. Finally, we demonstrated a kitchen assistive task by a humanoid robot.

Keywords. Humanoid, Task plannnig, Situation reasoning, Vision based behavior ontrol

1. Introduction

Development of robotic behaviors in human daily environments is one of the most desperately-needed application[1,2,3,4]. Most of humanoid research is focus on realizing individual action or sequence of actions, such as walking, carrying tray, pouring tea, washing dishes and so on, few research deals with how to select a goal of humanoid based on environmental situation and generatie sequence of actions from the goal. In this paper, we present humanoid robot system capable of providing scenario level description. High level task description as scenario help user to develop various daily assistive tasks of a humanoid robot. Previous researches on robot architecture with high level behavior description mainly focus on communication task of a robot[5,6]. On the other hand, we deal with daily assistive tasks of a humanoid robot.

Four key techniques are presented. 1) Three layered architecture integrating a vision based behavior control layer, a task planning layer, a situation based

[1]Corresponding Author: Graduate School of Information Science and Technology, The University of Tokyo, E-mail: k-okada@jsk.t.u-tokyo.ac.jp

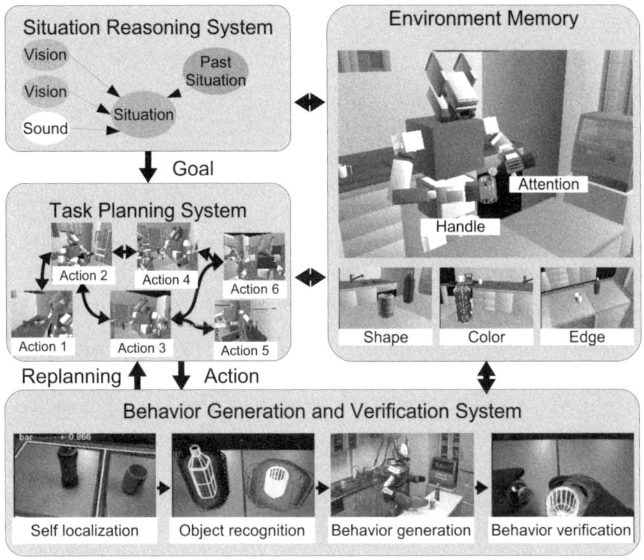

Figure 1. Action-recognition integrated humanoid system with task planning and situation reasoning for scenario level description and control

scenario generation layer. 2) Visual verification system for adapting symbol level description into real-world situation. 3) Task level planner providing high-level action primitives for scenario description, 4) Vision and auditory based situation interpretation for controlling the scenario.

2. Situation based Task Interpretation in Action-Recognition Integrated Humanoid System

Fig.1 presents action-recognition integrated humanoid system. It is composed of three layers: 1) Vision based behavior control layer, 2) Task planning layer and 3) Situation based scenario control layer. These layer required a environment memory which contains knowledges related to a manipulation and a recognition.

2.1. Vision based behavior control layer

This layer receives an action primitive from the task planning system and generate a humanoid behavior with a visual recognition and verification. For example, an action produced by the planning system is "Grasp Cup". This layer decompose this action into vision based cup detection, whole body reaching motion, grasping control and verification of these steps, which we call behavior.

In order to achieve vision based behavior control of humanoid manipulation action, we rely on knowledge based vision-guided robot system, which is achieved through the development of three components: 1) Manipulation knowledge based whole body motion generation system[7], 2) Visual feature knowledge based 3D object recognition system[8], 3) Vision based environment and behavior verification system by using both manipulation and visual feature knowledge[9].

The knowledge of humanoid robot system is illustrated as the "Environmental Memory" in the Fig.1. The system contains not only geometric shape information of objects and environment but also contains manipulation and visual recognition knowledge.

We have defined following manipulation knowledge: a `spot` represents a base coordinate when performing a task, a `handle` represents a position and constrains for grasping an object and an `attention` represents tool coordinates for manipulating an object as well as geometric shape information of objects and environment in the scene.

The sequence of an `attention` coordinates is the input of the whole body motion generator. In the case of pouring tea behavior, the sequence represents a rotating motion of the top of the bottle using an `attention` coordinate of a bottle.Then the motion generator calculates a motion of `handle` coordinates, which indicates the motion of the robot hand. Finally whole body motion is generated by calculating whole body joint angles from the motion of `handle`.

In order to recognize objects, we employ the Particle Filter algorithm[10,11] which is widely used because of its robust characteristics. Each particle represents the hypothesis that indicates the 3D position of the target object and is weighted by likelihood using multi visual cue integration method[8].

Visual feature knowledge for objects and events recognition is also described in the system: `Shape` for calculating 3D distance between this shape and visual 3D feature points, `Color` for calculating similarity between this histogram and the histogram taken from the view images and `Edge` on an object surface for calculating 2D edge distance on the image plane.

Three visual behavior controls are required to perform each motion: 1) Visual self localization, 2) Visual object localization, 3) Visual behavior verification[12]. Before performing each motion, the robot is assumed to be located on the `spot` position and positions of task relevant objects are known in advance. Thus the visual self localization and the visual object localization are required. After the motion, the robot verifies the behavior. We classified the verification process into two groups. One is an indirect verification and another is a direct verification.

Visual self localization, object recognition and behavior verification results are presented in the bottom of the Fig.1.

Thanks to the visual verification process, this layer is able to provides high level autonomous behavior command as "Move to Bar" "Hold Cup", "Hold Bottle" and "Pour Tea". User or other layers do not have to care about visual recognition and verification when executing these commands.

2.2. Task planning layer

Since vision based behavior control layer accepts high level action commands. It is easy to connect a behavior control layer and a high level task planner. We adopt the STRIP type operator for each action. For example the HOLD operator can be described as following:

```
HOLD:
 preconditions : (ON ?OBJECT ?SPOT) (AT ?SPOT)
 action        : (HOLD ?OBJECT ?ARM)
 effects       : (HOLD ?OBJECT ?ARM) ~(ON ?OBJECT ?SPOT)
```

Actions	Visual controls
Actions with self localization	
Move to counter	Recog. counter
Move to kitchen	Recog. sink
Actions with object localization	
Hold a cup	Recog. cup
Hold a bottle	Recog. bottle
Place a cup	Recog. cup
Place a bottle	Recog. bottle
Actionss with visual verification	
Pour tea	Recog. tea
Open tap	Recog. water
Close tap	Recog. water
Wash cup	—

Object	Knowledge
Cup	Shape
Bottle	Histogram, Shape
Counter	Edge
Sink	Edge
Search area	**Ttarget**
On counter	Cup, Bottle
Counter foot	Counter
Sink foot	Sink
Under tap	water flow
Event	**Knowledge**
Recog. tea	Color histogram
Recog. water	Water flow model

Table 1. Knowledge description in the kitchen experiment.

An `action` in the operator can be interpolated as a command to be issued to the behavior control layer. A `preconditions` in the operator consists of conjunctive logical expressions which must be true in order to apply the operator. Each expression evaluated not only the symbol level description of the system by but also using the real sensors. Conditions as `(ON ?X ?SPOT)`, `(AT ?SPOT)` `(WATER-FLOW)` are recognized using a vision sensor and a `(HOLD ?OBJECT ?ARM)` condition is recognized using torque sensors at a robot hand.

2.3. Vision and auditory integrated situation reasoning layer[13]

Estimating current situation gives adequate goal for task planning layer. We have developed situation reasoning method using the Dynamics Bayesian Network. Network node corresponds to the result of life sound recognition such as "tap water sound" or "table ware sound" and human location detected by 3D cylindrical model and 3D position of visual feature point.

Situation node have several states as "cooking", "washing" and "others". The brief node of the network is update continuously using vision and sound sources.

Scenario description can be written as the connection between the recognized situation and goal of the task planner. For example, when the situation reasoning layer interpret as "Washing" situation, then the goal status of the task planner is set to generate behavior sequence to wash a cup.

3. Tea Serving Task Experiment

3.1. Knowledge description for vision based behaviors

This section describes knowledges required to perform the experiment. Ten set of actions required for achieving this experiment are listed in the left table in the Table 1. First two actions require a vision based localization, next four requires an object detection and the last four requires visual verification.

```
HOLD:                                       OPEN-TAP:
 preconditions : (ON ?OBJECT ?SPOT)          preconditions : (AT ?SPOT) ~(WATER-FLOW)
                 (AT ?SPOT)                  action        : (OPEN-TAP)
 action        : (HOLD ?OBJECT ?ARM)         effects       : (WATER-FLOW)
 effects       : (HOLD ?OBJECT ?ARM)
                 ~(ON ?OBJECT ?SPOT)        CLOSE-TAP:
                                             preconditions : (AT ?SPOT) (WATER-FLOW)
PLACE:                                       action        : (CLOSE-TAP)
 preconditions : (HOLD ?OBJECT ?SPOT)        effects       : ~(WATER-FLOW)
                 (AT ?SPOT)
 action        : (PLACE ?OBJECT ?ARM)       WASH-CUP:
 effects       : ~(HOLD ?OBJECT)             preconditions : (HOLD CUP LARM) (AT SINK)
                 (ON ?OBJECT ?SPOT)                          (WATER-FLOW)
                                             action        : (WASH-CUP)
MOVE-TO:                                     effects       : (WASHED CUP)
 preconditions : (AT ?FROM)
 action        : (MOVE-TO ?FROM ?TO)        POUR-TEA:
 effects       : (AT ?TO) ~(AT ?FROM)        preconditions : (HOLD CUP LARM)
                                                             (HOLD BOTTLE RARM)
                                             action        : (POUR-TEA)
                                             effects       : (POURED CUP)
```

Table 2. Action operators in the kitchen experiment.

We defined four objects in the demo scene. For each object, we described associated visual recognition knowledges as shown in the right top table.

Three search areas are defined for detecting objects for grasping(cup and bottle). In this case, we used 2D search space definition since these objects are spinning objects, which has freedoms along with x and y axis and z position of the object is assumed to be the table height.

The right bottom table shows task relevant visual behavior verification knowledge. Recognizing tea is utilized for pouring tea behavior verifications. Recognizing water is applied to verify the open and close tap actions.

3.2. Action operators for task planning

We have defined seven operators for this experiment as presented in Table 2. Variable domain is defined as followings: `BAR`/`SINK` for `?SPOT`, `CUP`/`BOTTLE` for `?OBJECT` and `LARM`/`RARM` for `?ARM`.

Assuming that starting situation of this experiment can be described as `(AT BAR) (ON CUP BAR) (ON BOTTLE BAR)`, "Making tea" can be describe as setting goal status to be `(POURED CUP) (ON CUP BAR) (ON BOTTLE BAR)`. Then we obtain following behavior sequence by applying the task planner.

1. `(HOLD BOTTLE RARM)`
2. `(HOLD CUP LARM)`
3. `(POUR-TEA)`
4. `(PLACE BOTTLE RARM)`
5. `(PLACE CUP LARM)`

When we set goal status as `(WASHED CUP) (ON CUP SINK)`, then we obtain another task scenario "Cleaning up cup" automatically as `(HOLD CUP LARM) (MOVE-TO BAR SINK) (OPEN-TAP) (WASH-CUP) (PLACE CUP LARM) (CLOSE-TAP)`.

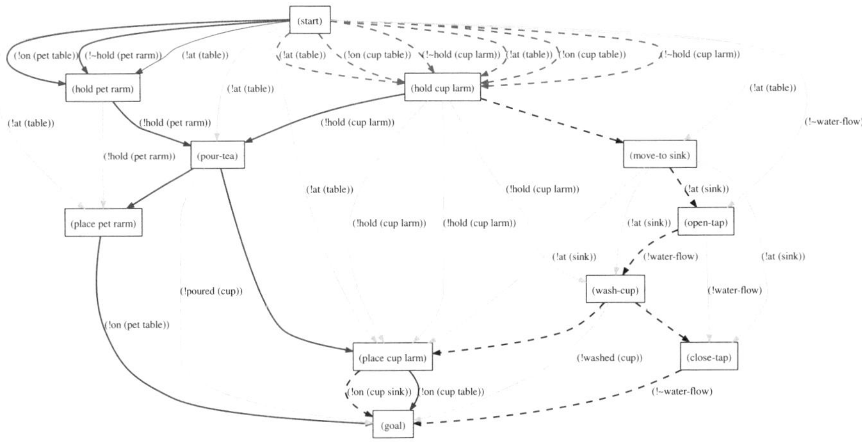

Figure 2. Task plan graph for tea serving experiment

We used POP planner (Partial Ordered Planner)[14] to solve this problem. Task graph to present these actions is illustrated in the Fig.2.

3.3. Tea serving task demonstration

We demonstrated the tea serving task experiment as shown in the Fig.3. The sequence of this experiment is as followings. The number on each line corresponds to the number in the figure.

1. The robot recognizes the cup (1) and holds (2).
2. The robot recognizes the bottle (3) and holds (4).
3. The robot pours tea into the cup from the bottle (5).
4. The robot places the cup (6) and the plastic bottle (7).
5. The human drink tea in the cup and place it (8-9).
6. The robot recognizes the cup (10) and holds.
7. The robot walks to the kitchen (11-12).
8. The robot localizes self position (13).
9. The robot opens the tap (14) and conform it (15).
10. The robot washes the cup (16).
11. The robot closes the tap and place the cup.

Sequence from 1 to 4 is generated by setting a goal situation as `(POURED CUP)` `(ON CUP BAR)` `(ON BOTTLE BAR)` and the rest of robot actions are generated by `(WASHED CUP)` `(ON CUP SINK)`.

3.4. Robustness

This demonstration is repeated very often at the lab on demand. Thanks to the robust object recognition using attention control, visual feature prediction, multi-cue integration and visual behavior verification, the task is rarely failed thus we believe the targeted robustness was reached. The rare case of the failure is when there are droplet on the cup or the bottle. It slips when the robot holds them.

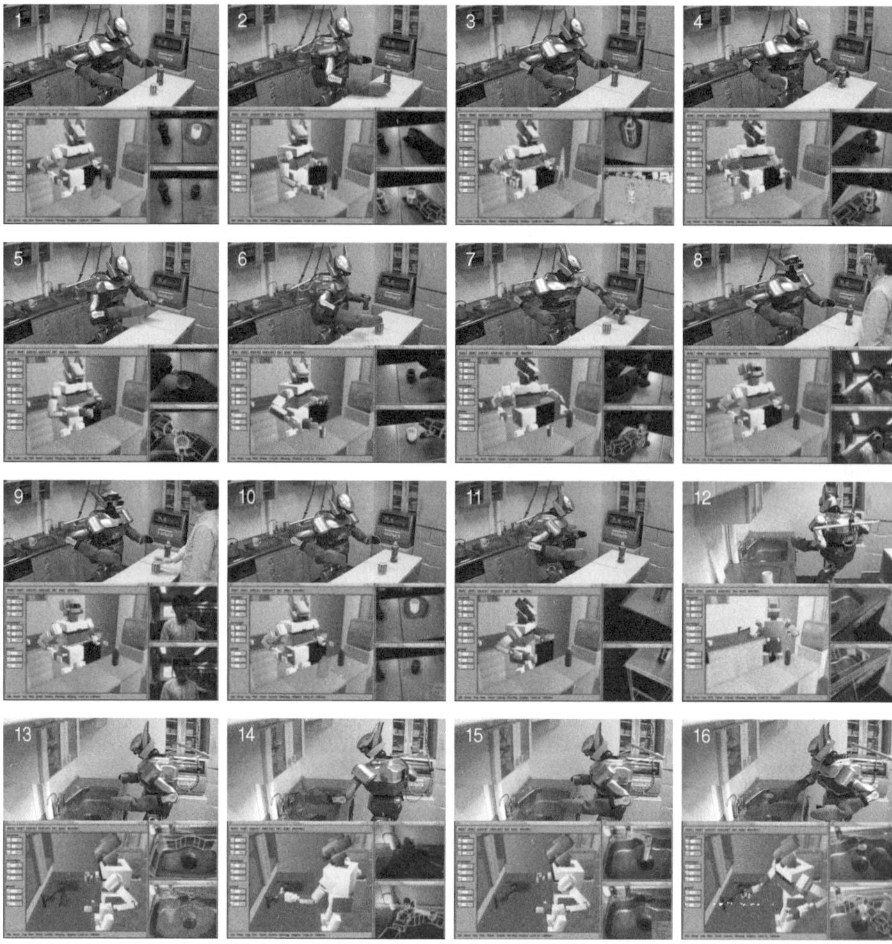

Figure 3. daily life support experiments using knowledge based on recognition system.

Changes of the lighting condition usually affects the recognition result, however our object recognition system is robust enough not to restrict using flash when taking photos. 3D feature points and 2D edges are robust to the illumination change and we use the HSV color space based histogram matching.

4. Conclusion

This paper describes scenario description and control method for a daily assistive humanoid robot. High level representation of a task scenario help user to develop various daily assistive tasks of a humanoid robot. Four key techniques are presented. 1) Three layered architecture integrating a vision based behavior control layer, a task planning layer, a situation based scenario control layer. 2) Visual verification system for adapting symbol level description into a real-world situation. 3) Task level planner which provides high-level action primitives for scenario description, 4) Vision and auditory based situation interpretation for controlling the scenario.

Finally we demonstrated a tea service task by a humanoid robot. Very robust vision based behavior control layer is able to connect a real humanoid robot to the high level task planner. Then task planner is able to provide high level goal description to generate a action sequence. Situation reasoning used to select goals by interpreting vision and auditory information.

References

[1] H. Inoue, S. Tachi, K. Tanie, K. Yokoi, S. Hirai, H. Hirukawa, K. Hirai, S. Nakayama, K. Sawada, T. Nishiyama, O. Miki, T. Itoko, H. Inaba, and M. Sudo. HRP: Humaonid Robotics Project of MITI. In *Proceedings of the First IEEE-RAS International Conference on Humanoid Robots (Humanoids 2000)*, 2000.

[2] Y. Sakagami, R. Watanabe, C. Aoyama, S. Matsunaga, N. Higaki, and K. Fujimura. The intelligent ASIMO: System overview and integration. In *Proceedings of the 2002 IEEE/RSJ International Conference on Intelligent Robots and Systems (IROS'02)*, pages 2478–2483, 2002.

[3] T. Asfour, K. Regenstein, P. Azad, J. Schroder, A. Bierbaum, N. Vahrenkamp, and R. Dillmann. ARMAR-III: An Integrated Humanoid Plattfrom for Sensory-Motor Control. In *2006 6th IEEE-RAS International Conference on Humanoid Robots (Humanoids 2006)*, pages 169–75, 2006.

[4] C.C. Kemp, A. Edsinger, and E. Torres-Jara. Challenges for robot manipulation in human environments. *IEEE Robotics & Automation Magazine*, 14(1):20–29, 2007.

[5] H. Ishiguro, T. Kanda, K. Kimoto, and T. Ishida. A robot architecture based on situated modules. In *Proceedings of the 1999 IEEE/RSJ Intl. Conference on Intelligent Robots and Systems (IROS'99)*, pages 1617–1624, 1999.

[6] Y. Hoshino, T. Takagi, U. Di Profio, and M. Fujita. Behavior description and control using behavior module for personal robot. In *Proceedings of The 2006 IEEE International Conference on Robotics and Automation*, pages 4165–4171, 2004.

[7] K. Okada, T. Ogura, A. Haneda, J. Fujimoto, F. Gravot, and M. Inaba. Humanoid Motion Generation System on HRP2-JSK for Daily Life Environment. In *2005 IEEE International Conference on Mechatronics and Automation (ICMA05)*, pages 1772–1777, 2005.

[8] K. Okada, M. Kojima, S. Tokutsu, T. Maki, Y. Mori, and M. Inaba. Multi-cue 3D Object Recognition in Knowledge-based Vision-guided Humanoid Robot System. In *Proceedings of the 2007 IEEE/RSJ Intl. Conference on Intelligent Robots and Systems (IROS'07)*, pages 3217–3222, 2007.

[9] K. Okada, M. Kojima, Y. Sagawa, T. Ichino, K. Sato, and M. Inaba. Vision based behavior verification system of humanoid robot for daily environment tasks. In *2006 6th IEEE-RAS International Conference on Humanoid Robots (Humanoids 2006)*, pages 7–12, 2006.

[10] Genshiro Kitagawa. Monte Carlo filter and smoother for non-Gaussian nonlinear state space models. *Journal of Computational and Graphical Statistics*, 5(1):1–25, March 1996.

[11] Michael Isard and Andrew Blake. Condensation – conditional density propagation for visual tracking. *International Journal of Computer Vision*, 29(1):5–28, 1998.

[12] K. Okada, M. Kojima, S. Tokutsu, Y. Mori, T. Maki, and M. Inaba. Task Guided Attention Control and Visual Verification in Tea Serving by the Daily Assistive Humanoid HRP2JSK. In *Proceedings of the 2008 IEEE/RSJ Intl. Conference on Intelligent Robots and Systems (IROS'08)*, page (Submitting), 2008.

[13] S. Tokutsu, K. Okada, and M. Inaba. Daily Life Sound Recognition based on Cepstrum for Environment Situation Reasoning in Humanoid Robot. In *Proceedings of the 2008 IEEE/RSJ Intl. Conference on Intelligent Robots and Systems (IROS'08)*, page (Submitting), 2008.

[14] S. Russell and P. Norvig. Planning and Acting. In *Artificial Intelligence: A Modern Approach, 2nd edition*, pages 445–448, 2002.

Intelligent Autonomous Systems 10
W. Burgard et al. (Eds.)
IOS Press, 2008

doi:10.3233/978-1-58603-887-8-406

A survey on mobile robots for industrial inspections

Andreas KROLL
Mechanical Engineering, University of Kassel, Germany

Abstract. Mobile service robots for carrying out inspections of industrial infrastructure gain importance e.g. due to aging infrastructure and demographics. This paper describes inspection tasks, non-destructive testing (NDT) methods and provides for a survey of autonomous mobile industrial service robot activities.

Keywords. Industrial infrastructure, inspection, mobile service robots, NDT

1. Introduction

Industrial robots are in general *immobile* and work on tasks such as assembly, welding or handling. *Mobile service robots* are typically deployed in entertainment, security, personal assistance, cleaning or civil construction. Mobile robots for industrial inspection are between these poles. They are motivated by: (i) replacing human missions in hazardous environments (e.g. radiation exposure) and on dangerous objects (e.g. in-service work on high voltage power lines), (ii) better access to tight or remote places (e.g. underground/underwater pipelines, high-up pipe-bridges), (iii) conducting repetitive, highly specialized and time-demanding tasks more reliably (e.g. inspect large structures for cracks), (iv) reducing downtime due to inspection, (v) speed up inspections and make the assessments more uniform, and to (vi) counteract consequences of demographical changes. Initial work addressed pipelines, nuclear power plants and waste depositories, aircraft surfaces, and underwater facilities. Recent applications include civil structures (building facades, concrete constructions, metal skeletons), electrical transmission and distribution infrastructure, and sewer systems.

In industrialized countries, grass root plant projects are rare and operational plants reach old age. E.g. 30+ year old refinery pipework is common in UK refineries, for which reason a governmental focus program on inspections was launched in 2004 [23]. Infrastructure inspection gains in importance and a robotized conduction can be advantageous, e.g.: The maintenance related downtime was cut by a factor of four in a refinery [3]. The downtime due to required inspections could be halved in a nuclear power plant [13]. Oil & gas offshore platform operational authorities currently investigate whether robots can take over not automated tasks wrt. operation, inspection and maintenance in order to operate platforms of tail-end fields unmanned [46].

This article aims at giving an introduction into the field of industrial inspection and at providing an overview on what type of inspection applications with surface-bound robots have already been addressed. The survey was compiled from results of a literature search. The next section gives some background on industrial inspection tasks and non-destructive testing methods. Section 3 surveys applications. Conclusions are drawn in section 4.

2. Inspection of industrial infrastructure

An inspection is an examination of an object or an activity. There are different types of industrial inspections: (i) maintenance and safety, (ii) security, and (iii) product-quality-related ones. This article focuses on type (i). These include routine inspections of plant components, which are commonly carried out using human "smell, noise and eyesight". In processing plants, special attention is paid to rotary machinery (e.g. bearings, eccentricities, lubricants), material health (e.g. cracks, corrosion, leaks, reserve of wear out) and instrumentation (e.g. calibration need, fouling) [12]. In addition, formal inspections - particularly of pressurized equipment - are carried out during major revisions, e.g. each 3rd to 5th year in case of large continuously operated plants.

Corrosion causes high costs, e.g. the direct cost was estimated as 276 billion USD p.a. in the USA [34]. As corrosion can have catastrophic consequences, the assessment of the structural health of infrastructure is a prime objective. Surprisingly, "simple" mechanical components such as pipes and vessels cause most frequently losses in processing plants [31,33]. Specialized non-destructive testing (NDT) methods are used by inspectors to determine the remaining wall thickness and to identify abnormalities such as cracks, see table 1 and refer to [7,24,39] for details. Leak detection is of particular interest, too: Escaped hazardous substances have to be detected and removed. Secondly, pressured equipment is often designed to show "leak-before-break" behaviour. This means that partial failures result in a detectable leak before a final fracture occurs [25]. Thirdly, leakages are a key maintenance issue. 23.5% of all damages in a large chemical plant appeared as leak. This made leaks the 2nd most frequent damage symptom (valued 1.59 MEUR p.a.); following just 2.2% behind the #1 symptom [40]. Table 2 records some methods for leak detection. Only few of the inspection methods can be applied remotely. A remote applicability is advantageous as high-up equipment (e.g. pipes in a pipe-bridge 10-20 m up) or high-rise components (e.g. storage tank surfaces) can be inspected from the ground without requiring climbing. Recently focal plan array (FPA) based IR-cameras became available permitting low-cost, portable thermographical and electro-optical measurements.

Table 1. NDT methods for inspecting structural health and corrosion

#	Method	Functional principle	Comment
1	CCD-camera / visual	Abnormal visual surface pattern are looked for. Aids include boroscopes, endoscopes or cameras.	Detects only surface defects; depends on illumination; clean surface required; suitable for area scans
2	Eddy current	Defects such as cracks change the electrical conductivity, which changes in turn the impedance of the measurement coil.	Inexpensive; applicable through normal paint; sensitive to small defects; time-consuming for large areas; only applicable to conductive materials; point-type measurement
3	Ultra-sonics	Ultrasonic sound waves are reflected at material defects (→defect detection) and interfaces (→wall thickness measurement); the	Coupling medium required; coupling may become superfluous when laser-ultrasound methods become

		response pattern is evaluated to determine time-of-flight.	deployable; point-type measurement
4	Magnetic flux leackage	Magnetic flux is "squeezed" out of a metal wall by any decrease of the wall thickness.	Inspected wall needs to be magnetized; provides only indirect and therefore qualitative thickness measurement; applicable to ferromagnetic materials
5	Thermo-graphy	Defects that change the thermal conductivity become visible during cooling / heating of objects.	Only defects detectable that affect the surface temperature; suitable for area scans
6	Magnetic particle inspection	If a magnetized object cracks, separate magnets result to both sides of the crack causing iron particles to accumulate at the crack.	Applicable to ferromagnetic materials
7	Radio-graphy	Absorption of radiation (X- or gamma rays) depends on thickness and density of the penetrated material.	Costly safety regulations; light-wide portable systems became recently available
8	Acoustic emission	Detects acoustic waves that are emitted during the growth of microscopic defects.	Static defects can not be detected
9	Liquid penetrant / dye penetrant	A liquid (often loaded with fluorescent dye) is applied to the object and seeps into surface defects. The seeped liquid is then drawn and made visible.	Easy applicable to large areas; complex geometries easy to handle; inexpensive; only surface breaking defects detectable; pre-cleaning required; requires application of chemicals

Table 2. Methods for leak detection

#	Method	Functional principle	Comment
1	Chemo-electrical	Many different principles including: thermo-catalytic, thermo-calorimetric, metal oxide semi-conductor, quartz crystal microbalance	Sensors often designed to be sensitive to a single gas or to a combustible mixture
2	Acoustic	A gas leak causes noise at the discharge point due to emerging vortices. A small liquid leak causes noise due to cavitation, which can be detected by hydrophone or transducer outside the pipe.	The noise wavelength is proportional to the leak diameter. Small leaks cause noise with frequency at the upper end of the acoustic range.
3	Thermo-graphy	Expansion of gas cools the vicinity of a leak. Liquid leaking from pipes with a different temperature from ambient can be detected.	Permits remote inspection
4	Electro-optical	Gases absorb electromagnetic radiation in characteristic ways and are thus observable by their spectral fingerprint.	Typically IR-laser-based systems; permits remote inspection
5	Model-based	Measured values are compared with predictions of thermodynamical model.	Used only for pipelines / main piping due to modelling effort

3. Applications of mobile inspection robots

Table 3 records work on inspection robots as identified in a literature scan. The survey does not aim at completeness but at indicating: (i) which application areas are addressed, (ii) what inspection sensorics are used, and (iii) how autonomously does the robot work. If several papers were identified from the same research group, the subset with the "best fit" wrt. inspection robots was selected. The boundary of the overview is difficult to draw. E.g. robots for building front inspection were excluded, as the major application will be for high-rise general buildings. On the other hand inspection robots for tunnels and sewers are included. Some activities wrt. leak/odour source localization have been included.

Table 3. Mobile inspection robot applications overview (ACFM: Alternating Current Field Measurement sensor; EC: Eddy current sensor; HS: Hall sensor; IR: infrared; MFL: Magnetic Flux Leakage; MO: Metal oxid gas sensor; SP/FP/OP/LP: Site-/Field-/Office-/Lab-tested Prototype; US: Ultrasonic sensor, VC: Video camera; -: not discussed: *not mentioned whether robot is used during tests or data gathered otherwise)

Ref.	Task	Locomotion type and platform	Inspection sensorics	System maturity	Comment
[17]	Underground nuclear storage facility: drum inspection for leaks & deterioration	Drive: wheeled (HelpMate/TRC)	6 VC; radiation detector	SP	No image processing; cordless
[9]	Ditto	Drive: 6-wheeled (K3A/Cybermotion)	VC with strobe light	SP; produc-tization planned	Image processing to detect deterioration; cordless
[51]	Nuclear power plant inspection	Guided motion: Monorail with suspended robot	Two robots with VC and two robots with acoustic & thermo-graphic sensors	SP	Abnormalities identi-fied by comparing video/ acoustic mea-surements with stored reference data; thermo-graphy analyzed by operator; inspection focus on steam / water leaks
[29]	Ditto	-	Stereo-VC system	Tested with pictures	Paper restricted to visual attention control
[13]	Ditto	Guided motion: Pole-guided manipulator	US	SP	Measurement interpretation by operator
[21]	Ditto	Guided motion: Pipe-clamped manipulator	US, VC	LP	Umbilical cord; scope limited to pipe intersect-tions; focus on inter-section weld inspection
[49]	Power plant inspection	Crawl: 2 sprocket chains with each 6 rollers for lateral and longitudinal movement	US, VC	LP	Measurement interpretation by operator
[19]	Ditto	-	VC	Tested with site photos*	Integration of robot with plant info. mgt' system
[3]	Metal storage tank internal inspection (wall, floor)	Drive: One robot with 4 magnetized wheels (permit climbing) and one Robot with 4 wheels	One robot with array of 8 US; one robot with array of 24 HS (MFL method)	Regular industrial use	In-service inspection; umbilical cord; tele-operated
[11]	Ditto	"Drive": 6 wheels; propeller-based adhesion for wall climbing	US; IR-camera	Water-tank tested prototype	In-service inspection; umbilical cord; ex-proof
[44]	Ditto	Drive: Tracks with adjustable rockers and switchable magnetic force	Colour VC; 8 US	Water-tank tested prototype	Ex-proof by pressurization; umbilical cord
[43]	Ditto	Drive & swim: 6-wheels and thrust based adhesion	US & EC array, ACFM	Water-tank tested prototype	Umbilical cord
[39]	Tunnel inspection	-	VC	OP for sensorics	

Ref.	Task	Locomotion type and platform	Inspection sensorics	System maturity	Comment
	(cracks)				
[37]	Storage tank external inspection	Crawl: 3 Segments with each 2 wheels and suction-cub-based adhesion	-	LP of single segments, complete system under construction	Umbilical cord
[16]	Ferromagne-tic material inspection on cracks	Drive: 2 independent-ly driven fixed wheels and 1 omni-directio-nal wheel; adhesion by magnetic wheels	HS (MFL method)	Tested conceptual prototype	Magnetic wheels magnetize inspection material as required by MFL method; umbilical cord
[27, 28]	Underground power cable inspection	Cable-guided motion on top of cable; 2-segments; hourglass-shaped wheels	IR, fringing el. field & acoustic sensor, VC	FP	
[22, 53]	High-voltage overhead line inspection	Line-guided motion: suspended, hourglass-shaped wheels	2 VC	Indoor and real outdoor power line tested prototype	Locomotion obstacle detection (e.g. isolator)
[35]	Ditto	Line-guided motion: suspended; wheels; grippers to help surpassing obstacles	-	Tested prototype	Knowledge-based strategy to overcome locomotion obstacles
[30]	Gas pipeline internal inspection	Drive: 4 radial distributed drive wheels	EC, VC	At service at Tokyo Gas	Off-service inspection; umbilical cord
[30]	Ditto	Drive: wheeled	VC	Field tested by Tokyo Gas	In-service inspection; cord-less
[30]	Ditto	-	MFL and EC method	In-service at Tokyo Gas	In-service inspection
[6]	Pipeline internal inspection (wall thickness)	Pipeline-guided, propelled by the transported fluid	HS (MFL method)	Commercial product	Cordless, in-service inspection
[36]	Ditto (weld condition)	Drive: 4-wheels	X-ray	Product	Cordless and cord-bound
[10]	Pipeline external inspection	Crawl: 3 set of omni-directional wheels at 0°, 120°, 240°	-	LP	Objective: detect corrosion under coatings: capable to handle bends
[1,2]	Sewer inspection	Drive: 4 specially oriented wheels	VC with fish-eye lens	(Dry) LP	Cordless; waterproof
[15]	Ditto	Drive: 4-wheels	Microwave backscatter and optical sensor	(Dry) LP	Sensor data fusion and fault detection & diagnostics
[14]	Ditto	Drive: Wheeled	Several VC for gas space, US for water space	FP	One-pipe sewer system
[8]	Ventilation duct inspection	Drive: Track-driven with permanent magnets for adhesion	VC	Prototype	Umbilical cord; tele-operated

Ref.	Task	Locomotion type and platform	Inspection sensorics	System maturity	Comment
[45]	Aircraft surface inspection	Crawl: Beam-crawler, suction cup-based adhesion (7 cups)	4 VC, EC	FP	Umbilical cord; tele-operated
[45]	Ditto	Drive: 4 wheels	Colour 3D-stereoscopic VC with dynamic lighting	FP	Tele-operated; only aircraft crown inspected
[41]	Ditto	-	EC, thermo-graphy	Lab-tested sensorics	Robot platform not yet considered
[50]	Ditto	Crawl: Traction-wheels and 2 suction cups for adhesion	-	LP	Umbilical cord
[5]	Ditto	Crawl: 16 suction cups for adhesion mounted on 2 legs for movement	-	LP	Umbilical cord; tele-operated
[47, 48]	Gas/Odour source localization	Drive: 4-wheeled, skid steered (RWII/Irobot)	8 MO (Figaro)	OP	Unventilated in-door environment; cordless
[42]	Ditto	Drive: 2 independent driven fixed wheels and 1 ball bearing	Wind direction (vane); quartz crystal micro-balance gas sensor	OP	In-door environment with constant air-airflow; umbilical cord
[38]	Ditto	Drive: 2 drive wheels and 2 additional supports	4 MO (Figaro)	OP	Focus on localization strategies
[26]	Ditto	Drive: Wheeled	VC, 4 MO (Figaro), 4 airflow sen-sors (ane-mometer)	OP	Focus on localization strategies
[32]	Material flow system	Propelled by the material flow system	-	Concept	
[4]	Steel skeleton	Climb: 2 Grippers	-	OP	
[18]	-	4-segment; tracks on all 4 sides of seg-ments; actuated joints	-	FP	"serpentine robot"

The following conclusions are drawn from analyzing the recorded work:

Locomotion: The prevailing approach is to construct robots that are capable of transporting the inspection sensors to close proximity of the inspection target at the cost of possibly complex locomotion mechanism. Wheeled robots are most often encountered. Guided motion-based robots partly result from the inspection object geometry (e.g. power lines, pipelines) and are partly used to avoid the additional requirements and complexity of "unconstraint" mobility (e.g. monorail inspection system for a section of a plant).

Inspection sensorics: Only the first four NDT methods in table 1 and the first method in table 2 are reportedly used with inspection robots. Except from video camera based inspection, they require the sensor to be close to the inspection target. Often only a single sensor is used. If several sensors are used, their measurements are evaluated separately; e.g. an US sensor is used for wall thickness measurement and MFL measurement for identifying cracks. Sensor data fusion is rarely considered. Some work focuses on the mechanics and the systems only feature "place holders" for inspection sensorics. No system was encountered that utilizes tele-measurements in tandem with a permissible simpler mechanical design.

Autonomy: Industrially deployed systems are typically tele-operated or motion is guided and therefore simple to automate. Measurements are manually interpreted and practitioners were sceptical in 1996 whether interpretation could be automated: "Aircraft inspectors and NDI supervisors are sceptical about the feasibility of automated flaw detection through image understanding algorithms [20]." 10Y later, the automatic interpretation of complex images taken in disturbed environments can not be regarded as solved. Autonomy is also restricted wrt. energy and communication: Some systems use an umbilical cord. Others are controlled wireless but store measurement data for external post mission analysis. Etc.

Maturity: Most of the work resulted in inspection robot prototypes. Some are rather basic, while others have been developed that far that site or field tests were carried out. Mobile robots for pipeline inspection have possibly the longest commercial deployment history as they have been deployed commercially since the 1960ies. However, they work in well-known environments and move along predefined paths. Such robots were referred to as "X-ray crawlers" and "intelligent pigs" [36] and results are typically not published in robotics journals or conferences.

Other: Typically, the research efforts target a single aspect of robotized inspection such as the locomotion mechanism or strategies for leak detection. Little work was identified where a team of experts of all relevant fields worked intensively on the design of an entire inspection robotic system. With few exceptions, the inspection robots work on their own and not in a multi-inspection-robot-team.

4. Conclusions

The idea of industrial inspection robots is not new as such, as mobile robots have been used for pipeline inspection in the Oil & Gas industry since the 1960ies [36]. However, autonomously conducted inspections of semi-structured industrial infrastructure such as processing plants provide for many not yet solved research problems e.g. regarding navigation, localization, measuring in disturbed environments, and measurement interpretation. The challenge of automating the interpretation of complex inspection measurements (currently carried out by particularly trained inspectors) is possible the most differentiating issue from other mobile service robot applications. In order to reliably take over tasks of human inspectors, the robots need to become much more intelligent. Only few initiatives behind the publications aim at designing an inspection robot with all required functions. Teams of specialists that cover all functional areas could possibly come up with novel system concepts. For instance, using IR-optical and optical measurements robots could possibly be designed for a subset of inspection tasks that conduct remote inspections (possibly in a team) from the plant floor. This would avoid requiring climbing up the inspection object in order to position the inspection sensorics in proximity to the test object. Sensor data fusion and more intelligent measurement interpretation methods are expected to shift the capabilities of inspection robots further towards reliable NDT results and towards the final objective of fully autonomous operation.

References

[1] A. Ahrary, M. Ishikawa, An automated fault detection system for sewer inspection robots, *Proc. Int. Symposium on Robotics ISR '05* (2005) 1-5.

[2] A. Ahrary, A.F. Nassiraei, M. Ishikawa. A study of an autonomous mobile robot for a sewer inspection system, *Artifical Life Robotics* **11** (2007) 23–27.

[3] J.L.G. Alonso, E.G. Ares, J. Torres, Autonomous integrated system for tank and pipe inspection, *Proc. 8th ECNDT Conf.*, Barcelona, British Institute of non destructive testing (2002).

[4] C. Balaguer, A. Gimenez, A. Jardon, Climbing Robots' Mobility for Inspection and Maintenance of 3D Complex Environments, *Autonomous Robots* **18** (2005) 157–169.

[5] Y. Bar-Cohen, P. Backes, Scanning large aerospace structures using open-architecture crawlers. National Space and Missile Materials Symposium, San Diego, CA, 27.2–2.3.2000.

[6] O.A. Barbian, Detection of corrsosion in pipelines with magnetic stray flux technique, *Materials and Corrosion* **46** (1995) 414-417.

[7] BDM, *Correction Detection Technologies: Sector study*. Final Report (1998).

[8] L. Bing Lam, A. Djordjevic, S.K. Tso, K.P. Liu, Development of a Range of Robot and Automation Prototypes for Service Applications, *Cutting Edge Robotics* (2005) 533-550.

[9] J.S. Byrd, An Intelligent Inspection and survey robot, Report of Dept. of Electrical and Computer Eng., University of South Carolina, USA (1996).

[10] P. Chatzakos, Y.P. Markopoulos, K. Hrissagis, A. Khalid, On the Development of a Modular External-pipe Crawling Omni-directional Mobile Robot. In: *Climbing and Walking Robots*, Springer (2006) 693-700.

[11] A.C. Cruz, M.S. Ribeiro, In service inspection robotized tool for tank filled with hazardous liquids – robotank inspect. In: *Climbing and Walking Robot,* Springer (2005) 1019-1031.

[12] J. Deinert, Erhöhung der Anlagenverfügbarkeit in einer Raffinerie, *Maintenance 2010*, Berlin (2005).

[13] F. Dirauf, B. Gohlke, E. Fischer, Innovative Robotics and Ultrasonic Technology at the Examination of Reactor Pressure Vessels in BWR and PWR Nuclear Power Stations. *NDT.net* **3,** 10 (1998).

[14] N. Elkmann, H. Althoff, Automated inspection systems for large underground concrete pipes partially filled with waste water, VDI-Berichte Nr. 1956, Robotik 2006, 15.-17.5.2006, München, Düsseldorf: VDI-Verlag (2006).

[15] C. Frey, H.-B. Kuntze, R. Munser, Anwendung von Neuro-Fuzzy-Methoden zur multisensoriellen Schadensdiagnose in Abwasserkanälen, *Proc. 12. WkShp Fuzzy-Control*, GMA FA 5.22, 13-15.11.2002, Dortmund, 81-89.

[16] M. Friedrich, L. Gatzoulis, G. Hayward, W. Galbraith, Small inspection vehicles for non-destructive testing applications, In: *Climbing and Walking Robots*, Springer (2006) 927-934.

[17] R. Fulbright, L. M. Stephens, SWAMI: An Autonomous Mobile Robot for Inspection of Nuclear Waste Storage Facilities. *Autonomous Robots* **2** (1995) 225-235.

[18] G. Granosiak, M.G. Hansen, J. Borenstein, The omnitread serpentine robot for industrial inspection and surveillance. *Industrial Robot* **32**, 2 (2005) 139-148.

[19] M. Gregori, L. Lombardi, M. Savini, A. Scianna, Autonomous plant inspection and anomaly detection, In: *Image Analysis and Processing - Lecture Notes in Computer Science 1311*, Springer, Berlin (1997).

[20] P. Gunatilake, M. Siegel, A. Jordan, G. Podnar, Image enhancement and understanding for remote visual inspection of aircraft surface, Nondestructive Evaluation of Aging Aircraft, airports, and Areospace hardware **2945**, December 1996, 416-427.

[21] X.G. Fu, G.Z. Yan, B. Yan, H. Liu, A new robot system for auto-inspection of intersected welds of pipes used in nuclear power stations, *Int. J. Adv. Manuf. Technol.* **28** (2006) 596–601.

[22] S. Fu, Y. Zhang. X. Zhao, Z. Liang, Z. Hou, A. Zou, M. Tan, W. Ye, L. Bo, Motion deblurring for a power transmission line inspection robot, *Computational Intelligence* **4114** (2006) 866-871.

[23] HSE, Chemical manufacture and storage – integrity of pipework systems project – UK refineries. Internet report, HSE, http://www.hse.gov.uk/chemicals/spctechgen33.htm (2004).

[24] HSE, Inspection/Non Destructive Testing. http://www.hse.gov.uk/comah/sragtech/techmeasndt.htm (2008).

[25] IAEA, Applicability of the leak before break concept, Report No. IAEA-TECDOC-710 (1993).

[26] H. Ishida, H. Tanaka, H. Taniguchi, T. Moriizumi, Mobile robot navigation using vision and olfaction to search for a gas/odor source, *Autonomous Robots* **20** (2006) 231–238.
[27] B. Jiang, A.P. Sample, A.V. Mamishev, Mobile Monitoring for Distributed Infrastructures, *Proc. IEEE Int. Conf. on Mechatronics & Automation*, Niagara Falls/Canada (2005) 138-143.
[28] B. Jiang, A.P. Sample, R.M. Wistort, A.V. Mamishev, Autonomous Robotic Monitoring of Underground Cable Systems, *Proc. Int. Conf. on Advanced Robotics* (2005) 673-679.
[29] N. Kita, Visual attention control for nuclear power plant inspection, *Proc. 15th Int. Conf. on Pattern Recognition* (2000) **4,** 118-123.
[30] M. Komori, K. Suyama, Inspection robots for gas pipelines of Tokyo Gas, *Advanced Robotics* **15**, 3 (2001) 365-370.
[31] A. Kroll, Priorisierung von Asset Managementaktivitäten für chemische Anlagenausrüstungen: Aussagen aus Ereignisstatistiken, *at – Automatisierungstechnik* **52** (2005) 228-238.
[32] B.T. Kuhlenkötter, T. Brutscheck, M. Bücker, Automation in der Instandhaltung, *GMA Kongress 2007*, Baden-Baden (2007) 331-340.
[33] B. Lafrenz, Schadensfälle in verfahrenstechnischen Anlagen – erhoben und ausgewertet nach Arbeitsschutzkriterien, *Forschungsbericht Fb 1022*, Bundesanstalt für Arbeitsschutz und Arbeitsmedizin, Dortmund (2004).
[34] M. Lettich, Is There a Cure for Corrosion Under Insulation?, *Insulation Outlook*, No. 11 (2005).
[35] W. Ludan, W. Hongguang, F. Lijin, Z. Mingyang, Research on obstacle-navigation control of a mobile robot for inspection of the power transmission lines based on expert system, *Proceedings of the 8th Int. Conf. on Climbing and Walking Robots* **4** (2006), 173-180.
[36] R. F. Lumb, Inspection of pipelines using nondestructive techniques, *Physics in Technology*, November 1977, 249-256.
[37] D. Longo, G. Muscato, A modular approach for the design of the Alicia climbing robot for industrial inspection, *Industrial Robot* **31**, 2 (2004) 148-158.
[38] C. Lytridis, E.E. Kadar, G.S. Virk, A systematic approach to the problem of odour source localisation, *Autonomous Robots* **20** (2006) 261–276.
[39] NACE, Business portal, www.nace.org (2008).
[40] W. Pohrer, Controlling in der Instandhaltung, *Maintenance 2010*, 9.-10.11.2005, Berlin.
[41] J.R. Rudlin, R. Fitch, G. Schweer, A. Boenisch, Eddy current and thermography application for robotic aircraft inspection, *Proc. BINDT Annual Conf.* (2002) 125-130.
[42] R.A. Russel, D. Thiel, R. Deveza, A. Mackay-Sim, A robotic system to locate harzadous chemical leaks, *Proceedings of IEEE International Conference on Robotics and Automation* (1995) 556-561.
[43] T.P. Sattar, H.E. Leon Rodriguez, J. Shang, B. Bridge, Automated NDT of floating production storage oil tanks with a swimming and climbing robot, In: *Climbing and Walking Robots*, Springer (2006) 935-942.
[44] H. Schempf, B. Chemel, N. Everett, Neptune: Above-Ground storage tank inspection robot, *Proc. IEEE Robotics & Automation*, Society Magazine **2**, 2, (1995) 9-15.
[45] M. Siegel, P. Gunatilake, Remote inspection technologies for aircraft skin inspection, *IEEE Workshop on Emergent Technologies and virtual Systems for Instrumentation and Measurement*, Niagara Falls/Canada (1997) 1-10.
[46] S. Vatland, M. Svenes, Tail integrated operations, *Project Newsletter 1*, ABB (2007).
[47] M. Wandel, Leckortung mit einem mobilen Roboter, Dissertation, Fakultät für Chemie und Pharmazie, Univ. Tübingen (2004).
[48] M. Wandel, U. Weimar, A. Lilienthal, A. Zell, Leakage localization with a mobile robot carrying chemical sensors, *Proceedings of Int. IEEE Conf on Electronics*, Circ, Malta (2001), 1247-1250.
[49] G. Wang, P. Ma, X. Cao, H. Sun, Y. Li, Application of wireless local area network technology in mobile robot for finned tube inspection, *Journal of Shanghai University* **8**, 2 (2004) 180-186.
[50] T.S. White, R. Alexander, G. Callow, A. Cooke, S. Harris, J. Sargent, A Mobile Climbing Robot for High Precision Manufacture and Inspection of Aerostructures, *International Journal of Robotics Research* **24**, 7 (2005) 589-598.
[51] S. Yamamoto, Development of inspection robot for nuclear power plant, *Proc. IEEE Int. Conf. on Robotics and Automation*, Nice, France (1992) 1559-1566.
[52] S.-N. Yu, J.-H. Jang, C.-S. Han, Auto inspection system using a mobile robot for detecting concrete cracks in a tunnel, *Automation in Construction* **16** (2007) 255-261.
[53] Y. Zhang, S. Fu, X. Zhao, Z. Liang, M. Tan, Y. Zhang, Visual Navigation for a power transmission line inspection robot, *Computational Intelligence* **4114** (2006) 887-892.

Intelligent Autonomous Systems 10
W. Burgard et al. (Eds.)
IOS Press, 2008

doi:10.3233/978-1-58603-887-8-415

PatrolGRAPH: a Distributed Algorithm for Multi-Robot Patrolling

M. Baglietto, G. Cannata, F. Capezio, A. Grosso, A. Sgorbissa, R. Zaccaria [1]
DIST - University of Genova, 16145 Genova, Italy

Abstract. This paper describes PatrolGRAPH, a novel real-time search algorithm which has been specifically designed to solve the "patrolling problem", i.e., the problem of repeatedly visiting a set of specific locations in the environment. The algorithm has extremely low requirements in terms of computational power, does not require inter-robot communication, and can even be implemented on memory-less robots. Moreover, the algorithm is proven to be statistically complete as well as easily implementable on real, marketable robot swarms for real-world applications.

Keywords. Multi-robot systems, RFID technology, real-time search.

1. Introduction

The work described in this paper has been done in the context of the European project ROBOSwarm. The project's main objective is building multi-robot systems to be used in different applications, ranging from autonomous cleaning and lawn-mowing to surveillance and patrolling, rescuing, humanitarian de-mining, etc. ROBOSwarm aims at finding solutions that are technologically plausible, with the purpose of designing robot swarms able to operate in the real-world *here and now*, not in *a future* when technology will be – hopefully – available. Finally, robots must be cheap, with low computational power and memory storage, and able to function even when wireless communication is not available or degraded, thus being unreliable for multi-robot coordination.

Towards this end, one of the core technologies of ROBOSwarm are passive RFID tags: these are low-cost, short-range, energetically self-sustained transponders that can be distributed in the environment and can store a limited amount of information. The use of RFIDs and other deployable sensor networks in robotics have been recently investigated, both to solve general robotic problems (i.e., navigation and localization [15][17][14]), as well as for specific applications (e.g., cleaning [12][13], exploration [10][11][2], or rescue applications in large scale disasters [18][16]). Both in the case that RFIDs are "manually" distributed during a setup phase which precedes execution, and in the case that robots themselves distribute RFIDs in the environment, they have been proven to provide a significant aid in different kinds of robotic tasks.

In this work, it is assumed that RFIDs are distributed in the environment prior to robot operation, and used to build a sort of "navigation graph": each RFIDs contains

[1]Corresponding Author: Antonio Sgorbissa, DIST - University of Genova, 16145 Genova, Italy; E-mail: antonio.sgorbissa@unige.it.

navigation directions to neighboring RFIDs, thus allowing robots to safely execute paths from and to different locations. In particular, we consider a swarm of robots which are required to repeatedly visit a set of specific locations, a variant of the dynamic coverage problem (see [1] and the references within) which is particularly relevant for surveillance and patrolling, continuous cleaning of crowded areas (e.g., malls, convention centers, restaurants, etc.), serving food or beverages (e.g., in hospitals or in a banquet), etc. As a consequence of the technological constraints, we are interested in distributed, real-time search algorithms (either biologically inspired or not [9] [8][7][4][5][6][3][2]), able to deal with situations in which a global representation of the environment, as well as global communication capabilities, are absent. However, in the perspective of designing and building real, marketable swarm of robots, we definitely privilege algorithms that can be formally proven to be (statistically) complete.

Section 2 describes *PatrolGRAPH*, a novel real-time search algorithm which allows to solve the patrolling problem, and gives formal proofs of statistical completeness. Section 3 compares *PatrolGRAPH* with *Node Count* (to our knowledge, the simpler algorithm for which formal proofs have been given [5]). Conclusions follow.

2. The PatrolGRAPH Algorithm

Navigation is based on a graph-like representation of the environment, in the following referred to as the *Navigation Graph* G_N. See Figure 1 on the left: G_N is a finite non-oriented graph of arbitrary order, and each vertex has an arbitrary number of incident edges. Cycles can be present. Differently from other scenarios, robots are not requested to visit repeatedly *all* vertices in G_N: instead, vertices are labeled as *terminal nodes* (i.e., vertices with only one incident edge) or *navigation nodes* (i.e., vertices with more incident edges). We are interested in real-time search algorithms which, as time passes, distribute robots in such a way that all *terminal nodes* receive the same number of visits. For example, *terminal nodes* can correspond to rooms or corridors to be cleaned, locations which must be surveyed (e.g., where a snapshot must be taken), etc. As usual, G_N can be represented as an oriented graph by simply duplicating all edges and assigning them a direction. The following constraints must be taken into account:

1. the algorithm must be implementable on very simple robots with limited computational power; memory requirements must be $O(1)$ with respect to the size of the graph. Robots cannot communicate with each other, neither globally nor locally.
2. robots do not know G_N, neither in advance nor during navigation. In fact, G_N is never stored in robots memory and, in general, it does not exist a central repository where G_N is stored. Instead, vertices as well as incident edges are stored into passive RFID tags with a very limited memory storage opportunely distributed in the environment. Notice that each RFID does not anything about adjacent vertices: it only knows the navigation directions to reach neighboring RFIDs (a RFID-based localization system is used, but it is not described here). Analogously, RFIDs do not know their own position in the environment; however, since RFIDs are physically located in the environment, this information is implicitly present: reaching a vertex in G_N correspond to reaching a well-defined region of the space. Figure 1 on the right shows the correspondence between vertices in

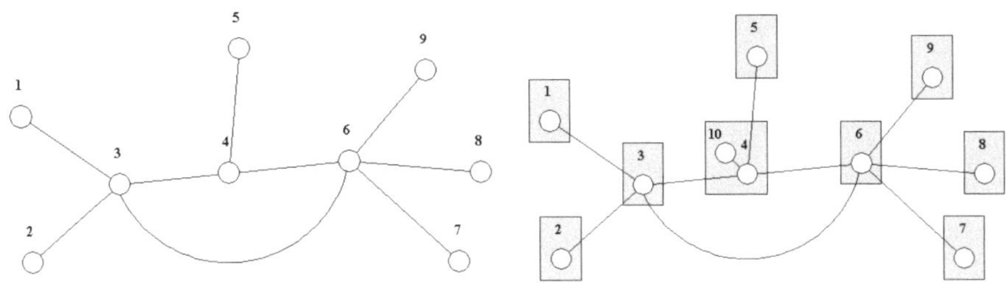

Figure 1. Left: navigation graph G_N. Right: implementation on RFID tags.

G_N and RFIDs (gray rectangles): vertex 4 is a *navigation node*: if one wants the corresponding RFID to be considered for patrolling, it is necessary to add a new *terminal node* 10. Notice that, in this case, it is not necessary to add a new RFID, since the same RFIDs can store information about vertices 4 and 10, as well as edges which connects them to other vertices in G_N and navigation directions.

3. RFIDs cannot communicate with each other. Robots can communicate indirectly by writing on/reading from RFIDs.

Consider Figure 1 on the left : vertices 1, 2, 5, 7, 8, 9, and 10 are labeled as *terminal nodes*, whereas the remaining ones are *navigation nodes*. In the following, two variants of the *PatrolGRAPH* algorithms able to solve the assigned problem while satisfying all constraints will be illustrated, and formal proofs will be given.

Suppose for sake of simplicity that only one robot is present. The general idea at the basis of *PatrolGRAPH* is the following: at time step $k = 0$ the robot starts from a *terminal node*, it moves from the current vertex to one of its adjacent vertices, and then it iterates the process (with the only constraints that it can never go back to the same vertex from where it came), until it finally reaches a *terminal node*. If G_N does not contain cycles (except for cycles with length 2, which are implicit since the graph is not oriented), it is guaranteed that one of the *terminal nodes* is reached in finite time. In fact, in this case, the rule which forbids backtracking allows to see the graph as a *tree* in which the start node is the *root* and other *terminal nodes* are *leaves*.

The two variants of *PatrolGRAPH* differ in how they choose the edge to be followed: whereas the random version chooses among the edges randomly (with equal probability), the deterministic version counts the number of times that each edge has been chosen, in order to guarantee that robots leaving from a vertex are equally distributed along all possible routes. In particular, we use the following notation: S denotes the finite set of vertices in G_N, and $s_{start} \in S$ denotes the start vertex. $A(s) \neq 0$ is the finite, nonempty set of (directed) edges that leave vertex $s \in S$. $succ(s, a)$ denotes the successor vertex that results from the traversal of edge $a \in A(s)$ in vertex $s \in S$. We also use two operators with the following semantics: the expression *is-terminal(s)* returns $true$ if and only if s is a *terminal node*; the expression *choose(s, p)* chooses one of the directed edges $a \in A(s)|succ(s, a) \neq p$. These operators are used to guarantee that the robot can never go back to the same vertex from where it came, except when it is in a *terminal node*. The random and the deterministic versions of the algorithm differ in how the operator *choose(s,p)* is implemented: whereas in the random version *choose(s,p)* calls the operator *choose-randomly (s, p)*, which chooses one of the leaving edges randomly (with equal probability), in the deterministic version *choose(s,p)* calls *choose-deterministic (s, p,*

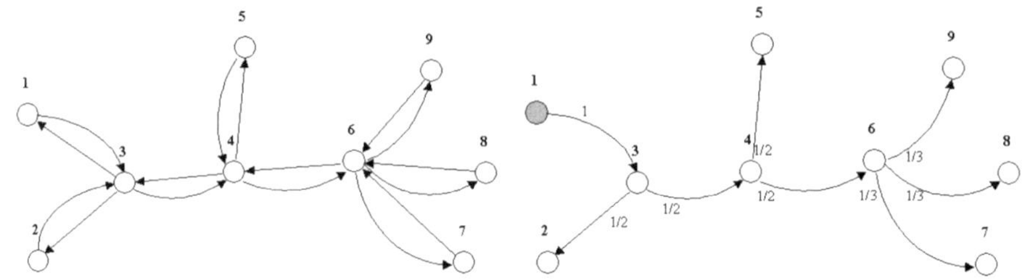

Figure 2. Left: navigation graph G_N. Right: graph $G_N, 1$.

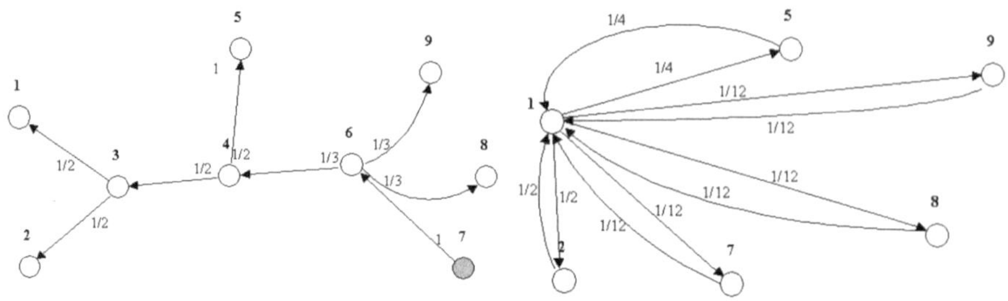

Figure 3. Left: graph $G_N, 7$. Right: graph G_T (only transitions from/to 1 are shown)

switch(s,p)), which selects edge a depending on p and the current value of $switch(s,p)$, and then move the switch to the next admissible edge in a circular fashion. Notice that, for each vertex s, the algorithm needs now to memorize a number of switches which equals the number of incident edges: each $switch(s,p)$ allows to uniformly distribute all robots coming from p along the remaining edges $a \in A(s)|succ(s,a) \neq p$.

Algorithm 1 PatrolGRAPH

s_{start} := choose an arbitrary terminal node.
a:= the only edge that leaves s_{start}.
$s := succ(s_{start}, a)$.
$p := s_{start}$.
while patrolling **do**
  if not *is-terminal(s)* **then**
    a:= *choose(s,p)* (either randomly or deterministically).
    $s := succ(s,a)$.
    $p := s$.
  else
    a := the only edge that leaves s.
    $s := succ(s,a)$.
    $p := s$.
  end if
end while

When more robots are present, each robot simply executes the algorithm: notice however that, when using the operator *choose-randomly (s, p)*, robots do not interfere with each other, but the same is not true in the case of *choose-deterministic (s, p, switch(s,p))*. In fact, since switches are shared among all robots (i.e., each RFID uniformly distributes visiting robots along available routes), it is in principle necessary to consider the single robot and the multi-robot case separately. For this reason, we start by demonstrating the properties of the random version of *PatrolGRAPH*, by showing how it meets all the constraints.

Assume, for sake of simplicity, that G_N has no cycles with length > 2 (see for example Figure 2 on the left). If, at some time, the robot starts from vertex 1, the graph available for navigation – as a consequence of the rule that forbids backtracking – reduces to $G_{N,1}$, a tree which root is 1. See Figure 2 on the right: edges leaving from a generic vertex s are labeled with transition probabilities, which – as stated – equals $1/(|A(s)|-1)$ for all leaving edges, where $|A(s)|$ is the number of incident edges in s and $|A(s)| - 1$ is the number of edges $a \in A(s)|succ(s,a) \neq p$. For example, the probability of ending up in 7 starting from 1 is $P(7|1) = 1/2 \times 1/2 \times 1/3 = 1/12$.

Suppose now that, starting from root 1, the robot ends up in 7: in the next phase, navigation will be performed on the basis of $G_{N,7}$, a tree which root is 7 (Figure 3 on the left). By considering Figures 2 on the right and 3 on the left, a very important property can be inferred: since the path from 1 to 7 requires to visit the same set of vertices $3, 4, 6$ as the path from 7 to 1, and the probability of choosing an edge depends only on the number of edges leaving that vertex, the probability of transition from 7 to 1 is the same than the probability of transition from 1 to 7. More formally, we have that, for each couple of *terminal nodes* (s_i, s_j):

$$P(s_i|s_j) = P(s_j|s_i) \tag{1}$$

It is straightforward to see that this is true even in presence of cycles with length > 2 in G_N, i.e., when each graph $G_{N,j}$ available for navigation without backtracking starting from s_j is no more a tree. In this case, there are infinite paths leading from s_j to s_i (since the robot is allowed to loop for an arbitrary number of times before reaching s_i), and $P(s_i|s_j)$ is the sum of the probabilities of all possible paths. However, also in this case, for each path in $G_{N,j}$ leading from s_j to s_i a path exists in $G_{N,i}$ from s_i to s_j which visits the same nodes but in the inverse sequence, and hence has the same probability. As a consequence, equation 1 still holds.

We now demonstrate the following theorem, which guarantees that – in the long term – all *terminal nodes* will be visited, in a statistical sense, the same number of times.

Theorem 1: If G_N is a finite non-oriented graph, the random version of *PatrolGRAPH* visits all terminal nodes with the same probability •

Proof: Since we are only interested in *terminal nodes* (i.e., we are not making any statement about *navigation nodes*), we consider a new system where states are the *terminal nodes* of G_N, and where the transition probability $P(s_j|s_i)$ from a generic state s_i to s_j equals the probability of ending in s_j starting from s_i after a random walk in G_N. See for example Figure 3 on the right (only the transitions from/to state 1 are shown for clarity): the graph G_T describes states (i.e., terminal nodes of G_N) and transitions between states (i.e., walks in G_N) corresponding to the navigation graph in Figure 2 on the left. The property described in Equation 1 is necessarily valid also in the new model. Also, since G_N is strongly connected by definition, G_T is strongly connected as well.

It is now possible to model the system as a Markov chain. This requires to write the transition matrix P (by assuming N terminal nodes):

$$P = \begin{bmatrix} P(1|1) & \cdots & P(1|N) \\ \vdots & \ddots & \vdots \\ P(N|1) & \cdots & P(N|N) \end{bmatrix} \tag{2}$$

P is a left stochastic matrix (i.e., $\forall i, j\ 0 \leq P(i|j) \leq 1$ and $\forall j \sum_i P(i|j) = 1$); moreover, since the graph G_T is strongly connected, P is irreducible (i.e., it exists $k > 1$ such that P^k is a positive matrix). Under these assumptions, the Perron-Frobenius theorem states that the system converges to an equilibrium state $\pi = P\pi$. Finally, since P is symmetric, it is also right stochastic ($\forall i \sum_j P(i|j) = 1$); this is sufficient to state that the equilibrium state is the uniform probability distribution:

$$\pi = [1/N \ldots 1/N] \tag{3}$$

which is exactly what we wanted to demonstrate □

The random version of *PatrolGRAPH* can be straightforwardly extended to the multi-robot case, since the random choice of each robot does not have any effect on the choices of other robots. However, the deterministic version requires a deeper investigation. In fact, since the choice of edges in G_N is now deterministic, it is not clear whether state transitions in G_T can still be modeled as a stochastic process, which allows to model the system as a Markov chain and to finally apply Theorem 1. This is especially true in the multi-robot case, since robots potentially interfere with each other by operating on switches; analyzing each robot separately could lead to unexpected results.

We therefore demonstrate the following:

Theorem 2: When deterministic *PatrolGRAPH* is used (either by a single robot or by many robots), the probability that a robot in s coming from p chooses an edge $a \in A(s)|succ(s, a) \neq p$ has a uniform distribution •

Proof: Consider that at time k the robot is in vertex $s \in G_N$, and assume that the number of incident edges is $|A(s)| = m$. The robot chooses one of the $m - 1$ leaving edges $a \in A(s)|succ(s, a) \neq p$ depending on the value of $switch(s, p)$ (where p is the vertex visited at time $k - 1$). In particular, since leaving edges are chosen in a circular fashion, the probability that $switch(s, p)$ returns a_i at step k depends only on the edge a_j chosen the last time that a robot visited vertex s coming from p. In particular, if $a_{prec(i)}$ is the leaving edges which precedes a_i according to the circular order established by $switch(s, p)$,

$$P(a_i|a_j = a_{prec(i)}) = 1, \; P(a_i|a_j \neq a_{prec(i)}) = 0 \tag{4}$$

It is now possible to model each $switch(s, p)$ as a stationary Markov process which states are $a_1, .., a_{m-1}$. The corresponding transition matrix can be easily written starting from the generic expression of the conditional probability $P(a_i|a_j)$ in Equation 4: it is straightforward to see that, once again, this is an irreducible doubly stochastic matrix which columns and rows sums up to 1. Therefore, the state necessarily converges to the uniform probability distribution

$$\pi = [P(a_1) \cdots P(a_{m-1})] = [1/(m-1) \cdots 1/(m-1)] \tag{5}$$

This is exactly what we wanted to demonstrate. From the statistical point of view, the same results in Theorem 1 can be applied to the deterministic case (which is nothing

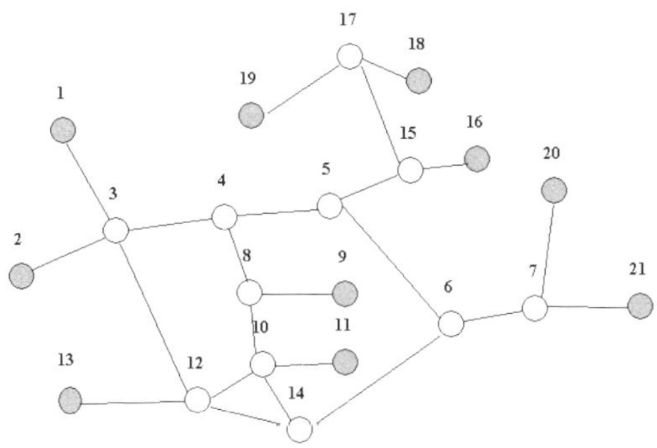

Figure 4. Navigation graph considered in experiments.

Vertex	1	2	3	4	5	6	7	8	9	10	11	12	13	14	15	16	17	18	19	20	21
1 robot	20	20	81	62	63	64	62	61	20	83	21	82	21	64	62	20	62	21	20	20	21
5 robot	103	103	411	310	310	311	310	309	103	409	102	413	104	310	311	104	312	104	104	103	104
10 robot	207	205	826	620	620	621	620	618	206	825	207	828	207	621	619	206	619	206	206	207	206
20 robot	413	412	1650	1241	1239	1241	1239	1240	413	1652	413	1652	413	1241	1239	413	1238	412	414	413	412

Figure 5. *PatrolGRAPH*: number of visits per vertex.

Vertex	1	2	3	4	5	6	7	8	9	10	11	12	13	14	15	16	17	18	19	20	21
1 robot	46	46	96	45	48	44	55	49	40	49	42	53	46	44	44	37	58	28	28	26	26
5 robot	230	230	480	225	240	220	275	245	200	245	210	265	230	220	220	185	290	140	140	130	130
10 robot	460	460	960	450	480	440	550	490	400	490	420	530	460	440	440	370	580	280	280	260	260
20 robot	920	920	1920	900	960	880	1100	980	800	980	840	1060	920	880	880	740	1160	560	560	520	520

Figure 6. *Node Count*: number of visits per vertex.

more than one of the possible sequences of events which can happen with probability π), thus yielding the same probability to be visited for all terminal nodes. □

3. Experimental results

The system has been extensively tested, both in the MatLab environment and in the Player/Stage architecture. Since the focus of the paper is on the properties of the *PatrolGRAPH* algorithm, only MatLab simulations are shown here. However, more realistic simulated experiments have been performed as well, with up to 20 Roomba robots equipped with an RFID antenna, as well as algorithms for RFID-based navigation, control and self-localization in presence of simulated sensor noise. In particular, MatLab experiments compare *PatrolGRAPH* with *Node Count*, the simplest real-time search algorithm for which formal proofs of complete coverage (in a statistical sense) have been given. *Node Count* differs from *PatrolGRAPH* in that it associates a value with each vertex of the graph, which counts how often each vertex has been already visited. When a robot enters a vertex, it increases the value of the vertex by one: it then moves to the neighboring vertex with the smallest value, that is, the neighboring vertex that has been visited the least number of times. *Node Count* and *PatrolGRAPH* have been quantitatively compared by running them on many different graphs. In the following, we discuss experiments performed in the graph in Figure 4: notice however that *only experiments performed with the deterministic version of PatrolGRAPH are shown*, which has similar

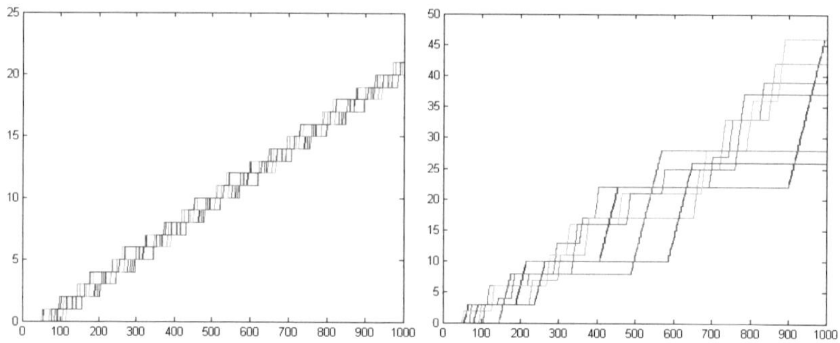

Figure 7. Number of visits versus time with 1 robot: left is *PatrolGRAPH*, right *Node Count*. Curves correspond to different *terminal nodes*

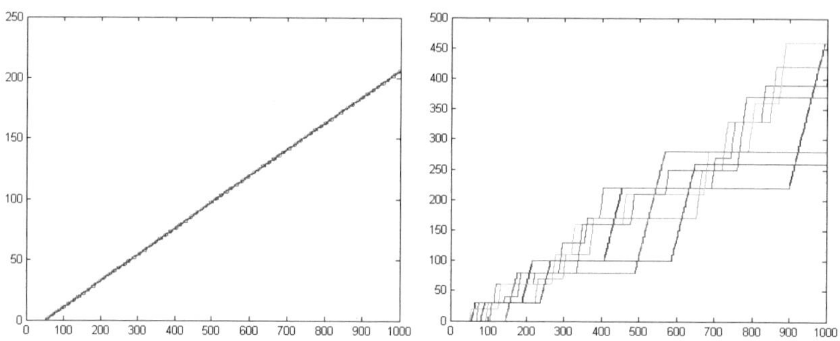

Figure 8. Number of visits versus time with 10 robot: left is *PatrolGRAPH*, right *Node Count*

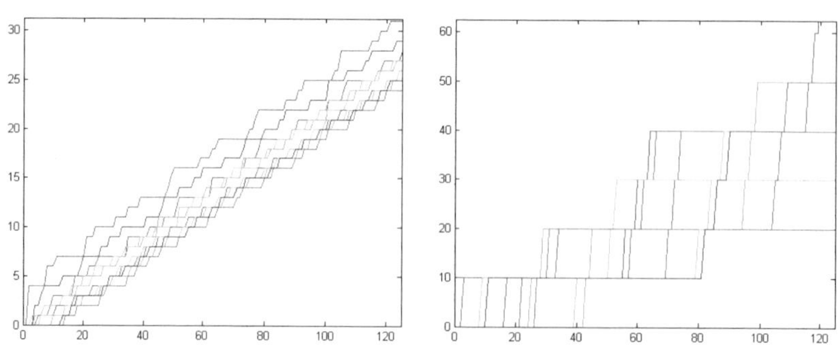

Figure 9. Transient behavior with 10 robot: left is *PatrolGRAPH*, right *Node Count*

requirements in terms of RFID memory as *Node Count*, and proves to behave more efficiently in the short term. Figure 5 shows the number of times that each vertex has been visited by *PatrolGRAPH* with an increasing number of robots, each moving for 1000 steps (the distance between vertices is ignored); Figure 6 shows the number of times that each vertex has been visited by *Node Count*. It is immediate to notice that *PatrolGRAPH* guarantees that all *terminal nodes* are visited the same number of times, whereas this property does not emerge in *Node Count*. However, *Node Count* appears to be more efficient, since – on average – *terminal nodes* receive a bigger number of visits, whereas *PatrolGRAPH* spends a lot of time in navigating from node to node. Both algorithms scale very well to an increasing number of robots M: the number of visits to each vertex is in fact proportional to M.

Figure 7 shows the number of visits to *terminal nodes* versus time when only one robot is present; plots correspond to *PatrolGRAPH* (to the left) and *Node Count* (to the right). Each curve corresponds to a different *terminal node*; finally, since we are interested into the regime behavior, the first 50 steps are not considered. It is easy to see that the peculiarity of *PatrolGRAPH* is "regularity": by considering a generic computation step, it never happens that the difference between the most visited and the least visited *terminal node* is greater than 2. Nothing similar happens in *Node Count*; in fact, it can happen that the same *terminal node* is visited 20 times before an "unlucky" *terminal node* is visited just 1 time.

Figure 8 and 9 plot analogous results when 10 robots are present. In particular, Figure 9 focuses on the transient behavior which corresponds to the initial steps: since the 10 robots start from the same *terminal node*, it can initially happen that some vertices are visited more than others: however, the difference between the most visited and the less visited *terminal node* is always ≤ 5 when *PatrolGRAPH* is used, and – most of all – this difference does not increase in time since the probability of visiting *terminal nodes* tends to the uniform probability. The situation is completely different with *Node Count*, where the difference between the most and the less visited *terminal node* goes up to 40 within the first 120 steps. Finally, the long term behavior of the random version of *PatrolGRAPH* is almost the same (not shown here), whereas bigger fluctuations are obviously present in the short term.

4. Conclusions

This paper has introduced two variants of *PatrolGRAPH*, a novel real-time search algorithm which allows a swarm of robots to patrol an area without global representations of the environment or global communication capabilities. The algorithm is suited to scenarios in which some areas (i.e., *terminal nodes*) must be repeatedly visited to perform some task, e.g., non stop surveillance and cleaning, table serving, etc.

A qualitative comparison with *Node Count* (and - to some extent - with other real time search algorithms) reveals that:

1. *PatrolGRAPH* has desirable properties in terms of complete coverage and uniformity in *terminal nodes* visits, whereas *Node Count* tends to result in more *terminal node* visits in a given time frame, and spends less time in *navigation nodes*. Basing on this observation, one could argue that *Node Count* is more efficient: however, this property is less evident in those domains in which the task to be performed in *terminal nodes* requires a time which is significantly longer than that required for navigation between nodes. If navigation time can be ignored, *PatrolGRAPH* uniformly distributes work among all *terminal nodes*, whereas other algorithms do not.
2. both algorithms work on very simple "memoryless" robots with limited computational power; neither algorithm require robots to explicitly communicate with each other. However, *Node Count* requires some information to be stored in the

vertices: this is true for the deterministic version of *PatrolGRAPH*, but not for the random version.

3. in neither case robots need to know the graph, since vertices as well as incident edges can be stored, for example, into passive RFID tags opportunely distributed in the environment. This allows to add new vertices to the graph in real-time (e.g., adding new RFIDs). Notice however that *Node Count* unrealistically requires that a robot in vertex s knows how many times neighboring vertices have been visited (which requires RFIDs to be very close to each other, or that they can communicate in some way), whereas neither version of *PatrolGRAPH* makes this very strong assumption. Moreover, if the robot is able to recognize distinctive places in the environment (i.e., T-junctions, crossroads, turns), the random *PatrolGRAPH* can be used to guarantee uniform coverage basing exclusively on the environment's topology, i.e., without needing RFID tags at all.

References

[1] H. Choset, Coverage for robotics - A survey of recent results, Annals Math. Artif. Intell., vol. 31, pp. 113126, 2001.

[2] M. A. Batalin and G. S. Sukhatme, The analysis of an efficient algorithm for robot coverage and exploration based on sensor network deployment, in Proc. IEEE Int. Conf. Robot. Autom. (ICRA), 2005.

[3] M. A. Batalin and G. S. Sukhatme, Spreading out: A local approach to multi-robot coverage, in Proc. 6th Int. Symp. Distrib. Autonomous Robot. Syst., 2002, pp. 373-382.

[4] R. T. Vaughan, K. Stoy, G. S. Sukhatme, and M. J. Mataric, Lost: Localization-space trails for robot teams, IEEE Trans. Robot. Autom.,Special Issue Multi-Robot Syst., vol. 18, no. 5, 2002.

[5] S. Koenig, B. Szymanski, and Y. Liu, Efficient and inefficient ant coverage methods, Annals Math. Artif. Intell., vol. 41, no. 76, pp. 31-31, 2001.

[6] J. Svennebring and S. Koenig, Trail-laying robots for robust terrain coverage, in Proc. IEEE Int. Conf. Robot. Autom. (ICRA), 2003, pp. 75-82.

[7] I. Wagner, M. Lindenbaum, and A. Bruckstein, Distributed covering by ant-robots using evaporating traces, IEEE Trans. Robot. Autom., vol. 15, no. 5, pp. 918-933, Oct. 1999.

[8] I.Wagner,M. Lindenbaum, and A. Bruckstein, Efficiently searching a dynamic graph by a smell-oriented vertex process, Annals Math. Artif. Intell., vol. 24, pp. 211-223, 1998.

[9] Balch, T. and Arkin, R. 1993. Avoiding the past: A simple, but effective strategy for reactive navigation. In International Conference on Robotics and Automation. 678-685.

[10] G. Dudek, M. Jenkin, E. Milios, and D. Wilkes, "Robotic exploration as graph construction," in IEEE Transactions on Robotics and Automation, 7-6, 1991.

[11] M. A. Bender, A. Fernandez, D. Ron, A. Sahai, and S. Vadhan, The power of a pebble: Exploring and mapping directed graphs, in Annual ACM Symposium on Theory of Computing (STOC '98), 1998.

[12] http://www.vorwerk-teppich.de/sc/vorwerk/rfid_en.html

[13] http://ebiquity.umbc.edu/blogger/2005/07/05/robot-vacuum-guided-by-rfid/

[14] Kanji Tanaka, Yoshihiko Kimuro, Kentaro Yamano, Mitsuru Hirayama, Eiji Kondo and Michihito Matsumoto, A Supervised Learning Approach to Robot Localization Using a Short-Range RFID Sensor, IEICE Transactions on Information and Systems 2007 E90-D(11):1762-1771.

[15] Myungsik Kim, Nak Young Chonga, RFID-based mobile robot guidance to a stationary target, Mechatronics, Volume 17, Issues 4-5, May-June 2007, Pages 217-229.

[16] V.A. Ziparo, A. Kleiner, B. Nebel, and D. Nardi. RFID-Based Exploration for Large Robot Teams In Proc. of the IEEE Int. Conf. on Robotics and Automation, 2007, pp. 4606-4613.

[17] D. Hhnel, W. Burgard, D. Fox, K. Fishkin, and M. Philipose. Mapping and Localization with RFID Technology Proc. of the IEEE International Conference on Robotics and Automation, 2004.

[18] A. Kleiner, J. Prediger and B. Nebel. RFID Technology-based Exploration and SLAM for Search And Rescue. In Proc. of the IEEE/RSJ Int. Conf. on Intelligent Robots and Systems, Beijing, 2006.

Subject Index

Intelligent Autonomous Systems 10
W. Burgard et al. (Eds.)
IOS Press, 2008

Author Index